Dynamic Systems: Modeling, Simulation, and Control

Dynamic Systems: Modeling, Simulation, and Control

Second Edition

Craig A. Kluever
University of Missouri-Columbia

VP AND EDITORIAL DIRECTOR	Laurie Rosatone
SENIOR DIRECTOR	Don Fowley
EDITOR	Jennifer Brady
EDITORIAL MANAGER	Judy Howarth
CONTENT MANAGEMENT DIRECTOR	Lisa Wojcik
CONTENT MANAGER	Nichole Urban
SENIOR CONTENT SPECIALIST	Nicole Repasky
PRODUCTION EDITOR	Vinolia Benedict Fernando
COVER PHOTO CREDIT	© atakan / Getty Images, (illustration) Courtesy of Craig A. Kluever

This book was set in 10/12 TimesLTStd-Roman by SPi Global and printed and bound by CPI Group (UK) Ltd, Croydon, CR0 4YY.

MATLAB® and Simulink are registered trademarks of The MathWorks, Inc. and are used with permission. See www.mathworks.com/trademarks for a list of additional trademarks. The MathWorks does not warrant the accuracy of the text or exercises in this book. This book's use or discussion of MATLAB® software or related products does not constitute endorsement or sponsorship by The MathWorks of a particular pedagogical approach or particular use of the MATLAB® software.

Founded in 1807, John Wiley & Sons, Inc. has been a valued source of knowledge and understanding for more than 200 years, helping people around the world meet their needs and fulfill their aspirations. Our company is built on a foundation of principles that include responsibility to the communities we serve and where we live and work. In 2008, we launched a Corporate Citizenship Initiative, a global effort to address the environmental, social, economic, and ethical challenges we face in our business. Among the issues we are addressing are carbon impact, paper specifications and procurement, ethical conduct within our business and among our vendors, and community and charitable support. For more information, please visit our website: www.wiley.com/go/citizenship.

ISBN: 978-1-119-66872-5 (PBK)
ISBN: 978-1-119-60189-0 (EVAL)

Library of Congress Cataloging-in-Publication Data

Names: Kluever, Craig A. (Craig Allan) author.
Title: Dynamic systems : modeling, simulation, and control / Craig A.
    Kluever, University of Missouri-Columbia.
Description: Second edition. | Hoboken, N.J. : John Wiley & Sons, Inc.,
    [2020] | Includes bibliographical references and index.
Identifiers: LCCN 2019048773 (print) | LCCN 2019048774 (ebook) | ISBN
    9781119668725 (paperback) | ISBN 9781119601982 (adobe pdf) | ISBN
9781119601869 (epub)
Subjects: LCSH: System analysis. | Dynamics–Mathematical models. |
    Mechanical engineering–Mathematical models.
Classification: LCC T57.6 .K48 2020 (print) | LCC T57.6 (ebook) | DDC
    620.001/185–dc23
LC record available at https://lccn.loc.gov/2019048773
LC ebook record available at https://lccn.loc.gov/2019048774

The inside back cover will contain printing identification and country of origin if omitted from this page. In addition, if the ISBN on the back cover differs from the ISBN on this page, the one on the back cover is correct.

10 9 8 7 6 5 4 3 2 1

Contents

OC Content available in eBook

SS Student solution available in interactive e-text

Preface

This textbook is intended for an introductory course in dynamic systems and control, typically required in undergraduate mechanical engineering and some aerospace engineering curricula. Such a course is usually taken in the junior or senior year, after the student has completed courses in mechanics, differential equations, and electrical circuits. The major topics of a dynamic systems and control course include (1) mathematical modeling, (2) system-response analysis, and (3) an introduction to feedback control systems. The primary objective of this textbook is a comprehensive yet concise treatment of these major topics with an emphasis on demonstrating physical engineering applications. It has been my experience that undergraduate students remain engaged in a system dynamics course when the concepts are presented in terms of real engineering systems (such as a hydraulic actuator) instead of academic examples. This textbook is a distillation of 25 years of course notes and strategies for teaching system dynamics in the Mechanical and Aerospace Engineering Department at the University of Missouri-Columbia. It is thus based on my extensive classroom experience and student feedback.

The primary revision for this second edition is the significant addition of new end-of-chapter problems. As with the first edition, Chapters 2–10 contain problems that are grouped into three categories: (1) conceptual problems, (2) MATLAB problems, and (3) engineering applications. This second edition contains over 100 new problems spread among these three categories. In many cases physical engineering systems (such as a suspension system, solenoid actuator, or filter circuit) are revisited throughout the textbook in the end-of-chapter problems. A new feature of the second edition is a group of nearly 50 "reserved" problems that are separated from their respective chapters. The instructor has access to these reserved problems and may use them for class discussions. Another new feature is a subset of nearly 70 end-of-chapter problems that have links to worked-out solutions in the enhanced ebook. Finally, the last new feature is the inclusion of nearly 30 videos of the author working example problems from Chapters 2–10 in the enhanced ebook and on the instructor's book companion website. It is my hope that the addition of new problems in various formats will greatly enhance the student's understanding of the course material.

Chapter 1 introduces dynamic systems and control, including definitions of the relevant terms and categories of dynamic systems. The following three chapters all deal with developing mathematical models of physical engineering systems. Chapter 2 introduces the fundamental techniques used to derive the modeling equations for mechanical systems. Mechanical systems, perhaps more intuitively understood by undergraduate students since they involve Newton's laws of motion, are treated first. Chapter 3 introduces the fundamental methods of developing the mathematical models for electrical and electromechanical systems. Models in this chapter are derived by applying Kirchhoff's and Faraday's laws and the electrical element laws that govern the interactions among electrical charge, current, magnetic flux, and voltage. Chapter 4 presents the fundamental techniques involved in developing models for fluid and thermal systems. It is the last chapter dedicated to the derivation of mathematical models. Fluid (hydraulic and pneumatic) models are based on the conservation of mass while thermal models are derived using the conservation of energy.

Chapter 5 presents standard forms for representing the mathematical models of dynamic systems: state-variable equations, state-space representation, input-output equations, transfer functions, and block diagrams. The linearization process is also described in this chapter. The key concept emphasized in Chapter 5 is that each standard form is simply a convenient representation of the system model (i.e., the differential equations) that lends itself to analysis of the system's dynamic response. Therefore Chapter 5 serves as a transition between developing the mathematical model (Chapters 2–4) and obtaining the system's response using either numerical simulation or analytical techniques (Chapters 6–10).

Chapter 6 presents numerical simulation methods for obtaining the response of dynamic systems. Here the MATLAB simulation software is used exclusively since it has become the standard computational platform for academia and industry. Simulating the response of linear systems using MATLAB commands is presented first. The graphical software Simulink is presented next, and it is the focus of this chapter and the primary simulation tool used throughout the remainder of the textbook. Simulink is used to simulate linear and nonlinear systems using the standard forms presented in Chapter 5.

The next three chapters involve the analytical solution of linear systems. Chapter 7 covers analytical methods for obtaining the system's response in the time domain with an emphasis on first- and second-order system response. Here the two key concepts are (1) the correlation between the roots of the characteristic equation and the form of the free (or transient) response; and (2) the equivalence of the characteristic roots, poles of the transfer function, and eigenvalues of the system matrix. Chapter 8 presents a brief overview of Laplace transform theory and its use in obtaining the response of linear dynamic systems. Chapter 9 involves frequency response, or the system response to periodic input functions. In this chapter, the emphasis is on the Bode diagram as a graphical depiction of the information required for complete frequency-response analysis.

Chapter 10 introduces feedback control systems where the PID controller (and its variants) is emphasized. Two graphical techniques, the root-locus method and the Bode diagram, are used to analyze the closed-loop response, design controllers, and assess stability. The chapter closes with a brief discussion of how controllers are implemented as discrete-time algorithms in a digital computer.

Serving as a capstone for the textbook, Chapter 11 presents case studies in dynamic systems and control. Mixed-discipline, integrated engineering systems, inspired by the research literature, serve as the case studies. The five case studies illustrate the major topics of the textbook: (1) developing mathematical models, (2) predicting the system's behavior using analytical and numerical methods, and (3) selecting the important system parameters in order to improve performance. Simulink is used extensively to obtain the dynamic response of these systems that often involve nonlinearities and other complexities.

Appendix A presents the basic and derived units used in this textbook. Appendix B provides a brief introduction to MATLAB, M-files, and the commands that pertain to solving problems involving dynamic systems and control. Appendix C is a primer on Simulink and expands on the brief Simulink tutorial covered in Chapter 6.

As previously mentioned, this textbook is an outgrowth of 25 years of teaching a system dynamics course. Chapters 1–10 can be covered in a single semester, and the instructor may choose to augment discussions with the case studies in Chapter 11. As previously noted, numerical simulation using MATLAB and Simulink (Chapter 6) is covered before analytical methods (Chapters 7–9). My experience is that presenting modeling and simulation of real engineering applications early in the semester engages the students in the subject material. Instructors who wish to present analytical methods before numerical simulation may simply cover Chapter 6 after Chapters 7 and 8. Furthermore, Chapter 6 is structured so that each instructor can choose their preferred depth of coverage.

Finally, I would like to thank my former students for their candid remarks and feedback that has helped to shape this book. In addition, many of them contributed by solving the new end-of-chapter problems, and for this I am grateful.

Craig A. Kluever
Mechanical & Aerospace Engineering
University of Missouri-Columbia
October, 2019

Chapter 1

Introduction to Dynamic Systems and Control

1.1 INTRODUCTION

In solving engineering problems, there is a need to understand and determine the dynamic response of a physical system that may consist of several components. These efforts involve *modeling*, *analysis*, and *simulation* of physical systems. Typically, building a prototype system and conducting experimental tests are either infeasible or too expensive for a preliminary design. Therefore, mathematical modeling, analysis, and simulation of engineering systems aid the design process immensely.

Dynamic systems and control involves the analysis, design, and control of physical engineering systems that are often composed of interacting mechanical, electrical, and fluid subsystem components. One example is an electrically controlled hydraulic actuator that is used to change the position of an aerodynamic surface (e.g., rudder) on an airplane. This system consists of several interacting components: an electromagnet circuit is used to open a mechanical valve that allows high-pressure hydraulic fluid to flow into a cylinder chamber; the fluid pressure causes a mechanical piston to move; and mechanical linkages connecting the hydraulic piston to the aerodynamic surface (e.g., rudder) cause the rudder to change position. Finally, an onboard digital computer (an "autopilot") uses feedback from sensors to adjust the operation of the hydraulic actuator so that the rudder position (and the subsequent response of the airplane) matches the desired value. This example demonstrates why it is advantageous for the engineer to understand the dynamic response of this interconnected system without relying on experimentation with a physical prototype.

Here are definitions of the important terms that we use throughout the book:

System: A combination of components acting together to perform a specified objective. The components or interacting elements have cause-and-effect (or input-output) relationships. One example of a system is a direct-current (DC) motor where a voltage input causes angular velocity (the output) of the mechanical load attached to the motor's shaft.

Dynamic system: A system where the current output variables (or *dynamic variables*) depend on the initial conditions (or stored energy) of the system and/or the previous input variables. The dynamic variables of the system (e.g., displacement, velocity, voltage, pressure) vary with time. For the DC motor example, the angular velocity of the motor is the dynamic variable and the circuit voltage is the input.

Modeling: The process of applying the appropriate fundamental physical laws in order to derive mathematical equations that adequately describe the physics of the engineering system. Dynamic systems are represented by differential equations. For the DC motor example, the electrical circuit is modeled by using Kirchhoff's voltage law, and the mechanical motion is modeled by using Newton's second law.

Mathematical model: A mathematical description of a dynamic system's behavior, which is usually a set of linear or nonlinear ordinary differential equations (ODEs). For the DC motor example, the mathematical model consists of a differential equation for the electrical current and a differential equation for mechanical motion.

Simulation: The process of obtaining the system's dynamic response by numerically solving the governing modeling equations. Simulation involves numerical integration of the model's differential equations and is performed by digital computers and simulation software.

System analysis: The use of analytical calculations or numerical simulation tools to determine the system response in order to assess its performance. Repeated analysis aids the design process where the system's configuration or parameters are altered to improve performance or meet desired constraints. For the DC motor example, we might apply a constant voltage input and determine the characteristics of the angular velocity response by using analytical calculations ("by hand") or numerical simulations. If the angular velocity response is inadequate, we could alter the system's parameters in order to improve performance.

1.2   CLASSIFICATION OF DYNAMIC SYSTEMS

In general, we can classify dynamic systems according to the following four categories: (1) distributed versus "lumped" systems, (2) continuous-time versus discrete-time systems, (3) time-varying versus time-invariant systems, and (4) linear versus nonlinear systems.

Distributed versus Lumped Systems

A *distributed system* requires an infinite number of "internal" variables, and, therefore, the system is governed by partial differential equations (PDEs). A *lumped system* involves a finite number of "internal" variables, and, therefore, the system is governed by ODEs. For example, if we want to model a hydraulic piston, we would "lump" all pressure distributions in a cylinder chamber into one single pressure term. Therefore, we would have one ODE for the time derivative of pressure (dP/dt or $\dot{P}$) for each "lump" of fluid in a particular chamber. In this textbook, we work exclusively with lumped systems and ODEs.

Continuous-Time versus Discrete-Time Systems

A *continuous-time system* involves variables and functions that are defined for all time, whereas a *discrete-time system* involves variables that are defined only at discrete time points. We may think of continuous-time systems consisting of variables in the "analog" domain, such as position $x(t)$. Discrete-time systems consist of variables in the "digital" domain, such as the sampled (measured) position $x(kT_s)$ that exists only at the discrete-time points $t = T_s, t = 2T_s, \ldots, t = kT_s$ where T_s is the sampling interval. Continuous-time systems are described by differential equations while discrete-time systems are described by difference equations. In this volume, we work with continuous-time systems and differential equations. We introduce discrete-time systems in Chapter 10 when we examine the role of digital computers in automatic control systems and the need to convert analog signals to digital signals and vice versa.

Time-Varying versus Time-Invariant Systems

In a *time-varying system*, the system parameters change with time (e.g., the friction coefficient changes with time). In a *time-invariant system*, the parameters remain constant. The reader should not confuse the variation of the system parameters with the variation of the dynamic variables. For the DC motor example, the system parameters would be electrical resistance of the circuit, inductance of the coil windings around the rotor, friction coefficient for the rotor bearings, and moment of inertia of the rotor. If these system parameters do not change with time (i.e., they are constants for the system model), then the DC motor is a time-invariant system. Of course, the dynamic variables associated with the DC motor (electrical current of the circuit and angular velocity of the output shaft) can change with time. We focus primarily on time-invariant systems in this text.

Linear versus Nonlinear Systems

Suppose we have a system or input–output relationship that is described by the function $y = f(u)$ where u is the input and y is the output. Linear systems obey the *superposition property*:

1. If $y_1 = f(u_1)$, then $ay_1 = f(au_1)$, where $a =$ any constant.
2. If $y_1 = f(u_1)$ and $y_2 = f(u_2)$, then $y_1 + y_2 = f(u_1 + u_2)$.

Consider again the DC motor example: suppose we apply 12 volts (V) to a motor and through measurements determine the steady-state (constant) angular velocity to be 1600 revolutions per minute (rpm). Next, if we apply 6 V to the motor and the measured steady-state angular velocity is 800 rpm then the system obeys the first superposition property and the DC motor system is linear. Of course, a physical system that demonstrates linearity (such as the DC motor) has a limited linear range of operation; that is, we cannot increase the input voltage by a factor of 100 and expect the corresponding angular velocity to increase by a factor of 100. Increasing the system input beyond a threshold may cause the output to saturate (i.e., reach a limit) and, therefore, the system is no longer linear.

The second superposition property shows that the total dynamic response of a linear system can be obtained by adding or superimposing the responses (or solutions) to individual input functions. Nonlinear systems do not obey either superposition property.

The following equations are examples of *linear* ODEs:

$$\ddot{x} + 3\dot{x} - 40x = 6u \tag{1.1}$$

$$2\ddot{x} + 0.4\dot{x} + 0.6e^{-2t}x = -8u \tag{1.2}$$

Equation (1.1) is a second-order linear ODE because the dynamic variable x and its derivatives appear as linear combinations (we will use the over-dot notation to denote derivatives with respect to time; hence, $\dot{x} = dx/dt$, $\ddot{x} = d^2x/dt^2$, etc.). Equation (1.1) involves constant coefficients and hence it is a *linear time-invariant* (LTI) differential equation. Equation (1.2) is linear as x and its derivatives appear in linear combinations. Because the coefficient $0.6e^{-2t}$ changes with time, Eq. (1.2) is a *linear time-varying* ODE. The following equation

$$2\ddot{x} + 3\dot{x} + 16x^2 = 5u \tag{1.3}$$

is a *nonlinear* ODE because of the x^2 term.

All physical systems are nonlinear. However, if we confine the input–output variables to a restricted (nominal) range, then we can often replace a nonlinear system with a *linear* model comprising linear differential equations. This important process is called *linearization*. Obtaining a linear model is extremely important and advantageous in system analysis because it is possible to obtain the analytical (closed-form) solution to linear ODEs. Nonlinear systems must be solved by using numerical methods to integrate the ODEs.

1.3 MODELING DYNAMIC SYSTEMS

A major focus of this book is mathematical modeling of dynamic systems. Developing an appropriate model is always the first step in system analysis because it is impossible to determine the system's response without a mathematical representation of the system dynamics. Mathematical models are obtained by applying the appropriate laws of physics to each element of a system. Some system parameters (such as friction characteristics) may be unknown, and these parameters are often determined through experimentation and observation, which lead to empirical relations. Engineering judgment must be used to trade model complexity with the accuracy of the analysis. Nonlinearities (such as gear backlash) are often ignored in preliminary design studies in order to derive linear models. Sometimes, low-order approximate linear models can be developed to accurately represent the system dynamics. These low-order linear models can be solved analytically ("by hand"), which gives the engineer an intuitive feel for the nature of the dynamic system. Furthermore, simulations are easier to construct with low-order linear models and therefore the time required to perform system analysis is reduced. Nonlinear models, on the other hand, require numerical solutions using simulation software. Extremely complex nonlinear models typically require small integration time steps to accurately solve the ODEs thus increasing computer-run time. Consequently, there is usually a trade-off between model complexity and analysis time.

Engineers must remember that the results obtained for a particular mathematical model are only approximate and are valid only to the extent of the assumptions used to derive the model. The model must be sufficiently sophisticated to demonstrate the significant features of the dynamic response without becoming too cumbersome for the available analysis tools. The validity of a mathematical model can often be verified by comparing the model solution (such as simulation results) with experimental results. The Shuttle Avionics Integration Laboratory (SAIL) was a hardware-in-the-loop test facility at NASA Johnson Space Center [1]. SAIL consisted of actual Space Shuttle hardware (such as the flight deck, cockpit displays, sensors, and electronic wiring) and mathematical models of the physical forces due to aerodynamics, gravity, and propulsion. Engineers and astronauts used SAIL to perform "real-time" simulations of Space Shuttle missions in order to test and validate the flight software. The simulation results from SAIL tests showed an excellent match with actual Shuttle flight data. SAIL tests, however, could occasionally be intermittent owing to their reliance on using very complex mathematical models (i.e., computer software) to interface with and drive all of the physical Shuttle hardware. The SAIL facility is an example of one extreme end of the mathematical modeling spectrum: a complex, "high-fidelity" simulation that was at times prone to sporadic testing. This trade-off was necessary in order to accurately model the Shuttle's flight dynamics.

Simulation Tools

Several commercial simulation tools have been developed to help engineers design and analyze dynamic systems. We briefly discuss a few of these software tools as examples of mathematical modeling and system analysis.

Simulink is a numerical simulation tool that is part of the MATLAB software package developed by MathWorks [2]. It uses a graphical user interface (GUI) to develop a block diagram representation of dynamic systems. Simulink is used by engineers in industry and academia. Constructing system models with Simulink is relatively easy and, therefore, it is often used to build simple models during the preliminary design stage. However, Simulink can be used to simulate complex, highly nonlinear systems. In this book, we use it extensively to simulate and analyze dynamic systems.

Caterpillar Inc. has developed the simulation tool Dynasty that allows engineers to construct complex models of large off-road vehicles [3]. The engineer can build software models of integrated machines by "dragging-and-dropping" subsystem models from a library. These subsystems include engines, linkages, drive trains, hydraulics, and controls. The underlying physics of each subsystem are contained within the mathematical model of the individual component. The Dynasty software simulates the dynamics of the integrated vehicle model and allows engineers to perform tests, analyze the dynamic response, and vary the subsystem components in order to improve overall system performance. Caterpillar engineers used Dynasty to analyze and design the 797B mining truck (its largest vehicle) and bring it to production in less than half the time it would take by building physical prototypes of the truck.

EASY5, originally developed by Boeing, is a graphics-based simulation tool for constructing virtual prototypes of engineering systems [4]. As in Dynasty, the user can select prebuilt components from libraries that include models of mechanical, electrical, hydraulic, pneumatic, and thermal subsystems. EASY5 can interface with Simulink and other computer-aided engineering software tools. Engineers have used EASY5 to analyze and design aerospace vehicles, for example.

In summary, it should be noted that all numerical simulation tools are constructed using the basic principles of mathematical modeling that are presented in this book. That is, the appropriate physical law (e.g., Newton's second law, Kirchhoff's voltage law) is applied to the particular system (mechanical, electrical, fluid, etc.) in order to develop the differential equations that describe the system dynamics. The differential equations are then solved using numerical integration methods. The solution to the differential equations is the system's dynamic response.

1.4 OBJECTIVES AND TEXTBOOK OUTLINE

The objective of this book is to present a comprehensive yet concise treatment of dynamic systems and control. In particular, on completing this volume, the reader should be able to accomplish the following tasks: (1) develop the mathematical models for mechanical, electrical, fluid, or thermal systems; (2) obtain the system's dynamic response (due to input functions and/or initial energy storage) by using numerical simulation tools and analytical techniques; and (3) analyze and design feedback control systems in order to achieve a desirable system response. This book primarily emphasizes lumped, continuous-time, LTI systems. Hence, all mathematical models involve ODEs and the majority have constant coefficients. Nonlinear systems are given considerable attention, and, therefore, we make frequent use of Simulink to obtain the dynamic response. Furthermore, we often utilize the linearization process in order to approximate the nonlinear dynamics with linear system dynamics. As is demonstrated, obtaining a linear mathematical model allows us to use a wealth of analytical tools for system analysis and graphical techniques for designing feedback control systems.

This book is organized according to its three major objectives. Chapters 2–4 deal with developing the mathematical models of physical engineering systems. In particular, Chapter 2 treats mechanical systems and the derivation of the modeling equations by applying Newton's laws of motion. Chapter 3 deals with mathematical models for electrical and electromechanical systems. Here we apply the element laws that govern the interaction between electrical charge, current, magnetic flux, and voltage. Mathematical models for fluid and thermal systems are developed in Chapter 4 by utilizing the conservation of mass and the conservation of energy, respectively. In Chapters 2–4, we focus on developing mathematical models of "real-world" physical engineering systems such as vehicle suspension systems, energy-transmission devices, and systems with mixed disciplines such as electromechanical, hydromechanical, and pneumatic actuators. Chapter 5 deals with the standard formats for representing the various mathematical models derived in the previous three chapters. These standard formats facilitate the second major topic of the book—obtaining the system response—whether we use numerical or analytical techniques.

Chapter 6 begins the system-analysis section of the book. This chapter introduces Simulink as the numerical simulation tool of choice for obtaining the response of linear and nonlinear dynamic systems. Chapter 7 presents analytical techniques for solving the mathematical modeling equations "by hand." Here we analyze the total system response (comprising the transient and steady-state responses) to input functions. In Chapter 8, we introduce the Laplace transformation method for obtaining the response of dynamic systems that are modeled by LTI differential equations. Chapter 9 deals with obtaining the response of a dynamic system that is driven by an oscillating or harmonic input function. This frequency-response analysis is aided by graphical techniques such as the Bode diagram.

Chapter 10 introduces the reader to the third major objective of the textbook: the analysis and design of feedback control systems. Here we investigate the use of feedback (from measurement sensors) to shape the system input function in order to achieve a desirable output response. Although different control schemes are discussed, this chapter emphasizes the proportional-integral-derivative (PID) controller (and its variants) because it is the most widely used control scheme in industry. Control-system design is aided by two graphical techniques, the root-locus method and the Bode diagram, that are discussed in this chapter.

Chapter 11 presents five engineering case studies that demonstrate the three major objectives of this textbook: modeling, analysis, and control of dynamic systems. These examples are inspired by research from the engineering literature and involve physical systems such as vehicle suspensions and actuators. The final chapter serves as a "capstone" for this book.

Appendix A presents units and Appendix B gives a brief overview of MATLAB usage, its commands, and programming with MATLAB. Only the MATLAB commands that pertain to solving problems in dynamic systems and control are presented in Appendix B. Appendix C is a tutorial on using Simulink to simulate linear and nonlinear dynamic systems.

REFERENCES

1. Melone, K., "SAILing Through Space," *Boeing Frontiers*, Vol. 9, September 2010, pp. 24–25.
2. http://www.mathworks.com/products/simulink/ (accessed 10 March 2014).
3. Dvorak, P., "Software Simulates Many Disciplines in One Model," *MachineDesign.com*, http://machinedesign.com/article/software-simulates-many-disciplines-in-one-model-1106 (accessed 10 March 2014).
4. http://www.mscsoftware.com/Products/CAE-Tools/Easy5.aspx (accessed 10 March 2014).

Chapter 2

Modeling Mechanical Systems

2.1 INTRODUCTION

The objective of this and the next two chapters is to develop the mathematical models of physical engineering systems. This chapter introduces the fundamental techniques for deriving the modeling equations for mechanical systems. These systems are composed of inertia, stiffness, and friction elements. Our mathematical models of mechanical systems are developed by applying Newton's laws of motion, which govern the interaction between force, mass, and acceleration. We utilize a lumped-system approach, and, therefore, the mathematical model consists of ordinary differential equations (ODEs). Mechanical systems with translational motion and rotational motion about a fixed axis are treated in this chapter.

The reader should keep in mind that the overall goal of this chapter is to derive the mathematical models that govern the behavior of mechanical systems. We do not (yet) attempt to obtain the mechanical system's response to force or motion inputs. Obtaining the system's response to known inputs is discussed in Chapters 6–9.

2.2 MECHANICAL ELEMENT LAWS

A mechanical system is composed of inertia, stiffness, and energy-dissipation elements. In addition, it may possess mechanical transformers, such as gears or levers. This section presents brief descriptions of the fundamental laws that govern these mechanical elements.

Inertia Elements

Inertia elements are either lumped masses (translational mechanical systems) or moments of inertia (rotational mechanical systems). They are easily identified in Newton's second law

$$\text{Force} = \text{mass} \times \text{acceleration} \qquad \text{(translational system)}$$

$$\text{Torque} = \text{moment of inertia} \times \text{angular acceleration} \qquad \text{(rotational system)}$$

Therefore, the inertia element is the ratio of force and acceleration (or torque and angular acceleration). A rigid body that has translational ("straight-line") motion has all of its mass lumped into a single element, m, with units of kg. A rigid body with purely rotational motion about an axis has all of its mass lumped into a moment of inertia, J, which is defined as

$$J = \int r^2 dm \qquad (2.1)$$

where dm is an incremental mass with radial distance r from the axis of rotation. Equation (2.1) shows that J has units of kg-m^2. Equations for moments of inertia can be derived for homogeneous, rigid bodies with standard shapes. One example is a cylindrical disk with radius R and a uniform mass distribution with total mass M. The moment of inertia about the axis of symmetry for a uniform disk is

$$J = \frac{1}{2} M R^2 \qquad (2.2)$$

Inertia elements can store potential energy due to position in a gravitational field, or kinetic energy due to motion. Potential energy ξ_P of a mass m in a uniform field with gravitational constant g is

$$\xi_P = mgh \tag{2.3}$$

where h is the vertical position of the mass measured from a reference height. Equation (2.3) shows that potential energy has the dimensions of force (mg) and length (h), or units of N-m or joule (J). Kinetic energy ξ_K of mass m moving with velocity $\dot{x} = dx/dt$ is

$$\xi_K = \frac{1}{2}m\dot{x}^2 \tag{2.4}$$

Kinetic energy of moment of inertia J rotating with angular velocity $\dot{\theta}$ is

$$\xi_K = \frac{1}{2}J\dot{\theta}^2 \tag{2.5}$$

As stated in Chapter 1, we adopt the over-dot convention throughout this book to indicate the derivative with respect to time; hence $\dot{\theta} = d\theta/dt$ and $\ddot{x} = d^2x/dt^2$. Equation (2.4) shows that translational kinetic energy has dimensions of mass (m) and velocity squared ($\dot{x}^2$), or units of kg-m^2/s^2, which is equivalent to N-m or joules. Equation (2.5) shows that rotational kinetic energy has dimensions of moment of inertia (J) and angular velocity squared ($\dot{\theta}^2$), or units of kg-m^2 rad^2/s^2, which is equivalent to N-m or joules. Clearly, all energy equation expressions must have the same units of N-m or joules.

Stiffness Elements

When a mechanical element stores energy due to a deformation or change in shape, it can be modeled as a stiffness element. In such cases, a fundamental relationship between force and the resulting deformation is required to model stiffness. The simplest force–deformation relationship is Hooke's law, which states that the force required to stretch or compress a spring is proportional to the displacement. Figure 2.1 shows a spring that is fixed at its left end, but free at the right end. Suppose a tensile force F is applied at the right (free) end and x is the corresponding displacement of the free end from its equilibrium (unstretched) position. The force required to produce displacement x is

$$F = kx \tag{2.6}$$

where k is called the *spring constant* and has units of N/m. Clearly, Eq. (2.6) is a *linear* relationship between force and displacement. Figure 2.1 shows that the positive convention for displacement x is to the right and, therefore, the positive convention for force F is also to the right. If force F is compressive, then both F and x are negative and Eq. (2.6) is still valid.

When both ends of a spring are free to move, then the force required to stretch or compress a spring depends on the *relative* displacement

$$F = k(x_2 - x_1) \tag{2.7}$$

Figure 2.2 shows the case where the tensile force F is applied at both ends of the spring k, and x_1 and x_2 are the absolute displacements of the free ends (positive displacement is to the right). The reference positions for x_1 and x_2 are shown in Fig. 2.2, and they represent the unstretched (equilibrium) positions when no force is applied to the spring. Hence, Eq. (2.7) and Fig. 2.2 show that if both ends of the spring are displaced by +0.1 m, then no force exists, $F = 0$.

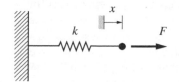

Figure 2.1 Force stretching the free end of a spring.

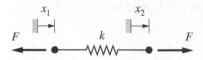

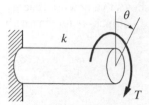

Figure 2.2 Force deflecting the free ends of a spring.

Figure 2.3 Torque rotating the free end of a torsional shaft.

If $x_2 = +0.25$ m and $x_1 = +0.15$ m, then a tensile force proportional to a relative displacement of 0.1 m exists. If $x_2 = +0.15$ m and $x_1 = +0.25$ m, then a compressive force proportional to a relative displacement of 0.1 m exists. Furthermore, Eq. (2.7) shows that the compressive force F is negative, which is in agreement with the convention for the positive force direction established by Fig. 2.2.

A rotational mechanical system exhibits stiffness when a relationship exists between an applied torque and the resulting angular displacement. A simple example is the torsional shaft depicted by Fig. 2.3, where the left end of the shaft is fixed and the right end is free. The positive angular displacement θ (in rad) is shown as clockwise and is measured relative to the untwisted (equilibrium) position of the free end. A positive applied torque T results in a positive angular displacement. The linear torque–displacement relation is

$$T = k\theta \tag{2.8}$$

where k is called the *torsional spring constant* and has units of N-m/rad. When both ends of the shaft are free to rotate, the torque depends on the *relative* angular displacement

$$T = k(\theta_2 - \theta_1) \tag{2.9}$$

Stiffness elements can store potential energy due to deformations or deflections. Potential energy ξ_P stored by an ideal translational spring is

$$\xi_P = \frac{1}{2}k\Delta x^2 \tag{2.10}$$

where $\Delta x = x_2 - x_1$, or the relative displacement between the free ends of the spring. Note that the units of energy in Eq. (2.10) are (N/m)-m^2, or N-m, which we recognize from basic mechanics as the same units for work (= force × displacement). Recall that energy can be defined as the capacity to perform work, and hence energy and work have the same units.

Potential energy of an ideal torsional spring is

$$\xi_P = \frac{1}{2}k\Delta\theta^2 \tag{2.11}$$

where $\Delta\theta = \theta_2 - \theta_1$ is the relative angular displacement between the free ends of the torsional spring. Potential energy stored in a torsional spring has units of (N-m/rad)-rad^2, or N-m, which are again the same units as work.

When we model mechanical systems that possess stiffness elements, we simply lump all of the system's stiffness into a "spring element," even though it may or may not be a physical mechanical spring. In some cases, such as a hydraulic valve with a return spring, a physical spring may exist in the mechanical system. In other cases, the "spring element" shown in Fig. 2.1 may be used to represent the inherent stiffness of a deformable body. In this book, we deal with lumped systems. Therefore, all stiffness is lumped into the spring element and all mass is lumped into the inertia element. Thus, we deal with "ideal" spring elements that possess no inertia and no energy-dissipation effects.

Spring elements may exhibit linear or nonlinear force–displacement relationships. Equations (2.6)–(2.9) represent linear spring elements. Practical mechanical springs exhibit nonlinear effects when subjected to extreme displacements; for example, the stiffness decreases as a spring is stretched beyond its yield point.

In some cases, it is possible to derive equations for the spring constant k. For example, the translational spring constant for a uniform rod in tension or compression is

$$k = \frac{EA}{L} \tag{2.12}$$

where E is Young's modulus of elasticity of the rod's material, A is the cross-sectional area, and L is the length of the rod. For a round shaft in torsion, the torsional spring constant is

$$k = \frac{\pi d^4 G}{32L} \tag{2.13}$$

where G is the shear modulus of elasticity, d is the diameter of the shaft, and L is the length. Expressions for k for mechanical springs can be found in machine-design textbooks, and these spring constants depend on material properties, geometric characteristics (such as coil radius and wire diameter), and the number of coils.

Friction Elements

When a mechanical element dissipates energy due to its motion, it can be modeled as a friction element. In such cases, a fundamental relationship between relative velocity and the resistive force is required to model friction. Just as we used a "spring element" to model stiffness in a mechanical system, we can use a "damper" (or "dashpot") element to model friction. Figure 2.4 shows a translational damper, which is a fluid-filled cylinder that encases a piston and a rod. The absolute velocity of the piston/rod is $\dot{x}_2$ (positive is to the right), and the absolute velocity of the cylinder is $\dot{x}_1$ (also positive to the right). If the damper exhibits a linear relationship between the resistive force and relative motion, then the damper force is

$$F = b(\dot{x}_2 - \dot{x}_1) \tag{2.14}$$

where b is the viscous friction coefficient, with dimensions of force/velocity, or units of N-s/m. Clearly, the damper force depends on the *relative* velocity between the piston/rod and cylinder.

Friction elements can only dissipate energy. From basic mechanics, we know that power is the time rate of change of energy, or force $\times$ velocity. Consider a translational damper where $\dot{x}$ is the velocity magnitude of the piston relative to the stationary cylinder. The rate of energy dissipation from the damper is

$$\dot{\xi}_f = -(b\dot{x})\dot{x} = -b\dot{x}^2 \tag{2.15}$$

Recall that energy is the scalar (or "dot") product of the force vector and displacement vector, and, therefore, a minus sign is inserted to indicate that the friction force $b\dot{x}$ is in the opposite direction of the velocity. Equation (2.15) shows that the rate of energy loss is always negative regardless of the direction of the velocity. Power or energy time rate of change has units of N-m/s or J/s or watts (W).

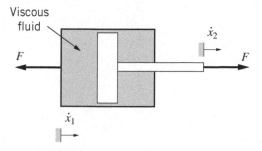

Figure 2.4 Translational damper.

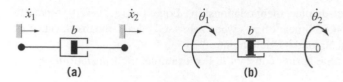

Figure 2.5 Generic symbols for damper elements: (a) translational damper and (b) rotational damper.

Next, consider a rotational or torsional damper, where the damping torque is proportional to the relative angular velocity

$$T = b(\dot\theta_2 - \dot\theta_1) \tag{2.16}$$

Here b is the torsional viscous friction coefficient, with dimensions of torque/angular velocity, or units of N-m-s/rad. An example of a rotational damper is a fluid drive, where two disks are separated by a viscous fluid, and a friction torque exists if the relative angular velocity $\dot\theta_2 - \dot\theta_1$ is not zero. The rate of energy dissipation from a rotational damper is the product of the friction torque and angular velocity, which is analogous to Eq. (2.15) and the rate of energy loss for a translational damper.

When we model mechanical systems that involve energy dissipation due to friction, we simply lump all of the system's friction into a "damper element," even though it may or may not be a physical piston–cylinder dashpot. In some cases, such as a shock absorber in an automobile, a physical piston–cylinder dashpot may exist in the mechanical system. Just as we used the generic spring symbol to represent stiffness, we can use the generic translational or rotational damper symbols shown in Fig. 2.5 to represent any friction in a mechanical system. Again, analogous to the ideal spring, the ideal damper element possesses no inertia or stiffness.

Friction in mechanical systems can involve nonlinear relationships between force and velocity. Common examples are dry (Coulomb) friction or square-law friction (such as aerodynamic drag). In the case of dry or Coulomb friction, the resistive force remains constant and opposes motion as long as relative velocity is not zero. For drag friction, the resistive force is proportional to the square of the relative velocity.

Mechanical Transformers

Mechanical devices that transform an input motion or a force are called mechanical transformers. Common examples include levers and gear trains. Figure 2.6 shows an ideal lever, which is a slender bar that rotates about a fixed pivot. An ideal lever is rigid, has neither inertia nor friction, and therefore cannot store or dissipate energy. The vertical displacements of the left and right ends of the lever in Fig. 2.6 are $L_1 \sin\theta$ and $L_2 \sin\theta$, respectively. For a small angular rotation, $\sin\theta \approx \theta$, and the vertical displacements are approximately $L_1\theta$ and $L_2\theta$. Because the ideal lever has no inertia, the moment about its pivot is zero, and, therefore, $f_1 L_1 \cos\theta = f_2 L_2 \cos\theta$; for small angles, $\cos\theta \approx 1$ and $f_1 L_1 = f_2 L_2$. If we consider force f_1 as the input force, then the output lever force is $f_2 = f_1 L_1 / L_2$, which is greater than the input force when length $L_1 > L_2$ (as in Fig. 2.6).

Figure 2.7 shows a gear train, which may be used to increase or decrease the angular velocity or torque from the input axis to the output axis. In an ideal gear train, the gears are assumed to have zero inertia, the gear teeth mesh perfectly without backlash, and energy is transmitted from the input to output axis without loss (no friction). Because

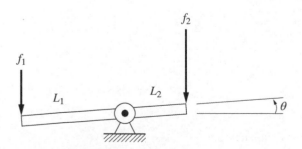

Figure 2.6 Ideal lever.

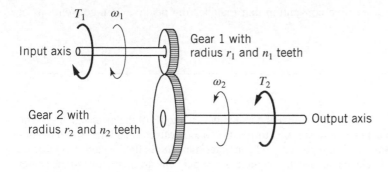

Figure 2.7 Ideal gear train.

the gear teeth mesh perfectly, the teeth are spaced equally on both gears, and therefore the ratio of the gear radii (r_2/r_1) is equal to the ratio of the number of gear teeth (n_2/n_1)

$$\frac{r_2}{r_1} = \frac{n_2}{n_1} = N \tag{2.17}$$

where N is called the gear ratio. We can determine the angular velocity input–output ratio for the gear train by noting that the velocity of the mesh point is the same for both gears, that is, $V_{mesh} = r_1\omega_1 = r_2\omega_2$. Therefore, the ratio of angular velocities is

$$\frac{\omega_1}{\omega_2} = \frac{r_2}{r_1} = N \tag{2.18}$$

If the input axis gear has a smaller radius than the output axis gear as in Fig. 2.7, then $N > 1$ and the output angular velocity ω_2 is less than the input angular velocity ω_1 and the gear train is a speed reducer. If $r_1 > r_2$, then $\omega_2 > \omega_1$ and the output shaft rotates faster than the input shaft.

Because friction is neglected in an ideal gear train, the energy is transmitted without loss. Work (or energy) for a rotational system is torque × angular displacement, and, therefore, power is torque × angular velocity. Equating power to the input shaft and power delivered to the output shaft yields $T_1\omega_1 = T_2\omega_2$, and, therefore, the input–output torque ratio is

$$\frac{T_1}{T_2} = \frac{\omega_2}{\omega_1} \tag{2.19}$$

Therefore, when the gear train is a speed reducer (as in Fig. 2.7), the output torque T_2 is greater than the input torque T_1.

2.3 TRANSLATIONAL MECHANICAL SYSTEMS

Mathematical models of mechanical systems can be derived using a systematic two-step process:

1. Draw a free-body diagram (FBD) of each inertia element with arrows denoting external forces (or torques) acting on each mass (or moment of inertia). Make use of Newton's third law to show equal-and-opposite reaction forces (or torques) on interconnected inertia elements. Carefully assign equations for each force (or torque) using the appropriate element law and the positive conventions for displacement variables.

2. Apply Newton's second law to each inertia element to obtain the mathematical model of the entire mechanical system.

Newton's second law for translational systems states that the sum of all external forces acting on a body is equal to the product of the mass m and acceleration $\ddot{x}$ of the body

$$\sum F = m\ddot{x} \tag{2.20}$$

The reader should be careful to sum forces with a positive convention that matches the positive sign convention for displacement (and, therefore, positive acceleration).

Example 2.1

A high-speed solenoid actuator–valve system is shown in Fig. 2.8. Derive the mathematical model of the mechanical system.

This type of actuator is used in hydraulic and pneumatic systems to position spool valves for metering fluid flow. Electric current flows through the coil of wire that surrounds the plunger and generates a magnetic field, which produces an attracting force on the armature, pulling it to the right. Hence, the electromagnetic force pulls the armature toward the center of the coil and closes the air gap. The armature is rigidly connected to the spool valve via the push rod, and therefore they can be considered as a single lumped mass. Motion of the armature–valve mass is purely translational and in the horizontal direction, and both the electromagnetic force and mass displacement are positive to the right. When the electromagnetic force displaces the armature–valve to the right from its seated position, the return spring is compressed and pushes on the spool valve to the left. Hence the return spring is used to return the armature–valve mass back to its seated position when the electromagnetic force is removed. For this example, let us assume that the return spring is undeflected when the armature–valve is in its seated position.

Figure 2.9 shows a schematic depiction of the mechanical components of the solenoid actuator using inertia, stiffness, and friction elements. Because the armature and spool valve are rigidly connected, we lump both masses into a single mass m. The position of the armature–valve mass is denoted by x, which is measured from the static equilibrium position (unstretched spring with mass at rest). Positive displacement x is to the right, as indicated in the figure. The electromagnetic force F_{em} is an external force that is applied directly to mass m. Let us assume that the friction caused by the valve's motion within the hydraulic fluid is modeled by a linear viscous friction coefficient b, or ideal damper element. Finally, the return spring is modeled by an ideal (linear) spring with stiffness coefficient k. The return spring is undeflected when the armature–valve mass is seated and $x = 0$. At this point, the reader should be able to identify the various mechanical elements in both the solenoid diagram (Fig. 2.8) and its equivalent mechanical system diagram (Fig. 2.9).

We derive the mathematical model for this system by applying our two-step process. First, the FBD of mass m (see Fig. 2.10) is drawn with the external forces F_{em} (applied electromagnetic force), spring force, and damper force. The applied force F_{em} is positive to the right, as given in the definition of the problem. The proper direction of the spring force can be determined by assuming a positive displacement of mass m. Figure 2.9 shows that if $x > 0$, the spring is compressed, and, therefore, it will "push" on mass m to the left with a force equal to kx. Figure 2.10 shows the proper direction of the spring force, with the proper equation. The direction of the damper force is determined in a similar manner: a positive velocity for mass m is assumed, which results in a resistive friction force that opposes motion. Hence the damper force $b\dot{x}$ is also to the left on the FBD. If mass m was moving to the left ($\dot{x} < 0$), then the damper force acts to the right and the FBD in Fig. 2.10 is still valid.

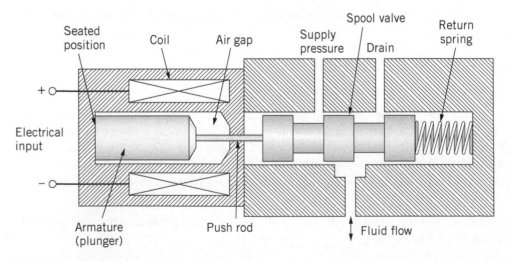

Figure 2.8 Solenoid actuator–valve system for Example 2.1.

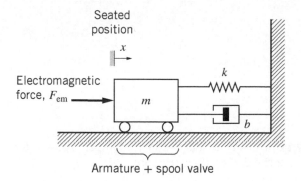

Seated position

x

Electromagnetic force, F_{em}

m

k

b

Armature + spool valve

Figure 2.9 Solenoid actuator as a mechanical system (Example 2.1).

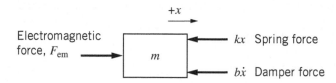

$+x$

Electromagnetic force, F_{em}

m

kx Spring force

$b\dot{x}$ Damper force

Figure 2.10 Free-body diagram for the solenoid actuator (Example 2.1).

Next, apply Newton's second law and sum all external forces on mass m with the sign convention as positive to the right

$$+ \to \sum F = F_{em} - kx - b\dot{x} = m\ddot{x}$$

It is common practice to place all "dynamic" or "output" variables and their derivatives on one side of the equal sign and all input variables on the other side. Therefore, rearranging the previous equation, we get

$$m\ddot{x} + b\dot{x} + kx = F_{em} \tag{2.21}$$

Equation (2.21) is the mathematical model of the mechanical component of the solenoid actuator. It is known as a *mass–spring–damper system* because it consists of an inertia, stiffness, and friction element. The mathematical model (2.21) is a linear, second-order ODE. In general, we obtain one second-order ODE for *each* inertia element of a mechanical system, provided that we use displacements of the inertia elements in the mathematical model. This fact arises from applying $F = m\ddot{x}$ to *each* mass, which results in a second-order ODE. The simple mass–spring–damper system in Fig. 2.9 is a one degree-of-freedom (1-DOF) system, as only one independent coordinate variable (x in this case) is needed to determine the position of the single inertia element. The electromagnetic force F_{em} is the input to the system. We revisit this solenoid example in Chapter 3 and discuss how the electromagnetic force is generated by the current in the coil.

Vertical Motion

Example 2.1 presents a mechanical system with horizontal translational motion. Many mechanical systems involve vertical translational motion where the gravitational forces must be accounted for in the dynamical equations. One example is the suspension system for an automobile, where shock absorbers and springs support the axle–wheel assembly and dampen vibrations from the road. Whether or not gravitational forces explicitly appear in the mathematical model depends on the choice of displacement coordinates, which we demonstrate with a simple 1-DOF example.

Example 2.2

The mechanical system shown in Fig. 2.11a is composed of a single mass m, a spring k, and a damper b. The vertical position of the mass is x, which is measured from the *undeflected* position of the spring.

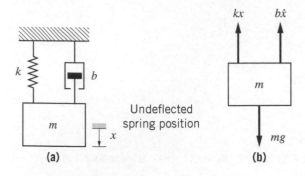

Figure 2.11 (a) Vertical mass–spring–damper system for Example 2.2 and (b) free-body diagram.

Figure 2.11b shows the FBD of mass m, which includes the spring force, damper force, and gravitational force mg. Summing all external forces with downward as the positive sign convention (see Fig. 2.11a) and applying Newton's second law, we obtain

$$+ \downarrow \sum F = -kx - b\dot{x} + mg = m\ddot{x}$$

Rearranging this equation with all dynamic variables on the left-hand side, we have

$$m\ddot{x} + b\dot{x} + kx = mg \qquad (2.22)$$

Equation (2.22) is the mathematical model of the vertical mass–spring–damper system for the case when displacement x is measured from the undeflected spring position. Note that the gravitational force mg appears on the right-hand side as an input to the mechanical system model.

We can rederive the mathematical model so that the gravitational force mg does not explicitly appear in the ODE. To begin, consider the case when the vertical mechanical system in Fig. 2.11a is at rest in static equilibrium, that is, $\ddot{x} = \dot{x} = 0$. Therefore, from the mathematical model (2.22), we obtain $kx = mg$ and the spring force balances the gravitational force. Define the static deflection of the spring as

$$d = \frac{mg}{k} \qquad (2.23)$$

Next, define z as the position of the mass relative to its static deflection, so that the total deflection is $x = d + z$. In other words, when $z = 0$, the mass is at its static deflection position, or $x = d$. We can compute the first and second time derivatives of $x = d + z$ to obtain $\dot{x} = \dot{d} + \dot{z}$ and $\ddot{x} = \ddot{d} + \ddot{z}$, which simplifies to $\dot{x} = \dot{z}$ and $\ddot{x} = \ddot{z}$ because the static deflection d is constant. Finally, we can substitute $\ddot{x} = \ddot{z}$, $\dot{x} = \dot{z}$, and $x = d + z$ into the mathematical model (2.22) to obtain

$$m\ddot{z} + b\dot{z} + k(d + z) = mg$$

The static deflection spring force cancels the gravitational force, or $kd = mg$, so our mathematical model becomes

$$m\ddot{z} + b\dot{z} + kz = 0 \qquad (2.24)$$

Equation (2.24) is the mathematical model of the simple mechanical system in terms of position variable z (deflection from static equilibrium). It is identical to modeling equation (2.22). Note that the gravitational force does not appear in Eq. (2.24). In general, the gravitational forces do not appear in the mathematical modeling equations for mechanical systems with vertical motion when all position variables are referenced to their respective static equilibrium positions. This convention is usually adopted because laboratory measurements of physical systems are typically referenced to a "zero" position when the system is at rest. The next example will further demonstrate this concept.

Example 2.3

Figure 2.12a shows a schematic diagram of a seat-suspension system, which is designed to attenuate (suppress) the road vibrations transmitted to the driver [1]. Derive the complete mathematical model.

Figure 2.12b shows the lumped mechanical model of the seat-suspension system. Mass m_1 is the total mass of the seat, while mass m_2 represents the mass of the driver. Ideal spring k_1 and viscous friction damper b_1 model a shock absorber

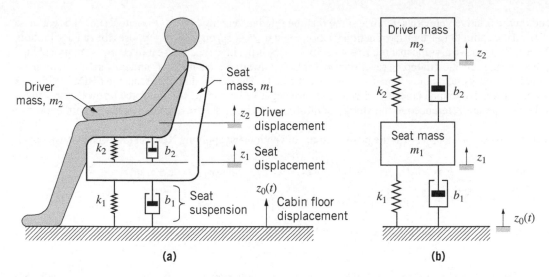

(a)   **(b)**

Figure 2.12 (a) Schematic diagram of the seat-suspension system for Example 2.3. (b) Mechanical model for the seat-suspension system.

connecting the seat to the vehicle's cabin floor. Spring constant k_2 and friction coefficient b_2 represent the stiffness and damping of the seat cushion, respectively. Finally, z_1 is the vertical displacement of the seat mass and z_2 is the displacement of the driver mass, and both are measured relative to their static equilibrium positions. The vertical displacement of the cabin floor (due to road vibrations) is $z_0(t)$ (note that upward is the positive sign convention for all displacements as defined by Reference 1). A complete description of the seat-suspension system analysis and design is presented in Chapter 11. However, all design problems involving dynamic systems begin with the development of mathematical models, which are presented in this example.

Figure 2.13 shows the FBDs of the two-mass mechanical system. The positive (upward) conventions for displacements z_1 and z_2 are also presented. Clearly, all spring and damper forces depend on the relative displacements and velocities between the seat mass and cabin floor, and the driver and seat masses, respectively. If we assume that relative displacement $z_1 - z_0$ is positive, then the suspension spring k_1 is in tension and the reaction force acts downward on seat mass m_1 as shown in Fig. 2.13. Similarly, if we assume that relative displacement $z_1 - z_2$ is positive, then the seat cushion is compressed and the reaction force acts downward on seat mass m_1 and upward on driver mass m_2 as shown by the equal-and-opposite spring force $k_2(z_1 - z_2)$ in the FBD.

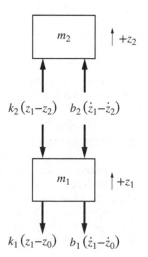

Figure 2.13 Free-body diagram for the seat-suspension system (Example 2.3).

Friction forces depend on the relative velocities. If we assume that the relative velocity $\dot{z}_1 - \dot{z}_0$ is positive (i.e., the seat mass m_1 is moving "away" from the cabin floor), then the reaction friction force $b_1(\dot{z}_1 - \dot{z}_0)$ on mass m_1 opposes the relative motion, as shown on the FBD. Similarly, if we assume that relative velocity $\dot{z}_1 - \dot{z}_2$ is positive (i.e., the seat mass m_1 is "approaching" the driver mass m_2), then the seat cushion damping reaction force acts downward on seat mass m_1 and upward on driver mass m_2 as shown by the equal-and-opposite damper force $b_2(\dot{z}_1 - \dot{z}_2)$ in the FBD. The reader should see that the FBDs in Fig. 2.13 remain valid if the assumed relative displacements and velocities are negative, in which case the force arrows are reversed. Finally, because displacements are referenced to the static equilibrium positions, the gravitational forces do not appear in the FBDs.

Summing all external forces with upward as the positive sign convention and applying Newton's second law, we obtain

$$\text{Mass 1: } + \uparrow \sum F = -k_2(z_1 - z_2) - b_2(\dot{z}_1 - \dot{z}_2) - k_1(z_1 - z_0) - b_1(\dot{z}_1 - \dot{z}_0) = m_1\ddot{z}_1$$

$$\text{Mass 2: } + \uparrow \sum F = k_2(z_1 - z_2) + b_2(\dot{z}_1 - \dot{z}_2) = m_2\ddot{z}_2$$

Rearranging these equations with the dynamic variables (z_1 and z_2) on the left-hand side and the input variable (z_0) on the right-hand side, we have

$$m_1\ddot{z}_1 + b_1\dot{z}_1 + b_2(\dot{z}_1 - \dot{z}_2) + k_1z_1 + k_2(z_1 - z_2) = b_1\dot{z}_0(t) + k_1z_0(t) \tag{2.25a}$$

$$m_2\ddot{z}_2 + b_2(\dot{z}_2 - \dot{z}_1) + k_2(z_2 - z_1) = 0 \tag{2.25b}$$

Equations (2.25a) and (2.25b) represent the mathematical model of the seat-suspension system. Because we have two inertia elements, the complete model consists of two second-order ODEs. Both ODEs are coupled, which means we cannot solve one ODE separately from the other. The model is linear because we have assumed linear stiffness and damper elements. Furthermore, the reader should note that all terms pertaining to the acceleration, velocity, and position of mass m_1 in Eq. (2.25a) have the same sign, that is, they are positive. Similarly, all terms associated with z_2 (and its derivatives) in Eq. (2.25b) have the same sign. In general, this condition holds for a system that is inherently stable. Intuition tells us that the seat-suspension system in Fig. 2.12 is stable and will always return to static equilibrium when input vibrations $z_0(t)$ have ceased. In Chapter 7, we discuss stability in detail when we investigate the response of dynamic systems.

Mechanical Systems with Nonlinearities

Many mechanical systems involve nonlinear effects such as Coulomb (or dry) friction, gear backlash, and stiffness elements that do not exhibit linear force–displacement characteristics. Another nonlinear effect is the presence of discontinuous forces, such as contact or damper forces that occur only when a mass remains in contact with a mechanical element that has stiffness and/or damping properties. We present mechanical systems with nonlinear effects in the following examples.

Example 2.4

Consider again the solenoid actuator–valve system shown in Fig. 2.8 and discussed in Example 2.1. Let us assume that Coulomb or dry friction acts on the armature–valve mass along with linear viscous friction. Derive the mathematical model of the mechanical system with this nonlinear friction effect.

Figure 2.14 shows the solenoid actuator with the return spring k, viscous friction coefficient b (due to motion of the spool valve in the hydraulic fluid), and dry friction force (due to sliding motion of the armature in the electric coil). The electromagnetic force F_{em} is the same as in Example 2.1.

Coulomb or dry friction is the kinetic friction force that exists when a mass is sliding relative to an unlubricated flat surface. From basic mechanics, we know that the magnitude of the dry friction force is $F_{dry} = \mu_k N$, where μ_k is the coefficient of kinetic friction and N is the normal force. Because the dry friction force always opposes the direction of motion, it is usually modeled as $F_{dry}\text{sgn}(\dot{x})$, where the operator "sgn" is the *signum function*, which returns the sign of its input value. In this case, the input to the signum function is velocity $\dot{x}$. Consequently, $\text{sgn}(\dot{x}) = 1$ when $\dot{x} > 0$, $\text{sgn}(\dot{x}) = -1$ when $\dot{x} < 0$, and $\text{sgn}(\dot{x}) = 0$ when $\dot{x} = 0$. Figure 2.15 shows the discontinuous nature of the dry friction force, which is clearly a nonlinear function of velocity $\dot{x}$.

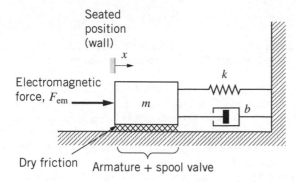

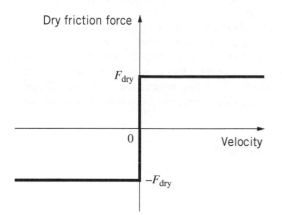

Figure 2.14 Solenoid actuator with dry friction (Example 2.4).

Figure 2.15 Dry friction force as a function of velocity.

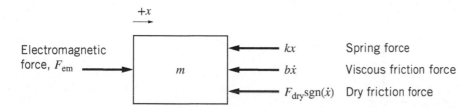

Figure 2.16 Free-body diagram for the solenoid actuator with dry friction (Example 2.4).

Figure 2.16 shows the FBD of the mechanical system with the spring, linear damper, dry friction, and electromagnetic forces. The reader should note that the dry friction force $F_{dry}\text{sgn}(\dot{x})$ in Fig. 2.16 will always oppose the motion of mass m regardless of the sign of velocity $\dot{x}$. Summing all external forces on mass m with the sign convention as positive to the right yields

$$+ \rightarrow \sum F = F_{em} - kx - b\dot{x} - F_{dry}\text{sgn}(\dot{x}) = m\ddot{x}$$

Rearranging this equation with all dynamic variables on the left-hand side yields

$$m\ddot{x} + b\dot{x} + F_{dry}\text{sgn}(\dot{x}) + kx = F_{em} \tag{2.26}$$

Equation (2.26) is the mathematical model of the mechanical component of the solenoid actuator system. It is nonlinear because of the inclusion of dry friction. If the dry friction force is ignored, Eq. (2.26) becomes the linear mathematical model of the actuator as derived in Example 2.1, or Eq. (2.21).

Example 2.5

Consider again the solenoid actuator shown in Fig. 2.8 and discussed in Examples 2.1 and 2.4. In most solenoid designs, the return spring will have a "preload" due to its compression when the valve is seated. Derive the mathematical model of the mechanical system with a preloaded return spring.

In Examples 2.1 and 2.4, the return spring was undeflected when the armature–valve mass was seated at the zero position. Therefore, when the electromagnetic force F_{em} is zero, the mathematical models (2.21) and (2.26) are satisfied when the armature–valve mass is at rest in the seated position, or $x = \dot{x} = \ddot{x} = 0$. However, a preloaded, compressed return spring will provide a force (to the left) when the valve is seated ($x = 0$) and at rest ($\dot{x} = \ddot{x} = 0$). Therefore, a wall-contact force acts on the left side of the armature mass to balance the preloaded spring when the system is in static equilibrium. We must incorporate the preload spring force and wall-contact force in our new mechanical model.

Figure 2.17 shows the mechanical system with the spring preload force F_{PL} and wall-contact force F_C. The mass displacement x is measured from the seated position. The seated position occurs when the left side of the armature mass is in contact with the wall. It is important to note that the wall-contact force can only "push" to the right and cannot "pull" the armature mass to the left. Consequently, the wall-contact discontinuity must be modeled correctly.

Figure 2.18 shows the FBD of the mechanical system with the spring preload and wall-contact force. The spring force kx, friction forces, and electromagnetic force F_{em} are the same as in Example 2.4. The spring force kx in Fig. 2.18 is due to the *additional* compression of the return spring when the mass is displaced to the right, or $x > 0$. Thus, the total spring force is $F_{PL} + kx$. Note that the return spring can never be in tension, as displacement x has a lower limit of zero because of wall contact and a compressive preload exists when $x = 0$. The wall-contact force will instantly become zero when the armature–valve mass is displaced from the seat, or $x > 0$. Summing all external forces on mass m with the sign convention as positive to the right yields

$$+ \rightarrow \sum F = F_{em} + F_C - kx - F_{PL} - b\dot{x} - F_{dry}\text{sgn}(\dot{x}) = m\ddot{x}$$

Rearranging this equation with all dynamic variables on the left-hand side yields

$$m\ddot{x} + b\dot{x} + F_{dry}\text{sgn}(\dot{x}) + kx = F_{em} - F_{PL} + F_C \qquad (2.27)$$

Equation (2.27) is the mathematical model of the mechanical system; however, we must account for the discontinuous wall-contact force F_C when the mass becomes unseated, or $x > 0$. Clearly, when the system is in static equilibrium and the mass

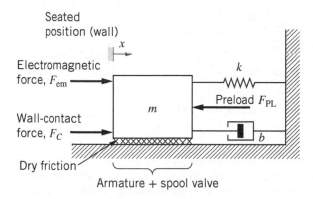

Figure 2.17 Solenoid actuator with preloaded return spring (Example 2.5).

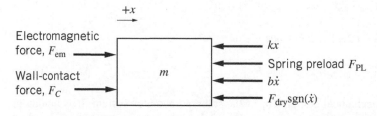

Figure 2.18 Free-body diagram for the solenoid actuator with preloaded return spring (Example 2.5).

is in the seated position (i.e., $\ddot{x} = \dot{x} = x = 0$), the right-hand side of Eq. (2.27) must equal zero. In this case, the contact force balances the difference between the spring preload and electromagnetic forces, or $F_C = F_{PL} - F_{em}$. However, when the electromagnetic force exceeds the spring preload force, a positive net force causes the mass to accelerate and eventually the mass is unseated and the contact force is zero. Therefore, the contact force is defined by

$$F_C = \begin{cases} F_{PL} - F_{em} & \text{if} \quad F_{em} < F_{PL} \\ 0 & \text{if} \quad F_{em} \geq F_{PL} \end{cases} \tag{2.28}$$

Equations (2.27) and (2.28) constitute the mathematical model of the solenoid actuator–valve system with a preloaded spring. We revisit this solenoid actuator in Chapter 3 when we discuss electromechanical systems and in Chapter 6 when we discuss numerical simulations.

Example 2.6

Figure 2.19 shows a schematic diagram of a piezoelectric actuator, which is designed to maintain contact with a slide mass and move it to a desired position [2]. Derive the complete mathematical model.

The actuator shown in Fig. 2.19 uses two sets of ceramic piezoelectric materials to provide external forces on the clamp mass m_1 (lead zirconate titanate or PZT is a common piezoelectric material that exhibits a mechanical deformation when a voltage is applied to a layer of ceramic materials). A vertical "stack" of PZT layers (not shown) provides a vertical force that "clamps" mass m_1 to the slide mass m_2 as shown in Fig. 2.19a. The horizontal stack of PZT layers extends when a voltage is applied and, therefore, pushes mass m_1 to the right as shown in Fig. 2.19b. The slide mass m_2 can be moved to a desired horizontal position by using the following sequence:

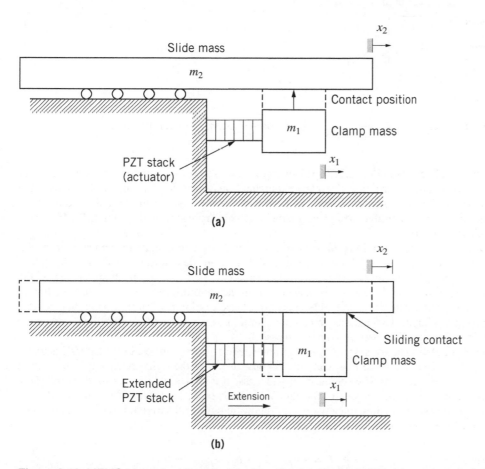

Figure 2.19 MEMS piezoelectric actuator operation for Example 2.6: (a) mass m_1 is "clamped" to slide mass m_2 and (b) PZT stack extends to move slide mass.

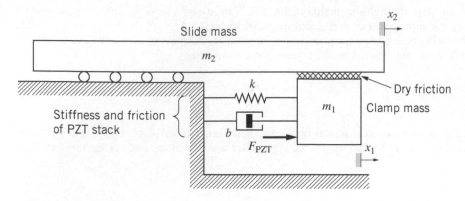

Figure 2.20 Mechanical model of the MEMS piezoelectric actuator (Example 2.6).

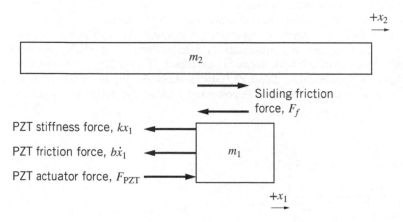

Figure 2.21 Free-body diagram of the MEMS piezoelectric actuator where mass m_1 slides relative to mass m_2 (Example 2.6).

(1) bring mass m_1 in contact with the slide mass (Fig. 2.19a), (2) energize and extend the actuator PZT stack in order to move masses m_1 and m_2 to the right (Fig. 2.19b), and (3) release the clamp mass and let it return to the undeflected (starting) position. This "clamp-extend-release" sequence is repeated until the slide mass is moved to the desired position. The actuator shown in Fig. 2.19 is proposed for manufacturing a very small microelectromechanical system (MEMS) where positioning a work piece with high accuracy is required.

Figure 2.20 shows the PZT actuator system as a lumped mechanical system with clamp mass m_1 and slide mass m_2. The stiffness and friction inherent in the horizontal PZT actuator are modeled by an ideal spring k and ideal damper b. Applying a voltage to the PZT stack extends the actuator and produces the force F_{PZT} that can only push on mass m_1 to the right as shown in Fig. 2.20. Position x_1 of the clamp mass m_1 is measured from the equilibrium or nonextended position of the actuator (i.e., zero applied voltage). The clamp mass and slide mass experience Coulomb (dry) friction at their contact surfaces when relative motion exists between the two masses. When the clamp mass is released from the slide mass at the end of the extension stroke, contact no longer exists between the two masses.

Figure 2.21 shows the FBD of the PZT actuator system. Stiffness and friction forces from the extended PZT actuator will act to the left as shown in Fig. 2.21. The PZT force F_{PZT} can only act on mass m_1 to the right (extension). The friction force F_f that results from contact and relative motion between mass m_1 and mass m_2 is shown in Fig. 2.21 for the case where the velocity of the clamp mass m_1 is greater than the velocity of the slide mass, or $\dot{x}_1 > \dot{x}_2$. This clamp-friction force F_f acts in equal-and-opposite pairs on the two masses according to Newton's third law. Summing all external forces on clamp mass m_1 and slide mass m_2 with the sign convention as positive to the right yields

Clamp mass: $\qquad + \rightarrow \sum F = -F_f - kx_1 - b\dot{x}_1 + F_{PZT} = m_1\ddot{x}_1$

Slide mass: $\qquad + \rightarrow \sum F = F_f = m_2\ddot{x}_2$

Rearranging these equations with all dynamic variables on the left-hand side yields

$$m_1\ddot{x}_1 + b\dot{x}_1 + kx_1 = F_{\text{PZT}} - F_f \tag{2.29}$$

$$m_2\ddot{x}_2 = F_f \tag{2.30}$$

When the clamp mass m_1 is in contact with mass m_2, the sliding friction force is

$$F_f = F_{\text{dry}}\,\text{sgn}(\dot{x}_1 - \dot{x}_2) \tag{2.31}$$

where the dry friction force $F_{\text{dry}} = \mu_k N_C$ is proportional to the clamping force N_C (produced by the vertical PZT stack) that acts normal to the slide mass m_2. The reader should note that the signum function sgn is used in Eq. (2.31) to determine the direction (or sign) of the friction force F_f. If $\dot{x}_1 - \dot{x}_2 > 0$ (e.g., clamp mass m_1 is moving to the right at a faster rate than mass m_2), then the arrows for friction force F_f are shown correctly in Fig. 2.21. However, Eq. (2.31) will correctly compute the friction force if $\dot{x}_1 - \dot{x}_2 < 0$ and mass m_2 is moving faster than mass m_1.

When the clamp mass m_1 is released ($N_C = 0$) and not in contact with the slide mass, the friction force is zero. Typically, the PZT actuator force F_{PZT} is set to zero during the release phase, and the stiffness force of the extended actuator returns the clamp mass m_1 to its equilibrium position ($x_1 = 0$). Using $F_{\text{PZT}} = 0$ and $F_f = 0$, Eqs. (2.29) and (2.30) become

$$\text{No contact (release):}\quad m_1\ddot{x}_1 + b\dot{x}_1 + kx_1 = 0 \tag{2.32}$$

$$m_2\ddot{x}_2 = 0 \tag{2.33}$$

In summary, the mathematical model of the PZT actuator system is composed of *two* sets of modeling equations

Sliding motion: $N_C > 0$

$$m_1\ddot{x}_1 + b\dot{x}_1 + kx_1 = F_{\text{PZT}} - \mu_k N_C\,\text{sgn}(\dot{x}_1 - \dot{x}_2) \tag{2.34a}$$

$$m_2\ddot{x}_2 = \mu_k N_C\,\text{sgn}(\dot{x}_1 - \dot{x}_2) \tag{2.34b}$$

No contact (release):

$N_C = 0$ and $F_{\text{PZT}} = 0$

$$m_1\ddot{x}_1 + b\dot{x}_1 + kx_1 = 0 \tag{2.35a}$$

$$m_2\ddot{x}_2 = 0 \tag{2.35b}$$

Equation sets (2.34) and (2.35) constitute the complete mathematical model of the PZT actuator. The complete system is nonlinear because of the dry friction force. Numerical integration of the system model is the only practical method for obtaining the system's response due to the nonlinear sliding friction. Furthermore, a numerical simulation must continuously monitor the normal clamping force N_C between the two masses and PZT actuator force F_{PZT} in order to switch to the appropriate set of ODEs that govern the system dynamics.

2.4 ROTATIONAL MECHANICAL SYSTEMS

Mathematical models of rotational mechanical systems can be derived using the same systematic two-step process that was used for translational systems:

1. Draw an FBD of each moment of inertia with arrows denoting external torques acting on each moment of inertia. Make use of Newton's third law to show equal-and-opposite reaction torques. Carefully assign equations for each torque using the appropriate element law and the positive conventions for angular displacement variables.

2. Apply Newton's second law to each moment of inertia to obtain the mathematical model of the entire mechanical system.

Newton's second law for rotational systems states that the sum of all external torques acting on a body is equal to the product of the moment of inertia J and angular acceleration $\ddot{\theta}$ of the body

$$\sum T = J\ddot{\theta} \tag{2.36}$$

The following examples demonstrate 1-DOF and 2-DOF rotational mechanical systems.

Example 2.7

Figure 2.22 shows a single-disk mechanical system, where the rotor is supported by bearings, and a motor provides input torque $T_{in}(t)$ directly to the rotor inertia J. Derive the mathematical model of the 1-DOF system.

In this example, we have a single displacement variable, angular position θ, which is measured clockwise from a fixed reference position as shown in Fig. 2.22. The rotor has friction due to the bearings and the fluid surrounding the rotor. Let us assume an ideal (linear) friction model, and lump the bearing and fluid friction into a single rotational friction coefficient b, with units of N-m-s/rad. Hence, the total friction torque is $b\dot{\theta}$, which always opposes the angular motion.

Figure 2.23 presents the FBD of the single moment of inertia J, showing the positive (clockwise) convention for angular displacement θ. The input torque $T_{in}(t)$ is in the positive direction. The direction of the friction torque is determined by assuming a positive angular velocity (clockwise). This positive angular velocity results in a resistive torque that opposes motion. Hence, $b\dot{\theta}$ is shown as counterclockwise on the FBD.

Summing all external torques on disk J with the sign convention as positive clockwise yields

$$\left(+\curvearrowright\right)\sum T = T_{in}(t) - b\dot{\theta} = J\ddot{\theta} \tag{2.37}$$

Rearranging Eq. (2.37), we obtain the mathematical model

$$J\ddot{\theta} + b\dot{\theta} = T_{in}(t) \tag{2.38}$$

Equation (2.38) is a linear, second-order model of the single-disk mechanical system. The dynamic variable is angular position θ, and the input is the applied torque $T_{in}(t)$.

Note that for this rotational mechanical system, angular position θ does not explicitly appear in the second-order model (2.38). Therefore, we can rewrite Eq. (2.38) as a *first-order* model by using angular velocity ω as the dynamic variable. Substituting $\omega = \dot{\theta}$ and $\dot{\omega} = \ddot{\theta}$ in Eq. (2.38) yields the first-order model

$$J\dot{\omega} + b\omega = T_{in}(t) \tag{2.39}$$

Solving the second-order model (2.38) will determine the angular position as a function of time, or $\theta(t)$, while solving the first-order model (2.39) will yield angular velocity $\omega(t)$.

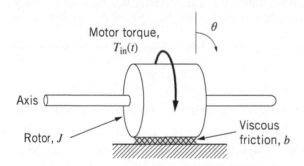

Figure 2.22 Single-disk mechanical system for Example 2.7.

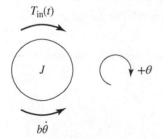

Figure 2.23 Free-body diagram for the rotor (Example 2.7).

Example 2.8

Figure 2.24 shows a wind turbine generator used for transforming mechanical energy into electrical energy. For this problem, let us assume that the turbine inertia J_1 and generator inertia J_2 are both rigidly connected to their respective gears in the gear train. Derive the mathematical model of this rotational mechanical system.

The turbine inertia J_1 consists of the wind turbine blades, turbine shaft, and gear 1 (with radius r_1), and the generator inertia J_2 consists of gear 2 (with radius r_2) and the generator rotor. Both the turbine and generator inertias experience viscous friction modeled by b_1 and b_2, respectively. The turbine blades extract energy from the wind and produce the aerodynamic torque T_{aero}, which is the input to the system and drives the gear train. The generator disk J_2 includes a coil of wire, and the rotational motion of the wire windings in a magnetic field generates electrical energy. In addition, the wire coil carrying current in the magnetic field experiences an induced force that opposes the motion of the generator, which is represented by the generator torque T_{gen} shown in Fig. 2.24. The details of the electromagnetic interactions are discussed in Chapter 3; for now, the reader should accept that T_{gen} is a known torque that opposes the positive rotation of the generator inertia J_2 as shown in Fig. 2.24.

Figure 2.25 shows the FBD of the wind turbine generator system. The positive convention for angular rotations θ_1 and θ_2 is shown on the figure. The contact force f_C at the mesh point between the gears is illustrated as a pair of equal-and-opposite forces, in accordance with Newton's third law. Because the aerodynamic torque T_{aero} provides a positive torque to the input shaft, the contact force f_C provides a positive transmitted torque to the output (generator) shaft, which is equal to $f_C r_2$. Summing the torques for each inertia element and applying Newton's second law yields

Turbine: (+ clockwise) $\left(+\searrow\right)\sum T = T_{\text{aero}} - b_1\dot{\theta}_1 - f_C r_1 = J_1\ddot{\theta}_1$ (2.40)

Generator: (+ counterclockwise) $\left(+\nearrow\right)\sum T = f_C r_2 - b_2\dot{\theta}_2 - T_{\text{gen}} = J_2\ddot{\theta}_2$ (2.41)

Our wind turbine generator system has *one* degree of freedom as angular rotations θ_1 and θ_2 are not independent because of the gear train. The velocity of both gears at their mesh point is $r_1\dot{\theta}_1 = r_2\dot{\theta}_2$, and the time derivative of the mesh point velocity yields $r_1\ddot{\theta}_1 = r_2\ddot{\theta}_2$. Therefore, Eqs. (2.40) and (2.41) are not independent. We can use Eq. (2.41) to determine the unknown contact force f_C

$$f_C = \frac{1}{r_2}(b_2\dot{\theta}_2 + T_{\text{gen}} + J_2\ddot{\theta}_2)$$

and substitute this expression into Eq. (2.40), which results in

$$J_1\ddot{\theta}_1 + b_1\dot{\theta}_1 = T_{\text{aero}} - \frac{r_1}{r_2}(b_2\dot{\theta}_2 + T_{\text{gen}} + J_2\ddot{\theta}_2)$$ (2.42)

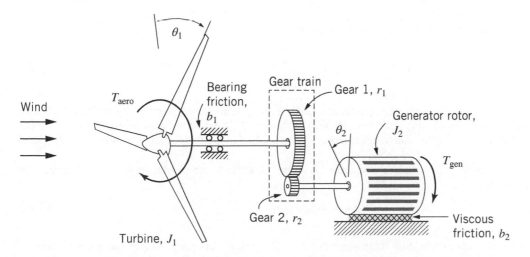

Figure 2.24 Wind turbine generator system for Example 2.8.

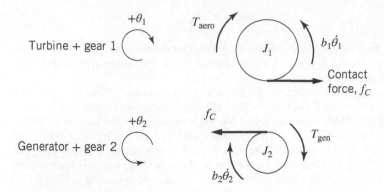

Figure 2.25 Free-body diagram of wind turbine generator system (Example 2.8).

We choose to write our system model in terms of turbine rotation angle θ_1; therefore, substitute $\dot{\theta}_2 = (r_1/r_2)\dot{\theta}_1$ for the generator shaft angular velocity and $\ddot{\theta}_2 = (r_1/r_2)\ddot{\theta}_1$ for the generator shaft angular acceleration in Eq. (2.42). Moving all dynamic variables to the left-hand side yields

$$J_1\ddot{\theta}_1 + \frac{r_1^2}{r_2^2}J_2\ddot{\theta}_1 + b_1\dot{\theta}_1 + \frac{r_1^2}{r_2^2}b_2\dot{\theta}_1 = T_{\text{aero}} - \frac{r_1}{r_2}T_{\text{gen}} \tag{2.43}$$

Finally, we can substitute the gear ratio $N = r_2/r_1$ into Eq. (2.43)

$$\left(J_1 + \frac{1}{N^2}J_2\right)\ddot{\theta}_1 + \left(b_1 + \frac{1}{N^2}b_2\right)\dot{\theta}_1 = T_{\text{aero}} - \frac{1}{N}T_{\text{gen}} \tag{2.44}$$

Equation (2.44) is the mathematical model of the wind turbine generator system. We can write the system model in a more compact form by defining the equivalent or "composite" inertia and friction coefficient as

$$J_{c1} = J_1 + \frac{1}{N^2}J_2$$

$$b_{c1} = b_1 + \frac{1}{N^2}b_2$$

Therefore, the complete system model using the composite coefficients is

$$J_{c1}\ddot{\theta}_1 + b_{c1}\dot{\theta}_1 = T_{\text{aero}} - \frac{1}{N}T_{\text{gen}} \tag{2.45}$$

Composite terms J_{c1} and b_{c1} represent the equivalent inertia and friction coefficients experienced by the turbine shaft. The generator inertia "reflected" back to the turbine shaft through the gear train is J_2/N^2, and the generator friction coefficient "reflected" to the turbine shaft is b_2/N^2. The equivalent external torque applied to the turbine shaft is the sum of T_{aero} and $-T_{\text{gen}}/N$.

Finally, note that we can rewrite the complete mathematical model (2.45) in terms of angular velocity of the turbine shaft using $\omega_1 = \dot{\theta}_1$ and $\dot{\omega}_1 = \ddot{\theta}_1$

$$J_{c1}\dot{\omega}_1 + b_{c1}\omega_1 = T_{\text{aero}} - \frac{1}{N}T_{\text{gen}} \tag{2.46}$$

Equation (2.46) is a *first-order* model of the wind turbine generator, and its solution will only yield angular velocity information. The reader should note the similarity between the wind turbine generator model (2.46) and the single-disk mechanical model (2.39) from Example 2.7.

Example 2.9

Figure 2.26 shows a dual-disk mechanical system that has been proposed as an efficient generator for hybrid vehicles [3]. Derive the complete mathematical model.

The mechanical system represented by Fig. 2.26 is composed of a toroidal-segment piston (disk J_1) matched with a toroidal-segment cylinder (disk J_2). Both disks rotate about a common axis. The disks are connected by a torsional spring, represented by rotational spring constant k. Angular displacements θ_1 (piston disk) and θ_2 (cylinder disk) are measured from their equilibrium positions, with positive rotation as clockwise when viewing the system from the left. Both disks experience friction, which is modeling by viscous friction coefficient b. A gas-pressure torque from a diesel engine, $T_{in}(t)$, drives the two-disk system in equal-and-opposite pairs as indicated in Fig. 2.26. Angular rotation of the disks relative to stationary magnets generates electrical current and reaction torques, but these effects are not included in this example. During the normal operating mode, the input engine torque $T_{in}(t)$ is pulsed so that the elastic system deflects in a manner such that both disks vibrate about a mean angular motion in one direction. At this early stage of the textbook, it is important that the reader is able to derive the mathematical model given the mechanical system with its displacement and input variables defined in Fig. 2.26.

As with the previous examples, we start with an FBD of the coupled rotational mechanical system, shown in Fig. 2.27. Both disks are shown in the FBDs, with the positive (clockwise) angular displacements θ_1 and θ_2. Input torque $T_{in}(t)$ is shown opposing the positive rotation for piston disk J_1 and in the same direction as positive rotation for cylinder disk J_2, which is in agreement with the torque arrows given in Fig. 2.26. Both friction torques depend only on the angular velocities of the respective disks, and both torques oppose the positive rotation directions. The torque from twisting the torsional spring k depends on the relative angular displacement $\theta_1 - \theta_2$. If we assume that piston angle θ_1 is greater than cylinder angle θ_2, then the twist in the torsional spring k will impart a negative reaction torque on piston disk J_1 and an equal-and-opposite positive torque on cylinder disk J_2 as shown in the FBD. Of course, piston angle θ_1 can be less than cylinder angle θ_2. The reader should see that the spring torque arrows and corresponding equations shown in Fig. 2.27 remain valid in this case.

Summing all external torques with clockwise as the positive sign convention and applying Newton's second law for a rotational system, we obtain

$$\text{Disk 1:} \quad \left(+\curvearrowright\right)\sum T = -k\left(\theta_1 - \theta_2\right) - b\dot{\theta}_1 - T_{in}(t) = J_1\ddot{\theta}_1$$

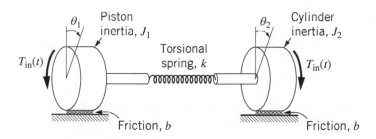

Figure 2.26 Mechanical model of the dual-disk generator for Example 2.9.

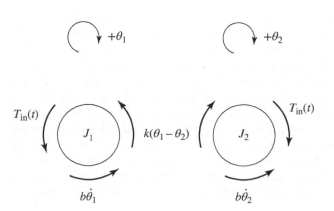

Figure 2.27 Free-body diagram for the dual-disk generator system (Example 2.9).

$$\text{Disk 2:} \qquad \left(+\right) \sum T = k(\theta_1 - \theta_2) - b\dot{\theta}_2 + T_{\text{in}}(t) = J_2\ddot{\theta}_2$$

Rearranging these equations with the dynamic variables (θ_1 and θ_2) on the left-hand side and the input variable $T_{\text{in}}(t)$ on the right-hand side, we have

$$J_1\ddot{\theta}_1 + b\dot{\theta}_1 + k(\theta_1 - \theta_2) = -T_{\text{in}}(t) \tag{2.47a}$$

$$J_2\ddot{\theta}_2 + b\dot{\theta}_2 + k(\theta_2 - \theta_1) = T_{\text{in}}(t) \tag{2.47b}$$

Equations (2.47a) and (2.47b) represent the mathematical model of the dual-disk generator system. Because we have two inertia elements, the complete model consists of two coupled, second-order ODEs. The model is linear, as we have assumed linear stiffness and damper elements. Note that all terms pertaining to the angular acceleration, velocity, and position of disk J_1 in Eq. (2.47a) have the same sign; similarly, all terms associated with θ_2 (and its derivatives) in Eq. (2.47b) have the same sign. This characteristic holds for inherently stable systems, and the reader is encouraged to use this sign check to verify that their FBD and algebra steps have led to the correct mathematical model.

We may rewrite the mathematical model in terms of the *relative* angular displacement between the cylinder disk J_2 and the piston disk J_1, that is, $\Delta\theta = \theta_2 - \theta_1$. Subtracting Eq. (2.47a) from Eq. (2.47b) yields

$$J_2\ddot{\theta}_2 - J_1\ddot{\theta}_1 + b(\dot{\theta}_2 - \dot{\theta}_1) + 2k(\theta_2 - \theta_1) = 2T_{\text{in}}(t)$$

Because the piston and cylinder inertias are equal (the dual-disk system is balanced), we can substitute $J = J_1 = J_2$. Furthermore, we can substitute the relative angular displacement variable $\Delta\theta = \theta_2 - \theta_1$ and its derivatives $\Delta\dot{\theta} = \dot{\theta}_2 - \dot{\theta}_1$ and $\Delta\ddot{\theta} = \ddot{\theta}_2 - \ddot{\theta}_1$ to yield

$$J\Delta\ddot{\theta} + b\Delta\dot{\theta} + 2k\Delta\theta = 2T_{\text{in}}(t) \tag{2.48}$$

Equation (2.48) is a single linear, second-order ODE modeling equation for the dual-disk mechanical system. In the case of equal disk inertias it represents the same system dynamics as Eqs. (2.47a) and (2.47b). However, Eq. (2.48) uses the relative angular displacement $\Delta\theta$ as the dynamic variable and hence its solution only provides information about the relative angle between the two disks. While the use of the relative angular twist $\Delta\theta$ as the single dynamic variable may at first seem restrictive, it can be used to compute the system's important performance measures. For example, consider the net input power: the power input to a mechanical system is the product of the input force (or torque) and velocity (or angular velocity). In this case, input torque $T_{\text{in}}(t)$ is defined as positive when it is in the same direction as positive rotation of the cylinder disk θ_2 and negative rotation of the piston disk θ_1 (see Fig. 2.26). Therefore, the net input power is

$$P_{\text{in}} = T_{\text{in}}(t)\dot{\theta}_2 - T_{\text{in}}(t)\dot{\theta}_1 = T_{\text{in}}(t)\Delta\dot{\theta} \tag{2.49}$$

Equation (2.49) shows that the net input power can be computed from the relative angular velocity between the two disks, which can be determined from the solution of the single modeling equation (2.48).

SUMMARY

In this chapter, we introduced a systematic approach for developing the mathematical model of mechanical systems. First, we discussed the physical characteristics of the inertia, stiffness, and energy-dissipation elements that make up a mechanical system. Next, we began the modeling process by drawing all forces on an FBD for each inertia element. Newton's third law is used to draw the equal-and-opposite reaction forces that exist between inertia elements. The summation of all forces in the assumed positive direction is equated to the product of mass and acceleration (Newton's second law). Thus, each inertia element in a mechanical system requires a second-order ODE because acceleration is the second time derivative of position. For example, a mechanical system composed of three lumped masses, each of which has an independent displacement variable, is modeled by three second-order ODEs. In the special case where the modeling equation does *not* depend on the displacement variable, velocity can be used as the dynamic variable and, therefore, we may represent a single lumped inertia with a *first*-order ODE. For a mechanical system with translational motion, the horizontal or vertical displacement is the dynamic variable. For rotational motion, we draw an FBD showing the applied torques on each moment of inertia and angular displacement is the dynamic variable.

REFERENCES

1. Choi, S.B., Choi, J.H., Lee, Y.S., and Han, M.S., "Vibration Control of an ER Seat Suspension for a Commercial Vehicle," *ASME Journal of Dynamic Systems, Measurement, and Control*, Vol. 125, Mar. 2003, pp. 60–68.

2. Salisbury, S.P., Mrad, R.B., Waechter, D.F., and Prasad, S.E., "Design, Modeling, and Closed-Loop Control of a Complementary Clamp Piezoworm Stage," *IEEE/ASME Transactions on Mechatronics*, Vol. 14, Dec. 2009, pp. 724–732.

3. Dunne, J.F., "Dynamic Modelling and Control of Semifree-Piston Motion in a Rotary Diesel Generator Concept," *ASME Journal of Dynamic Systems, Measurement, and Control*, Vol. 132, Sept. 2010, pp. 051003/1–051003/12.

4. Jayanth, G.R., "The Fundamental Bandwidth Limit of Piezoelectrically Actuated Nanopositioners With Motion Amplification," *ASME Journal of Dynamic Systems, Measurement, and Control*, Vol. 139, Nov. 2017, pp. 114501/1–114501/3.

5. Wilbanks, J.J., and Leamy, M.J., "Two-Scale Command Shaping for Reducing Powertrain Vibration During Engine Restart," *ASME Journal of Dynamic Systems, Measurement, and Control*, Vol. 139, Sept. 2017, pp. 091004/1–091004/11.

6. Zhu, H., Hu, C., and Liu, Y., "Optimum Design of a Passive Suspension System of a Semisubmersible for Pitching Reduction," *ASME Journal of Dynamic Systems, Measurement, and Control*, Vol. 138, Dec. 2016, pp. 121003/1–121003/8.

7. Liu, J.-J., and Yang, Y.-P., "Disk Wobble Control in Optical Disk Drives," *ASME Journal of Dynamic Systems, Measurement, and Control*, Vol. 127, Sept. 2005, pp. 508–514.

8. O'Connor, D.N., Eppinger, S.D., Seering, W.P., and Wormley, D.N., "Active Control of a High-Speed Pantograph," *ASME Journal of Dynamic Systems, Measurement, and Control*, Vol. 119, Mar. 1997, pp. 1–4.

9. Genin, J., Ginsberg, J.H., and Ting, E.C., "Longitude Train-Track Dynamics: A New Approach," *ASME Journal of Dynamic Systems, Measurement, and Control*, Vol. 96, Dec. 1974, pp. 466–469.

Chapter 3

Modeling Electrical and Electromechanical Systems

3.1 INTRODUCTION

Electrical circuits and electromechanical devices are used extensively by mechanical engineers in instrumentation and in the conversion between electrical energy and mechanical energy. This chapter introduces the fundamental techniques for developing the modeling equations for electrical systems. These systems are composed of resistor, capacitor, and inductor elements. Our mathematical models of electrical systems are developed by applying Kirchhoff's voltage and current laws for electrical circuits, as well as the element laws that govern the interaction between electrical charge, current, magnetic flux linkage, and voltage. Electromechanical systems involve the interaction between electrical and mechanical energy as demonstrated by motors, generators, and actuators. These systems require analysis of the electrical circuit in conjunction with free-body diagrams of the mechanical components.

As with mechanical systems in Chapter 2, we utilize a lumped-parameter approach, and, therefore, the mathematical model of electrical systems consist of ordinary differential equations (ODEs). While our goal in this chapter is to develop the methods for modeling electrical and electromechanical systems, the reader should bear in mind that we intend to emphasize engineering applications of these systems. Hence, we develop models for devices such as motors and actuators, and these examples are used throughout the remainder of the book. As with the previous chapter, we develop the mathematical models, but we do not obtain the solutions to the governing ODEs. Methods for obtaining the system response are presented in Chapters 6–9.

3.2 ELECTRICAL ELEMENT LAWS

An electrical system is composed of electrical elements, which can be grouped into two categories: (1) passive elements and (2) active elements. Passive elements cannot introduce energy into a system; they can only store or dissipate energy. Resistors, capacitors, and inductors are passive electrical elements. The three basic mechanical elements (inertia, stiffness, and friction) are also passive elements because they can only store energy (inertia and stiffness elements) or dissipate energy (friction elements). Therefore, it is possible to draw analogies between the passive electrical and mechanical elements. Active elements, on the other hand, can introduce energy into an electrical system. Voltage and current sources are active elements, and they are analogous to the force or motion inputs for a mechanical system.

This section presents brief descriptions of the fundamental laws that govern these electrical elements. We use the basic concepts of electricity and magnetism that are developed in university physics courses. Current I is the rate of flow of electrical charge q (in coulombs, C), or $I = \dot{q}$. Therefore, current I has units C/s or ampere (A). Voltage e (in volts, V) is the electrical potential difference between two points or the ends of a two-terminal element. At times we show that part of a circuit is connected to the "ground" where the voltage is zero.

Resistor

Resistors are electrical elements that hinder (resist) the flow of current. Resistors dissipate electrical energy by converting it into heat and hence they are analogous to friction elements in a mechanical system. Figure 3.1 shows the symbol for a two-terminal resistor element with resistance R. In Fig. 3.1, current I flows through the resistor element R and e_R is

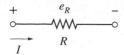

Figure 3.1 Resistor element.

the voltage potential across the two terminals (the plus sign in Fig. 3.1 denotes the higher electrical potential). Ohm's law defines the "voltage drop" e_R across an ideal resistor as

$$e_R = RI \tag{3.1}$$

where R is the resistance in V/A or ohms (Ω). Equation (3.1) is the voltage–current relationship for a linear resistor; resistors may exhibit a nonlinear relationship between current and voltage drop. Power is the time-rate of energy ξ and power for an electrical element is voltage $\times$ current in watts (W). Therefore, the power dissipated by a resistor is

$$\dot{\xi}_R = -e_R I = -RI^2 \tag{3.2}$$

Equation (3.2) has a minus sign to indicate that resistors always dissipate energy. The reader should note that power dissipated by a resistor is analogous to the power dissipated by a mechanical damper as presented by Eq. (2.15): $\dot{\xi}_f = -b\dot{x}^2$. Electrical resistance R is analogous to friction coefficient b and current I ($= \dot{q}$) is analogous to velocity $\dot{x}$.

Capacitor

Two conductors separated by a nonconducting medium form a capacitor. One example is metallic parallel plates separated by a thin dielectric material. Capacitors store energy in the electric field that results from a voltage potential across the two conductors. Figure 3.2 shows the symbol for a two-terminal capacitor with current I and voltage potential e_C across the two terminals. Ideal (linear) capacitors obey the charge–voltage relationship

$$q = Ce_C \tag{3.3}$$

where C is the capacitance in C/V or farads (F). Capacitance is a measure of the charge that can be stored for a given voltage across the conductors. Capacitance C depends on material and geometric properties, such as the area of the parallel plates and the distance between the two plates. We can relate capacitance to current by taking the time derivative of Eq. (3.3)

$$\dot{q} = I = C\dot{e}_C \tag{3.4}$$

The voltage drop across a capacitor can be obtained by integrating Eq. (3.4)

$$e_C(t) = e_C(0) + \frac{1}{C}\int_0^t I(\tau)d\tau \tag{3.5}$$

Capacitors can store energy due to their voltage

$$\xi_C = \frac{1}{2}Ce_C^2 \tag{3.6}$$

The time derivative of Eq. (3.6) yields the power

$$\dot{\xi}_C = Ce_C\dot{e}_C \tag{3.7}$$

Substituting Eq. (3.4) for $C\dot{e}_C$ in Eq. (3.7), we see that power is voltage $\times$ current.

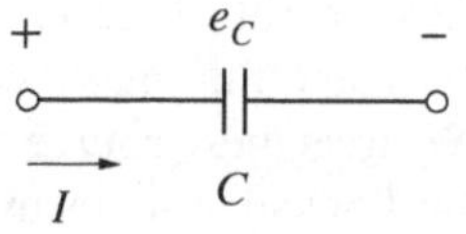

Figure 3.2 Capacitor element.

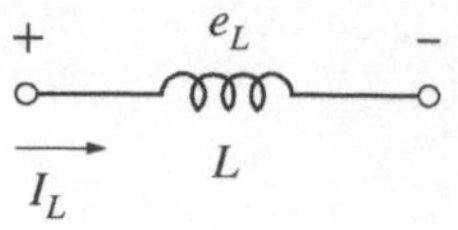

Figure 3.3 Inductor element.

Inductor

A simple coil of wire forms an inductor. Inductors store energy in the magnetic field that results from current flowing through the coil of wire. Figure 3.3 shows the symbol for a two-terminal inductor with current I_L and voltage potential e_L across the two terminals. Ideal inductors exhibit a linear relationship between current I_L and magnetic flux linkage λ

$$\lambda = LI_L \tag{3.8}$$

where L is the inductance in webers/ampere (Wb/A) or henries (H). Magnetic flux linkage λ has units of webers (Wb), and it is the product of magnetic flux density (Wb/m^2), coil area (m^2), and the number of turns (or loops) in the coil of wire. Inductance L depends on material and geometric properties, such as the number of loops (turns) and area of the coil. If the coil is wrapped around a ferromagnetic core, the inductance becomes a nonlinear function.

Faraday's law of magnetic induction states that a coil of wire will have a voltage difference induced across it if the magnetic flux changes with time. The time derivative of flux linkage is equal to the voltage across the inductor

$$\dot{\lambda} = e_L \tag{3.9}$$

For a fixed inductor with constant inductance L, we can substitute the time derivative of Eq. (3.8) into Eq. (3.9) to yield

$$e_L = L\dot{i}_L \tag{3.10}$$

Inductors can store energy in their magnetic field due to current

$$\xi_L = \frac{1}{2}LI_L^2 \tag{3.11}$$

The time derivative of Eq. (3.11) yields the power

$$\dot{\xi}_L = LI_L\dot{i}_L \tag{3.12}$$

Substituting Eq. (3.10) for $L\dot{i}_L$ in Eq. (3.11), we see that power is voltage × current.

Sources

We utilize two types of ideal sources for electrical systems: voltage and current sources. Figure 3.4a shows an ideal voltage source that provides the specified input voltage $e_{in}(t)$ to the circuit regardless of the amount of current being drawn from it. The positive terminal of the voltage source shown in Fig. 3.4a indicates the positive direction of current flow (current is assumed to flow from the higher potential to the lower potential). Figure 3.4b shows an ideal current source that provides the specified input current $I_{in}(t)$ to the circuit regardless of the amount of voltage that may be required. The arrow symbol in the current source denotes the positive convention for current flow. We treat these sources as the known inputs to the electrical system just as we treated force and displacement inputs for mechanical systems in Chapter 2.

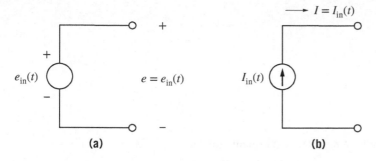

Figure 3.4 Ideal electrical sources: (a) voltage source and (b) current source.

3.3 ELECTRICAL SYSTEMS

In Chapter 2, we derived mathematical models of mechanical systems by drawing free-body diagrams and applying Newton's second law to each inertia element. In all cases, the "dynamic variables" of the mechanical model were displacement variables, such as the position for a translational system or angular displacement for a rotational system. It is clear that knowledge of these displacements and their derivatives (velocities) completely determine the total (kinetic + potential) energy of the mechanical system. For electrical systems, the important "dynamic variables" are voltage and current: Eqs. (3.4) and (3.10) show that capacitor voltage e_C and inductor current I_L are governed by first-order ODEs. Furthermore, the energy stored by capacitors and inductors depends on the capacitor voltage e_C and inductor current I_L; see Eqs. (3.6) and (3.11). Therefore, the mathematical models of electrical systems can be written in terms of the important "dynamic variables" (e_C and I_L) of the governing first-order ODEs for the energy-storage elements. The voltages and currents in an electrical circuit that are not e_C or I_L (e.g., voltage drop across a resistor) can be written in terms of the important dynamic variables by using Kirchhoff's laws.

Kirchhoff's Voltage Law

Kirchhoff's voltage (or "loop") law states that the algebraic sum of all voltages across the elements for any closed path (loop) is equal to zero. Figure 3.5 shows a circuit that consists of a single loop with voltage source $e_{in}(t)$ and three passive elements. These three elements can be any combination of resistors, capacitors, or inductors. The positive current flow I is shown in the figure, where current flows through each passive element from the positive terminal to the negative terminal. The convention is to assign a *negative* sign for a "voltage drop" (moving with the current across a passive element, or moving from + to − across an active voltage source) and a *positive* sign for a "voltage rise" (moving against

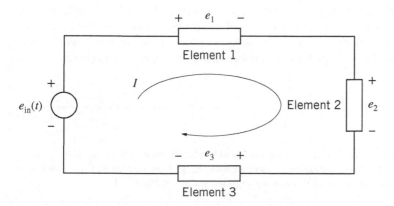

Figure 3.5 Example of Kirchhoff's voltage (loop) law.

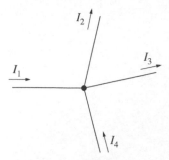

Figure 3.6 Example of Kirchhoff's current (node) law.

the current across a passive element, or moving from − to + across a voltage source). Summing the voltages in the loop (moving with the current in a clockwise direction) results in

$$\text{Clockwise:} \quad -e_1 - e_2 - e_3 + e_{\text{in}}(t) = 0 \tag{3.13}$$

Of course, we could sum voltages around the loop moving in a counterclockwise (against the current) direction

$$\text{Counterclockwise:} \quad -e_{\text{in}}(t) + e_3 + e_2 + e_1 = 0 \tag{3.14}$$

which yields the same algebraic result as Eq. (3.13).

Kirchhoff's Current Law

Kirchhoff's current (or "node") law states that the algebraic sum of all currents entering and leaving node is equal to zero. A node is defined as a junction of three or more wires, and we choose to assign a positive sign convention for current entering a node and a negative convention for current leaving a node. Figure 3.6 shows four wires carrying currents I_1, I_2, I_3, and I_4 where all wires meet at a single node. Applying Kirchhoff's node law yields

$$I_1 - I_2 - I_3 + I_4 = 0 \tag{3.15}$$

Mathematical Models of Electrical Systems

Mathematical models of electrical systems can be derived using a systematic two-step process:

1. Write the corresponding first-order ODE for each energy-storage element (capacitor or inductor). The dynamic variables of the ODEs will be either voltage e_C (for a capacitor) or current I_L (for an inductor).

2. Use Kirchhoff's laws to express the unknown voltages and currents in terms of either the dynamic variables associated with the energy-storage elements (e_C or I_L) or the sources (input voltage e_{in} or input current I_{in}).

In general, we write a first-order ODE for each energy-storage element; that is, each capacitor and each inductor. For example, a circuit with two capacitors and one inductor will result in a third-order mathematical model (i.e., three first-order ODEs). It is important that the complete model of the electrical system be in terms of the dynamic variables associated with the energy-storage elements and the source input variable. The two-step modeling process is best illustrated by the following examples.

Example 3.1

Figure 3.7 shows a series RL circuit with a voltage source. Derive the mathematical model of the electrical system.

This circuit contains a single energy-storage element (inductor L) and a single loop. Consequently, the model will consist of one first-order ODE, which is the dynamic equation for current through the inductor, Eq. (3.10)

$$L\dot{I}_L = e_L \tag{3.16}$$

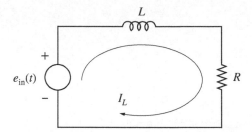

Figure 3.7 Series RL circuit for Example 3.1.

Next, we must express inductor voltage e_L in terms of the dynamic variable I_L and/or the source voltage $e_{in}(t)$. To do so, we apply Kirchhoff's voltage law around the loop in a clockwise direction

$$-e_L - e_R + e_{in}(t) = 0 \qquad (3.17)$$

Substituting Ohm's law for the voltage across the resistor ($e_R = RI_L$) in Eq. (3.17), the inductor voltage is

$$e_L = e_{in}(t) - RI_L \qquad (3.18)$$

Substituting Eq. (3.18) into the first-order ODE (3.16) yields

$$L\dot{I}_L = e_{in}(t) - RI_L$$

Finally, we move all terms involving the dynamic variable I_L to the left-hand side

$$L\dot{I}_L + RI_L = e_{in}(t) \qquad (3.19)$$

Equation (3.19) is the mathematical model of this simple series RL circuit. It is a linear, time-invariant first-order ODE.

For this simple example, we could have started with Kirchhoff's voltage law, Eq. (3.17), repeated below

$$-e_L - e_R + e_{in}(t) = 0$$

Next, we simply substitute the appropriate element laws for voltage across an inductor ($e_L = L\dot{I}_L$) and voltage across a resistor ($e_R = RI_L$) to yield the mathematical model (3.19). For simple single-loop circuits, it may be easier to derive the mathematical model by starting with the loop equation from Kirchhoff's voltage law.

Example 3.2

Figure 3.8 shows a series RLC circuit with a voltage source. Derive the mathematical model of the electrical system.

This circuit contains two energy-storage elements: inductor L and capacitor C. Therefore, the model consists of two first-order ODEs in terms of capacitor voltage e_C and inductor current I_L

$$\text{Capacitor voltage:} \quad C\dot{e}_C = I_L \qquad (3.20)$$

$$\text{Inductor current:} \quad L\dot{I}_L = e_L \qquad (3.21)$$

The two important dynamic variables are capacitor voltage e_C and inductor current I_L, and the system input is source voltage $e_{in}(t)$. Hence, Eq. (3.20) is already complete, because its right-hand side is expressed in terms of the current I_L. Next, we must express the inductor voltage e_L in Eq. (3.21) in terms of e_C, I_L, or $e_{in}(t)$. To do so, we apply Kirchhoff's voltage law around the single loop in a clockwise direction

$$-e_R - e_L - e_C + e_{in}(t) = 0 \qquad (3.22)$$

Substituting Ohm's law ($e_R = RI_L$) into Eq. (3.22) and solving for inductor voltage e_L yields

$$e_L = -RI_L - e_C + e_{in}(t) \qquad (3.23)$$

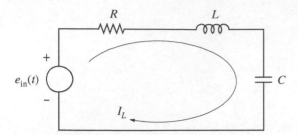

Figure 3.8 Series RLC circuit for Example 3.2.

Equation (3.23) can be substituted into Eq. (3.21) to yield

$$L\dot{I}_L = -RI_L - e_C + e_{in}(t) \tag{3.24}$$

Finally, moving all terms involving the dynamic variables e_C and I_L in Eqs. (3.20) and (3.24) to the left-hand sides yields

$$C\dot{e}_C - I_L = 0 \tag{3.25}$$

$$L\dot{I}_L + RI_L + e_C = e_{in}(t) \tag{3.26}$$

Equations (3.25) and (3.26) are the mathematical modeling equations for the series RLC electrical system. The complete system is linear and second order because it consists of two first-order, linear, coupled ODEs.

Because this electrical system is simple, we can derive another form of the mathematical model directly from Kirchhoff's voltage law, Eq. (3.22), repeated below

$$-e_R - e_L - e_C + e_{in}(t) = 0 \tag{3.27}$$

Next, substitute the appropriate expressions for the voltage drops across each of the three passive elements

$$-RI_L - L\dot{I}_L - e_C(0) - \frac{1}{C}\int I_L dt + e_{in}(t) = 0 \tag{3.28}$$

Equation (3.28) is an integro-differential equation as it involves both the derivative and integral of current I_L. We can take the time derivative of Eq. (3.28) to eliminate the integral term (in addition, move the input variable $e_{in}(t)$ to the right-hand side)

$$L\ddot{I}_L + R\dot{I}_L + \frac{1}{C}I_L = \dot{e}_{in}(t) \tag{3.29}$$

Equation (3.29) is the mathematical model of the RLC circuit. It consists of a single linear second-order ODE with dynamic variable I_L and input variable $e_{in}(t)$. The reader should note that the single second-order modeling equation (3.29) is *equivalent* to the two first-order modeling equations (3.25) and (3.26). To show this, we can take the time derivative of Eq. (3.26)

$$L\ddot{I}_L + R\dot{I}_L + \dot{e}_C = \dot{e}_{in}(t) \tag{3.30}$$

Next, we substitute Eq. (3.25) for the time derivative of capacitor voltage ($\dot{e}_C = I_L/C$) to yield

$$L\ddot{I}_L + R\dot{I}_L + \frac{1}{C}I_L = \dot{e}_{in}(t) \tag{3.31}$$

which is equivalent to the second-order model (3.29).

In summary, we may use either mathematical model to represent the dynamics of the RLC circuit. If we choose the two first-order equations (3.25) and (3.26), our solution will be in terms of dynamic variables e_C and I_L. If we use the single second-order ODE (3.29), our single dynamic variable is I_L and the input is the time derivative of the source voltage $e_{in}(t)$.

Example 3.3

Figure 3.9 shows a parallel RLC circuit driven by a current source. Derive the mathematical model of the electrical system.

We begin the model development as we did in Example 3.2: this circuit contains two energy-storage elements, inductor L and capacitor C. Therefore, the model consists of two first-order ODEs in terms of capacitor voltage e_C and inductor current I_L

$$\text{Capacitor voltage:} \quad C\dot{e}_C = I_C \tag{3.32}$$

$$\text{Inductor current:} \quad L\dot{I}_L = e_L \tag{3.33}$$

Because the subsequent model involves only dynamic variables e_C and I_L and source current $I_{\text{in}}(t)$, we must express capacitor current I_C and inductor voltage e_L in terms of these variables. We can apply Kirchhoff's current law to the common node that connects the wires containing the current source, resistor, capacitor, and inductor. Figure 3.9 shows that currents I_R, I_L, and I_C are flowing out of the top node, while source current $I_{\text{in}}(t)$ is flowing into the node. Hence, Kirchhoff's current law yields

$$-I_R - I_L - I_C + I_{\text{in}}(t) = 0 \tag{3.34}$$

which can be solved for capacitor current

$$I_C = I_{\text{in}}(t) - I_R - I_L \tag{3.35}$$

Resistor current I_R must now be expressed in terms of e_C, I_L, or $I_{\text{in}}(t)$. We can apply Kirchhoff's voltage law to any loop that contains two passive elements. For example, moving clockwise around the loop containing resistor R and inductor L yields

$$-e_L + e_R = 0 \tag{3.36}$$

In a similar fashion, moving clockwise around the right-end loop containing inductor L and capacitor C yields

$$-e_C + e_L = 0 \tag{3.37}$$

Clearly, Eqs. (3.36) and (3.37) show that all voltage drops are equal: $e_R = e_L = e_C$ (we should recall from a university physics or elementary circuits course that two or more elements connected in a parallel circuit have the same voltage across their shared terminals). Using Ohm's law, we can express resistor current as $I_R = e_R/R = e_C/R$, and, therefore, the capacitor current in Eq. (3.35) becomes

$$I_C = I_{\text{in}}(t) - \frac{e_C}{R} - I_L \tag{3.38}$$

In addition, we can substitute for inductor voltage ($e_L = e_C$) in Eq. (3.33). Substituting Eq. (3.38) for I_C in Eq. (3.32) and capacitor voltage for e_L in Eq. (3.33) yields

$$C\dot{e}_C + \frac{1}{R}e_C + I_L = I_{\text{in}}(t) \tag{3.39}$$

$$L\dot{I}_L - e_C = 0 \tag{3.40}$$

Equations (3.39) and (3.40) are the mathematical modeling equations for the parallel RLC circuit. The complete system is linear and second order as it consists of two first-order, linear, coupled ODEs.

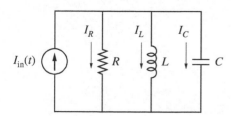

Figure 3.9 Parallel RLC circuit for Example 3.3.

We can express the electrical model as a *single* second-order ODE in terms of capacitor voltage e_C by taking a time derivative of Eq. (3.39)

$$C\ddot{e}_C + \frac{1}{R}\dot{e}_C + \dot{i}_L = \dot{I}_{in}(t) \tag{3.41}$$

Next, solve Eq. (3.40) for the time-rate of inductor current, $\dot{i}_L = e_C/L$, and substitute this result into Eq. (3.41) to yield

$$C\ddot{e}_C + \frac{1}{R}\dot{e}_C + \frac{1}{L}e_C = \dot{I}_{in}(t) \tag{3.42}$$

Equation (3.42) is the mathematical model of the parallel RLC circuit, and it is equivalent to the model represented by the two, coupled first-order equations (3.39) and (3.40).

Example 3.4

Figure 3.10 shows a dual-loop electrical system driven by a current source. Derive the mathematical model of the electrical system.

Because this system has only one energy-storage element, we begin with the basic modeling equation for a capacitor

$$\text{Capacitor voltage:}\quad C\dot{e}_C = I_C \tag{3.43}$$

Next, we use Kirchhoff's current law at the node marked "A" in Fig. 3.10

$$I_{in}(t) - I_C - I_2 = 0 \tag{3.44}$$

Therefore, substituting Eq. (3.44) for capacitor current in Eq. (3.43) we obtain

$$C\dot{e}_C = I_{in}(t) - I_2 \tag{3.45}$$

Hence, we need an expression for current through resistor R_2. Applying Kirchhoff's voltage law to the right-hand loop in Fig. 3.10 (moving clockwise) yields

$$-e_{R_2} + e_C + e_{R_1} = 0 \tag{3.46}$$

We can express both resistor voltage drops in Eq. (3.46) using Ohm's law

$$-R_2 I_2 + e_C + R_1 I_C = 0 \tag{3.47}$$

Substituting $I_C = I_{in}(t) - I_2$ in Eq. (3.47) yields

$$-R_2 I_2 + e_C + R_1(I_{in}(t) - I_2) = 0 \tag{3.48}$$

Grouping terms in Eq. (3.48) that involve current through resistor R_2, we obtain

$$(R_1 + R_2)I_2 = e_C + R_1 I_{in}(t) \tag{3.49}$$

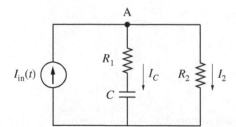

Figure 3.10 Electrical system for Example 3.4.

Finally, we can solve Eq. (3.49) for current I_2 and substitute the result into Eq. (3.45) to obtain the dynamic equation for the capacitor

$$C\dot{e}_C = I_{\text{in}}(t) - \frac{e_C}{R_1 + R_2} - \frac{R_1 I_{\text{in}}(t)}{R_1 + R_2} \tag{3.50}$$

Multiplying Eq. (3.50) by $R_1 + R_2$ and rearranging yields

$$(R_1 + R_2)C\dot{e}_C + e_C = R_2 I_{\text{in}}(t) \tag{3.51}$$

Equation (3.51) is the mathematical modeling equation for the electrical system. The system is a linear first-order ODE as the circuit consists of a single capacitor. The reader should note that all terms in Eq. (3.51) are voltages.

Example 3.5

Figure 3.11 shows a dual-loop electrical system driven by a voltage source. Derive the mathematical model of the electrical system.

We begin the model development as we did in Examples 3.2 and 3.3: this network contains two energy-storage elements, inductor L and capacitor C. Therefore, the model consists of two first-order ODEs in terms of capacitor voltage e_C and inductor current I_L

$$\text{Capacitor voltage:} \quad C\dot{e}_C = I_C \tag{3.52}$$

$$\text{Inductor current:} \quad L\dot{I}_L = e_L \tag{3.53}$$

We must express capacitor current I_C and inductor voltage e_L in terms of dynamic variables I_L and e_C and input voltage $e_{\text{in}}(t)$. To begin, apply Kirchhoff's current law to node "A" in Fig. 3.11

$$I_1 - I_C - I_L = 0 \tag{3.54}$$

Therefore, the capacitor current required in Eq. (3.52) is $I_C = I_1 - I_L$. Current through resistor R_1 can be computed from Ohm's law, $I_1 = e_{R_1}/R_1$, if we can determine the voltage drop across the resistor. Voltage drop across R_1 is determined from Kirchhoff's voltage law around the left-hand side loop (moving clockwise):

$$-e_{R_1} - e_C + e_{\text{in}}(t) = 0 \tag{3.55}$$

Therefore, the voltage drop for R_1 is

$$e_{R_1} = e_{\text{in}}(t) - e_C \tag{3.56}$$

and the current through resistor R_1 is

$$I_1 = \frac{e_{\text{in}}(t) - e_C}{R_1} \tag{3.57}$$

Finally, substituting Eq. (3.57) for resistor R_1 current and using Eq. (3.54) for capacitor current in Eq. (3.52), the modeling equation for capacitor voltage is

$$C\dot{e}_C = \frac{e_{\text{in}}(t) - e_C}{R_1} - I_L \tag{3.58}$$

Equation (3.58) is complete because it is in terms of e_C, I_L, and $e_{\text{in}}(t)$.

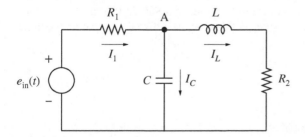

Figure 3.11 Electrical system for Example 3.5.

Next, we must determine an expression for the inductor voltage e_L in the dynamic equation for the inductor, Eq. (3.53). Applying Kirchhoff's voltage law to the right-hand side loop in Fig. 3.11 (moving clockwise) yields

$$-e_L - e_{R_2} + e_C = 0 \tag{3.59}$$

Therefore, inductor voltage drop is $e_L = e_C - e_{R_2}$. Voltage drop across resistor R_2 is determined by Ohm's law, $e_{R_2} = R_2 I_L$, and, therefore, Eq. (3.53) becomes

$$L\dot{I}_L = e_C - R_2 I_L \tag{3.60}$$

Finally, we can multiply Eq. (3.58) by R_1 and place all dynamic variables (e_C and I_L) in Eqs. (3.58) and (3.60) on the left-hand sides and the system input $e_{\text{in}}(t)$ on the right-hand sides to yield

$$R_1 C \dot{e}_C + e_C + R_1 I_L = e_{\text{in}}(t) \tag{3.61}$$

$$L\dot{I}_L + R_2 I_L - e_C = 0 \tag{3.62}$$

Equations (3.61) and (3.62) are the mathematical modeling equations for the dual-loop electrical system. The complete system is linear and second order because it consists of two first-order, linear, coupled ODEs. The reader should note that all terms in Eqs. (3.61) and (3.62) are voltages.

3.4 OPERATIONAL-AMPLIFIER CIRCUITS

An operational amplifier ("op-amp") is a modern electronic device that is used to amplify ("gain") an input voltage signal. They can also be used in circuits to construct *filters* that remove a desired range of frequencies from the input signal. Op amps were initially developed in the 1940s and during their evolution have utilized vacuum tubes, transistors, and integrated circuits. We do not investigate the inner-working details of an op amp; instead this section focuses on basic op-amp circuits.

Figure 3.12 shows the schematic diagram of an op amp that has two terminals on the input (left) side and one output terminal (right side). The input terminals with the negative and positive signs are known as the *inverting* and *noninverting* terminals, respectively. The output voltage e_O of the op amp shown in Fig. 3.12 is

$$e_O = K(e_B - e_A) \tag{3.63}$$

where K is the "voltage gain" of the op amp, which is usually very large and on the order of 10^5 V/V.

The analysis of op-amp circuits is greatly simplified by utilizing what is known as an *ideal* op amp. An ideal op amp has the following characteristics:

1. The input terminals of the op amp draw negligible current.
2. The voltage difference at the input terminals $e_B - e_A$ is zero.
3. The gain K is infinite.

These ideal op-amp characteristics show that it is difficult to determine the output voltage e_O using the configuration in Fig. 3.12 and Eq. (3.63) as the input $e_B - e_A \approx 0$ and the gain K is infinite. We can see that using a "negative feedback" circuit connection from the output terminal to the inverting (negative) input terminal (not shown in Fig. 3.12) causes the second idealized condition. All of the op-amp circuits that we consider in this chapter utilize this negative feedback configuration, which we demonstrate in the following examples.

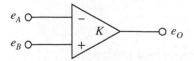

Figure 3.12 Operational amplifier.

Example 3.6

Figure 3.13 shows an op-amp circuit with input (source) voltage $e_{in}(t)$ and output voltage e_O. Derive the relationship between input and output voltages.

This circuit contains resistor R_1 between the input voltage and the negative input terminal of the op amp and a second resistor R_2 between the output voltage and the negative input terminal of the op amp (negative feedback). The positive input terminal of the op amp is directly connected to the ground and hence voltage $e_B = 0$. Because the input voltage difference is zero for an ideal op amp with negative feedback (i.e., $e_B - e_A = 0$) and $e_B = 0$, the input voltage e_A to the inverting terminal is also zero.

Kirchhoff's current law is applied at the top node between the two resistors in Fig. 3.13:

$$I_1 - I_A - I_2 = 0 \tag{3.64}$$

However, an ideal op amp draws negligible current ($I_A = 0$) and, therefore, $I_1 = I_2$ and the currents through the two resistors are equal. We can write expressions for I_1 and I_2 by using Ohm's law and dividing the respective voltage drop by the resistance

$$\frac{e_{in}(t) - e_A}{R_1} = \frac{e_A - e_O}{R_2} \tag{3.65}$$

The left-side term in Eq. (3.65) is current I_1 and the right-side term is current I_2. We can rewrite Eq. (3.65) as

$$R_2(e_{in}(t) - e_A) = R_1(e_A - e_O) \tag{3.66}$$

Rearranging Eq. (3.66) with the op-amp input voltage e_A on the right-hand side, we obtain

$$R_2 e_{in}(t) + R_1 e_O = (R_1 + R_2)e_A \tag{3.67}$$

or, solving for e_A

$$e_A = \frac{R_2}{R_1 + R_2} e_{in}(t) + \frac{R_1}{R_1 + R_2} e_O \tag{3.68}$$

Next, substitute Eq. (3.68) into the amplifier gain equation (3.63)

$$e_O = K(e_B - e_A) = -\frac{KR_2}{R_1 + R_2} e_{in}(t) - \frac{KR_1}{R_1 + R_2} e_O \tag{3.69}$$

Note that $e_B = 0$ as the positive op-amp input terminal in Fig. 3.13 is directly connected to the ground. Equation (3.69) is rearranged with all output voltage terms on the left-hand side to yield

$$e_O \left(1 + \frac{KR_1}{R_1 + R_2} \right) = -\frac{KR_2}{R_1 + R_2} e_{in}(t) \tag{3.70}$$

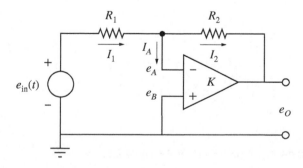

Figure 3.13 Op-amp circuit for Example 3.6.

Equation (3.70) can be simplified by multiplying both sides by $R_1 + R_2$

$$e_O(R_1 + R_2 + KR_1) = -KR_2 e_{in}(t) \tag{3.71}$$

Finally, the output voltage is

$$e_O = \frac{-KR_2}{R_1 + R_2 + KR_1} e_{in}(t) \tag{3.72}$$

Because the gain K is extremely large, we can take the limit of Eq. (3.72) as $K \to \infty$ to obtain the output voltage relationship for an ideal op-amp circuit

$$e_O = \frac{-R_2}{R_1} e_{in}(t) \tag{3.73}$$

Equation (3.73) shows that the output voltage of the op-amp circuit can be controlled by selecting the values of the two resistors R_1 and R_2. Note that the particular value of the op-amp gain K does not factor into the output voltage—the only requirement is that the gain K is very high. Because the output voltage e_O has the opposite sign of the input voltage, the circuit in Fig. 3.13 is called an *inverting amplifier*. The input–output voltage relationship of Eq. (3.73) is an algebraic equation and is not an ODE because the circuit does not contain any energy-storage elements.

Note that the op-amp input voltage e_A can be obtained by substituting the output voltage defined by Eq. (3.73) into Eq. (3.68)

$$e_A = \frac{R_2}{R_1 + R_2} e_{in}(t) + \frac{R_1}{R_1 + R_2} \frac{-R_2}{R_1} e_{in}(t) = 0 \tag{3.74}$$

Hence, the negative feedback connection in Fig. 3.13 between the op-amp output and input terminals results in $e_A = 0$. Because $e_B = 0$, the voltage difference at the input terminals is $e_B - e_A = 0$, which is the second characteristic of an ideal op amp.

As a final note to this example, consider the op-amp circuit in Fig. 3.13 with the following numerical values: $e_{in}(t) = 1.5$ V, $R_1 = 2\,\Omega$, and $R_2 = 4\,\Omega$. Hence, using Eq. (3.73) the output voltage is $e_O = -3$ V. Therefore, the current across the first resistor R_1 is $I_1 = 1.5$ V$/2\,\Omega = 0.75$ A as current flows from the higher potential ($e_{in} = 1.5$ V) to the lower potential ($e_A = 0$). Current across the second resistor R_2 is $I_2 = 3$ V$/4\,\Omega - 0.75$ A as current flows from the higher potential ($e_A = 0$) to the lower potential ($e_O = -3$ V).

Example 3.7

Figure 3.14 shows an op-amp circuit with input voltage $e_{in}(t)$, output voltage e_O, and a capacitor C in a branch connecting output and input terminals. Derive the relationship between input and output voltages.

Because this op-amp circuit contains a negative feedback connection between output and input terminals, we use the ideal op-amp characteristics and set $e_A = 0$ (positive input terminal $e_B = 0$ as it is connected to the ground). Furthermore, the op amp draws no current ($I_A = 0$), so applying Kirchhoff's current law at the top node yields

$$I_1 = I_2 + I_3 \tag{3.75}$$

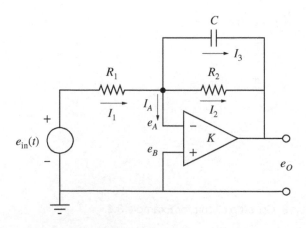

Figure 3.14 Op-amp circuit for Example 3.7.

We can substitute Ohm's law for currents I_1 and I_2 and the governing equation for a capacitor, Eq. (3.4) for current I_3 into Eq. (3.75) to yield

$$\frac{e_{\text{in}}(t) - e_A}{R_1} = \frac{e_A - e_O}{R_2} + C\frac{d}{dt}(e_A - e_O) \tag{3.76}$$

Setting $e_A = 0$ because of the negative feedback connection between output and input terminals, Eq. (3.76) becomes

$$\frac{e_{\text{in}}(t)}{R_1} = \frac{-e_O}{R_2} + C\frac{d}{dt}(-e_O) \tag{3.77}$$

which can be rewritten as

$$R_2 C\dot{e}_O + e_O = \frac{-R_2}{R_1}e_{\text{in}}(t) \tag{3.78}$$

Equation (3.78) is a first-order ODE model of the op-amp circuit in Fig. 3.14. We obtain a dynamic model (an ODE) as the circuit includes an energy-storage element (capacitor C). Clearly, if the capacitor is removed, the circuit becomes the inverting amplifier in Example 3.6 and Eq. (3.78) becomes Eq. (3.73).

3.5 ELECTROMECHANICAL SYSTEMS

As stated in Section 3.1, a principal objective of this chapter is to develop mathematical models of electromechanical systems, which are created by combining mechanical and electrical elements. Mechanical and aerospace engineers use electromechanical systems to convert electrical energy into mechanical energy, such as displacement and/or velocity of a mechanical element. These devices are called *actuators*, and common examples include motors and solenoids. Engineers use similar concepts to develop instrumentation devices that convert mechanical energy into electrical signals for measurements. Examples of electromechanical sensors include accelerometers, linear variable differential transformers (LVDTs), and rotary encoders. We present derivations of the mathematical models for a rotational direct current (DC) motor, a translational solenoid actuator, and an electrostatic actuator for micromechanical systems.

Current–Magnetic Field Interaction

Electromechanical systems utilize the interaction between an electrical current and a magnetic field in order to establish a mechanical force. These current–magnetic field interactions are described by Faraday's laws of induction and Lorentz's force law. For the purposes of this chapter, our treatment of electromagnetism is based on principles that are typically presented in a university-level physics course. Therefore, we can state three basic relationships between electrical current and magnetism:

1. An electrical current establishes a magnetic field.
2. A current-carrying wire in a magnetic field has a force exerted on it.
3. A wire moving relative to a magnetic field will have a voltage induced between the ends of the wire.

Figure 3.15 illustrates the first basic current–magnetism relationship: a wire carrying current I establishes concentric-circle magnetic field lines around the wire. The Biot–Savart law describes the magnetic field $\mathbf{B}$ shown in Fig. 3.15, where the direction of the field lines is defined by applying the "right-hand rule."

Figure 3.16 illustrates the second current–magnetism relationship by showing a stationary wire carrying current I in magnetic field $\mathbf{B}$. The magnetic field $\mathbf{B}$ is a vector with direction from the north pole (N) to the south pole (S) of the permanent magnet and has magnetic flux density in Wb/m^2 or tesla. The force induced on the stationary wire is governed by the vector (or "cross") product

$$\mathbf{F} = I\boldsymbol{\ell} \times \mathbf{B} \tag{3.79}$$

where $\boldsymbol{\ell}$ is a vector with direction along the wire (in the direction of current flow) with magnitude equal to the length of the wire in the field. Figure 3.16 shows that the induced force vector $\mathbf{F}$ follows the "right-hand rule" of the cross product

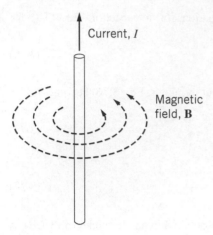

Current, I

Magnetic field, $\mathbf{B}$

Figure 3.15 A current-carrying wire establishes a magnetic field.

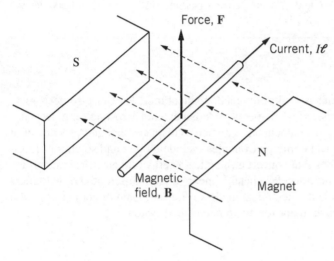

Force, $\mathbf{F}$

Current, $I\boldsymbol{\ell}$

S

Magnetic field, $\mathbf{B}$

N

Magnet

Figure 3.16 Force induced on a current-carrying wire in a magnetic field.

and is perpendicular to vectors $\mathbf{B}$ and $\boldsymbol{\ell}$. If the stationary wire is perpendicular to the magnetic field vector $\mathbf{B}$, then the magnitude of the induced force is

$$F = B\ell I \tag{3.80}$$

where ℓ is the length of the wire in the magnetic field and B is the magnetic flux density. If the wire remains perpendicular to the field $\mathbf{B}$, we may use the scalar equation (3.80) to compute the magnitude of the induced force. The induced or electromagnetic force is the basis of electromechanical actuators such as DC motors and solenoids.

Figure 3.17 illustrates the third basic current–magnetism relationship by showing a wire that is *moving* relative to magnetic field $\mathbf{B}$. The voltage induced on the moving wire is

$$e_b = (\mathbf{v} \times \mathbf{B}) \cdot \boldsymbol{\ell} \tag{3.81}$$

where $\mathbf{v}$ is the velocity vector of the wire. The induced voltage e_b is the "dot" or scalar product of the velocity–magnetic field cross product and vector $\boldsymbol{\ell}$. The cross product $\mathbf{v} \times \mathbf{B}$ establishes the direction of the positive (+) polarity of the induced voltage e_b shown in Fig. 3.17, or the direction of current caused by the induced voltage. We can better understand the induced-voltage effect with the stationary current-carrying wire in Fig. 3.16. Equation (3.79) and Fig. 3.16 show that current vector $I\boldsymbol{\ell}$ and magnetic field $\mathbf{B}$ induce the force $\mathbf{F}$, which in turn will cause the wire to move with velocity $\mathbf{v}$ shown in Fig. 3.17. Equation (3.81) and Fig. 3.17 show that velocity and magnetic field interaction induces a voltage e_b that *opposes* the current vector $I\boldsymbol{\ell}$ in Fig. 3.16 that originally established the induced force $\mathbf{F}$. Hence, the induced voltage e_b is traditionally called the *back electromotive force*, or *back emf*. The reader should note that "electromotive force" (emf) is a rather old term that is synonymous

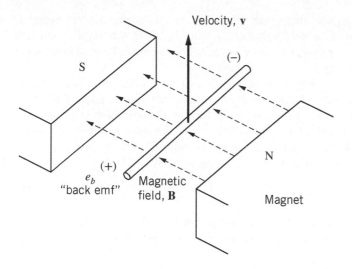

Figure 3.17 Voltage induced on a wire moving relative to a magnetic field.

with voltage. It should also be noted that motion of the wire can be caused by an external applied force (rather than the induced electromagnetic force as described here), such as steam-powered turbines used in electrical power-generation plants.

If the moving wire is perpendicular to the magnetic field vector **B**, the magnitude of the induced voltage is

$$e_b = B\ell v \tag{3.82}$$

If the wire remains perpendicular to the field **B**, we may use the scalar equation (3.82) to compute the magnitude of the back emf.

DC Motor

A DC motor is an electromechanical system that converts electrical energy (voltage source) to mechanical energy (rotational motion) by utilizing the basic current–magnetism relationships. A DC motor consists of an armature (or *rotor*) and a magnetic field established by the *stator*. The armature or rotor disk is wrapped with a coil of wire, or *armature windings*. Figure 3.18 shows a cross-sectional view of a simple DC motor: the stator is a permanent magnet (with north and south poles) that establishes the radial magnetic field **B** and the armature (rotor disk) rotates on an axis fixed in the center of the magnetic field. Armature windings are wrapped on the periphery of the rotor so that the direction of current flow is perpendicular to the plane (or "page") shown in Fig. 3.18 (in other words, the windings are parallel to the rotor's axis). Commutator brushes (not shown) maintain electrical contact between the armature-circuit voltage source

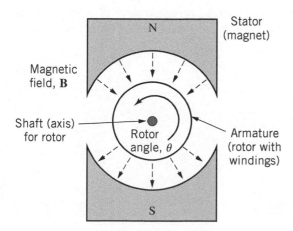

Figure 3.18 Stator and armature of a DC motor.

(not shown) and the rotating armature. The reader should note that while simple DC motors use permanent magnets (such as those shown in Fig. 3.18), the stator may consist of an electromagnet produced by "field windings" around an iron core connected to an auxiliary voltage source. In either case, the stator is stationary and establishes the magnetic field **B** that interacts with the current in the armature windings.

The armature windings in the "top half" of the rotor in Fig. 3.18 carry current flowing "into" the plane (page), while windings in the "bottom half" carry current flowing "out" of the page. Furthermore, the radial magnetic field lines **B** remain perpendicular to the armature windings, and therefore the cross product in Eq. (3.79) results in an induced force on each individual wire that is tangent to the rotor's surface and in the direction commensurate with the positive rotation angle θ shown in Fig. 3.18. Hence, the *total* electromagnetic torque on the rotor (T_m) is the product of the summation of the induced forces (on each wire) and the moment arm (i.e., the radius of the rotor)

$$T_m = Fr = B\ell Ir \tag{3.83}$$

where r is the rotor's radius and ℓ is the total length of the armature (wire) windings that are in the radial magnetic field. If the magnetic flux density B is constant, the three terms $B\ell r$ can be lumped into a single constant, $K_m = B\ell r$, and Eq. (3.83) becomes

$$T_m = K_m I \tag{3.84}$$

In other words, net electromagnetic torque on the rotor T_m is linearly proportional to the current I in the armature windings. The constant K_m is typically called the *motor-torque constant* and has units of N-m/A. Manufacturers of DC motors often provide the motor-torque constant K_m in catalogs summarizing motor specifications.

Equation (3.84) accounts for the net torque applied to the mechanical part of the DC motor system. A positive torque will produce positive rotational motion in the direction shown in Fig. 3.18. Hence, the armature windings will move relative to the radial magnetic field and this motion will result in an induced voltage (back emf) that opposes the armature current. Because the velocity vectors of the windings remain tangent to the rotor, the individual velocity vectors are always perpendicular to the radial magnetic field **B**. Hence, the *total* induced voltage (e_b) is

$$e_b = B\ell v = B\ell r\dot{\theta} \tag{3.85}$$

where $\dot{\theta}$ is the angular velocity of the rotor. Note that the circumferential velocity of every armature winding on the rotor is $v = r\dot{\theta}$. If the magnetic field is constant, we can define a new constant $K_b = B\ell r$ and Eq. (3.85) becomes

$$e_b = K_b\dot{\theta} \tag{3.86}$$

Induced voltage (back emf) e_b is linearly proportional to the angular velocity of the rotor. The constant K_b is typically called the *back-emf constant* and has units of V-s/rad. Although it is not apparent, the units for K_m (N-m/A) and K_b (V-s/rad) are equivalent as $1\,\text{V} = 1\,\text{kg-m}^2/(\text{s}^3\text{-A}) = 1\,\text{N-m}/(\text{s-A})$. Therefore, K_m and K_b have identical numerical values when expressed using the basic SI units. Manufacturers of DC motors also often provide the back-emf constant K_b in catalogs.

Figure 3.19 shows a schematic diagram of the DC motor. The armature circuit is composed of the voltage source $e_{in}(t)$, armature coil inductance L_a (due to the windings), armature resistance R_a, and back-emf e_b. Note that the back

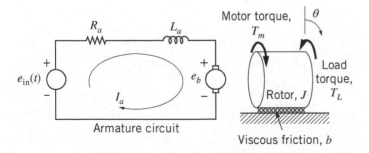

Figure 3.19 Schematic diagram of a DC motor.

emf is represented by a modified voltage-source symbol with positive and negative terminals that oppose the positive (clockwise) flow of armature current I_a. The mechanical component of the DC motor is shown to the right of the armature circuit and includes the moment of inertia for the rotor J, viscous friction coefficient b, motor torque T_m (from the current–magnetism interaction), and load torque T_L. Note that positive angular rotation of the rotor is clockwise and corresponds to positive armature current I_a and positive motor torque T_m.

We can derive the complete mathematical model of the DC motor by applying Kirchhoff's laws to the armature circuit and Newton's laws to the mechanical rotor. To begin, we use Kirchhoff's voltage law around the loop, moving clockwise

$$-e_R - e_L - e_b + e_{\text{in}}(t) = 0$$

The reader should note that the first three voltage terms are voltage drops as the assumed positive current flow is from the positive to negative terminals for the resistor, inductor, and back emf. Next, we substitute the appropriate element laws for voltage drop across a resistor ($e_R = R_a I_a$), voltage across an inductor ($e_L = L_a \dot{I}_a$), and back emf ($e_b = K_b \dot{\theta}$) to yield

$$L_a \dot{I}_a + R_a I_a + K_b \dot{\theta} = e_{\text{in}}(t) \tag{3.87}$$

The mathematical model of the mechanical component is derived using the methods developed in Chapter 2. Figure 3.20 shows the free-body diagram of the mechanical armature rotor with motor torque ($T_m = K_m I_a$), viscous friction torque ($b\dot{\theta}$), and load torque (T_L). Summing torques on the rotor (with a positive clockwise convention) and applying Newton's second law yields

$$\left(+\searrow \right) \sum T = K_m I_a - T_L - b\dot{\theta} = J\ddot{\theta} \tag{3.88}$$

The *complete* mathematical model of the DC motor consists of the electrical system equation (3.87) and the mechanical system equation (3.88)

$$L_a \dot{I}_a + R_a I_a = e_{\text{in}}(t) - K_b \dot{\theta} \tag{3.89a}$$

$$J\ddot{\theta} + b\dot{\theta} = K_m I_a - T_L \tag{3.89b}$$

Therefore, we see that the mathematical model of the DC motor is third order: one first-order ODE for the armature circuit (one energy-storage element, L_a) and one second-order ODE for the mechanical rotor. The dynamic variables are armature current I_a and rotor angle θ, and the system input variables are armature voltage $e_{\text{in}}(t)$ and load torque T_L. Equations (3.89a) and (3.89b) are linear and coupled because they cannot be solved separately. The right-hand side of the mechanical system equation (3.89b) shows that a positive armature current produces a positive motor torque that in turn accelerates the armature rotor. However, the right-hand side of the armature equation (3.89a) shows that positive angular velocity of the rotor creates a negative induced voltage (the back emf) that in turn reduces the net voltage of the circuit.

Because the rotor's angular position θ does not appear in Eqs. (3.89a) and (3.89b), we may substitute $\omega = \dot{\theta}$ and $\dot{\omega} = \ddot{\theta}$ in order to develop a reduced-order model:

$$L_a \dot{I}_a + R_a I_a = e_{\text{in}}(t) - K_b \omega \tag{3.90a}$$

$$J\dot{\omega} + b\omega = K_m I_a - T_L \tag{3.90b}$$

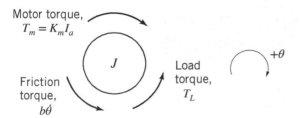

Figure 3.20 Free-body diagram of the DC motor armature rotor.

Now, the mathematical model of the DC motor is second order and consists of two coupled first-order ODEs. The solution to the second-order model will yield information for dynamic variables current $I_a(t)$ and angular velocity $\omega(t)$ but not angular position $\theta(t)$.

Solenoid Actuator

A solenoid actuator is an electromechanical device that converts electrical energy (voltage source) to mechanical energy (translational motion) by utilizing the same basic current–magnetism principles that govern the operation of a DC motor. Solenoids can deliver a translational force to either push or pull a mechanical load such as a valve in hydraulic or pneumatic systems. A solenoid actuator consists of a coil of wire with an iron core (the armature or plunger) that moves in and out of the center of the coil. Figure 3.21 shows the components of a push-type solenoid actuator. Energizing the voltage source $e_{in}(t)$ causes current to flow through the coil that in turn establishes a magnetic field. The energized coil acts as an electromagnet and applies a force to the armature (plunger), drawing it toward the center of the coil (to the right in Fig. 3.21). The push-type solenoid shown in Fig. 3.21 uses a push-pin to move a load (e.g., valve mass) to the right. A return spring is typically used to deliver a force to the displaced load mass so that it returns to its original position when the current is zero.

The coil inductance of a solenoid actuator is a nonlinear function of the armature position. Inductance (and hence magnetic flux) decreases the farther the armature is moved out of the coil and increases the closer the armature is drawn into the center of the coil. Figure 3.22 shows the solenoid actuator–valve system from Example 2.1 (recall that we derived the mechanical model for this system in Chapter 2). Note that Fig. 3.22 shows a push-type solenoid where

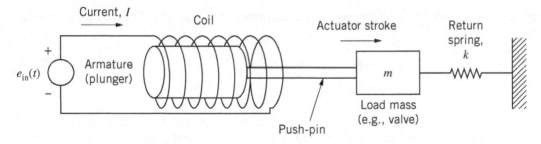

Figure 3.21 Armature and coil for a solenoid actuator.

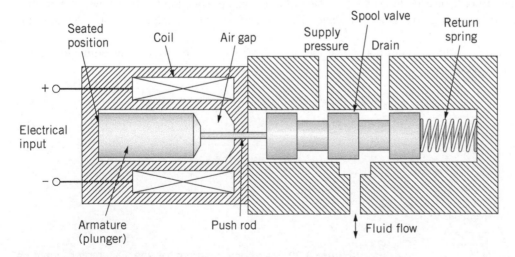

Figure 3.22 Solenoid actuator–valve system.

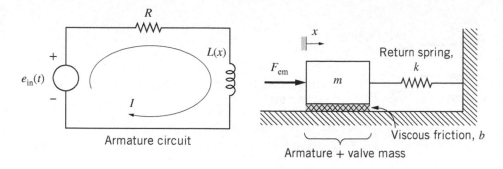

Figure 3.23 Schematic diagram of a solenoid actuator.

energizing the coil draws the armature toward the center of the coil and therefore pushing the valve to the right. One accepted method for modeling the coil inductance is to use the nonlinear expression [1, 2]

$$L(x) = \frac{c}{d-x} = \frac{L_0}{1-(x/d)} \tag{3.91}$$

where x is the armature displacement (measured positive to the right from the seated position; see Fig. 3.22). The constants c and d depend on the geometry and material properties of the solenoid coil. Note that coil inductance $L(x)$ is minimum when $x = 0$ (seated armature) and increases when $x > 0$ and the armature moves to the right to close the air gap. The inductance when $x = 0$ is

$$L_0 = \frac{\mu A N^2}{l} \tag{3.92}$$

where N is the number of turns of the coil, A is the area of the air gap, l is the coil length, and μ is the magnetic permeabilities of air and the iron core. The minimum coil inductance L_0 is a known constant given the values for A, N, l, and μ.

Figure 3.23 shows a schematic diagram of the solenoid actuator. The armature circuit (coil) is composed of the voltage source $e_{in}(t)$, armature coil inductance $L(x)$, and armature resistance R. A single lumped mass m represents the sum of the armature (plunger) and the load (valve) masses. The energized solenoid coil produces the electromagnetic force F_{em} that pushes on mass m to the right. Displacement of the mass is x (positive to the right), and the return spring k and viscous friction b act on the armature–valve mass m.

As with the DC motor, we develop the complete mathematical model of the solenoid actuator by applying Kirchhoff's laws to the armature circuit and Newton's laws to the single inertia element. To begin, we apply Kirchhoff's voltage law around the loop

$$-e_R - e_L + e_{in}(t) = 0 \tag{3.93}$$

Computing the solenoid inductor voltage e_L is more complicated than the inductor voltage for the DC motor because solenoid inductance changes with the plunger position. Therefore, we use Eq. (3.9) to equate the solenoid inductor voltage to the time derivative of magnetic flux linkage

$$\dot{\lambda} = e_L \tag{3.94}$$

where flux linkage is defined by Eq. (3.8) as the product of inductance and current, or $\lambda = L(x)I$. Because both inductance and current can change with time, the time derivative of the flux linkage is

$$\dot{\lambda} = \frac{dL}{dt}I + L\frac{dI}{dt} \tag{3.95}$$

Using the chain rule to expand dL/dt, Eq. (3.95) becomes

$$\dot{\lambda} = \frac{dL}{dx}\frac{dx}{dt}I + L\frac{dI}{dt} \tag{3.96}$$

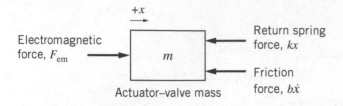

Figure 3.24 Free-body diagram of the solenoid actuator.

or, using the compact notation

$$\dot{\lambda} = L_x \dot{x} I + L(x)\dot{I} \tag{3.97}$$

where L_x is the short-hand notation for the derivative dL/dx. Using Eq. (3.91), the derivative dL/dx is

$$L_x \equiv \frac{dL}{dx} = \frac{L_0}{d\left(1 - (x/d)\right)^2} \tag{3.98}$$

Finally, we can substitute Eq. (3.97) into Kirchhoff's loop equation (3.93) for inductor voltage e_L along with resistor voltage drop $e_R = RI$ to yield the mathematical model of the solenoid circuit

$$L(x)\dot{I} + RI = e_{in}(t) - L_x \dot{x} I \tag{3.99}$$

Note that the right-hand-side term $L_x \dot{x} I$ in Eq. (3.99) acts like the back-emf term in the DC motor modeling equation (3.89a): when the actuator mass moves with positive velocity toward the center of the coil, it induces a negative voltage that decreases the net voltage in the circuit. Furthermore, the induced voltage $L_x \dot{x} I$ for the solenoid is *nonlinear* whereas the back emf of the DC motor is a linear term ($K_b \dot{\theta}$).

   The mathematical model of the mechanical component of the solenoid actuator is derived using the methods developed in Chapter 2. Figure 3.24 shows the free-body diagram of the armature–valve mass with electromagnetic force (F_{em}), viscous friction force ($b\dot{x}$), and return spring force (kx). We have assumed that no preload force (initial compression) exists in the return spring and hence there is no wall-contact force. Summing forces on the mass and applying Newton's second law yields

$$+ \rightarrow \sum F = F_{em} - kx - b\dot{x} = m\ddot{x}$$

Grouping all terms involving displacement x yields

$$m\ddot{x} + b\dot{x} + kx = F_{em} \tag{3.100}$$

In order to complete the model, we need an expression for the electromagnetic force F_{em}, which is generated by the energy stored in the solenoid coil. From work and energy principles, we know that the product of the electromagnetic force and an incremental displacement dx is equal to an incremental change in energy $d\xi$

$$F_{em} dx = d\xi$$

or, solving for the electromagnetic force

$$F_{em} = \frac{d\xi}{dx} \tag{3.101}$$

Equation (3.11) states that the energy stored in an inductor is due to the inductance and the current

$$\xi = \frac{1}{2} L I^2$$

Therefore, taking the derivative of energy with respect to displacement x and substituting the result into Eq. (3.101) yields an expression for the electromagnetic force

$$F_{em} = \frac{1}{2} \frac{dL}{dx} I^2 \tag{3.102}$$

We see that the electromagnetic force is a nonlinear function of current and displacement as Eq. (3.98) shows that the derivative L_x is a nonlinear function of x.

The *complete* mathematical model of the solenoid actuator consists of the electrical system equation (3.99) and the mechanical system equation (3.100) with Eq. (3.102) used to define the electromagnetic force

$$L(x)\dot{I} + RI = e_{\text{in}}(t) - L_x \dot{x}I \tag{3.103}$$

$$m\ddot{x} + b\dot{x} + kx = \frac{1}{2}L_x I^2 \tag{3.104}$$

We see that the mathematical model of the solenoid actuator is third order: one first-order ODE for the solenoid circuit and one second-order ODE for the mechanical mass. The dynamic variables are the coil current I and plunger displacement x, and the system input variable is armature voltage $e_{\text{in}}(t)$. Equations (3.103) and (3.104) are coupled nonlinear differential equations. The reader should recall that Eqs. (3.91) and (3.98) are also needed to define inductance $L(x)$ and its derivative L_x, both of which are nonlinear functions of plunger displacement.

Electrostatic Microactuator

Motors and solenoid actuators are driven by the current–magnetism interaction that is governed by Faraday's laws. However, for tiny devices at the microscale, there is no room for coils and electromagnetic induction [3]. Microelectromechanical systems (MEMS) often use electrostatic forces for actuation, and applications include surgical microgrippers and optical microshutters [3–5]. The electrostatic driving force is the electrical force of repulsion or attraction between charged particles.

Figure 3.25 shows a MEMS device commonly known as a *comb-drive* actuator [3–5]. Comb-drive actuators consist of two interlocking "finger" structures that have the appearance of two intertwined combs. The interlocking fingers are misaligned parallel plates that act as a series of capacitors where the movable drive and stationary closure plates carry opposite charges. An input voltage $e_{\text{in}}(t)$ is applied to the comb structure that establishes the electrostatic force that attempts to realign the parallel plates of the interconnected comb. Hence, the electrostatic force attracts the drive arm toward the stationary closure arm. Movement of the comb structure could actuate the extension arms of a microgripper device. Deflections for these MEMS devices are on the order of microns, where $1\,\mu\text{m} = 10^{-6}\,\text{m}$. A stiffness element (modeled by spring constant k) is used to retract the drive arm when the electrostatic force is removed.

Figure 3.26 shows a schematic diagram for a comb-drive actuator where the intertwined parallel plates are replaced by a lumped capacitor $C(x)$. The comb circuit consists of the voltage source $e_{\text{in}}(t)$, comb capacitance $C(x)$, and resistance R. A single lumped mass m represents the movable drive-comb structure that experiences stiffness and friction forces modeled by kx and $b\dot{x}$, respectively. Charging the misaligned parallel plates of the comb produces an electrostatic force F_{es} that pulls the drive arm to the left in order to realign the plates.

The equivalent capacitance of the comb is

$$C(x) = \frac{n\varepsilon_0 A}{d} = \frac{n\varepsilon_0(x_0 + x)w}{d} \tag{3.105}$$

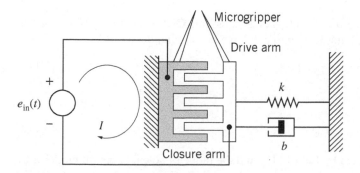

Figure 3.25 MEMS comb-drive actuator.

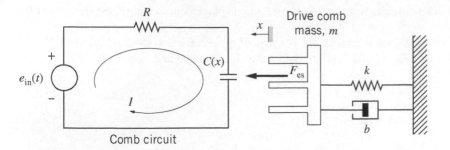

Figure 3.26 Schematic diagram of a MEMS comb-drive actuator.

where n is the number of fingers, ε_0 is the dielectric constant in air (in F/m), A is the overlapping area of the fingers, and d is the gap spacing between fingers [3–5]. Overlapping area A is the product of finger width w and overlap distance $x_0 + x$, where x_0 is the initial overlap between fingers when the comb is uncharged and undeflected (zero electrostatic force). Figures 3.25 and 3.26 and Eq. (3.105) show that pulling the drive comb to the left ($x > 0$) increases the overlap area and therefore increases the capacitance $C(x)$.

As with the electromagnetic actuators, we develop the complete mathematical model of the comb-drive actuator by applying Kirchhoff's laws to the comb circuit and Newton's laws to the single inertia element. However, computing the comb capacitor voltage e_C is somewhat complicated because capacitance changes with position x. We begin with the capacitor's basic charge–voltage relation (3.3):

$$q = C(x)e_C \tag{3.106}$$

The time derivative of the charge is current, I

$$\dot{q} = \frac{dC}{dt}e_C + C\frac{de_C}{dt} = I \tag{3.107}$$

Using the chain rule to expand dC/dt, Eq. (3.107) becomes

$$\dot{q} = \frac{dC}{dx}\frac{dx}{dt}e_C + C\frac{de_C}{dt} = I \tag{3.108}$$

or, using the compact notation

$$\dot{q} = C_x \dot{x}e_C + C(x)\dot{e}_C = I \tag{3.109}$$

where C_x is short-hand notation for the derivative dC/dx. Using Eq. (3.105), the derivative dC/dx is

$$C_x \equiv \frac{dC}{dx} = \frac{n\varepsilon_0 w}{d} \tag{3.110}$$

Hence, the change in capacitance because of position x is a constant. We can substitute Ohm's law $I = e_R/R$ for current in Eq. (3.109) and determine resistor voltage e_R by applying Kirchhoff's voltage law around the loop

$$-e_R - e_C + e_{in}(t) = 0 \tag{3.111}$$

Substituting $I = (e_{in}(t) - e_C)/R$ into Eq. (3.109) yields

$$RC(x)\dot{e}_C + e_C = e_{in}(t) - RC_x\dot{x}e_C \tag{3.112}$$

Note the similarity between the capacitor circuit modeled by Eq. (3.112) and the inductor solenoid circuit modeled by Eq. (3.103). Both circuits contain energy storage elements that change with position, $C(x)$ or $L(x)$, and both circuits contain nonlinear back-emf (voltage) terms that depend on velocity, $RC_x\dot{x}e_C$ and $L_x\dot{x}I$.

The mathematical model of the mechanical component of the microactuator is derived using a free-body diagram that is essentially identical to the solenoid free-body diagram shown in Fig. 3.24 except that the electrostatic force F_{es} pulls the drive comb to the left (see Fig. 3.26). Applying Newton's second law yields the familiar mass–spring–damper modeling equation

$$m\ddot{x} + b\dot{x} + kx = F_{es} \tag{3.113}$$

In order to complete the model, we need an expression for the electrostatic force F_{es}, which is generated by the energy stored in the capacitor. We use the same work and energy principles that we employed to obtain the electromagnetic force F_{em} for the solenoid, that is,

$$F_{es} = \frac{d\xi}{dx} \tag{3.114}$$

Equation (3.6) shows that the energy stored in a capacitor is due to its capacitance and its voltage

$$\xi = \frac{1}{2} C e_C^2$$

Therefore, taking the derivative of energy with respect to displacement x and substituting the result into Eq. (3.114) yields an expression for the electrostatic force

$$F_{es} = \frac{1}{2} \frac{dC}{dx} e_C^2 \tag{3.115}$$

We see that the electrostatic force is a nonlinear function of voltage e_C. Equation (3.110) shows that dC/dx is a constant.

The *complete* mathematical model of the comb-drive actuator consists of the electrical system equation (3.112) and the mechanical system equation (3.113) with Eq. (3.115) used to define the electrostatic force

$$RC(x)\dot{e}_C + e_C = e_{in}(t) - RC_x\dot{x}e_C \tag{3.116}$$

$$m\ddot{x} + b\dot{x} + kx = \frac{1}{2} C_x e_C^2 \tag{3.117}$$

We see that the mathematical model of the microactuator is third order: one first-order ODE for the comb circuit and one second-order ODE for the mechanical mass. The dynamic variables are the capacitor voltage e_C and drive-comb displacement x, and the system input variable is source voltage $e_{in}(t)$. Equations (3.116) and (3.117) are coupled nonlinear differential equations. The reader should recall that Eqs. (3.105) and (3.110) are also needed to define capacitance $C(x)$ and its derivative C_x.

As a final note, we can compute the electrostatic force for a "typical" MEMS actuator [3, 5] with $n = 100$ fingers, $\varepsilon_0 = 8.85 \ (10^{-12})$ F/m, gap $d = 2 \ \mu m$, and finger width $w = 2 \ \mu m$. Using Eq. (3.110) we see that the capacitance derivative with position is $C_x = 8.85 \ (10^{-10})$ F/m. Equation (3.115) shows that the electrostatic force is $F_{es} = 1.77(10^{-7})$ N or $0.177 \ \mu N$ if the capacitor voltage is 20 V.

SUMMARY

This chapter has illustrated the development of the mathematical models for electrical and electromechanical systems. First, we presented the physical laws that govern the interaction between charge, current, and voltage of electrical elements such as resistors, capacitors, and inductors. It is important for the reader to remember that only capacitors and inductors can store electrical energy and that each energy-storage element requires a single first-order ODE. Voltage across a capacitor and current through an inductor are the two dynamic variables of interest for capacitor and inductor elements, respectively. For example, an electrical system composed of two inductors and one capacitor is modeled by three first-order ODEs: two first-order ODEs for the time-rates of the two inductor currents (I_L) and one first-order ODE for the time-rate of capacitor voltage (e_C). The voltages and currents in an electrical circuit that are not e_C or I_L (e.g., voltage drop across a resistor) can be expressed in terms of the important dynamic variables by applying Kirchhoff's voltage and/or current laws to the electrical circuit. Furthermore, we showed how to model electrical systems that contained an operational amplifier. We ended the chapter with a discussion of electromechanical systems that involve the transfer between electrical and mechanical energy. Electromechanical systems employ the interaction between current and magnetism in order to convert electrical current into a mechanical force or mechanical motion into electrical voltage. MEMS actuators use the interaction between voltage and charge in order to convert electrical voltage into an electrostatic force.

REFERENCES

1. Yuan, Q., and Li, P.Y., "Self-Calibration of Push-Pull Solenoid Actuators in Electrohydraulic Valves," ASME Paper No. 2004-62109, Nov. 2004.
2. Chladny, R.R., Koch, C.R., and Lynch, A.F., "Modeling Automotive Gas-Exchange Solenoid Valve Actuators," *IEEE Transactions on Magnetics*, Vol. 41, No. 3, Mar. 2005, pp. 1155–1162.
3. Hsu, T.-R., *MEMS and Microsystems: Design, Manufacture, and Nanoscale Engineering*, 2nd ed., Wiley, Hoboken, NJ, 2008.
4. Tang, W.C., Lim, M.G., and Howe, R.T., "Electrostatic Comb Drive Levitation and Control Method," *IEEE Journal of Microelectromechanical Systems*, Vol. 1, No. 4, 1992, pp. 170–178.
5. Legtenberg, R., Groeneveld, A.W., and Elwenspoek, M., "Comb-Drive Actuators for Large Displacements," *Journal of Micromechanics and Microengineering*, Vol. 6, 1996, pp. 320–329.
6. Vaughan, N.D., and Gamble, J.B., "The Modeling and Simulation of a Proportional Solenoid Valve," *ASME Journal of Dynamic Systems, Measurement, and Control*, Vol. 118, Mar. 1996, pp. 120–125.
7. Yeh, T.-J., Chung, Y.-J., Wu, W.-C., "Sliding Control of Magnetic Bearing Systems," *ASME Journal of Dynamic Systems, Measurement, and Control*, Vol. 123, Sept. 2001, pp. 353–362.

Chapter 4

Modeling Fluid and Thermal Systems

4.1 INTRODUCTION

Pressurized fluids (liquids and gases) are used by mechanical engineers in designing devices that deliver forces and torques to mechanical loads. *Hydraulic systems* use a liquid as the working fluid while *pneumatic systems* use air or other gases. Hydraulic actuators are used by construction and farm machinery to lift loads, move soil, and drive rotational augers. Hydraulic actuators are also used on aerospace vehicles to position aerodynamic surfaces (rudders, elevators, ailerons, and flaps), deploy landing gear, and swivel rocket engines. In addition, hydraulic and pneumatic actuators are used to maneuver robot manipulators and activate automotive braking systems. Like the electromechanical systems investigated in Chapter 3, these fluid systems convert energy from a power source to mechanical energy (position and velocity). In the case of fluid systems, the power source is a pressurized fluid, whether it is liquid (hydraulic system) or gas (pneumatic system). Thermal systems involve the transfer of heat energy, and temperature is typically the dynamic variable of interest.

This chapter introduces the fundamental techniques for developing the modeling equations for fluid and thermal systems. Our mathematical models of fluid systems are developed by utilizing the conservation of mass while thermal system models are developed by applying the conservation of energy. When a fluid system is used in an actuator to move or interact with a mechanical load, we also make use of free-body diagrams of the mechanical components and Newton's laws in order to develop the complete mathematical model. As with mechanical systems in Chapter 2 and electrical systems in Chapter 3, we utilize a lumped-parameter approach, and, therefore, the mathematical models derived in this chapter consist of ordinary differential equations (ODEs). Our goal is to develop models of systems and devices that utilize fluid and/or thermal components that are of practical use to mechanical and aerospace engineers. In particular, we focus our discussion on fluid-system actuators. This chapter is the final chapter devoted to modeling physical systems; the subsequent chapters develop the techniques for obtaining and analyzing the system's dynamic response.

4.2 HYDRAULIC SYSTEMS

A general fluid system is composed of fundamental elements: (1) a pump that provides a high-pressure fluid; (2) a fluid capacitance due to fluid energy stored in a reservoir or tank; and (3) hoses, pipes, and valves that connect the various reservoirs and control flow. If the fluid system is a translational actuator (such as a hydraulic servomechanism), it typically involves a cylindrical reservoir where the pressurized fluid moves a piston connected to a mechanical load to perform work.

This section presents brief descriptions of the fundamental relationships that govern hydraulic systems where a liquid is the working fluid. We use the basic concepts of fluid mechanics that are developed in university physics courses. The fundamental variables of a hydraulic system are pressure P (in N/m^2 or pascals, Pa), mass-flow rate $w = \dot{m}$ (in kg/s), and volumetric-flow rate $Q = \dot{V}$ (in m^3/s), where m and V are the fluid mass and volume, respectively. It is important to use *absolute* pressure, or pressure relative to a perfect vacuum, in theoretical calculations (absolute pressure = gage pressure + atmospheric pressure). We use volumetric-flow rate Q to describe the flow of liquids in hydraulic systems. Density ρ is a physical property of a fluid, and it is the amount of mass per unit volume (in kg/m^3).

Fluid Bulk Modulus

A fluid is said to be *incompressible* if its density remains constant and *compressible* if its density changes with pressure. Under relatively low pressures, liquids can be considered as incompressible fluids, while high-pressure hydraulic fluids are compressible. The *fluid bulk modulus* β measures the fluid's resistance to compressibility and is defined as

$$\beta = \rho_0 \frac{dP}{d\rho} \tag{4.1}$$

where ρ_0 is a reference fluid density at a nominal pressure and temperature. The derivative $dP/d\rho$ in Eq. (4.1) is computed at a constant temperature. Note that fluid bulk modulus β has the same units as pressure (N/m^2 or Pa) because the term $\rho_0/d\rho$ in Eq. (4.1) is nondimensional. Fluid bulk modulus is the *inverse* of the compressibility of a fluid, and, therefore, fluids that exhibit small changes in density under high pressure (i.e., $d\rho/dP$ is "small") have an extremely large value for β. For example, a typical hydraulic oil used in industrial applications has a fluid bulk modulus of $\beta = 10^9$ Pa (or 1 GPa) and a nominal density of about $\rho_0 = 860 \text{ kg/m}^3$. Using these nominal values, we can apply Eq. (4.1) to obtain the density change due to an increase in pressure

$$\frac{d\rho}{dP} = \frac{\rho_0}{\beta} = 8.6(10^{-7}) \text{ kg/m}^3\text{-Pa}$$

Therefore, the first-order change in fluid density for a hydraulic pressure increase of 20 MPa is about 17.2 kg/m^3 or a 2% increase in the nominal density. Fluid bulk modulus can be thought of as the fluid analog to the elastic modulus of a solid, and, therefore, a very large β means that the fluid is extremely "stiff."

Resistance of Hydraulic Systems

A fluid resistance element is any component that resists flow and dissipates energy, and, therefore, they are analogous to electrical resistors. In general, the flow can be characterized as *laminar* or *turbulent*. Figure 4.1 shows laminar flow through a pipe where the streamlines are smooth and parallel. Laminar flow exists when the pipe diameter is "large" and the flow velocity is "small" (note that flow velocity in Fig. 4.1 is $v = Q/A$). Laminar fluid resistance exhibits a linear relationship between the pressure drop $\Delta P = P_1 - P_2$ and the volumetric-flow rate Q

$$\Delta P = R_L Q \tag{4.2}$$

where R_L is the laminar fluid resistance (in Pa-s/m^3). Note that laminar resistance (4.2) is analogous to electrical resistance $e_R = RI$, where the voltage drop across the resistor e_R is analogous to the pressure drop ΔP, and current flow I is analogous to fluid flow Q. Laminar flow exists when the pressure difference ΔP is "small" and consequently the fluid flow Q is "small" (or, low velocity). For laminar pipe flow where the pipe length L is significantly larger than the pipe diameter d (as in Fig. 4.1), the laminar fluid resistance can be computed using the Hagen−Poiseuille law

$$R_L = \frac{128\mu L}{\pi d^4} \tag{4.3}$$

where μ is the dynamic (or absolute) viscosity of the fluid, in Pa-s.

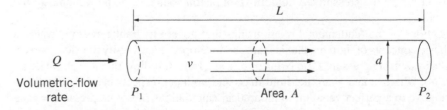

Figure 4.1 Laminar pipe flow.

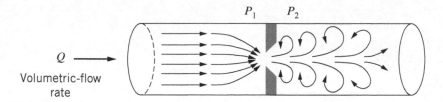

Figure 4.2 Turbulent flow through an orifice.

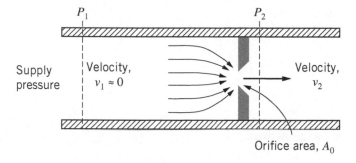

Figure 4.3 Hydraulic fluid flowing through a sharp-edged orifice.

Figure 4.2 shows turbulent flow through a sharp-edged opening (or orifice) in a pipe. Turbulent flow is character-ized by irregular, eddy-like, swirling flow where the streamlines are not uniform or parallel. Turbulent fluid resistance exhibits a nonlinear relationship between the pressure drop $\Delta P = P_1 - P_2$ across the orifice and the volumetric-flow rate Q

$$\Delta P = R_T Q^2 \tag{4.4}$$

where R_T is the nonlinear (turbulent) fluid resistance (in Pa-s^2/m^6). Turbulent flow exists when the pressure difference ΔP across the orifice is "large" and consequently the fluid velocity is large. We can rewrite the turbulent flow equation (4.4) in terms of volumetric-flow rate Q

$$Q = K_T \sqrt{\Delta P} \tag{4.5}$$

where $K_T = \sqrt{1/R_T}$ is a turbulent flow coefficient.

Most industrial hydraulic systems involve high-pressure flow through valve openings or small restrictions and hence the resulting flow is typically turbulent. Let us now develop an approximate model for turbulent hydraulic flow through an orifice or sharp-edged valve opening. Figure 4.3 shows hydraulic oil flowing through a sharp-edged orifice (such as a valve opening) with area A_0. The fluid has very high pressure P_1 at station 1, which is upstream from the orifice, and lower pressure P_2 at station 2, which is immediately downstream from the orifice. Let us assume that the velocity of the oil at station 1 is small and can be neglected, that is, $v_1 \approx 0$. Next, we can apply Bernoulli's equation to the fluid flow between stations 1 and 2:

$$\frac{1}{2}\rho v_1^2 + P_1 = \frac{1}{2}\rho v_2^2 + P_2 \tag{4.6}$$

Bernoulli's equation assumes incompressible, steady, frictionless flow and consequently the total energy is conserved along a streamline between stations 1 and 2. Assuming $v_1 \approx 0$, we can solve Eq. (4.6) for the flow velocity v_2 through the orifice

$$v_2 = \sqrt{\frac{2}{\rho}(P_1 - P_2)} \tag{4.7}$$

We can multiply the velocity through the orifice v_2 by orifice area A_0 to get the volumetric-flow rate through the orifice Q

$$Q = A_0 \sqrt{\frac{2}{\rho}(P_1 - P_2)} \tag{4.8}$$

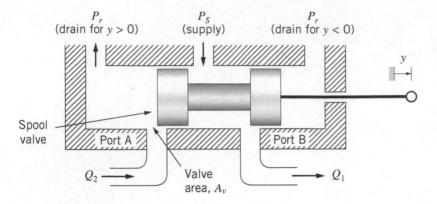

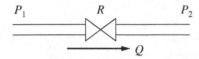

Figure 4.4 Hydraulic flow through a spool valve.

Figure 4.5 Fluid resistance element.

Equation (4.8) is the idealized volumetric-flow rate where the fluid pressure difference $P_1 - P_2$ is converted into kinetic energy without any losses. Real hydraulic flow through the orifice will incur frictional losses, which can be accounted for by multiplying the right-hand side of Eq. (4.8) by the "discharge coefficient" $C_d < 1$

$$Q = C_d A_0 \sqrt{\frac{2}{\rho}(P_1 - P_2)} \tag{4.9}$$

The reader should note that the orifice-flow equation (4.9) is equivalent to the turbulent-flow equations (4.4) and (4.5) where the turbulent flow coefficient is $K_T = C_d A_0 \sqrt{2/\rho}$. Let us use Eq. (4.9) to model hydraulic flow through an orifice or a valve opening. A discharge coefficient $C_d = 0.62$ is typically used for the high-pressure valve flow found in industrial hydraulic systems. Area A_0 may be constant or variable: it is constant for a fixed-geometry orifice and variable for a valve. Figure 4.4 shows a four-way spool valve that is used to meter flow in a hydraulic circuit. When the spool valve is moved to the right ($y > 0$, as shown in Fig. 4.4), port B is opened and consequently the high-pressure oil from the supply pressure P_S flows through port B to provide volumetric-flow rate Q_1. In addition, oil with volumetric-flow rate Q_2 flows through port A to the low-pressure drain P_r as shown in Fig. 4.4. When the valve is moved to the left ($y < 0$), the flow is reversed. Valve area in Fig. 4.4 is $A_v = h|y|$ where h is the height of the valve opening.

Figure 4.5 shows a generic "valve" symbol used to depict fluid resistance. This symbol represents the lumped fluid resistance in a system that could be due to laminar flow through a long pipe, turbulent flow through a sharp-edged orifice, or turbulent flow through a valve opening. This valve symbol is analogous to the damper (or dashpot) symbol used to represent friction in a mechanical system.

Fluid Capacitance

The capacitance of a fluid reservoir or tank is a measure of its ability to store energy due to fluid pressure. For hydraulic systems (liquids), the fluid capacitance C is usually defined as the ratio of the change in volume V to the change in pressure P

$$C = \frac{dV}{dP} \tag{4.10}$$

Hydraulic fluid capacitance has units of m³/Pa or m⁵/N. Figure 4.6 shows a reservoir (tank) with constant circular cross-sectional area A partially filled with a liquid. The pressure at the base of the tank is determined from the hydrostatic equation

$$P = \rho g h + P_{atm} \tag{4.11}$$

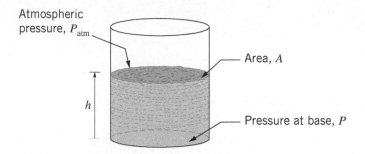

Figure 4.6 Hydraulic reservoir.

where g is the gravitational acceleration, h is the height of the liquid, and P_{atm} is the atmospheric pressure. Base pressure is the sum of the atmospheric pressure (acting at the surface) and the weight of the column of liquid (ρgAh) divided by its area (A). We can compute the fluid capacitance of the tank using Eq. (4.10) by noting that volume is $V = Ah$, and hence the differential is $dV = Adh$. Next, we can compute the differentials of both sides of the hydrostatic equation (4.11) for an incompressible fluid ($\rho = $ constant) to yield

$$dP = \rho g dh \tag{4.12}$$

Using Eqs. (4.10), (4.12), and the differential $dV = Adh$, the fluid capacitance of the tank is

$$C = \frac{dV}{dP} = \frac{Adh}{\rho g dh} = \frac{A}{\rho g} \tag{4.13}$$

Therefore, the fluid capacitance dV/dP of a hydraulic reservoir with a constant cross-sectional area containing an incompressible fluid is a constant.

The fluid capacitance definition (4.10) can be rewritten by separating variables

$$CdP = dV \tag{4.14}$$

Dividing both sides of Eq. (4.14) by dt yields

$$C\dot{P} = \dot{V} = Q \tag{4.15}$$

Equation (4.15) is analogous to the fundamental equation of an electrical capacitor, $C\dot{e}_C = I$, which states that the product of electrical capacitance and the time-rate of voltage is equal to the current flow through the capacitor. Thus, in the hydraulic tank system, pressure is analogous to electric potential (voltage) and volumetric-flow rate Q is analogous to electrical current. Let us revisit Eq. (4.15) when we apply the conservation of mass to a control volume (CV) in a fluid system.

Fluid Inertance

Fluid inertance (or fluid inductance) is the effect due to the fluid's inertia as it accelerates along a pipe. For fluid systems, the inertance L_f is defined as the ratio of the change in pressure P to the change in the time-rate of volumetric-flow rate $\dot{Q}$

$$L_f = \frac{dP}{d\dot{Q}} \tag{4.16}$$

Hydraulic fluid inertance has units of Pa-s^2/m^3. We can use Eq. (4.16) to write an expression relating pressure drop to flow acceleration

$$\Delta P = L_f \dot{Q} \tag{4.17}$$

Equation (4.17) is analogous to the electrical inductor relationship $e_L = L\dot{i}$, where pressure drop ΔP is analogous to the inductor voltage drop e_L, and volumetric-flow rate Q is analogous to current I. While inertia effects are important in modeling mechanical systems, fluid inertia effects are usually insignificant and can be ignored in modeling fluid systems.

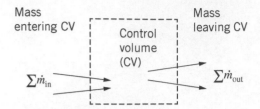

Figure 4.7 Control volume.

Fluid Sources

We utilize two types of ideal sources for fluid systems: pressure and flow sources. These ideal fluid sources are analogous to the ideal voltage and current sources for electrical systems. A pump driven by a motor is typically used to provide a pressurized fluid or a desired fluid flow rate. We do not consider the details that govern the operation of a pump in this chapter but instead assume that a desired pressure source or flow source is a known input to the fluid system.

Conservation of Mass

Mathematical models of fluid systems can be derived by applying the conservation of mass to a CV. Figure 4.7 shows a CV where mass could be entering (positive $\dot{m}$) and leaving (negative $\dot{m}$) through various paths. In addition, mass could be accumulating in the CV. The conservation of mass for the CV yields

$$\dot{m}_{CV} = \sum \dot{m}_{in} - \sum \dot{m}_{out} \tag{4.18}$$

where $\dot{m}_{CV}$ is the net rate of change of the total fluid mass in the CV. If mass does not accumulate in the CV (i.e., steady flow through the CV), then $\dot{m}_{CV} = 0$. We may rewrite the mass continuity equation (4.18) using the symbol $w = \dot{m}$ for mass-flow rate

$$w_{CV} = \sum w_{in} - \sum w_{out} \tag{4.19}$$

The right-hand side terms of the mass continuity equation (4.19) are due to fluid flowing into and out of the CV through pipes, orifices, or valves. The left-hand side of Eq. (4.19) is the time derivative of the total mass contained in the CV, $m_{CV} = \rho V$

$$w_{CV} = \dot{m}_{CV} = \frac{d}{dt}(\rho V) = \dot{\rho}V + \rho\dot{V} \tag{4.20}$$

Therefore, the net mass-flow rate in the CV is affected by the changes in fluid density ρ and volume V. Let us now discuss different hydraulic systems that involve incompressible fluids ($\dot{\rho} = 0$) and compressible fluids ($\dot{\rho} \neq 0$).

Modeling Hydraulic Tank Systems

As a first example of a simple fluid system, we derive the mathematical model of a hydraulic reservoir (tank) system. Because the fluid pressure in a tank is due to its weight (i.e., the hydrostatic equation (4.11)) pressure is relatively low, and, therefore, the fluid can be considered to be incompressible ($\dot{\rho} = 0$). Using Eqs. (4.19) and (4.20), the net rate of change of fluid mass is due solely to its changing volume

$$w_{CV} = \rho\dot{V} = \sum w_{in} - \sum w_{out} \tag{4.21}$$

We can divide Eq. (4.21) by density ρ and use input and output volumetric-flow rates instead of mass-flow rates.

$$\dot{V} = \sum Q_{in} - \sum Q_{out} \tag{4.22}$$

Substituting Eq. (4.15) for the left-hand side of Eq. (4.22) yields

$$C\dot{P} = \sum Q_{in} - \sum Q_{out} \tag{4.23}$$

Equation (4.23) is our fundamental modeling equation for a hydraulic system with incompressible flow. Note that the left side of Eq. (4.23) is $\dot{V} = (dV/dP)(dP/dt)$ where $dV/dP = C$. We must apply Eq. (4.23) to *each* hydraulic reservoir (or CV) if we have a system of connected tanks. The following example demonstrates the modeling steps for a single-tank hydraulic system.

Example 4.1

Figure 4.8 shows a single hydraulic tank with input volumetric-flow rate Q_{in}.

 (a) Derive the mathematical model of the hydraulic system assuming laminar flow through the valve.
 (b) Derive the mathematical model of the hydraulic system assuming turbulent flow through the valve.
 (c) Repeat problems (a) and (b) with the model expressed in terms of liquid height h.

 (a) Because the hydraulic system consists of a single tank, the mathematical model is derived from the conservation of mass (4.23) for a single CV

$$C\dot{P} = Q_{in} - Q_{out} \tag{4.24}$$

where P is the pressure at the base of the tank, Q_{in} is the (given) input volumetric flow, and Q_{out} is the output volumetric flow through the valve. We use Eq. (4.2) to represent the laminar output flow rate

$$Q_{out} = \frac{\Delta P}{R_L} \tag{4.25}$$

where R_L is the laminar fluid resistance, and the pressure drop across the valve is $\Delta P = P - P_{atm}$ because the pressure at the outlet of the valve is atmospheric pressure P_{atm}. Using Eq. (4.25) in Eq. (4.24) yields

$$C\dot{P} = Q_{in} - \frac{P - P_{atm}}{R_L} \tag{4.26}$$

Multiplying Eq. (4.26) by R_L and grouping terms involving the dynamic variable P on the left-hand side yields

$$R_L C\dot{P} + P = R_L Q_{in} + P_{atm} \tag{4.27}$$

Equation (4.27) is the mathematic model of the hydraulic tank system with laminar flow through the valve. The system is modeled by a single first-order linear ODE because we have a single fluid capacitance that can store energy. The fluid capacitance of the tank is a constant: $C = A/(\rho g)$. The system inputs are Q_{in} and P_{atm} and consequently they are grouped on the right-hand side of Eq. (4.27).

 (b) For turbulent flow through the valve, we use Eq. (4.5) for the output volumetric-flow rate

$$Q_{out} = K_T \sqrt{P - P_{atm}} \tag{4.28}$$

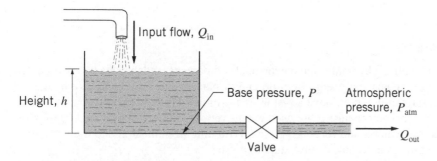

Figure 4.8 Hydraulic tank system for Example 4.1.

where K_T is a turbulent flow coefficient and $\Delta P = P - P_{atm}$. Substituting Eq. (4.28) into the continuity equation (4.24) yields

$$C\dot{P} + K_T \sqrt{P - P_{atm}} = Q_{in} \qquad (4.29)$$

Equation (4.29) is the mathematic model of the hydraulic tank system with turbulent flow through the valve. The system is modeling by a single first-order *nonlinear* ODE as we have a single fluid capacitance that can store energy.

(c) We want to write the linear and nonlinear mathematical models (4.27) and (4.29) in terms of liquid height h instead of base pressure P. Pressure and height are related by the hydrostatic equation (4.11)

$$P = \rho g h + P_{atm} \qquad (4.30)$$

The time derivative of hydrostatic pressure (4.30) is

$$\dot{P} = \rho g \dot{h} \qquad (4.31)$$

Next, we substitute Eq. (4.31) for $\dot{P}$ and Eq. (4.30) for $P - P_{atm}$ in the linear hydraulic tank model (4.27) to yield

$$R_L C \rho g \dot{h} + \rho g h = R_L Q_{in} \qquad (4.32)$$

Finally, we can divide all terms in Eq. (4.32) by ρg

$$R_L C \dot{h} + h = \frac{R_L}{\rho g} Q_{in} \qquad (4.33)$$

Equation (4.33) is the mathematical model of the hydraulic tank with laminar valve flow where liquid height h is the dynamic variable. Equations (4.27) and (4.33) are equivalent dynamic models of the hydraulic tank with laminar flow through the valve.

We can obtain the nonlinear (turbulent flow) model in terms of liquid height h by substituting Eqs. (4.30) and (4.31) for $P - P_{atm}$ and $\dot{P}$ in the nonlinear hydraulic tank model (4.29) to yield

$$C \rho g \dot{h} + K_T \sqrt{\rho g h} = Q_{in} \qquad (4.34)$$

Equation (4.34) is the mathematical model of the hydraulic tank with turbulent valve flow where liquid height h is the dynamic variable. The reader should note that if we substitute fluid capacitance $C = A/(\rho g)$ in Eq. (4.34), the first left-hand side term is clearly the time-rate of volume ($A\dot{h} = \dot{V}$) and exhibits a dimensional match with Q_{in} and $Q_{out} = K_T \sqrt{\Delta P}$. Equations (4.29) and (4.34) are equivalent dynamic models of the hydraulic tank with turbulent valve flow.

In summary, hydraulic tank systems require a first-order ODE for each fluid capacitance (reservoir) and the ODE can be expressed with either pressure P or liquid height h as the dynamic variable. Therefore, if we have two interconnected tanks, the complete mathematical model will involve two first-order ODEs. Flow through valves (resistances), whether it is laminar (linear) or turbulent (nonlinear), will always be a function of the pressure drop ΔP across the valve. Because fluid-system actuators are the primary focus of this chapter, we do not pursue additional examples of hydraulic tank systems.

Modeling Hydromechanical Systems

Hydromechanical systems are created by combining hydraulic and mechanical components, and they are used to convert the energy stored in the pressurized fluid to mechanical energy (motion and displacement). For example, a hydraulic actuator (or servomechanism) uses a pressurized liquid to move a piston in a cylinder that is connected to a mechanical load. As previously indicated, hydraulic servomechanisms are used extensively in industry to provide large forces for heavy machinery such as harvesters, excavators, and forging presses. Hydraulic accumulators store energy in a pressurized reservoir and often use a spring-mass mechanical subsystem to do so.

Figure 4.9 shows a simple hydraulic actuator that consists of a piston and cylinder with circular cross-sectional area A. In this elementary example, hydraulic oil is flowing into the left chamber of the cylinder (Q_{in}), and, therefore,

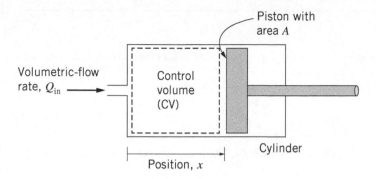

Figure 4.9 Hydraulic piston and cylinder.

we establish a CV on the left side of the cylinder (most hydraulic servomechanisms involve pressurized fluid on both sides of the cylinder so that the piston can move left or right). Conservation of mass (4.19) is applied to the CV

$$w_{CV} = w_{in} - w_{out} \tag{4.35}$$

The left-hand side of Eq. (4.35) is the time derivative of total mass in the CV

$$w_{CV} = \dot{m}_{CV} = \dot{\rho}V + \rho\dot{V} \tag{4.36}$$

We consider the high-pressure hydraulic fluid to be compressible, and therefore $\dot{\rho} \neq 0$ (the reader should note that most hydraulic oils will experience a density change less than 2% despite a pressure increase on the order of 20 MPa). By expressing the input mass-flow rate as $w_{in} = \rho Q_{in}$ and noting that the output mass-flow rate w_{out} is zero, Eq. (4.35) becomes

$$\dot{\rho}V + \rho\dot{V} = \rho Q_{in} \tag{4.37}$$

The time derivative of density can be expressed by using the chain rule

$$\dot{\rho} = \frac{d\rho}{dP}\frac{dP}{dt} \tag{4.38}$$

The definition of fluid bulk modulus, Eq. (4.1), can be used to solve for the change in density due to change in pressure, or $d\rho/dP = \rho/\beta$. Therefore, the time-rate of density is $\dot{\rho} = \rho\dot{P}/\beta$ and the mass-continuity equation (4.37) becomes

$$\frac{\rho\dot{P}V}{\beta} + \rho\dot{V} = \rho Q_{in} \tag{4.39}$$

Dividing out density and rearranging Eq. (4.39) yields

$$\dot{P} = \frac{\beta}{V}(Q_{in} - \dot{V}) \tag{4.40}$$

Equation (4.40) is the fundamental modeling equation for the time-rate of pressure for a hydraulic cylinder with a *compressible* fluid. The hydraulic actuator shown in Fig. 4.9 is known as a *single-acting* cylinder because hydraulic fluid flows to one side of the piston. If we compare Eq. (4.40) with the hydraulic capacitance equation (4.15), we see that V/β is the fluid capacitance of a hydraulic actuator. Equation (4.40) shows that fluid flowing into the CV ($Q_{in} > 0$) increases the fluid pressure while an expanding CV ($\dot{V} > 0$) decreases pressure. The instantaneous volume of the CV is $V = Ax$ (where x is the position of the piston in Fig. 4.9), and, therefore, the time-rate of the volume is $\dot{V} = A\dot{x}$. It is clear that the motion of the piston (x and $\dot{x}$) must be accounted for in the hydraulic servo equation (4.40), and consequently we must include a model of the mechanical system (piston and load). Finally, it should be noted that Eq. (4.40) is nonlinear as the coefficient $1/V$ involves the inverse of the dynamic variable x.

Example 4.2

Figure 4.10 shows a simple hydraulic actuator with an input flow rate Q_{in} to the cylinder and a piston connected to the load mass. Derive the mathematical model of the hydromechanical system.

We begin with Eq. (4.40), which models the pressure change in the left-hand side of the cylinder

$$\dot{P} = \frac{\beta}{V}(Q_{in} - \dot{V}) \tag{4.41}$$

where P and V are the pressure and volume of the left-side cylinder, respectively. The instantaneous volume is

$$V = V_0 + Ax \tag{4.42}$$

where piston position x is measured from the static equilibrium position (no spring deflection) and V_0 is the left-side chamber volume when $x = 0$. If we use Eq. (4.42), the time-rate of chamber volume is $\dot{V} = A\dot{x}$ and hence Eq. (4.41) becomes

$$\dot{P} = \frac{\beta}{V_0 + Ax}(Q_{in} - A\dot{x}) \tag{4.43}$$

Next, we derive the mechanical model that governs the position and velocity of the piston and load mass. Figure 4.11 shows a free-body diagram of the mechanical system where m is the total mass of the piston, connecting rod, and load mass. The hydraulic-pressure force PA acts on the left side of the piston, while the atmospheric pressure force $P_{atm}A$ acts on the right side of the piston (note that although the area of the right side of the piston is less than A due to the connecting rod area A_{rod}, the atmospheric pressure distribution on the load mass results in an incremental force $P_{atm}A_{rod}$ acting to the left and hence the *net* atmospheric pressure force on the piston/load mass is $P_{atm}A$). The spring, friction, and load forces act on the load mass as shown in Fig. 4.11. Next, apply Newton's second law and sum all external forces on piston/load mass m with the sign convention as positive to the right:

$$+ \rightarrow \sum F = PA - P_{atm}A - kx - b\dot{x} - F_L = m\ddot{x} \tag{4.44}$$

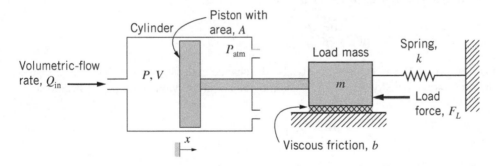

Figure 4.10 Hydraulic actuator for Example 4.2.

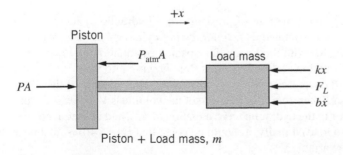

Figure 4.11 Free-body diagram of the hydraulic actuator (Example 4.2).

Rearranging Eq. (4.44) yields

$$m\ddot{x} + b\dot{x} + kx = (P - P_{\text{atm}})A - F_L \tag{4.45}$$

Equations (4.43) and (4.45) constitute the mathematical model of the hydromechanical system. The reader should note that the total system involves a first-order ODE for the fluid capacitance and a second-order ODE for the mechanical inertia. Equations (4.43) and (4.45) are coupled as pressure P is required in the mechanical model (4.45), and position and velocity (x and $\dot{x}$) are required in the fluid model (4.43). The system is nonlinear because the fluid model equation (4.43) is a nonlinear ODE. Finally, the reader should note that the dynamic variables are chamber pressure P and position x, while the system input variables are input flow rate Q_{in}, atmospheric pressure P_{atm}, and load force F_L.

Example 4.3

Figure 4.12 shows an accumulator in a hydraulic circuit where Q_{in} is the input volumetric-flow rate. Derive the mathematical model of the complete system.

Accumulators are placed downstream from hydraulic sources (e.g., pumps) in order to attenuate (reduce) ripples or spikes in flow rate or pressure. The system shown in Fig. 4.12 consists of a hose (with constant volume V_h) connecting a pump to a hydraulic load. The load might be a hydraulic motor that is part of a hydrostatic transmission, which may require a prescribed flow rate Q_h for proper operation. The accumulator consists of a chamber with pressure P_c and volume V_c and a movable plate (mass m) restrained by spring k. High-pressure fluid is allowed to flow from the hose to the accumulator through a sharp-edged orifice with constant area A_0. The accumulator can store fluid during rapid increases in hose pressure P_h and release fluid to the hose during rapid decreases in hose pressure. Consequently, accumulators are used to dampen out fluctuations in hose pressure or flow-rate.

We begin by applying the pressure-rate equation (4.40) to each fluid CV: the hose CV and the accumulator CV

$$\text{Hose:} \quad \dot{P}_h = \frac{\beta}{V_h}(Q_{\text{in}} - Q_c - Q_h - \dot{V}_h) \tag{4.46}$$

$$\text{Accumulator:} \quad \dot{P}_c = \frac{\beta}{V_c}(Q_c - \dot{V}_c) \tag{4.47}$$

Note that in Eq. (4.46) the net volumetric-flow rate into the hose CV is $Q_{\text{in}} - Q_c - Q_h$. Furthermore, the hose volume V_h is constant, and, therefore, $\dot{V}_h = 0$ in Eq. (4.46). We can use Eq. (4.9) to represent the turbulent flow through the sharp-edged orifice

$$Q_c = C_d A_0 \sqrt{\frac{2}{\rho}(P_h - P_c)} \tag{4.48}$$

where ρ is the nominal fluid density and C_d is the discharge coefficient. The reader should note that Eq. (4.48) indicates flow from the hose to the accumulator that occurs when $P_h > P_c$ as depicted in Fig. 4.12. However, it is possible for the

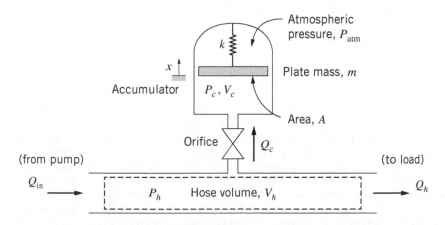

Figure 4.12 Hydraulic accumulator for Example 4.3.

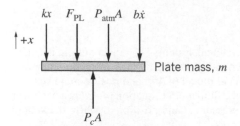

Figure 4.13 Free-body diagram of the hydraulic accumulator plate (Example 4.3).

accumulator pressure to exceed the hose pressure, which would reverse the direction (and sign) of flow Q_c. For this case, $P_c > P_h$ and we cannot use Eq. (4.48) because the radicand would be negative. Therefore, we can modify Eq. (4.48) to produce a general equation that accommodates positive or negative orifice flow

$$Q_c = C_d A_0 \text{sgn}(P_h - P_c)\sqrt{\frac{2}{\rho}|P_h - P_c|} \tag{4.49}$$

Note that the radicand is always positive (due to the absolute value of the pressure difference) and that the signum function operation $\text{sgn}(P_h - P_c)$ results in either $+1$, -1, or zero, and, therefore, determines whether orifice flow is positive (hose to accumulator), negative (accumulator to hose), or zero.

The accumulator volume is

$$V_c = V_{c0} + Ax \tag{4.50}$$

where V_{c0} is the accumulator volume when $x = 0$. Plate position x is measured from the static equilibrium position where the spring force balances the nominal pressure force. It is clear that the time-rate of the accumulator CV is $\dot{V}_c = A\dot{x}$.

Next, we derive the mechanical model for the accumulator plate mass by utilizing the free-body diagram shown in Fig. 4.13. Note that we have assumed that linear viscous friction acts on the plate $(b\dot{x})$ and that atmospheric pressure P_{atm} acts on the top part of the plate. Furthermore, the accumulator spring is initially compressed in order to balance the nominal pressure force and hence the spring preload force F_{PL} is included in the free-body diagram. The reader should note that when the plate is at rest at position $x = 0$, the accumulator is in static equilibrium and the preload F_{PL} balances the pressure forces. Summing all external forces on plate mass m yields

$$+\uparrow \sum F = P_c A - kx - F_{\text{PL}} - P_{\text{atm}}A - b\dot{x} = m\ddot{x} \tag{4.51}$$

or, after rearranging Eq. (4.51)

$$m\ddot{x} + b\dot{x} + kx = (P_c - P_{\text{atm}})A - F_{\text{PL}} \tag{4.52}$$

Equation (4.52) clearly shows that when the spring preload force F_{PL} balances the differential pressure force $(P_c - P_{\text{atm}})A$ the plate mass can be at static equilibrium (i.e., $\ddot{x} = \dot{x} = x = 0$).

The complete mathematical model of the accumulator system is composed of Eqs. (4.46), (4.47), and (4.52), which are repeated below with the proper substitutions for the volumes and their time-rates:

$$\text{Hose CV:} \quad \dot{P}_h = \frac{\beta}{V_h}(Q_{\text{in}} - Q_c - Q_h) \tag{4.53}$$

$$\text{Accumulator CV:} \quad \dot{P}_c = \frac{\beta}{V_{c0} + Ax}(Q_c - A\dot{x}) \tag{4.54}$$

$$\text{Accumulator plate:} \quad m\ddot{x} + b\dot{x} + kx = (P_c - P_{\text{atm}})A - F_{\text{PL}} \tag{4.55}$$

Equations (4.53), (4.54), and (4.55) represent the mathematical model of the hydromechanical system. Equation (4.49) is also required to define the orifice flow rate Q_c. The complete model consists of two first-order ODEs and one second-order ODE and hence the model is fourth order. The ODEs are coupled because knowledge of the mechanical motion is required in the accumulator CV equation (4.54) and pressure information is required in the orifice flow equation (4.49) and the mechanical equation (4.55). The system is clearly nonlinear because of the orifice flow equation (4.49). In summary, the dynamic variables of the system are hose pressure P_h, accumulator pressure P_c, and plate position x, while the system input variables are input flow rate Q_{in}, hose flow rate Q_h, atmospheric pressure P_{atm}, and spring preload force F_{PL}.

Example 4.4

Consider again the hydraulic system from Example 4.3. Compute the fluid capacitance C of the hose and initial volumetric-flow rate Q_c through the orifice if the system has the following characteristics at time $t = 0$: hose volume $V_h = 0.003$ m^3, hose pressure $P_h = 2(10^7)$ Pa, accumulator pressure $P_c = 1.95(10^7)$ Pa, fluid density $\rho = 875$ kg/m^3, bulk modulus $\beta = 0.8(10^9)$ Pa, discharge coefficient $C_d = 0.62$, and orifice area $A_0 = 5(10^{-6})$ m^2.

Fluid capacitance of the fixed-volume hose is $C = V_h/\beta = 0.003$ m^3/$0.8(10^9)$ Pa, or $C = 3.75(10^{-12})$ m^3/Pa.

Equation (4.48) determines the initial volumetric-flow rate through the orifice

$$Q_c = C_d A_0 \sqrt{\frac{2}{\rho}(P_h - P_c)}$$

where the initial pressure difference across the orifice is $P_h - P_c = 2(10^7) - 1.95(10^7) = 5(10^5)$ Pa. Using the given numerical values for the hydraulic fluid and orifice yields an initial volumetric-flow rate of $Q_c = 1.048(10^{-4})$ m^3/s.

4.3 PNEUMATIC SYSTEMS

This section presents brief descriptions of the fundamental relationships that govern pneumatic systems where a gas (often air) is the working fluid. In the previous section, we noted that very large pressure changes (on the order of 20 MPa) produced a 2% increase in the density of typical hydraulic oils. While pneumatic systems involve much lower operating pressures when compared to hydraulic systems, they nearly always involve compressible gases where the density changes significantly with pressure. Consequently, pneumatic systems are less "stiff" when compared to hydraulic systems, and, therefore, exhibit a slower response to changes in the operating state. Another difference between hydraulic and pneumatic systems involves the inclusion of thermodynamic effects. Although temperature changes can affect the properties of liquids (such as viscosity and bulk modulus), these effects are small compared to the pressure variations and hence were not considered in the development of hydraulic system models. Pneumatic systems, on the other hand, exhibit a functional relationship between pressure, temperature, and density as demonstrated by the ideal gas law:

$$P = \rho R T \tag{4.56}$$

where P is the absolute pressure, ρ is the gas density, R is the gas constant, and T is the absolute temperature (in kelvin, K). Including thermodynamic effects complicates the analysis of pneumatic systems immensely.

The fundamental variables of a pneumatic system are pressure P and mass-flow rate w. Because gases are highly compressible, we cannot simply relate mass-flow rate to volumetric-flow rate. Furthermore, unlike gases, liquids "fill" a vessel with a recognizable liquid height. Hence we use mass-flow rate w for pneumatic systems instead of volumetric-flow rate Q.

Resistance of Pneumatic Systems

In rare cases where the gas is incompressible (i.e., very low-speed flow), the pneumatic resistance can be modeled by either the linear laminar equation (4.2) or the nonlinear turbulent equation (4.4). The corresponding laminar or turbulent resistance coefficients R_L and R_T can be approximated from experimental results, such as a plot of mass-flow rate w versus pressure drop.

In most industrial applications (such as pneumatic actuators), the working gas flows through valves and orifices at a high speed, and therefore the gas is compressible. Compressible gas flow is a complex phenomenon, and, therefore, we do not develop the flow equations here. Instead, we present the results for gas flow through a sharp-edged orifice, which we can use to model compressible flow in a pneumatic system (see [1] for a detailed discussion of compressible gas flow).

Figure 4.14 shows compressible gas flow through a sharp-edged orifice with area A_0 at the throat (minimum area). Expressions for the mass-flow rate of the gas can be derived by assuming that the expansion of an ideal gas through the orifice is isentropic (i.e., frictionless and adiabatic). In addition, we need to consider two cases: (1) "unchoked" flow and (2) "choked" flow. The flow is said to be "choked" when it achieves sonic conditions (the speed of sound, or Mach = 1) at the throat. The ratio of the downstream-to-upstream pressures, P_2/P_1, determines whether or not the flow is choked. Clearly, if the upstream and downstream pressures are nearly equal ($P_2/P_1 \approx 1$), then no gas flows through the orifice.

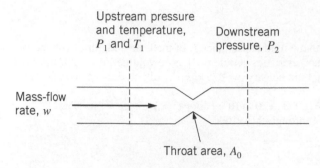

Figure 4.14 Gas flow through a sharp-edged orifice.

Gas begins to flow through the orifice at an increasing speed as the pressure ratio P_2/P_1 decreases from unity. When the pressure ratio P_2/P_1 is greater than the critical ratio C_r, the gas flow is subsonic and "unchoked" and the corresponding mass-flow rate is

$$\text{"unchoked"} \qquad w = C_d A_0 P_1 \sqrt{\frac{2\gamma}{(\gamma-1)RT_1}\left[\left(\frac{P_2}{P_1}\right)^{\frac{2}{\gamma}} - \left(\frac{P_2}{P_1}\right)^{\frac{\gamma+1}{\gamma}}\right]} \quad \text{if } P_2/P_1 > C_r \qquad (4.57)$$

where γ is the ratio of specific heats ($=1.4$ for air) and C_d is the discharge coefficient for losses associated with flow through the orifice. Equation (4.57) is a highly nonlinear function of pressure ratio P_2/P_1, temperature T_1, orifice area A_0, discharge coefficient C_d, gas constant R, and the ratio of specific heats γ. The unchoked flow equation clearly shows that mass-flow rate is zero if the pressure ratio P_2/P_1 is exactly unity. If the downstream pressure P_2 becomes low enough, the flow speed increases until it reaches the sonic (Mach 1) condition at the throat and the flow becomes choked. Decreasing downstream pressure P_2 below this critical point will not alter the sonic conditions at the throat. In this case, the choked mass-flow rate is

$$\text{"choked"} \qquad w = C_d A_0 P_1 \sqrt{\frac{\gamma}{RT_1} C_r^{\frac{\gamma+1}{\gamma}}} \quad \text{if } P_2/P_1 \leq C_r \qquad (4.58)$$

The critical pressure ratio that divides the unchoked and choked flow regimes is a function of γ

$$C_r = \left(\frac{2}{\gamma+1}\right)^{\frac{\gamma}{\gamma-1}} \qquad (4.59)$$

For air, $\gamma = 1.4$, and, therefore, the critical pressure is $C_r = 0.528$. Figure 4.15 shows the mass-flow rate for air flowing through a sharp-edged orifice with the following conditions: upstream pressure $P_1 = 6(10^5)$ Pa (about 87 psia), orifice area $A_0 = 4\,\text{mm}^2$, upstream temperature $T_1 = 298$ K, and discharge coefficient $C_d = 0.8$. The downstream pressure P_2 is varied between a very small value (near vacuum) and an upper limit equal to P_1. Choked flow, as computed by Eq. (4.58), is easily identified by the constant mass-flow rate when pressure ratio $P_2/P_1 \leq 0.528$. The flow becomes unchoked (subsonic) when $P_2/P_1 > 0.528$ and decreases to zero when $P_2/P_1 = 1$.

Pneumatic Capacitance

Because pneumatic systems often involve gas flowing into a constant-volume vessel, the fluid capacitance C is usually defined as the ratio of the change in mass m to the change in pressure P

$$C = \frac{dm}{dP} \qquad (4.60)$$

that has units of kg/Pa or kg-m^2/N. The mass of a gas in a vessel is $m = \rho V$, and, therefore, the differential of mass for a constant-volume vessel is $dm = Vd\rho$. Consequently, the pneumatic capacitance (4.60) for a constant-volume container becomes

$$C = V\frac{d\rho}{dP} \qquad (4.61)$$

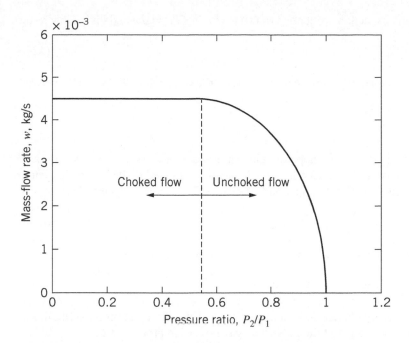

Figure 4.15 Mass-flow rate for air through a sharp-edged orifice.

Therefore, pneumatic capacitance depends on the compressibility of the gas with respect to change in pressure. Recall that from basic thermodynamics, the expansion process of a gas may be at a constant temperature (isothermal), constant pressure (isobaric), or constant entropy (isentropic). We consider here isothermal and isentropic (adiabatic and reversible) processes, which we can model with the polytropic process

$$\frac{P}{\rho^n} = a \quad \text{or} \quad P = a\rho^n \tag{4.62}$$

where a is a constant and n is the polytropic exponent. For an isothermal process $n = 1$; for an isentropic process $n = \gamma$. Taking differentials of both sides of Eq. (4.62) yields

$$dP = an\rho^{n-1}d\rho \tag{4.63}$$

Solving Eq. (4.63) for $d\rho/dP$, we obtain

$$\frac{d\rho}{dP} = \frac{\rho^{1-n}}{an} = \frac{\rho\rho^{-n}}{an} \tag{4.64}$$

Next, solve Eq. (4.62) for the constant $a = P\rho^{-n}$ and substitute into Eq. (4.64) to yield

$$\frac{d\rho}{dP} = \frac{\rho}{nP} \tag{4.65}$$

Using the ideal gas law (4.56) to substitute for pressure P in Eq. (4.65), the derivative $d\rho/dP$ becomes

$$\frac{d\rho}{dP} = \frac{1}{nRT} \tag{4.66}$$

Finally, substituting Eq. (4.66) into Eq. (4.61) yields the pneumatic capacitance for a constant-volume vessel:

$$C = \frac{V}{nRT} \tag{4.67}$$

Note that the pneumatic capacitance of a fixed vessel can vary depending on the gas temperature, type of gas, and thermodynamic process.

We can separate variables in the pneumatic capacitance equation (4.60) to obtain $C\,dP = dm$ and divide this result by dt to yield the differential equation

$$C\dot{P} = \dot{m} = w \tag{4.68}$$

Equation (4.68) is analogous to the fundamental equation of an electrical capacitor and the hydraulic reservoir derived in Section 4.2.

Modeling Pneumatic Systems

Mathematical models of pneumatic systems can be derived by applying the conservation of mass and the CV approach demonstrated in Section 4.2. The derivation is only slightly different from the steps we applied to the hydraulic system with a compressible fluid. To begin, recall the mass-continuity equation (4.19) for a CV, which is repeated here

$$w_{CV} = \sum w_{in} - \sum w_{out} \tag{4.69}$$

The total mass of the gas in the CV at any instant is $m_{CV} = \rho V$, and, therefore, its time derivative is

$$w_{CV} = \dot{m}_{CV} = \dot{\rho}V + \rho\dot{V} \tag{4.70}$$

We used the fluid bulk modulus to define the density change $\dot{\rho}$ for a hydraulic system with a compressible fluid. For pneumatic systems, we find $\dot{\rho}$ by taking the time derivative of the polytropic gas expansion process (4.62)

$$\dot{P} = an\rho^{n-1}\dot{\rho} = \frac{n}{\rho}a\rho^n\dot{\rho} \tag{4.71}$$

Substituting the polytropic process model $P = a\rho^n$ in Eq. (4.71) yields an expression for the time derivative of density

$$\dot{\rho} = \frac{\rho}{nP}\dot{P} \tag{4.72}$$

Substituting Eq. (4.72) into the mass-flow rate equation (4.70) and using the ideal gas law to replace density ($\rho = P/RT$) we obtain

$$w_{CV} = \frac{V}{nRT}\dot{P} + \frac{P}{RT}\dot{V} = \sum w_{in} - \sum w_{out} \tag{4.73}$$

Because we are interested in the pressure variation of the pneumatic system, we solve Eq. (4.73) for the time-rate of pressure to obtain

$$\dot{P} = \frac{nRT}{V}\left(w_{net} - \frac{P}{RT}\dot{V}\right) \tag{4.74}$$

where $w_{net} = \sum w_{in} - \sum w_{out}$ is the net rate of change of the total fluid mass in the CV. Equation (4.74) is our fundamental modeling equation for a pneumatic system. We see that for a constant-volume vessel ($\dot{V} = 0$), Eq. (4.74) is identical to the differential equation for pressure (4.68) that was derived from the pneumatic capacitance definition. Furthermore, the pneumatic pressure-rate equation (4.74) has a structure that is very similar to the compressible hydraulic pressure-rate equation (4.40): both equations demonstrate an increase in pressure for a positive net flow into the CV and a decrease in pressure when the CV expands (i.e., $\dot{V} > 0$). For the case of constant volume ($\dot{V} = 0$), the fluid capacitance of the hydraulic system in Eq. (4.40) is V/β while the pneumatic capacitance in Eq. (4.74) is V/nRT, which matches Eq. (4.67).

Example 4.5

Figure 4.16 shows a simple pneumatic system that consists of a chamber with fixed volume V connected to an air-supply tank with constant pressure P_S. Derive the mathematical model of the pneumatic system assuming compressible flow through a sharp-edged orifice (valve) with throat area A_0.

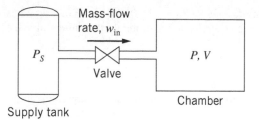

Figure 4.16 Pneumatic system for Example 4.5.

The rate of pressure change in the chamber is governed by Eq. (4.74), which is the fundamental modeling equation of a pneumatic vessel

$$\dot{P} = \frac{nRT}{V}\left(w_{in} - \frac{P}{RT}\dot{V}\right) \tag{4.75}$$

where w_{in} is the mass-flow rate from the supply tank through the valve. Because chamber volume is constant, we set $\dot{V} = 0$ in Eq. (4.75) to obtain

$$C\dot{P} = w_{in} \tag{4.76}$$

where the pneumatic capacitance for the fixed-volume vessel is

$$C = \frac{V}{nRT} \tag{4.77}$$

The mass-flow rate w_{in} for a compressible gas is governed by Eqs. (4.57) and (4.58) depending on whether the flow is unchoked (subsonic) or choked. Therefore, using the compressible flow equations the pressure-rate equation (4.76) becomes

$$C\dot{P} = C_d A_0 P_S \sqrt{\frac{2\gamma}{(\gamma-1)RT}\left[\left(\frac{P}{P_S}\right)^{\frac{2}{\gamma}} - \left(\frac{P}{P_S}\right)^{\frac{\gamma+1}{\gamma}}\right]} \quad \text{if} \quad P/P_S > C_r \tag{4.78}$$

$$C\dot{P} = C_d A_0 P_S \sqrt{\frac{\gamma}{RT}C_r^{\frac{\gamma+1}{\gamma}}} \quad \text{if} \quad P/P_S \leq C_r \tag{4.79}$$

Equations (4.78) and (4.79) are the mathematical modeling equations of the simple pneumatic tank system. Equation (4.77) is used to define the pneumatic capacitance C and Eq. (4.59) is needed to define the critical pressure ratio ($C_r = 0.528$ for air). The mathematical model consists of the appropriate first-order ODE (i.e., unchoked or choked flow) and is clearly nonlinear for unchoked flow. The single dynamic variable is chamber pressure P and the system input is supply pressure P_S. System parameters include chamber volume V, polytropic exponent n, orifice area A_0, upstream gas temperature T, and discharge coefficient C_d.

Example 4.6

Consider again the pneumatic system from Example 4.5. Compute the pneumatic capacitance C and initial mass-flow rate w_{in} through the valve if the system has the following characteristics at the instant the valve is opened: chamber volume $V = 3(10^{-4})$ m^3, upstream air temperature $T = 298$ K, upstream supply pressure $P_S = 6(10^5)$ Pa, downstream chamber pressure $P = 1.2(10^5)$ Pa, discharge coefficient $C_d = 0.8$, and valve area $A_0 = 2(10^{-6})$ m^2. Assume an isothermal expansion process with air as the working gas.

Equation (4.67) defines the pneumatic capacitance for a fixed-volume vessel:

$$C = \frac{V}{nRT}$$

where $n = 1$ (isothermal process) and $R = 287$ N-m/kg-K is the gas constant for air. Using the pneumatic characteristics given for this problem, the pneumatic capacitance is $C = 3.5077(10^{-9})$ kg-m^2/N.

The initial input mass-flow rate w_{in} is determined by either Eq. (4.57) for unchoked flow or Eq. (4.58) for choked flow. Hence, we must compute the initial downstream-to-upstream pressure ratio P/P_S and compare this ratio with the critical pressure ratio, where $C_r = 0.528$ for air (see Eq. (4.59)). The initial pressure ratio is $P/P_S = 1.2(10^5)/6(10^5) = 0.2$, which is less than the critical pressure ratio C_r. Therefore, the initial mass-flow rate is *choked*, and we compute its numerical value using Eq. (4.58)

$$w_{in} = C_d A_0 P_S \sqrt{\frac{\gamma}{RT} C_r^{\frac{\gamma+1}{\gamma}}}$$

where $\gamma = 1.4$ for air. Using the numerical values for the gas and valve characteristics in the above equation yields $w_{in} = 0.002248 \, \text{kg/s}$.

4.4   THERMAL SYSTEMS

Thermal systems involve the storage and flow of heat energy. Temperature T (in kelvin, K) is the principal dynamic variable of interest and heat flow rate q is also a fundamental variable in thermal models. Heat energy has units of joules (J), and, therefore, the rate of heat transfer q has units J/s or watts (W), which has the same units as power. Thermal system models are derived by applying the conservation of energy to the system boundary and noting the heat energy rates into and out of the system. Figure 4.17 shows an *open thermal system* with a boundary that encloses a thermal capacitance C. The boundary could be an insulating material that impedes the flow of heat energy. Heat energy could flow into the capacitance (e.g., a heating element) or heat energy could flow out of the capacitance (e.g., heat flowing from a hot chamber to the colder ambient surroundings). Furthermore, heat energy can enter or leave the system because of mass transferring across the boundary (e.g., fluid flow). Applying the conservation of energy to the system boundary in Fig. 4.17 yields

$$\dot{\xi}_h = \sum \dot{h}_{in} - \sum \dot{h}_{out} + \sum q_{in} - \sum q_{out} \tag{4.80}$$

where $\dot{\xi}_h$ is the net time-rate of heat energy stored within the system boundary. The time-rate of enthalpy $\dot{h}$ is because of mass transferring across the system boundary. Therefore, the rate of enthalpy change is zero for a *closed thermal system* (i.e., no mass transfer). Equation (4.80) assumes that the system does not generate heat within the boundary and no work is done on the system or by the system. The energy balance equation (4.80) is essentially the first law of thermodynamics expressed as the time-rate of energy contained within the system boundary.

Thermal systems are generally more difficult to model compared to mechanical, electrical, or fluid systems. Temperature typically exhibits a spatial variation; that is, temperature usually varies between different points in a body. Therefore, temperature of a body could be represented as $T(x, y, z, t)$, which states that the temperature varies with the Cartesian coordinate location (x, y, z) within the body as well as time t. Therefore, thermal systems are more accurately modeled as *distributed systems*, which require partial differential equations (PDEs) instead of ODEs as the modeling equations. In order to derive simplified, approximate thermal models, we assume that all points in a "thermal body" possess the same (average) temperature. This assumption allows us to derive lumped-parameter models where each "thermal body" (or thermal capacitance) has a single, uniform temperature. Therefore, our lumped-parameter thermal models are similar to our lumped-parameter fluid models, where each fluid capacitance (i.e., chamber or vessel) possesses a single pressure at each instant of time (i.e., there is no pressure variation within a fluid capacitance).

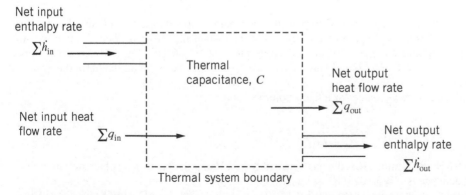

Figure 4.17   Thermal system boundary for an open system.

Thermal Resistance

Heat can be transferred in three ways: conduction, convection, and radiation. Conduction involves the diffusion of heat energy between two bodies that are in physical contact, such as heat transfer through a solid material. Convection involves the transfer of heat energy through the motion of a fluid. Radiation involves the transfer of heat through the absorption of electromagnetic radiation such as infrared waves and solar energy. Conductive or convective heat transfer can be approximated by a linear function of temperature difference, while heat transfer via radiation is a highly nonlinear function of the temperature difference. Therefore, we consider only conduction and convection in this section.

Thermal resistance elements hinder (resist) the flow of heat energy despite a temperature difference. Thermal insulating materials can be modeled as thermal resistance elements. For conduction or convection, the rate of heat transfer q can be approximated by a linear function of the temperature difference ΔT

$$q = \frac{1}{R}\Delta T \tag{4.81}$$

where R is the thermal resistance (in K-s/J or K/W). Another way to express Eq. (4.81) is $\Delta T = Rq$, which is analogous to Ohm's law for an electrical resistor ($e_R = RI$), where heat-flow rate q is analogous to electrical current I and temperature difference ΔT is analogous to voltage drop e_R. Equation (4.81) makes sense intuitively if we consider two extreme cases. When thermal resistance is infinite, $R \to \infty$ (i.e., perfect insulation), the heat transfer rate $q \to 0$ despite the magnitude of the temperature difference ΔT between the two bodies. At the other extreme, when $R \to 0$ (i.e., zero insulation) the heat transfer rate $q \to \infty$, which implies that the heat is transferred instantaneously between two bodies.

Thermal resistance R for conduction is proportional to the thickness of the material (x) and inversely proportional to area normal to the heat flow (A) and the material's thermal conductivity coefficient (k):

$$\text{Conduction:} \quad R = \frac{x}{kA} \tag{4.82}$$

For example, copper has a thermal conductivity coefficient k that is on the order of 18 times greater than the conductivity coefficient for stainless steel. Consequently, copper is an excellent conductor of heat and a poor insulating material.

Thermal resistance R for convection is inversely proportional to the area A and the convection coefficient H:

$$\text{Convection:} \quad R = \frac{1}{HA} \tag{4.83}$$

The convection coefficient for water is 50–100 times greater than the convection coefficient for air, and hence air is a better insulator compared to water.

Thermal Capacitance

Thermal capacitance is a measure of a body's ability to store heat energy due to its mass and thermal properties. Thermal capacitance C is the product of the body's mass m and its specific heat capacity at constant pressure c_p (with units J/kg-K)

$$C = mc_p \tag{4.84}$$

Therefore, thermal capacitance has units of stored energy per unit temperature, or J/K. For example, 1 kg of water has more than four times the thermal capacitance of 1 kg of air.

Modeling Thermal Systems

Thermal system models are based on the energy balance equation (4.80) where the rate of energy stored by the thermal capacitance is $\dot{\xi}_h = C\dot{T}$ as capacitance has units of J/K and the time-rate of temperature has units of K/s. Substituting the time-rate of enthalpy $\dot{h} = \dot{m}c_pT$ into Eq. (4.80), we obtain

$$C\dot{T} = \sum \dot{m}_{\text{in}}c_pT_{\text{in}} - \sum \dot{m}_{\text{out}}c_pT_{\text{out}} + \sum q_{\text{in}} - \sum q_{\text{out}} \tag{4.85}$$

where T_{in} and T_{out} are the fluid temperatures of the streams flowing into and out of the thermal capacitance, respectively, and T is the uniform temperature of the lumped thermal capacitance.

Models of thermal systems can be derived using the following steps:

1. Draw a thermal system boundary around each thermal capacitance, identifying whether the system is a closed or an open system.

2. Label the input and output heat transfer rates q_i between thermal capacitances or their surroundings. For open systems, label the enthalpy rates $\dot{h}_i$ due to mass flowing into or out of the thermal capacitance.

3. Apply the energy balance equation (4.85) to each thermal capacitance.

It is important to note that the resulting mathematical model will consist of one first-order differential equation for each thermal capacitance where the dynamic variable is the temperature of each capacitance. We illustrate this modeling process with the following examples.

Example 4.7

Figure 4.18 shows an interior office room with a baseboard heater, which can be modeled by a lumped-capacitance thermal system (e.g., see [2]). The air in the room has total thermal capacitance C and temperature T and the baseboard heater provides heat input q_{BH}. The four walls and ceiling and floor surfaces are modeled by six different thermal resistances (R_i, $i = 1, 2, \dots 6$) due to the different materials, dimensions, and existence of a window or door for that surface. Derive the model of the thermal system.

Figure 4.19 shows the boundary of the thermal system and the heat flow rates. The system boundary is the rectangular volume enclosed by the four walls and ceiling and floor surfaces. Because the rectangular room has six surfaces (each modeled by a discrete thermal resistance), we show six outgoing heat flow rates q_i, $i = 1, 2, \dots, 6$ that are normal to each surface. The input heat from the baseboard heater is q_{BH}. Because there is no mass crossing the system boundary, we use the energy balance equation (4.85) without the enthalpy-rate terms

$$C\dot{T} = \sum q_{in} - \sum q_{out} \tag{4.86}$$

where the single input heat flow rate is clearly q_{BH} and the output heat flow rates from the room to its surroundings are q_i, $i = 1, 2, \dots, 6$. Each of the six output heat flow rates can be expressed using Eq. (4.81) and the temperature difference across the corresponding thermal resistance surface and its thermal resistance R_i

$$q_i = \frac{T - T_a}{R_i} \quad i = 1, 2, \dots, 6 \tag{4.87}$$

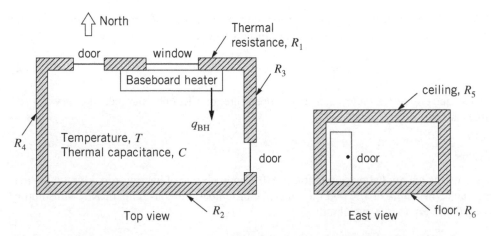

Figure 4.18 Thermal system: interior room with baseboard heater (Example 4.7).

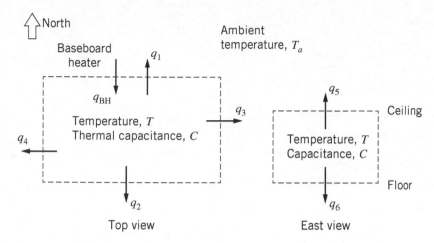

Figure 4.19 Thermal system boundary and heat flow rates (Example 4.7).

where T_a is the ambient (exterior) temperature. Substituting Eq. (4.87) for the six output heat flow rates into the energy balance equation (4.86), we obtain

$$C\dot{T} = q_{BH} - \frac{T - T_a}{R_1} - \frac{T - T_a}{R_2} - \frac{T - T_a}{R_3} - \frac{T - T_a}{R_4} - \frac{T - T_a}{R_5} - \frac{T - T_a}{R_6} \tag{4.88}$$

Equation (4.88) is an acceptable form for the thermal system model. However, we can simplify Eq. (4.88) using the following compact notation

$$C\dot{T} = q_{BH} - \sum_{i=1}^{6} \frac{T - T_a}{R_i} \tag{4.89}$$

Equation (4.89) can be further simplified as

$$C\dot{T} = q_{BH} - \frac{T - T_a}{R_{EQ}} \tag{4.90}$$

where R_{EQ} is the equivalent (or combined) thermal resistance defined by

$$\frac{1}{R_{EQ}} = \frac{1}{R_1} + \frac{1}{R_2} + \cdots + \frac{1}{R_6} = \sum_{i=1}^{6} \frac{1}{R_i} \tag{4.91}$$

The reader should note that the equivalent thermal resistance is defined in the same manner as the equivalent electrical resistance for parallel resistors. Finally, we can move all terms involving the dynamic variable T in Eq. (4.90) to the left-hand side to yield

$$R_{EQ}C\dot{T} + T = R_{EQ}q_{BH} + T_a \tag{4.92}$$

Equation (4.92) is the mathematical model of the thermal system and is equivalent to Eq. (4.88). Room temperature T is the dynamic variable and baseboard heater input q_{BH} and ambient temperature T_a are the two system inputs.

Example 4.8

Figure 4.20 shows a schematic diagram of a double-pipe heat exchanger, which is used to transfer heat from the "hot" fluid flowing through the tube to the surrounding "cold" fluid flowing between the outer shell and the tube. Both flows are steady, and therefore the input and output mass-flow rates of the tube and shell flows are $\dot{m}_1$ and $\dot{m}_2$, respectively. Derive the mathematical model of the heat exchanger.

The tube pipe shown in Fig. 4.20 is inside the shell pipe. A "hot" fluid solution flows through the tube pipe and a "cold" fluid (usually water) flows through the annular region between the shell and tube walls (the two fluids do not mix).

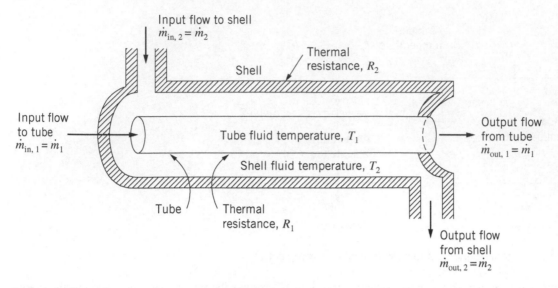

Figure 4.20 Double-pipe heat exchanger (Example 4.8).

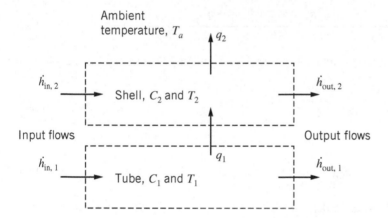

Figure 4.21 Heat exchanger thermal boundaries and heat flow rates (Example 4.8).

The temperatures of the input flows to the tube and shell are $T_{in,1}$ and $T_{in,2}$, respectively. Let us assume that the temperatures of the output flows from the tube and shell are equal to the temperatures of their lumped capacitances, that is, $T_{out,1} = T_1$ and $T_{out,2} = T_2$. Note that the walls of the tube and shell pipes have thermal resistances R_1 and R_2, respectively.

Figure 4.21 shows the boundaries of the thermal capacitances of the heat exchanger and the heat flow rates. Note that we have two thermal capacitances C_1 (tube fluid) and C_2 (shell fluid) and heat flow rate q_1 from the tube to the shell fluid and heat flow rate q_2 from the shell fluid to the ambient air outside the shell. Because the heat exchanger is an open system (i.e., mass is transferred across the system boundaries), we include the enthalpy rates in Fig. 4.21 and in the energy balance equation (4.85):

$$\text{Tube fluid:}\quad C_1\dot{T}_1 = \dot{h}_{in,1} - \dot{h}_{out,1} - q_1 \tag{4.93}$$

$$\text{Shell fluid:}\quad C_2\dot{T}_2 = \dot{h}_{in,2} - \dot{h}_{out,2} + q_1 - q_2 \tag{4.94}$$

Next, use the thermal resistance equation (4.81) to define each heat-transfer rate as $q_1 = (T_1 - T_2)/R_1$ (tube to shell fluid) and $q_2 = (T_2 - T_a)/R_2$ (shell fluid to surroundings). Furthermore, substitute the enthalpy rate equation $\dot{h} = \dot{m}c_p T$ into Eqs. (4.93) and (4.94) for each fluid stream to yield

$$C_1\dot{T}_1 = \dot{m}_1 c_{p,1} T_{in,1} - \dot{m}_1 c_{p,1} T_1 - \frac{T_1 - T_2}{R_1} \tag{4.95}$$

$$C_2\dot{T}_2 = \dot{m}_2 c_{p,2} T_{\text{in},2} - \dot{m}_2 c_{p,2} T_2 + \frac{T_1 - T_2}{R_1} - \frac{T_2 - T_a}{R_2} \tag{4.96}$$

Note that we must use two distinct specific heats $c_{p,1}$ and $c_{p,2}$ because the tube fluid is a chemical solution and the shell fluid is water. Finally, we can move all dynamic variables (T_1 and T_2) to the left-hand sides of Eqs. (4.95) and (4.96) to obtain

$$R_1 C_1 \dot{T}_1 + R_1 \dot{m}_1 c_{p,1} T_1 + T_1 - T_2 = R_1 \dot{m}_1 c_{p,1} T_{\text{in},1} \tag{4.97}$$

$$R_1 R_2 C_2 \dot{T}_2 + R_1 R_2 \dot{m}_2 c_{p,2} T_2 + (R_1 + R_2) T_2 - R_2 T_1 = R_1 R_2 \dot{m}_2 c_{p,2} T_{\text{in},2} + R_1 T_a \tag{4.98}$$

Equations (4.97) and (4.98) constitute the mathematical model of the heat exchanger system. Because we have two thermal capacitances, the complete model consists of two first-order differential equations. The two modeling equations are coupled. If the mass-flow rates $\dot{m}_1$ and $\dot{m}_2$ are constant parameters, then the thermal system model is linear in the temperature variables. Our two dynamic variables are fluid temperatures T_1 and T_2, and the system inputs are input fluid temperatures $T_{\text{in},1}$ and $T_{\text{in},2}$, mass-flow rates $\dot{m}_1$ and $\dot{m}_2$, and the ambient temperature T_a.

SUMMARY

In this chapter, we demonstrated how to model fluid and thermal systems. We began each modeling section with a discussion on the physical characteristics of resistance elements (i.e., fluid and thermal resistance) and energy-storage elements (i.e., fluid and thermal capacitance). Fluid energy is stored because of the pressure of each lumped fluid capacitance, while thermal energy is stored as a result of the temperature of each lumped thermal capacitance. Hydraulic systems involve liquids as the working fluid while pneumatic systems involve gases (usually air). Fluid-system models are derived by applying the conservation of mass to a CV where the uniform fluid pressure inside the CV is the dynamic variable of interest. Consequently each CV (or fluid capacitance) in a fluid system will require a first-order ODE with pressure as the dynamic variable. Fluid-system models are often nonlinear as turbulent valve flow is a nonlinear function of pressure. Thermodynamic effects and the interaction between pressure, density, and temperature are required to fully model pneumatic systems, which complicates the modeling process. Thermal-system models are developed by applying the conservation of energy to a thermal boundary and by accounting for the heat-transfer rates across the system boundary. Each lumped thermal capacitance will require a first-order ODE with temperature as the dynamic variable.

REFERENCES

1. Blackburn, J.F., Reethof, G., and Shearer, J.L., *Fluid Power Control*, MIT Press, Cambridge, MA, 1960, pp. 61–69, 214–219.

2. Pfafflin, J.R., "Space Heating Dynamics," *IEEE Transactions on Industry Applications*, Vol. IA-19, No. 5, 1983, pp. 844–847.

3. Doria, A., and Lucchini, M., "An Experimentally Validated Model of a Motorcycle Shock Absorber for Studying Suspension Dynamics," DETC2011-47214, Proceedings of the ASME 2011 International Design Engineering Technical Conferences & Computers and Information in Engineering Conference, Aug. 28–31, 2011, Washington DC.

4. Langjord, H., and Johansen, T.A., "Dual-Mode Switched Control of an Electropneumatic Clutch Actuator," *IEEE/ASME Transactions on Mechatronics*, Vol. 15, No. 6, 2010, pp. 969–981.

5. Franklin, G.F., Powell, J.D., and Emami-Naeini A., *Feedback Control of Dynamic Systems*, 4th ed., Prentice Hall, Upper Saddle River, NJ, 2002, pp. 59–61.

Chapter 5

Standard Models for Dynamic Systems

5.1 INTRODUCTION

Chapters 2–4 discussed deriving the mathematical models for mechanical, electrical, electromechanical, fluid, and thermal systems. In each case, the complete mathematical model consists of a collection of first-order differential equations (such as an electrical circuit with a single inductor) and/or second-order differential equations (such as a mechanical system with a single inertia element). When elements are interconnected to form a system with multiple dynamic variables, the complete model consists of a collection of coupled ordinary differential equations (ODEs).

In this chapter, we present standard forms for representing the complete mathematical model. The objective is to take the collection of differential equations (i.e., the complete modeling equations) and present them in a convenient form for analyzing the dynamic system's response. System analysis may involve analytical methods or numerical methods such as MATLAB and Simulink; in either case, the complete system must be represented in a convenient standard format. The reader should remember that the starting point for system analysis is always the derivation of the mathematical modeling equations from the fundamental laws (such as Newton's second law, Kirchhoff's laws), and that the standard models are simply representations of these modeling equations.

5.2 STATE-VARIABLE EQUATIONS

One standard method for representing a system is to use *state variables*, which are a set of the dynamic variables that completely define all characteristics of a system. The state variables are usually the physical variables of the system, such as displacement and velocity for mechanical systems, current for electrical systems, pressure for fluid systems, and temperature for thermal systems. State variables can therefore be used to determine the energy stored in a system. The *state* of a system is the smallest set of dynamic variables that can completely define all characteristics of the system. Therefore, the state variables are the dynamic variables that make up the state of the system. For example, we typically choose position z and velocity $\dot{z}$ as the state variables of a single-mass mechanical system. Hence $(z, \dot{z})$ is the state of the system. Potential (or kinetic) energy cannot be selected as an additional state variable because it can be derived from the position (or velocity) state variable. Any characteristic of the system, such as energy or momentum, can be derived from the knowledge of the state variables.

The standard convention is to use $x_1, x_2, \ldots, x_n$ as the *state variables* and $u_1, u_2, \ldots, u_r$ as the *input* (or control) variables. Integer n is the total number of state variables, or the order of the system, and integer r is the total number of input variables. The *state-variable equations* are a collection of n differential equations that are the first-order derivatives of each state variable:

$$\dot{x}_1 = f_1(x_1, x_2, \ldots, x_n, u_1, u_2, \ldots, u_r)$$

$$\dot{x}_2 = f_2(x_1, x_2, \ldots, x_n, u_1, u_2, \ldots, u_r)$$

$$\vdots$$

$$\dot{x}_n = f_n(x_1, x_2, \ldots, x_n, u_1, u_2, \ldots, u_r) \tag{5.1}$$

The right-hand side functions $f_1, f_2, \ldots, f_n$ may be linear or nonlinear, and they can only depend on the state variables x_i or the input variables u_j. If all right-hand side functions f_i are linear, then the state-variable equations (5.1) may be written in a convenient matrix-vector format called the *state-space representation* (SSR), which is described in Section 5.3. If any single function f_i is nonlinear, we have two options: (1) use a numerical integration method (such as Runge–Kutta) to obtain the system response, or (2) derive a linear approximation of the system (as described in Section 5.4), which can then lead to an SSR. In either case, developing the state-variable equations is the starting point. The following three examples demonstrate how to develop the state-variable equations.

Example 5.1

Determine the state-variable equations for the system modeled by the following ODEs, where z and w are the dynamic variables and v is the input

$$2\ddot{z} + 0.8z - 0.4w + 0.2\dot{z}w = 0 \tag{5.2}$$

$$4\dot{w} + 3w + 0.1w^3 - 6z = 8v \tag{5.3}$$

The first step is to determine the order of the system. Equation (5.2) is a second-order nonlinear ODE in dynamic variables z and w, while Eq. (5.3) is a first-order nonlinear ODE in dynamic variables w and z and input v. Hence the complete system is third order, and we need three state variables. We define the state variables $x_1 = z$, $x_2 = \dot{z}$, and $x_3 = w$ and the single input $u = v$. Next, we write the three first-order time derivatives of the three state variables

$$\dot{x}_1 = \dot{z}$$
$$\dot{x}_2 = \ddot{z} = 0.5(-0.8z + 0.4w - 0.2\dot{z}w)$$
$$\dot{x}_3 = \dot{w} = 0.25(-3w - 0.1w^3 + 6z + 8v) \tag{5.4}$$

Note that we have substituted the second-order ODE (5.2) for $\ddot{z}$, and the first-order ODE (5.3) for $\dot{w}$. Because we want all three right-hand sides of Eq. (5.4) to be functions of the states x_i and input u, we substitute $x_1 = z$, $x_2 = \dot{z}$, $x_3 = w$, and $u = v$ to yield

$$\dot{x}_1 = x_2$$
$$\dot{x}_2 = -0.4x_1 + 0.2x_3 - 0.1x_2x_3$$
$$\dot{x}_3 = -0.75x_3 - 0.025x_3^3 + 1.5x_1 + 2u \tag{5.5}$$

Equations (5.5) are the state-variable equations of the system described by Eqs. (5.2) and (5.3). All three right-hand sides of the state-variable equations are functions of the states x_i and input u. Two of the three state-variable equations are nonlinear, due to the nonlinear terms $-0.1x_2x_3$ (second state-variable equation) and $-0.025x_3^3$ (third state-variable equation). The reader should note that the assignment of the state variables is arbitrary; for example, we could have swapped the definitions of states x_1 and x_3 and used $x_1 = w$ and $x_3 = z$.

Example 5.2

Consider the simple single-mass mechanical system shown in Fig. 5.1, which involves both stiffness and damping elements. Determine the state-variable equations.

The position of the mass is denoted by z, which is measured from the static equilibrium position when applied force $F_a(t) = 0$. Stiffness is modeled by a nonlinear spring, which exhibits the following nonlinear force–displacement relationship:

$$f_k(z) = k_1 z + k_3 z^3 \tag{5.6}$$

Damping force is assumed to be a linear function of viscous friction coefficient b and velocity. The applied force $F_a(t)$ is the input to the system.

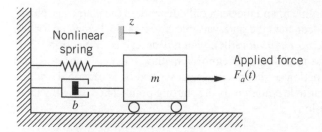

Figure 5.1 Mechanical system for Example 5.2.

We start with the mathematical model, which is derived using a free-body diagram and the methods of which are presented in Chapter 2. The modeling equation is identical to the mass–spring–damper model in Example 2.1, except that the nonlinear spring force (5.6) must be used:

$$m\ddot{z} + b\dot{z} + k_1 z + k_3 z^3 = F_a(t) \tag{5.7}$$

Equation (5.7) is a second-order (nonlinear) ODE. Hence $n = 2$ and it requires two state variables. We select position z and velocity $\dot{z}$ as the two state variables because knowledge of these variables will determine the total (potential + kinetic) energy of the system. The applied force is the single system input. Therefore, using our convention where x_i are state variables and u_j are input variables, we have $x_1 = z$, $x_2 = \dot{z}$, and $u = F_a(t)$.

Once we have defined the state variables, we simply write the first-order differential equations by taking time derivatives of each state variable. The state-variable equations are

$$\dot{x}_1 = \dot{z} \tag{5.8}$$

$$\dot{x}_2 = \ddot{z} = \frac{1}{m}\left(-b\dot{z} - k_1 z - k_3 z^3 + F_a(t)\right) \tag{5.9}$$

Notice that we have solved the mathematical model (5.7) for acceleration $\ddot{z}$ and substituted it into the second state-variable equation (5.9). Our standard convention for state-variable equations presented by Eq. (5.1) shows that the right-hand side functions must solely involve states x_i and inputs u_j. Therefore, we substitute $x_1 = z$, $x_2 = \dot{z}$, and $u = F_a(t)$ into Eqs. (5.8) and (5.9) to produce the final form of the state-variable equations:

$$\dot{x}_1 = x_2 \tag{5.10}$$

$$\dot{x}_2 = -\frac{k_1}{m}x_1 - \frac{k_3}{m}x_1^3 - \frac{b}{m}x_2 + \frac{1}{m}u \tag{5.11}$$

The first state-variable equation (5.10) is linear, but the second state-variable equation (5.11) is nonlinear due to the term involving x_1^3. The reader should note that we could have reversed the indices and selected $x_1 = \dot{z}$ and $x_2 = z$ as the state variables, in which case Eqs. (5.10) and (5.11) would be reversed. However, the usual practice is to define the successive state variable as the time derivative of the preceding state variable.

Example 5.3

Figure 5.2 shows the solenoid actuator–valve system described in Chapters 2 and 3. Obtain a set of state-variable equations for this system.

The solenoid actuator consists of a coil circuit and a valve mass constrained by a return spring. When a voltage is applied to the armature circuit, an electromagnetic force is produced, which pushes on the valve to meter hydraulic flow. Recall that in Chapter 3 we showed that the inductance $L(z)$ of the solenoid coil is a nonlinear function of armature displacement z. We also showed that the electromagnetic force F_{em} is a nonlinear function of current I and position z, or $F_{em} = 0.5KI^2$, where $K = dL/dz$. In addition, the motion of the plunger core in the coil produces a "back-emf" voltage $KI\dot{z}$, which is a nonlinear function of current, velocity, and position. In order to simplify matters, let us assume nominal (constant) values for inductance L and its gradient $K = dL/dz$; furthermore, let us assume linear friction and zero spring preload for the mechanical subsystem. Therefore, the complete mathematical model of the solenoid actuator is

$$L\dot{I} + RI + KI\dot{z} = e_{in}(t) \tag{5.12}$$

$$m\ddot{z} + b\dot{z} + kz = 0.5KI^2 \tag{5.13}$$

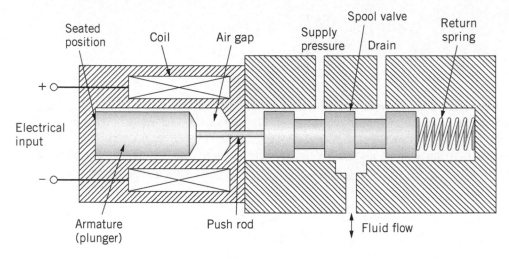

Figure 5.2 Solenoid actuator–valve system for Example 5.3.

Equation (5.12) is a first-order nonlinear ODE, and Eq. (5.13) is a second-order nonlinear ODE. Hence $n = 3$ and the complete system requires three state variables. We select current I, and armature position z and velocity $\dot{z}$ as the three state variables, and applied voltage $e_{\text{in}}(t)$ is the single system input. Therefore, we have $x_1 = I$, $x_2 = z$, $x_3 = \dot{z}$, and $u = e_{\text{in}}(t)$.

Once we have defined the state variables, we write the n first-order differential equations by taking a time derivative of each state variable

$$\dot{x}_1 = \dot{I} = \frac{1}{L}\left(-RI - KI\dot{z} + e_{\text{in}}(t)\right) \tag{5.14}$$

$$\dot{x}_2 = \dot{z} \tag{5.15}$$

$$\dot{x}_3 = \ddot{z} = \frac{1}{m}\left(-b\dot{z} - kz + 0.5KI^2\right) \tag{5.16}$$

Notice that we have substituted the mathematical modeling equations (5.12) and (5.13) for $\dot{I}$ and $\ddot{z}$, respectively. Finally, we substitute $x_1 = I$, $x_2 = z$, and $x_3 = \dot{z}$ and $u = e_{\text{in}}(t)$ in the three first-order differential equations to yield

$$\dot{x}_1 = -\frac{R}{L}x_1 - \frac{K}{L}x_1 x_3 + \frac{1}{L}u \tag{5.17}$$

$$\dot{x}_2 = x_3 \tag{5.18}$$

$$\dot{x}_3 = -\frac{k}{m}x_2 - \frac{b}{m}x_3 + \frac{K}{2m}x_1^2 \tag{5.19}$$

Equations (5.17), (5.18), and (5.19) are the state-variable equations for the solenoid actuator. The reader should note that each right-hand side of the state-variable equations involves only the states x_i and the input u.

5.3 STATE-SPACE REPRESENTATION

If the mathematical modeling equations representing a system are linear, then the resulting state-variable equations (5.1) will be linear first-order ODEs. In this case, we can write the state-variable equations in a convenient matrix-vector format called the *state-space representation* (SSR). The SSR is well-suited for implementation in a numerical computer simulation using MATLAB or Simulink as discussed in Chapter 6.

A few definitions are in order before the SSR is presented. Recall that an nth-order system will require n state variables $x_1, x_2, \ldots, x_n$. We define the *state vector* **x** as the $n \times 1$ column vector composed of the state variables x_i

$$\mathbf{x} = \begin{bmatrix} x_1 \\ x_2 \\ \vdots \\ x_n \end{bmatrix}$$

It should be noted that the state vector does not represent a physical vector (such as a three-component force vector in mechanics), but rather a convenient collection of all n state variables. The *state space* is defined as the n-dimensional "geometric space" that contains the state vector $\mathbf{x}$.

A complete SSR includes two equations in a matrix-vector format: the *state equation* and the *output equation*. The output variables are denoted by $y_1, y_2, \ldots, y_m$, and they are functions of the state and input variables:

$$y_1 = h_1(x_1, x_2, \ldots, x_n, u_1, u_2, \ldots, u_r)$$

$$y_2 = h_2(x_1, x_2, \ldots, x_n, u_1, u_2, \ldots, u_r)$$

$$\vdots$$

$$y_m = h_m(x_1, x_2, \ldots, x_n, u_1, u_2, \ldots, u_r) \tag{5.20}$$

The output equations (5.20) may be linear or nonlinear; however, the output equations must be linear in order to use the matrix-vector SSR. Output variables usually represent sensor measurements of a system's response. For example, if a 1-DOF rotational mechanical system has state variables $x_1 =$ angular position and $x_2 =$ angular velocity and a tachometer is measuring angular velocity, the single output equation is $y = x_2$.

For example, if our system is linear time invariant (LTI) and third order ($n = 3$) with two inputs ($r = 2$), then our state-variable equations will have the general form:

$$\dot{x}_1 = a_{11}x_1 + a_{12}x_2 + a_{13}x_3 + b_{11}u_1 + b_{12}u_2$$

$$\dot{x}_2 = a_{21}x_1 + a_{22}x_2 + a_{23}x_3 + b_{21}u_1 + b_{22}u_2$$

$$\dot{x}_3 = a_{31}x_1 + a_{32}x_2 + a_{33}x_3 + b_{31}u_1 + b_{32}u_2$$

Note that the first time derivatives of the states are linear combinations of all three states (x_1, x_2, x_3) and both inputs (u_1, u_2). In this case where $n = 3$ and $r = 2$, we will have a total of $n^2 = 9$ a_{ij} coefficients and $n \times r = 6$ b_{ij} coefficients. If the system has two sensors that produce two measurements ($m = 2$) that are linear functions of the state and input variables, our output equations will have the general form:

$$y_1 = c_{11}x_1 + c_{12}x_2 + c_{13}x_3 + d_{11}u_1 + d_{12}u_2$$

$$y_2 = c_{21}x_1 + c_{22}x_2 + c_{23}x_3 + d_{21}u_1 + d_{22}u_2$$

In this case where $n = 3$, $r = 2$, and $m = 2$, we will have a total of $m \times n = 6$ c_{ij} coefficients and $m \times r = 4$ d_{ij} coefficients. For a time-invariant system, all coefficients a, b, c, and d are constants.

For a general, nth-order LTI system with r inputs and m outputs, the state equations will have the form

$$\dot{x}_1 = a_{11}x_1 + a_{12}x_2 + \cdots + a_{1n}x_n + b_{11}u_1 + b_{12}u_2 + \cdots b_{1r}u_r$$

$$\dot{x}_2 = a_{21}x_1 + a_{22}x_2 + \cdots + a_{2n}x_n + b_{21}u_1 + b_{22}u_2 + \cdots b_{2r}u_r$$

$$\vdots$$

$$\dot{x}_n = a_{n1}x_1 + a_{n2}x_2 + \cdots + a_{nn}x_n + b_{n1}u_1 + b_{n2}u_2 + \cdots b_{nr}u_r$$

and the output equations will have the form

$$y_1 = c_{11}x_1 + c_{12}x_2 + \cdots + c_{1n}x_n + d_{11}u_1 + d_{12}u_2 + \cdots d_{1r}u_r$$

$$y_2 = c_{21}x_1 + c_{22}x_2 + \cdots + c_{2n}x_n + d_{21}u_1 + d_{22}u_2 + \cdots d_{2r}u_r$$

$$\vdots$$

$$y_m = c_{m1}x_1 + c_{m2}x_2 + \cdots + c_{mn}x_n + d_{m1}u_1 + d_{m2}u_2 + \cdots d_{mr}u_r$$

Because the state-variable and output equations are linear combinations of the state and input variables, we can assemble both equations in a compact matrix-vector format. To begin, we assemble the r input variables into an input vector $\mathbf{u}$ and the m output variables into an output vector $\mathbf{y}$, in the same manner as the definition of the state vector $\mathbf{x}$:

$$\mathbf{u} = \begin{bmatrix} u_1 \\ u_2 \\ \vdots \\ u_r \end{bmatrix} \qquad \mathbf{y} = \begin{bmatrix} y_1 \\ y_2 \\ \vdots \\ y_m \end{bmatrix}$$

Therefore, all three vectors $\mathbf{x}$, $\mathbf{u}$, and $\mathbf{y}$ are column vectors. We note that the first time derivative of the state vector is still an $n \times 1$ vector:

$$\dot{\mathbf{x}} = \begin{bmatrix} \dot{x}_1 \\ \dot{x}_2 \\ \vdots \\ \dot{x}_n \end{bmatrix}$$

We can assemble the a, b, c, and d coefficients of the state and output equations into four matrices:

$$\mathbf{A} = \begin{bmatrix} a_{11} & a_{12} & \cdots & a_{1n} \\ a_{21} & a_{22} & \cdots & a_{2n} \\ \vdots & \vdots & \ddots & \vdots \\ a_{n1} & a_{n2} & \cdots & a_{nn} \end{bmatrix} \qquad \mathbf{B} = \begin{bmatrix} b_{11} & b_{12} & \cdots & b_{1r} \\ b_{21} & b_{22} & \cdots & b_{2r} \\ \vdots & \vdots & \ddots & \vdots \\ b_{n1} & b_{n2} & \cdots & b_{nr} \end{bmatrix}$$

$$\mathbf{C} = \begin{bmatrix} c_{11} & c_{12} & \cdots & c_{1n} \\ c_{21} & c_{22} & \cdots & c_{2n} \\ \vdots & \vdots & \ddots & \vdots \\ c_{m1} & c_{m2} & \cdots & c_{mn} \end{bmatrix} \qquad \mathbf{D} = \begin{bmatrix} d_{11} & d_{12} & \cdots & d_{1r} \\ d_{21} & d_{22} & \cdots & d_{2r} \\ \vdots & \vdots & \ddots & \vdots \\ d_{m1} & d_{m2} & \cdots & d_{mr} \end{bmatrix}$$

The $n \times n$ (square) matrix $\mathbf{A}$ is the *state* or *system* matrix; the $n \times r$ matrix $\mathbf{B}$ is the *input* matrix; the $m \times n$ matrix $\mathbf{C}$ is the *output* matrix; and the $m \times r$ matrix $\mathbf{D}$ is the *direct-link* matrix. Finally, we can use these matrix and vector definitions to present a compact matrix-vector representation of the state and output equations

$$\dot{\mathbf{x}} = \mathbf{A}\mathbf{x} + \mathbf{B}\mathbf{u} \tag{5.21}$$

$$\mathbf{y} = \mathbf{C}\mathbf{x} + \mathbf{D}\mathbf{u} \tag{5.22}$$

Equation (5.21) is the *state equation*, and Eq. (5.22) is the *output equation*, and together they constitute a complete SSR. The reader should note that the state equation (5.21) represents the system dynamics; that is, the linear coefficients from the differential equations that comprise the mathematical model are contained in matrices $\mathbf{A}$ and $\mathbf{B}$. The output equation (5.22) is an algebraic linear mapping between the state and input variables and the outputs or measurements.

A final note is in order to explain matrix-vector multiplication. All right-hand side terms in the state and output equations involve a matrix multiplied by a column vector, and both left-hand side terms are column vectors. When multiplying a matrix and a column vector, the number of columns of the matrix must equal the number of rows of the column vector. For example, consider a case with four states ($n = 4$) and two inputs ($r = 2$). The dimensions of the matrices and vectors of the state equation for this case are

$$\begin{bmatrix} \dot{x}_1 \\ \dot{x}_2 \\ \dot{x}_3 \\ \dot{x}_4 \end{bmatrix} = \begin{bmatrix} a_{11} & a_{12} & a_{13} & a_{14} \\ a_{21} & a_{22} & a_{23} & a_{24} \\ a_{31} & a_{32} & a_{33} & a_{34} \\ a_{41} & a_{42} & a_{43} & a_{44} \end{bmatrix} \begin{bmatrix} x_1 \\ x_2 \\ x_3 \\ x_4 \end{bmatrix} + \begin{bmatrix} b_{11} & b_{12} \\ b_{21} & b_{22} \\ b_{31} & b_{32} \\ b_{41} & b_{42} \end{bmatrix} \begin{bmatrix} u_1 \\ u_2 \end{bmatrix}$$

$$4 \times 1 = [4 \times 4][4 \times 1] + [4 \times 2][2 \times 1]$$

Here, we see that the state matrix **A** must have four columns in order to multiply with the 4×1 state vector **x**, and the input matrix **B** must have two columns in order to multiply with the 2×1 input vector **u**. Because the left-hand side is the first derivative of 4×1 state vector **x**, both matrices **A** and **B** must have four rows. The reader should verify that any single state-variable equation can be written by simply using the coefficients from the appropriate rows in matrices **A** and **B**. For example, the third state-variable equation is

$$\dot{x}_3 = a_{31}x_1 + a_{32}x_2 + a_{33}x_3 + a_{34}x_4 + b_{31}u_1 + b_{32}u_2$$

which uses the coefficients from the third rows of matrices **A** and **B**. Similar arguments apply to the matrix-vector multiplication for the output equation.

An SSR does not change the system dynamics; it is simply a compact matrix-vector format for representing the mathematical model (the ODEs) and the desired output variables. As previously stated, this compact format is well-suited for representing complex systems with multiple inputs and multiple outputs in a computer simulation environment such as MATLAB and Simulink. It should be stressed that an SSR can be obtained only if the mathematical modeling equations are linear. The following examples illustrate how to develop a complete SSR.

Example 5.4

Given the state-variable equations of a third-order system, obtain the SSR matrices if the two output variables are $y_1 = x_1$ and $y_2 = x_2 - x_3$.

$$\dot{x}_1 = -6.2x_1 - 2.3x_2 + 8.4x_3$$

$$\dot{x}_2 = -x_2 + 2.7x_3 + 3u_1$$

$$\dot{x}_3 = -4.1x_1 - 1.5x_2 + 3.9x_3 + 4u_2 \tag{5.23}$$

We can develop the complete SSR equations and matrices *only* if the system is linear, and we see that the three state-variable equations (5.23) are indeed linear combinations of the states x_i and inputs u_j. Let us develop the state equation first: Eq. (5.21) presents the general format of the state equation. In this case, the state vector **x** is a 3×1 column vector, and the input vector **u** is a 2×1 column vector (note that Eq. (5.23) includes two inputs u_1 and u_2). The rows of the state matrix **A** and input matrix **B** contain the state and input coefficients from the three respective state-variable equations. The state equation is

$$\dot{\mathbf{x}} = \begin{bmatrix} \dot{x}_1 \\ \dot{x}_2 \\ \dot{x}_3 \end{bmatrix} = \begin{bmatrix} -6.2 & -2.3 & 8.4 \\ 0 & -1 & 2.7 \\ -4.1 & -1.5 & 3.9 \end{bmatrix} \begin{bmatrix} x_1 \\ x_2 \\ x_3 \end{bmatrix} + \begin{bmatrix} 0 & 0 \\ 3 & 0 \\ 0 & 4 \end{bmatrix} \begin{bmatrix} u_1 \\ u_2 \end{bmatrix} \tag{5.24}$$

The reader should be able to perform the matrix-vector multiplication in Eq. (5.24) and reproduce the three individual state-variable equations in Eq. (5.23).

The system has two output variables, $y_1 = x_1$ and $y_2 = x_2 - x_3$. Therefore, the output vector **y** is a 2×1 column vector. Equation (5.22) presents the general format of the output equation, which, takes the form

$$\mathbf{y} = \begin{bmatrix} y_1 \\ y_2 \end{bmatrix} = \begin{bmatrix} 1 & 0 & 0 \\ 0 & 1 & -1 \end{bmatrix} \begin{bmatrix} x_1 \\ x_2 \\ x_3 \end{bmatrix} + \begin{bmatrix} 0 & 0 \\ 0 & 0 \end{bmatrix} \begin{bmatrix} u_1 \\ u_2 \end{bmatrix} \tag{5.25}$$

Again, the reader should be able to carry out the matrix-vector multiplication in Eq. (5.25) and reproduce the desired output variables, $y_1 = x_1$ and $y_2 = x_2 - x_3$. The complete SSR is

$$\dot{\mathbf{x}} = \mathbf{A}\mathbf{x} + \mathbf{B}\mathbf{u}$$

$$\mathbf{y} = \mathbf{C}\mathbf{x} + \mathbf{D}\mathbf{u}$$

where the state and input matrices are

$$\mathbf{A} = \begin{bmatrix} -6.2 & -2.3 & 8.4 \\ 0 & -1 & 2.7 \\ -4.1 & -1.5 & 3.9 \end{bmatrix} \quad \mathbf{B} = \begin{bmatrix} 0 & 0 \\ 3 & 0 \\ 0 & 4 \end{bmatrix}$$

and the output and direct-link matrices are

$$\mathbf{C} = \begin{bmatrix} 1 & 0 & 0 \\ 0 & 1 & -1 \end{bmatrix} \quad \mathbf{D} = \begin{bmatrix} 0 & 0 \\ 0 & 0 \end{bmatrix}$$

In summary, we have a third-order system ($n = 3$) with two inputs and two outputs. Hence, the state matrix $\mathbf{A}$ is 3×3, the input matrix $\mathbf{B}$ is 3×2, the output matrix $\mathbf{C}$ is 2×3, and the direct-link matrix $\mathbf{D}$ is a 2×2 null matrix. Even though the direct-link matrix $\mathbf{D}$ contains all zeros, it must be defined in order to perform numerical simulations with MATLAB and Simulink, as we see in Chapter 6.

Example 5.5

Consider again the simple single-mass mechanical system shown in Fig. 5.1 that was described in Example 5.2. Obtain a complete SSR if the stiffness is modeled by an ideal (linear) spring element. A single sensor measures the translational displacement of the mass.

We are able to develop the SSR *only* if the state and output equations are linear. Therefore, let us assume a linear spring force and hence $k_1 = k$ (linear spring constant) and $k_3 = 0$. Recall that the state variables are $x_1 = z$ (position) and $x_2 = \dot{z}$ (velocity), and the input variable is $u = F_a(t)$ (applied force). The state vector is a column vector with two elements

$$\mathbf{x} = \begin{bmatrix} x_1 \\ x_2 \end{bmatrix}$$

The linear state-variable equations are

$$\dot{x}_1 = x_2 \tag{5.26}$$

$$\dot{x}_2 = -\frac{k}{m}x_1 - \frac{b}{m}x_2 + \frac{1}{m}u \tag{5.27}$$

First, let us construct the matrix-vector state equation. The first rows of the $\mathbf{A}$ and $\mathbf{B}$ matrices will involve the coefficients associated with the first state-variable equation (5.26). Because the first state-variable equation is $\dot{x}_1 = x_2$, the first row of the state matrix $\mathbf{A}$ consists of a zero coefficient for the first column (which multiplies x_1) and a unity coefficient for the second column (which multiplies x_2). The first row of the input matrix $\mathbf{B}$ consists of a zero coefficient as the first state-variable equation does not include the input u. The second rows of the $\mathbf{A}$ and $\mathbf{B}$ matrices will involve the coefficients from the second state-variable equation (5.27). The state equation then takes the form

$$\dot{\mathbf{x}} = \begin{bmatrix} \dot{x}_1 \\ \dot{x}_2 \end{bmatrix} = \begin{bmatrix} 0 & 1 \\ -k/m & -b/m \end{bmatrix} \begin{bmatrix} x_1 \\ x_2 \end{bmatrix} + \begin{bmatrix} 0 \\ 1/m \end{bmatrix} u \tag{5.28}$$

The reader should be able to perform the matrix-vector multiplication in Eq. (5.28) and arrive at the two state-variables equations (5.26) and (5.27).

In general, the output equation depends on which variable or variables are either measured or defined to be the system output. In this case, a single sensor measures the translational position of the mass. Hence, the output is $y = z = x_1$. The output equation is also a matrix-vector equation, which for this problem takes the form

$$y = \begin{bmatrix} 1 & 0 \end{bmatrix} \begin{bmatrix} x_1 \\ x_2 \end{bmatrix} + [0]u \tag{5.29}$$

The complete SSR is

$$\dot{\mathbf{x}} = \mathbf{A}\mathbf{x} + \mathbf{B}u$$

$$y = \mathbf{C}\mathbf{x} + Du$$

where the state and input matrices are

$$\mathbf{A} = \begin{bmatrix} 0 & 1 \\ -k/m & -b/m \end{bmatrix} \qquad \mathbf{B} = \begin{bmatrix} 0 \\ 1/m \end{bmatrix}$$

and the output and direct-link matrices are

$$\mathbf{C} = \begin{bmatrix} 1 & 0 \end{bmatrix} \qquad D = 0$$

In summary, we have a second-order system ($n = 2$) with one input and one output. Hence, the state matrix $\mathbf{A}$ is 2×2, the input matrix $\mathbf{B}$ is 2×1 (a column vector), the output matrix $\mathbf{C}$ is 1×2 (a row vector), and the direct-link matrix D is a scalar (it is zero in this case).

Example 5.6

Obtain a complete SSR of the electromechanical DC motor system that was presented in Chapter 3. A tachometer is used to measure the angular velocity of the rotor ($\dot{\theta}$), and an ammeter is used to measure current in the armature circuit (I).

The mathematical model of the DC motor was developed in Chapter 3, and the governing system model is repeated below

$$L\dot{I} + RI = e_{\text{in}}(t) - K_b\dot{\theta} \tag{5.30}$$

$$J\ddot{\theta} + b\dot{\theta} = K_m I - T_L \tag{5.31}$$

Equation (5.30) is a first-order linear ODE, and Eq. (5.31) is a second-order linear ODE. Consequently, $n = 3$ and the system requires three state variables. We select current I, angular displacement θ, and angular velocity $\dot{\theta}$ as the three state variables. The applied armature voltage $e_{\text{in}}(t)$ and load torque T_L are the *two* inputs to the system. Therefore, we have states $x_1 = I$, $x_2 = \theta$, $x_3 = \dot{\theta}$ and inputs $u_1 = e_{\text{in}}(t)$, $u_2 = T_L$.

Next, we write the three first-order state equations by taking a time derivative of each state variable, and substitute Eq. (5.30) for the time derivative of current I and Eq. (5.31) for the time derivative of angular velocity $\dot{\theta}$

$$\dot{x}_1 = \dot{I} = \frac{1}{L}\left(-RI + e_{\text{in}}(t) - K_b\dot{\theta}\right) \tag{5.32}$$

$$\dot{x}_2 = \dot{\theta} \tag{5.33}$$

$$\dot{x}_3 = \ddot{\theta} = \frac{1}{J}\left(-b\dot{\theta} + K_m I - T_L\right) \tag{5.34}$$

We substitute the states $x_1 = I, x_2 = \theta, x_3 = \dot{\theta}$ and the inputs $u_1 = e_{\text{in}}(t), u_2 = T_L$ in the three first-order differential equations to yield

$$\dot{x}_1 = -\frac{R}{L}x_1 + \frac{1}{L}u_1 - \frac{K_b}{L}x_3 \tag{5.35}$$

$$\dot{x}_2 = x_3 \tag{5.36}$$

$$\dot{x}_3 = -\frac{b}{J}x_3 + \frac{K_m}{J}x_1 - \frac{1}{J}u_2 \tag{5.37}$$

Finally, we put Eqs. (5.35)–(5.37) in the standard matrix-vector format to construct the state equation. The rows of the $\mathbf{A}$ and $\mathbf{B}$ matrices will involve the coefficients associated with each state-variable equation. The state equation is

$$\dot{\mathbf{x}} = \begin{bmatrix} \dot{x}_1 \\ \dot{x}_2 \\ \dot{x}_3 \end{bmatrix} = \begin{bmatrix} -R/L & 0 & -K_b/L \\ 0 & 0 & 1 \\ K_m/J & 0 & -b/J \end{bmatrix} \mathbf{x} + \begin{bmatrix} 1/L & 0 \\ 0 & 0 \\ 0 & -1/J \end{bmatrix} \mathbf{u} \tag{5.38}$$

The reader should be able to multiply each row of state matrix **A** and input matrix **B** by column vectors **x** and **u**, respectively, and reproduce the three state equations (5.35)–(5.37).

The system has two measurements, angular velocity and current, and therefore the output variables are $y_1 = \dot{\theta}$ and $y_2 = I$. Both outputs are state variables: $y_1 = \dot{\theta} = x_3$ and $y_2 = I = x_1$. Thus, the output equation is

$$\mathbf{y} = \begin{bmatrix} y_1 \\ y_2 \end{bmatrix} = \begin{bmatrix} 0 & 0 & 1 \\ 1 & 0 & 0 \end{bmatrix} \mathbf{x} + \begin{bmatrix} 0 & 0 \\ 0 & 0 \end{bmatrix} \mathbf{u} \tag{5.39}$$

The complete SSR is

$$\dot{\mathbf{x}} = \mathbf{A}\mathbf{x} + \mathbf{B}\mathbf{u}$$

$$\mathbf{y} = \mathbf{C}\mathbf{x} + \mathbf{D}\mathbf{u}$$

where the state and input matrices are

$$\mathbf{A} = \begin{bmatrix} -R/L & 0 & -K_b/L \\ 0 & 0 & 1 \\ K_m/J & 0 & -b/J \end{bmatrix} \quad \mathbf{B} = \begin{bmatrix} 1/L & 0 \\ 0 & 0 \\ 0 & -1/J \end{bmatrix}$$

and the output and direct-link matrices are

$$\mathbf{C} = \begin{bmatrix} 0 & 0 & 1 \\ 1 & 0 & 0 \end{bmatrix} \quad \mathbf{D} = \begin{bmatrix} 0 & 0 \\ 0 & 0 \end{bmatrix}$$

In summary, we have a third-order system ($n = 3$) with two inputs and two outputs. Consequently, the state matrix **A** is 3×3, the input matrix **B** is 3×2, the output matrix **C** is 2×3, and the direct-link matrix **D** is a 2×2 null matrix.

Example 5.7

Note that the SSR of the DC motor in Example 5.6 does not depend on angular displacement of the rotor (θ). Thus, this state variable can be eliminated. Obtain a "reduced-order" SSR of the DC motor in terms of two state variables: current I and angular velocity $\dot{\theta}$. The two outputs (measurements) are the two state variables.

In Example 5.6, the second state variable ($x_2 = \theta$) does not appear in the state equations (5.32)–(5.34), nor is it an output variable. Another way to recognize the system's lack of dependence on angular displacement θ is to note the three zeros in the second column of the state matrix **A** and the two zeros in the second column of the output matrix **C** (elements contained in the second columns of **A** and **C** are coefficients of the second state variable x_2). Therefore, we can eliminate the second state equation (5.33) of our third-order SSR in Example 5.6. Furthermore, we can substitute $\omega = \dot{\theta}$ and $\dot{\omega} = \ddot{\theta}$ in the electrical and mechanical modeling equations, or the first and third state equations (5.32) and (5.34). The armature circuit and mechanical modeling equations thus become

$$\dot{I} = \frac{1}{L}\left(-RI + e_{in}(t) - K_b\omega\right) \tag{5.40}$$

$$\dot{\omega} = \frac{1}{J}\left(-b\omega + K_m I - T_L\right) \tag{5.41}$$

Note that both equations are first-order ODEs. Consequently, we need two state variables ($n = 2$). We choose the states $x_1 = I$ and $x_2 = \omega$ and inputs $u_1 = e_{in}(t)$, $u_2 = T_L$. Substituting for the states and input variables in Eqs. (5.40) and (5.41) and using the matrix-vector format yields the state equation

$$\dot{\mathbf{x}} = \begin{bmatrix} \dot{x}_1 \\ \dot{x}_2 \end{bmatrix} = \begin{bmatrix} -R/L & -K_b/L \\ K_m/J & -b/J \end{bmatrix} \mathbf{x} + \begin{bmatrix} 1/L & 0 \\ 0 & -1/J \end{bmatrix} \mathbf{u} \tag{5.42}$$

Note that the **A** and **B** matrices for this reduced-order SSR are identical to the third-order state and input matrices in Eq. (5.38) with the second row removed (the state equation for angular displacement θ) and the second column removed from the third-order state matrix in Eq. (5.38). The output equation for the reduced-order SSR is

$$\mathbf{y} = \begin{bmatrix} y_1 \\ y_2 \end{bmatrix} = \begin{bmatrix} 0 & 1 \\ 1 & 0 \end{bmatrix} \mathbf{x} + \begin{bmatrix} 0 & 0 \\ 0 & 0 \end{bmatrix} \mathbf{u} \tag{5.43}$$

It should be noted that the reduced-order output matrix **C** is identical to the third-order output matrix in Eq. (5.39) with the second column removed.

A final note is in order. If we obtain the system response of the DC motor for a given input voltage $e_{in}(t)$ and load torque T_L (say, using MATLAB's Simulink), we will obtain identical system output responses for $y_1(t)$ and $y_2(t)$ (angular velocity ω and current I) whether we use the third-order SSR derived in Example 5.6 or the second-order SSR presented in this example. Both state-space models are based on the same governing system dynamics, and both use the same input and output variables. The third-order SSR in Example 5.6 carries along an additional state variable (θ) that does not contribute to either the system dynamics or the output, and hence this state variable and its state equation can therefore be eliminated. However, the angular position information $\theta(t)$ of the DC motor is lost when we use the second-order SSR presented in this example.

Example 5.8

Figure 5.3 shows the seat-suspension system presented by Example 2.3 in Chapter 2. Obtain a complete SSR, where the two sensor measurements are displacement and acceleration of the mass that represents the driver.

The governing mathematical modeling equations are repeated

$$m_1\ddot{z}_1 + b_1\dot{z}_1 + b_2(\dot{z}_1 - \dot{z}_2) + k_1 z_1 + k_2(z_1 - z_2) = b_1\dot{z}_0(t) + k_1 z_0(t) \tag{5.44a}$$

$$m_2\ddot{z}_2 + b_2(\dot{z}_2 - \dot{z}_1) + k_2(z_2 - z_1) = 0 \tag{5.44b}$$

Recall that the system variables are the vertical displacements of the seat mass (z_1) and driver mass (z_2), and that both are measured relative to their static equilibrium positions. The vertical displacement of the cabin floor (due to road vibrations) is $z_0(t)$, which is considered to be an input to the system. The remaining parameters are the seat and driver masses (m_1 and m_2), and the passive friction and stiffness coefficients (b_i and k_i, respectively). The two measurements associated with the test rig are driver displacement z_2 and driver acceleration $\ddot{z}_2$. Displacement is measured by a linear variable differential transformer (LVDT), an electromechanical device used to measure translational displacement. Acceleration is measured by an accelerometer.

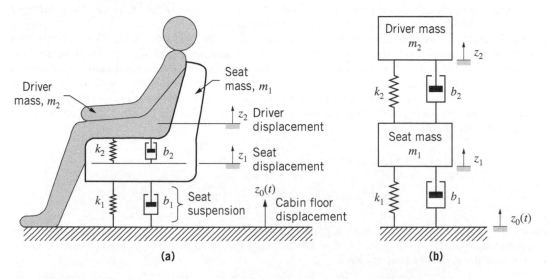

(a) **(b)**

Figure 5.3 (a) Schematic diagram of the seat-suspension system for Example 5.8.
(b) Mechanical model for the seat-suspension system.

The objective is to develop a complete SSR given the modeling equations (5.44a) and (5.44b). Clearly, this system is linear and fourth-order ($n = 4$). At first glance, displacements z_1 and z_2 and their derivatives are the obvious choices for state variables. However, if we choose $x_1 = z_1$ and $x_2 = \dot{z}_1$, we see from Eq. (5.44a) that the second state equation ($\dot{x}_2 = \ddot{z}_1$) will involve the term $\dot{z}_0(t)$, which is the derivative of the input $u = z_0(t)$. Our standard state equation (5.21) does not contain a matrix-vector term involving $\dot{\mathbf{u}}$, so we cannot utilize this choice of state variables along with a single input defined as $u = z_0(t)$. In general, when the mathematical model involves derivatives of the input variables, such as Eq. (5.44a), the choice of state variables becomes more complicated and less intuitive. We show two solutions to this state-space problem when the time derivative of the input u appears in the system dynamics. The first solution method is more intuitive and easier to apply. It is pursued here in this example; the second method is less intuitive and is reserved as Problem 5.32 at the end of this chapter.

The intuitive or easier solution is to simply define an additional input variable that is the derivative of z_0. Hence, we define *two* input variables: $u_1 = z_0(t)$ and $u_2 = \dot{z}_0(t)$. Now, we are able to define the following four state variables:

$$x_1 = z_1 \tag{5.45}$$

$$x_2 = \dot{z}_1 \tag{5.46}$$

$$x_3 = z_2 \tag{5.47}$$

$$x_4 = \dot{z}_2 \tag{5.48}$$

Taking the first time derivative of each state variable and substituting the system dynamics (5.44a) and (5.44b) for $\ddot{z}_1$ and $\ddot{z}_2$ yields

$$\dot{x}_1 = \dot{z}_1 \tag{5.49}$$

$$\dot{x}_2 = \ddot{z}_1 = \frac{1}{m_1}[-b_1\dot{z}_1 - b_2(\dot{z}_1 - \dot{z}_2) - k_1 z_1 - k_2(z_1 - z_2) + b_1\dot{z}_0(t) + k_1 z_0(t)] \tag{5.50}$$

$$\dot{x}_3 = \dot{z}_2 \tag{5.51}$$

$$\dot{x}_4 = \ddot{z}_2 = \frac{1}{m_2}[-b_2(\dot{z}_2 - \dot{z}_1) - k_2(z_2 - z_1)] \tag{5.52}$$

Next, we substitute for the physical variables using the definitions of the states ($x_1 = z_1$, $x_2 = \dot{z}_1$, $x_3 = z_2$, and $x_4 = \dot{z}_2$) and the two inputs ($u_1 = z_0(t)$, $u_2 = \dot{z}_0(t)$) to yield

$$\dot{x}_1 = x_2 \tag{5.53}$$

$$\dot{x}_2 = \frac{-k_1 - k_2}{m_1}x_1 + \frac{-b_1 - b_2}{m_1}x_2 + \frac{k_2}{m_1}x_3 + \frac{b_2}{m_1}x_4 + \frac{k_1}{m_1}u_1 + \frac{b_1}{m_1}u_2 \tag{5.54}$$

$$\dot{x}_3 = x_4 \tag{5.55}$$

$$\dot{x}_4 = \frac{k_2}{m_2}x_1 + \frac{b_2}{m_2}x_2 - \frac{k_2}{m_2}x_3 - \frac{b_2}{m_2}x_4 \tag{5.56}$$

The complete state equation in matrix-vector format is the collection of Eqs. (5.53)–(5.56)

$$\begin{bmatrix} \dot{x}_1 \\ \dot{x}_2 \\ \dot{x}_3 \\ \dot{x}_4 \end{bmatrix} = \begin{bmatrix} 0 & 1 & 0 & 0 \\ -(k_1 + k_2)/m_1 & -(b_1 + b_2)/m_1 & k_2/m_1 & b_2/m_1 \\ 0 & 0 & 0 & 1 \\ k_2/m_2 & b_2/m_2 & -k_2/m_2 & -b_2/m_2 \end{bmatrix} \begin{bmatrix} x_1 \\ x_2 \\ x_3 \\ x_4 \end{bmatrix} + \begin{bmatrix} 0 & 0 \\ k_1/m_1 & b_1/m_1 \\ 0 & 0 \\ 0 & 0 \end{bmatrix} \begin{bmatrix} u_1 \\ u_2 \end{bmatrix} \tag{5.57}$$

Recall that the two system outputs or measurements are specified as $y_1 = z_2$ and $y_2 = \ddot{z}_2$. The first output equation is simply $y_1 = x_3$, and the second output equation is obtained from the fourth state equation (5.56), or the bottom-row terms in Eq. (5.57)

$$\begin{bmatrix} y_1 \\ y_2 \end{bmatrix} = \begin{bmatrix} 0 & 0 & 1 & 0 \\ k_2/m_2 & b_2/m_2 & -k_2/m_2 & -b_2/m_2 \end{bmatrix} \begin{bmatrix} x_1 \\ x_2 \\ x_3 \\ x_4 \end{bmatrix} + \begin{bmatrix} 0 & 0 \\ 0 & 0 \end{bmatrix} \begin{bmatrix} u_1 \\ u_2 \end{bmatrix} \tag{5.58}$$

Equations (5.57) and (5.58) constitute the complete SSR. *A strong word of caution is in order*: we have greatly simplified the state-space derivation by defining *two independent* input variables $u_1 = z_0(t)$ and $u_2 = \dot{z}_0(t)$. In reality, there is only one independent system input, and it is $u = z_0(t)$. In our two-input SSR summarized by Eqs. (5.57) and (5.58), there is no constraint requiring that the second input u_2 is the time derivative of input u_1. Therefore, we must be careful to enforce the constraint $u_2 = \dot{u}_1$ when we analyze or simulate the dynamics of the two-input SSR derived in this example. We revisit this system and this modeling issue in Chapter 11, when we investigate the ability of the seat suspension to suppress road vibrations transmitted by the cabin-floor input $z_0(t)$.

5.4   LINEARIZATION

Most real-world dynamic systems are nonlinear; that is, they are modeled by nonlinear differential equations. There is no general, unifying theory for obtaining the solution of a nonlinear system, and in most cases we must rely on numerical integration schemes to obtain the system response. On the other hand, a wealth of analysis tools exists for obtaining the solution of a linear system. Furthermore, there are many control system design techniques that can be applied only to linear system dynamics. Therefore, it is desirable to have a linear system model for analysis and design purposes.

Linearization is a method for converting a nonlinear equation (or model) into a linear model. Linearization relies on a Taylor series expansion about a nominal (or reference) operating point, where only the first-order terms are retained. Because second- and higher-order terms are neglected in the Taylor-series expansion, the resulting linear model is accurate only if the system's state does not deviate too far from the nominal operating point. Linearization relies on three basic steps:

1. Choose (or solve for) the nominal operating point (or states) about which to linearize the system. The operating point may be given, or it may be an equilibrium point that can be obtained from the governing nonlinear model. In many cases, the nominal operating point will be a static state. We use x^* and u^* to denote the nominal state and input variables, respectively.

2. Redefine the nonlinear modeling equations in terms of the *nominal* variables and the *perturbation* variables (or deviations) with respect to the nominal values. We use the convention $\delta x = x - x^*$ for the perturbation (deviation) from the nominal state x^* and $\delta u = u - u^*$ for the perturbation from nominal input u^*.

3. Expand the nonlinear modeling equations in a Taylor series about the nominal operating point and retain only the first-order (linear) terms. The resulting linearized model will be in terms of the perturbation variables δx and δu.

As previously stated, the linearized model is reasonably accurate and representative of the (true) nonlinear model as long as the dynamic variables do not deviate "too far" from the operating point (e.g., δx remains "small" at all times).

The three-step linearization process is demonstrated by the following example. Suppose we have a nonlinear system with one state variable x, and one input variable u

$$\dot{x} = f(x, u) \tag{5.59}$$

Step 1: The nominal input u^* (which is likely given in the problem) results in the nominal state solution x^*, which is the nominal operating point.

Step 2: The perturbation variables are $\delta x = x - x^*$ and $\delta u = u - u^*$. Therefore, we can substitute $x = \delta x + x^*$ and $u = \delta u + u^*$ into the nonlinear model (5.59)

$$\delta \dot{x} + \dot{x}^* = f(\delta x + x^*, \delta u + u^*) \tag{5.60}$$

Step 3: Expand the right-hand side of the nonlinear equation (5.60) in Taylor series about x^* and u^*

$$\delta \dot{x} + \dot{x}^* = f(x^*, u^*) + \left.\frac{\partial f}{\partial x}\right|_* \delta x + \frac{1}{2!}\left.\frac{\partial^2 f}{\partial x^2}\right|_* \delta x^2 + \cdots + \left.\frac{\partial f}{\partial u}\right|_* \delta u + \frac{1}{2!}\left.\frac{\partial^2 f}{\partial u^2}\right|_* \delta u^2 + \cdots \tag{5.61}$$

Because the nominal state solution is $\dot{x}^* = f(x^*, u^*)$, these two terms cancel each other in Eq. (5.61). It should be noted that all partial derivatives in Eq. (5.61) are evaluated at the nominal state and input, x^* and u^*, respectively. Finally, eliminating all Taylor-series terms higher than first-order yields

$$\delta\dot{x} = \frac{\partial f}{\partial x}\bigg|_{*} \delta x + \frac{\partial f}{\partial u}\bigg|_{*} \delta u \qquad (5.62)$$

Equation (5.62) is the linearized model of the original nonlinear system (5.59). The two first-order derivatives $\partial f/\partial x$ and $\partial f/\partial u$ are constants as long as x^* and u^* are constants. It is important to note that the linear model (5.62) is in terms of the *perturbation variables* δx and δu. Therefore, the solution of the linearized model (5.62) yields $\delta x(t)$, which is the time history of the state deviation from the operating point x^*. If we want to estimate the state history from the linear solution, we use $x(t) = \delta x(t) + x^*$.

Example 5.9

Derive the linear model of the following nonlinear state-variable equation. Perform the linearization about the static equilibrium state x^* that results from the nominal input $u^* = 2$.

$$\dot{x} = -2x - 0.4x^3 + 0.3u = f(x, u) \qquad (5.63)$$

The first step is to obtain the nominal state x^* given the nominal input $u^* = 2$. We assume that a static equilibrium state exists when the nominal input is $u^* = 2$; that is, $\dot{x} = 0$ and hence x remains constant. Solving Eq. (5.63) for x with $\dot{x} = 0$ and $u = u^* = 2$ yields a third-order polynomial in x

$$\dot{x} = -2x - 0.4x^3 + 0.6 = 0 \qquad (5.64)$$

We can use MATLAB's `roots` command to obtain the three roots

```
>> roots([ -0.4 0 -2 0.6 ])
```

where the row vector in the square brackets contains the coefficients of the third-order polynomial in Eq. (5.64) in descending powers of x. The three roots include one real root ($x = 0.2949$), and two complex conjugate roots ($x = -0.1474 \pm j2.2506$; recall that j is the imaginary number, $j = \sqrt{-1}$). Because the equilibrium state must be a real number, the nominal state is $x^* = 0.2949$.

Next, we define the perturbation variables $\delta x = x - x^*$ and $\delta u = u - u^*$ and write the linearized equation (5.62) that results from the first-order Taylor-series expansion

$$\delta\dot{x} = \frac{\partial f}{\partial x}\bigg|_{*} \delta x + \frac{\partial f}{\partial u}\bigg|_{*} \delta u \qquad (5.65)$$

The two partial derivatives are easily evaluated using the right-hand-side function $f(x, u)$ defined in Eq. (5.63). We evaluate the partial derivatives at the nominal state x^* and nominal input u^*

$$\frac{\partial f}{\partial x}\bigg|_{*} = -2 - 1.2x^2|_{*} = -2 - 1.2(0.2949^2) = -2.1044$$

$$\frac{\partial f}{\partial u}\bigg|_{*} = 0.3$$

Note that the value of the nominal input u^* was not needed for evaluating either of the two partial derivatives. In addition, the partial derivative $\partial f/\partial u$ is 0.3, which is the same input coefficient in the nonlinear equation (5.63) because the term involving u is already linear. Finally, substituting the numerical values of the two partial derivatives into the first-order Taylor-series equation (5.65) yields

$$\delta\dot{x} = -2.1044\delta x + 0.3\delta u \qquad (5.66)$$

Equation (5.66) is the *linearized* version of the original nonlinear equation (5.63), where the linearization has been performed about the nominal state $x^* = 0.2949$. This value is the equilibrium state for the nominal input $u^* = 2$. Note that the linearized equation is written in terms of the *perturbation* variables δx and δu. We show in Chapter 7 that it is relatively easy to obtain the analytical solution to a linear first-order ODE such as Eq. (5.66). However, the solution of Eq. (5.66) produces the perturbation variable $\delta x(t)$; if we wish to estimate the *nonlinear* solution of Eq. (5.63), we must add the perturbation variable to the nominal state, or $x(t) = x^* + \delta x(t)$.

Example 5.10

Consider the simple hydraulic system shown in Fig. 5.4, which consists of a single tank with constant cross-sectional area A being filled with a liquid with input volumetric-flow rate Q_{in}. Output flow through the valve is turbulent. Our objective is to derive a linear model of the hydraulic tank dynamics given a nominal input volumetric-flow rate Q_{in}^*.

We see in Fig. 5.4 that the output volumetric-flow rate through the valve is Q_{out}. The state variable for this system is pressure P at the base of the tank, and atmospheric pressure at the outlet of the valve is P_{atm}. The model for this simple hydraulic system can be derived from the conservation of mass, as demonstrated in Chapter 4

$$\frac{dV}{dP}\frac{dP}{dt} = C\dot{P} = Q_{in} - Q_{out} \tag{5.67}$$

where $C = dV/dP$ is the fluid capacitance of the tank. This capacitance is fixed because of its constant cross-sectional area. Turbulent valve flow is represented by the nonlinear equation

$$Q_{out} = K_T \sqrt{P - P_{atm}} \tag{5.68}$$

Therefore, the nonlinear modeling equation for the hydraulic tank system is

$$C\dot{P} = Q_{in} - K_T \sqrt{P - P_{atm}} \tag{5.69}$$

We begin by rewriting the nonlinear model (5.69) as a state-variable equation

$$\dot{P} = \frac{1}{C}Q_{in} - \frac{K_T}{C}\sqrt{P - P_{atm}} = f(P, Q_{in}) \tag{5.70}$$

The first step is to determine the operating point P^* given a nominal (constant) input Q_{in}^*. We can expect that over time the outflow Q_{out} will balance the constant input flow Q_{in}^*. Therefore, $\dot{P} = 0$ and pressure reaches a constant value P^* (the liquid-level height in the tank will also reach a constant value, as pressure and liquid-level height are related by the hydrostatic equation). Solving Eq. (5.69) for the constant pressure when $C\dot{P} = 0$ yields

$$P^* = \frac{Q_{in}^{*2}}{K_T^2} + P_{atm} \tag{5.71}$$

which is the nominal pressure or operating point of the hydraulic system. The perturbation variables are defined as $\delta P = P - P^*$ and $\delta Q_{in} = Q_{in} - Q_{in}^*$. Next, we use Eq. (5.62), the first-order Taylor-series expansion

$$\delta\dot{P} = \left.\frac{\partial f}{\partial P}\right|_* \delta P + \left.\frac{\partial f}{\partial Q_{in}}\right|_* \delta Q_{in} \tag{5.72}$$

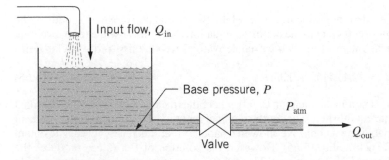

Figure 5.4 Hydraulic tank system for Example 5.10.

where the first-order partial derivatives can be determined using Eq. (5.70)

$$\frac{\partial f}{\partial P}\bigg|_* = \frac{-K_T}{2C\sqrt{P^* - P_{\text{atm}}}} \qquad \frac{\partial f}{\partial Q_{\text{in}}}\bigg|_* = \frac{1}{C}$$

Substituting Eq. (5.71) for the nominal P^* yields the first-order derivative

$$\frac{\partial f}{\partial P}\bigg|_* = \frac{-K_T^2}{2CQ_{\text{in}}^*}$$

This derivative is a known constant, given numerical values for turbulent flow coefficient K_T, fluid capacitance C, and nominal input flow rate Q_{in}^*. Finally, the linear hydraulic model is

$$\delta\dot{P} = \frac{-K_T^2}{2CQ_{\text{in}}^*}\delta P + \frac{1}{C}\delta Q_{\text{in}} \tag{5.73}$$

The solution to the linear model will produce the pressure perturbation $\delta P(t)$. We revisit this problem in Chapter 6 when we compare the system responses of the linear and nonlinear hydraulic models using numerical simulations.

The linearization method can be generalized and applied to the nth-order vector of nonlinear state equations

$$\dot{\mathbf{x}} = \mathbf{f}(\mathbf{x}, \mathbf{u}) \tag{5.74}$$

The nominal input vector time history $\mathbf{u}^*(t)$ will produce a nominal state vector history $\mathbf{x}^*(t)$, sometimes called the state trajectory. For example, a predefined program for motor torque inputs for a robotic system will produce nominal trajectories for the positions and velocities of the robot linkages. The three-step linearization process previously described can be applied to the nonlinear vector system (5.74), and the result is

$$\delta\dot{\mathbf{x}} = \frac{\partial \mathbf{f}}{\partial \mathbf{x}}\bigg|_* \delta\mathbf{x} + \frac{\partial \mathbf{f}}{\partial \mathbf{u}}\bigg|_* \delta\mathbf{u} \tag{5.75}$$

where $\delta\mathbf{x} = \mathbf{x} - \mathbf{x}^*$ and $\delta\mathbf{u} = \mathbf{u} - \mathbf{u}^*$. Finally, the linearized system (5.75) may be written in the compact state equation format

$$\delta\dot{\mathbf{x}} = \mathbf{A}\delta\mathbf{x} + \mathbf{B}\delta\mathbf{u} \tag{5.76}$$

where the $\mathbf{A}$ and $\mathbf{B}$ matrices are the first-order partial derivatives of the nonlinear state equations (5.74)

$$\mathbf{A} = \begin{bmatrix} \partial f_1/x_1 & \partial f_1/x_2 & \cdots & \partial f_1/x_n \\ \partial f_2/x_1 & \partial f_2/x_2 & \cdots & \partial f_2/x_n \\ \vdots & \vdots & \ddots & \vdots \\ \partial f_n/x_1 & \partial f_n/x_2 & \cdots & \partial f_n/x_n \end{bmatrix} \qquad \mathbf{B} = \begin{bmatrix} \partial f_1/u_1 & \partial f_1/u_2 & \cdots & \partial f_1/u_r \\ \partial f_2/u_1 & \partial f_2/u_2 & \cdots & \partial f_2/u_r \\ \vdots & \vdots & \ddots & \vdots \\ \partial f_n/u_1 & \partial f_n/u_2 & \cdots & \partial f_n/u_r \end{bmatrix}$$

These matrices are evaluated at the nominal state and input vectors. Therefore, the linearized system matrices will be time varying if $\mathbf{x}^*(t)$ and $\mathbf{u}^*(t)$ vary with time. The following example demonstrates how to develop the linearized state and input matrices for a case with constant input u^* and a constant nominal state vector $\mathbf{x}^*$.

Example 5.11

Consider again the nonlinear system from Example 5.1 and the corresponding state-variables equations. Derive the linear state equation for a nominal constant input $u^* = 0.15$.

$$\dot{x}_1 = f_1(\mathbf{x}, u) = x_2 \tag{5.77}$$

$$\dot{x}_2 = f_2(\mathbf{x}, u) = -0.4x_1 + 0.2x_3 - 0.1x_2x_3 \tag{5.78}$$

$$\dot{x}_3 = f_3(\mathbf{x}, u) = -0.75x_3 - 0.025x_3^3 + 1.5x_1 + 2u \tag{5.79}$$

The linearization is performed about the nominal state vector. We must determine the equilibrium state vector $\mathbf{x}^*$ given constant input u^*. The first state-variable equation (5.77) shows that setting $\dot{x}_1 = 0$ for equilibrium requires that state $x_2 = 0$. Using $x_2 = 0$ in Eq. (5.78) with $\dot{x}_2 = 0$ results in the equilibrium condition $-0.4x_1 + 0.2x_3 = 0$, or $x_1 = 0.5x_3$. The final state-variable equation (5.79) with condition $x_1 = 0.5x_3$ and nominal input $u = u^* = 0.15$ yields the third-order polynomial in x_3

$$\dot{x}_3 = -0.75x_3 - 0.025x_3^3 + 1.5(0.5x_3) + 2(0.15) = 0 \tag{5.80}$$

or

$$-0.025x_3^3 + 0.3 = 0 \tag{5.81}$$

We can use MATLAB's `roots` command to obtain the three roots

```
>> roots([ -0.025 0 0 0.3 ])
```

The three roots include one real root ($x_3 = 2.2894$) and two complex conjugate roots ($x_3 = -1.1447 \pm j1.9827$). Because the equilibrium state must be a real number, the nominal value of the third state is $x_3^* = 2.2894$. Hence the nominal value of the first state is $x_1^* = 1.1447$ and the nominal state vector is

$$\mathbf{x}^* = \begin{bmatrix} x_1^* \\ x_2^* \\ x_3^* \end{bmatrix} = \begin{bmatrix} 1.1447 \\ 0 \\ 2.2894 \end{bmatrix} \tag{5.82}$$

Equations (5.75) and (5.76) show that the system matrix $\mathbf{A}$ is composed of the first-order partial derivatives of the three right-hand side functions $f_i(\mathbf{x}, u)$ in Eqs. (5.77)–(5.79). The first-order partial derivatives are

$$\mathbf{A} = \begin{bmatrix} \partial f_1/\partial x_1 & \partial f_1/\partial x_2 & \partial f_1/\partial x_3 \\ \partial f_2/\partial x_1 & \partial f_2/\partial x_2 & \partial f_2/\partial x_3 \\ \partial f_3/\partial x_1 & \partial f_3/\partial x_2 & \partial f_3/\partial x_3 \end{bmatrix}_{\mathbf{x}^*} = \begin{bmatrix} 0 & 1 & 0 \\ -0.4 & -0.1x_3 & 0.2 - 0.1x_2 \\ 1.5 & 0 & -0.75 - 0.075x_3^2 \end{bmatrix}_{\mathbf{x}^*} \tag{5.83}$$

Evaluation of each matrix element at the nominal state vector $\mathbf{x}^*$ yields the system matrix

$$\mathbf{A} = \begin{bmatrix} 0 & 1 & 0 \\ -0.4 & -0.2289 & 0.2 \\ 1.5 & 0 & -1.1431 \end{bmatrix}$$

The input matrix $\mathbf{B}$ is composed of the first-order partial derivatives of $f_i(\mathbf{x}, u)$ with respect to input u

$$\mathbf{B} = \begin{bmatrix} \partial f_1/\partial u \\ \partial f_2/\partial u \\ \partial f_3/\partial u \end{bmatrix}_{\mathbf{x}^*} = \begin{bmatrix} 0 \\ 0 \\ 2 \end{bmatrix}$$

Because input u appears linearly in Eqs. (5.77)–(5.79), the input matrix $\mathbf{B}$ contains the linear coefficients of u. The linear state equation is

$$\delta\dot{\mathbf{x}} = \begin{bmatrix} 0 & 1 & 0 \\ -0.4 & -0.2289 & 0.2 \\ 1.5 & 0 & -1.1431 \end{bmatrix} \delta\mathbf{x} + \begin{bmatrix} 0 \\ 0 \\ 2 \end{bmatrix} \delta u \tag{5.84}$$

Equation (5.84) is the linearized version of the nonlinear system (5.77)–(5.79). The linearized system is in terms of the perturbation variables $\delta\mathbf{x} = \mathbf{x} - \mathbf{x}^*$ and $\delta u = u - u^*$. The nonlinear system has been linearized about the equilibrium state vector shown in Eq. (5.82).

5.5 INPUT–OUTPUT EQUATIONS

In the previous sections, we developed state-variable equations and, in the case of linear systems, the SSR. We also presented a linearization process that can take the nonlinear state-variable equations and develop a linear state-space model. In general, the state-variable and state-space equations will involve a collection of first-order, *coupled* differential equations, which means that they must be solved simultaneously. In this section, we develop *input–output* (I/O) equations that are solely a function of the desired output and input variables and their derivatives.

Consider a single-input, single-output (SISO) dynamic system shown in Fig. 5.5, represented by the generic "block diagram" or "black-box" diagram. For an SISO system, an I/O equation involves only input variable u and output variable y and their derivatives:

$$a_n y^{(n)} + a_{n-1} y^{(n-1)} + \cdots + a_2 \ddot{y} + a_1 \dot{y} + a_0 y = b_m u^{(m)} + \cdots + b_1 \dot{u} + b_0 u \tag{5.85}$$

where $y^{(n)} = d^n y/dt^n$, $y^{(n-1)} = d^{n-1} y/dt^{n-1}$, and so on. In general, the highest derivative of the input variable is less than or equal to the highest derivative of the output variable, or $m \leq n$. For a time-invariant system, the coefficients a_i and b_i are constants. Equation (5.85) is the general form of an I/O equation for an SISO system. For systems with two or more inputs, the right-hand side of the I/O equation will involve additional input terms. If we have a system with p output variables, we will have p I/O equations, one for each output variable. Therefore, unlike the coupled state-variable equations, we can solve each I/O equation independently of the others. The following examples illustrate the derivations of I/O equations.

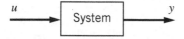

Figure 5.5 Single-input, single-output system.

Example 5.12

Figure 5.6 shows an electrical system comprising a series RLC circuit and input voltage source $e_{\text{in}}(t)$. Derive the I/O equation with output $y = I$ (loop current) and $u = e_{\text{in}}(t)$.

The mathematical model of the RLC circuit can be derived by applying Kirchhoff's voltage law around the single loop, which yields

$$-e_L - e_R - e_C + e_{\text{in}}(t) = 0 \tag{5.86}$$

Substituting for the voltages across each element, we obtain

$$L\dot{I} + RI + \frac{1}{C}\int I\,dt = e_{\text{in}}(t) \tag{5.87}$$

Taking a time derivative of Eq. (5.87) in order to eliminate the integral term yields

$$L\ddot{I} + R\dot{I} + \frac{1}{C}I = \dot{e}_{\text{in}}(t) \tag{5.88}$$

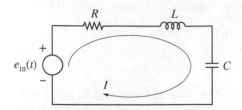

Figure 5.6 Series RLC circuit for Example 5.12.

Let current I be the output variable y, and source voltage $e_{in}(t)$ be the input variable u. Therefore, the I/O equation is obtained directly from Eq. (5.88)

$$\ddot{y} + \frac{R}{L}\dot{y} + \frac{1}{LC}y = \frac{1}{L}\dot{u} \tag{5.89}$$

which matches the basic form of the I/O equation (5.85). Here, the leading coefficient a_2 becomes unity by dividing the entire equation by inductance L.

Differential Operator

Deriving the I/O equation is relatively easy when the governing mathematical model consists of a single differential equation with one dynamic (output) variable and one input variable, which is the case for Example 5.12. When the mathematical model consists of two or more differential equations with multiple output and input variables, obtaining the I/O equations becomes significantly more complicated, as each output variable must be separated into an independent I/O equation. One way to simplify the analysis is to define the differential operator or "D operator" as

$$D \equiv \frac{d}{dt}$$

Therefore, time derivatives can be written as powers of operator D: for example, $Dy = \dot{y}$, $D^2y = \ddot{y}$, etc. We can use the D operator to manipulate the governing dynamical equations in order to obtain the desired I/O equation, which is demonstrated in the following example.

Example 5.13

Figure 5.7 shows a simple two-mass mechanical system. We wish to derive the I/O equation with the displacement of mass m_1 as the output variable, or $y = z_1$, and the applied force $F_a(t)$ as the input variable.

Displacements z_1 and z_2 are measured from their respective equilibrium (unstretched) positions. The mathematical model can be derived using a free-body diagram and the methods from Chapter 2. The result is

$$m_1\ddot{z}_1 + b\dot{z}_1 + k(z_1 - z_2) = 0$$
$$m_2\ddot{z}_2 + k(z_2 - z_1) = F_a(t) \tag{5.90}$$

Therefore, the mathematical model (5.90) becomes

$$m_1\ddot{y} + b\dot{y} + k(y - z_2) = 0$$
$$m_2\ddot{z}_2 + k(z_2 - y) = u \tag{5.91}$$

Clearly, we need to eliminate z_2 from the differential equation containing $\ddot{y}$ in order to obtain the I/O equation, that is, in terms of output y only. By applying the D operator, we can rewrite Eq. (5.91) as

$$m_1D^2y + bDy + k(y - z_2) = 0 \tag{5.92}$$
$$m_2D^2z_2 + k(z_2 - y) = u \tag{5.93}$$

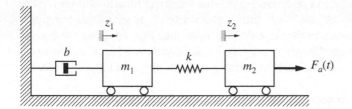

Figure 5.7 Two-mass mechanical system for Example 5.13.

We can solve Eq. (5.93) for displacement z_2

$$z_2 = \frac{u + ky}{m_2 D^2 + k}$$

and substitute into Eq. (5.92). The result is

$$m_1 D^2 y + bDy + ky - \frac{k(u + ky)}{m_2 D^2 + k} = 0$$

Multiplying this equation by $m_2 D^2 + k$ in order to clear the fraction yields

$$m_1 D^2 (m_2 D^2 + k)y + bD(m_2 D^2 + k)y + k(m_2 D^2 + k)y - k(u + ky) = 0$$

Finally, we can convert this equation from operator form to a differential equation by replacing the D^k terms with the respective time derivatives

$$m_1 m_2 y^{(4)} + bm_2\,\dddot{y} + (m_1 k + m_2 k)\ddot{y} + bk\dot{y} = ku \tag{5.94}$$

where $y^{(4)} = d^4 y / dt^4$. Equation (5.94) is the I/O equation of the dual-mass system with output $y = z_1$. Note that the governing mathematical model (5.90) is fourth order (i.e., we would need four state variables for an SSR). Consequently, the I/O equation (5.94) is of order four.

In general, deriving the I/O equation is simple when the system is modeled by a single differential equation with one dynamic variable. Deriving the I/O equations is considerably more difficult than developing the state-variable equations when the system involves two or more differential equations with multiple input and output variables. Furthermore, most numerical simulation methods (such as MATLAB and Simulink) require that the dynamic system consist of a collection of first-order differential equations (i.e., state-variable equations) as the standard strategy is to numerically integrate n differential equations simultaneously. For these reasons, we typically employ I/O equations for SISO systems and we use state-variable models for both SISO and multiple-input, multiple-output (MIMO) systems.

5.6   TRANSFER FUNCTIONS

Transfer functions are a convenient way to represent and analyze the I/O (or cause-and-effect) relationship of an LTI dynamic system. They are frequently used in simulation diagrams ("block diagrams," to be discussed in the next section) to characterize a desired input/output equation. Numerical algorithms such as MATLAB and Simulink use transfer functions to represent dynamic systems that have a single input and a single output.

Traditionally, transfer functions are introduced using Laplace transform methods. The *Laplace transform* maps the function $f(t)$ from the time domain to the domain of the complex variable s and is defined by

$$\mathscr{L}\{f(t)\} = F(s) = \int_0^\infty f(t)e^{-st}dt \tag{5.95}$$

The reader may recall that Laplace transforms can be used to solve linear differential equations with constant coefficients, where the Laplace transformation converts a differential equation into an algebraic equation in variable s.

If a forcing function $f(t)$ exists in the differential equation, it can be converted into the s-domain function $F(s)$ using Eq. (5.95). Hence, a table comprising the Laplace transforms of "standard" time functions (such as sinusoidal functions) can be constructed. The solution to the differential equation in the time domain is obtained by applying the inverse Laplace transform to the algebraic equation in the s-domain. Chapter 8 presents a brief overview of Laplace transform theory, including transforms of common time functions, Laplace transform properties, and the solution of differential equations using Laplace transform methods.

In this book, we do not rely solely on Laplace transform methods to solve linear differential equations. Although the Laplace-based solution method is systematic, it involves the often tedious process of computing the partial-fraction expansion so that tables of common transforms may be used to obtain the inverse Laplace transform. Instead, our focus is on the direct analysis of the time-domain differential equations for low-order SISO systems, and numerical simulation techniques for complex, MIMO systems. However, we use transfer functions to *represent* and analyze single-input, single-output dynamic systems without formally using Laplace transforms. It should be noted that the simulation software Simulink uses transfer functions to represent an I/O differential equation, but it does not use Laplace transform theory to obtain the system response (it uses direct numerical integration of the time-domain differential equations). We show two techniques for developing the transfer function without using Laplace transforms.

The following example demonstrates the transfer function in the s-domain without dealing with Laplace transforms. To begin, consider the following third-order I/O equation

$$a_3\ddddot{y} + a_2\ddot{y} + a_1\dot{y} + a_0 y = b_1\dot{u} + b_0 u \tag{5.96}$$

Next, consider an exponential input, $u(t) = U(s)e^{st}$, where $s = \sigma + j\omega$ is a complex variable and $U(s)$ is a complex function. The exponential function is

$$e^{st} = e^{\sigma t}e^{j\omega t} = e^{\sigma t}(\cos \omega t + j \sin \omega t) \tag{5.97}$$

where the last substitution comes from applying Euler's formula $e^{j\theta} = \cos \theta + j \sin \theta$. Equation (5.97) shows that e^{st} is complex; it consists of real and imaginary parts. Because input $u(t)$ is strictly real, $U(s)$ will be the complex conjugate of e^{st}. Recall from differential equations that if the input is $u(t) = U(s)e^{st}$, the particular solution (i.e., the response due to the input) will also be an exponential function, $y(t) = Y(s)e^{st}$. Hence, we can substitute these exponential input and output functions and their time derivatives ($\dot{y} = sY(s)e^{st}$, $\ddot{y} = s^2Y(s)e^{st}$, $\dot{u} = sU(s)e^{st}$, etc.) into the I/O differential equation (5.96) to yield

$$(a_3 s^3 + a_2 s^2 + a_1 s + a_0)Y(s)e^{st} = (b_1 s + b_0)U(s)e^{st} \tag{5.98}$$

Finally, use Eq. (5.98) to form the ratio $Y(s)/U(s)$

$$\frac{Y(s)}{U(s)} = \frac{b_1 s + b_0}{a_3 s^3 + a_2 s^2 + a_1 s + a_0} = G(s) \tag{5.99}$$

We define the *transfer function* $G(s)$ as the ratio shown in Eq. (5.99). The formal mathematical definition of the transfer function of a linear, time-invariant I/O equation is the Laplace transform of the output $Y(s)$ divided by the Laplace transform of the input $U(s)$ with the assumption of zero initial conditions (we repeat this definition in Chapter 8). The formal definition of the transfer function is identical to the result demonstrated by this example.

As a second example, we can derive the "time-domain" transfer function by applying the differential D operator to the third-order I/O equation (5.96)

$$a_3 D^3 y + a_2 D^2 y + a_1 Dy + a_0 y = b_1 Du + b_0 u \tag{5.100}$$

We can factor out $y(t)$ and $u(t)$ from both sides of Eq. (5.100) and form the ratio of output to input:

$$\frac{y(t)}{u(t)} = \frac{b_1 D + b_0}{a_3 D^3 + a_2 D^2 + a_1 D + a_0} \tag{5.101}$$

Equation (5.101) is identical to the transfer function $G(s)$ in Eq. (5.99) if we simply replace the differential operator D with the complex variable s. Note that in Laplace transform theory, the differentiation theorem states that the Laplace transform of $\dot{y}$ is equal to $sY(s) - y(0)$, and the Laplace transform of $\ddot{y}$ is $s^2 Y(s) - sy(0) - \dot{y}(0)$, and so on for higher-order derivatives. Because all initial conditions $y(0)$, $\dot{y}(0)$, and $\ddot{y}(0)$ are assumed to be zero for a transfer function, we can conclude that multiplying by the kth power of Laplace variable s in the Laplace domain is equivalent to the kth derivative in the time domain.

It is this author's opinion that using the Laplace transform method to obtain the system response is unduly tedious and time consuming. Therefore, we focus on the analysis of the time-domain differential equations and numerical simulations using MATLAB and Simulink. Furthermore, it should be reemphasized that while MATLAB and Simulink very often rely on transfer functions (which use the complex variable s) to *represent* I/O differential equations, their numerical solution techniques are based on the time-domain differential equations and *not* Laplace transform theory. For the sake of *representing* dynamic systems, we can simply interchange the D and s symbols to produce transfer functions. Furthermore, as we see in Chapters 7, 9, and 10, we use transfer functions to analyze the system's response without relying on Laplace transform theory (however, we present Laplace transform methods in Chapter 8 for the sake of completeness). The following two examples demonstrate the relationship between the time-domain ODE and the transfer function.

Example 5.14

Equation (5.89) is the I/O equation of the series RLC circuit from Example 5.12 (Fig. 5.6), where current is the output variable y and source voltage is the input variable u. Derive the system transfer function $G(s)$ that relates output to input and use the following numerical values for the electrical system parameters: resistance $R = 2\,\Omega$, inductance $L = 0.25\,\text{H}$, and capacitance $C = 0.4\,\text{F}$.

Using the prescribed values for the electrical parameters, the I/O equation is

$$\ddot{y} + 8\dot{y} + 10y = 4\dot{u} \tag{5.102}$$

Applying the D operator, Eq. (5.102) becomes

$$D^2 y + 8Dy + 10y = 4Du \tag{5.103}$$

which can be solved for the output-to-input ratio $y(t)/u(t)$

$$\frac{y(t)}{u(t)} = \frac{4D}{D^2 + 8D + 10} \tag{5.104}$$

Replacing the differential operator D with s yields the transfer function

$$G(s) = \frac{4s}{s^2 + 8s + 10} \tag{5.105}$$

The reader should also be able to convert a transfer function into an I/O differential equation.

Example 5.15

Equation (5.106) is the transfer function for output variable $y(t)$ and input variable $u(t)$

$$G(s) = \frac{7s + 4}{2s^2 + 10s + 28} = \frac{Y(s)}{U(s)} \tag{5.106}$$

Derive the time-domain ODE for the I/O equation of this system.

To begin, rewrite the transfer function $G(s)$ with Laplace variable s replaced by the differential operator D

$$\frac{7D + 4}{2D^2 + 10D + 28} = \frac{y(t)}{u(t)} \tag{5.107}$$

Cross multiplying the denominator term in Eq. (5.107) $(2D^2 + 10D + 28)$ by the output y and the numerator term $(7D + 4)$ by the input u yields

$$(2D^2 + 10D + 28)y = (7D + 4)u \tag{5.108}$$

Finally, apply the differential operator to produce the time derivative terms $D^2y = \ddot{y}$, $Dy = \dot{y}$, and $Du = \dot{u}$, which yields the differential equation

$$2\ddot{y} + 10\dot{y} + 28y = 7\dot{u} + 4u \tag{5.109}$$

Equation (5.109) is the desired I/O equation, the equivalent to the transfer function shown in Eq. (5.106). We can divide the I/O equation by 2 if we desire a unity coefficient on the highest derivative term, which yields the equivalent I/O equation

$$\ddot{y} + 5\dot{y} + 14y = 3.5\dot{u} + 2u \tag{5.110}$$

With enough practice, the reader should be able to develop transfer functions directly from I/O equations (and vice versa) without relying on the intermediate step with the D operator.

5.7   BLOCK DIAGRAMS

Block diagrams are standard pictorial or graphical representations of interconnected systems. Each dynamic system that has an I/O relationship is a "block," which is usually a single transfer function. Other types of blocks include multiplying factors ("gains"), time differentiation, and integration with time ("integrators"). Blocks are connected by signal paths, and these paths represent the flow of input and output signals and computations. Signal flow into a block represents a math operation, usually multiplication. Simulink is based on block diagrams, which are constructed by connecting the icons for various types of blocks in the Simulink environment (discussed in Chapter 6 and Appendix C).

Standard Block-Diagram Components

Figure 5.8 shows the multiplication of input signal u by the constant (or *gain*) K to produce the output signal y. Simulink uses a triangle-shaped block for a gain, since it is the traditional symbol for an operational amplifier ("op amp") used to boost electrical signals.

Figure 5.9 shows three blocks that represent the time integration of input signal u. Figure 5.9b shows integration as the inverse of the differential or D operator, while Fig. 5.9c uses the transfer function $1/s$ to denote integration. Simulink uses the $1/s$ block in Fig. 5.9c to represent integration. The initial value of the integrator output, $y(0)$, can be set in the Simulink environment (see Chapter 6 and Appendix C).

Figure 5.10 shows a transfer function block that represents an I/O differential equation. For example, if the transfer function $G(s)$ in Fig. 5.10 is

$$G(s) = \frac{3s + 2}{s^2 + 4s + 20} \tag{5.111}$$

the corresponding I/O equation is

$$\ddot{y} + 4\dot{y} + 20y = 3\dot{u} + 2u \tag{5.112}$$

It is important to note that the transfer function $G(s)$ represents the I/O relationship or mathematical model of a dynamic system and is independent of the nature of the input function $u(t)$. For example, if we apply an arbitrary input signal (such as a constant or a sinusoidal function) to the block diagram shown in Fig. 5.10, the output $y(t)$ will be determined by the I/O equation (5.112).

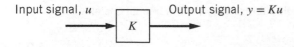

Input signal, u Output signal, $y = Ku$

Figure 5.8 Gain block.

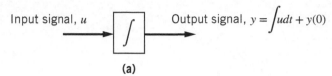

(a)

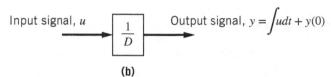

(b)

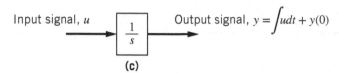

(c)

Figure 5.9 Integrator blocks: (a) integral symbol, (b) *D*-operator symbol, and (c) transfer-function symbol.

Figure 5.10 Transfer function block.

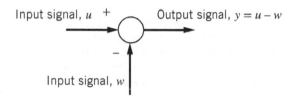

Figure 5.11 Summing junction.

We often need to represent the addition and subtraction of dynamic variables in a block diagram. Figure 5.11 shows a common component known as a *summing junction*. We must include a plus or minus sign with each input signal to indicate addition or subtraction.

Example 5.16

Figure 5.12 shows a simple series RL circuit that has voltage source $e_{in}(t)$ as the input. The desired output, perhaps measured by a voltmeter, is the voltage across the resistor R. We wish to draw the block diagrams for this system using (1) a transfer function block and (2) an integrator block.

As with all problems, we start with the mathematical model. Applying Kirchhoff's voltage law around the loop yields

$$-e_L - e_R + e_{in}(t) = 0$$

where the voltage across the inductor is $e_L = L\dot{i}$ and the voltage across the resistor is $e_R = RI$. By placing all terms involving the dynamic variable (current, I) on the left-hand side, we obtain the mathematical model

$$L\dot{i} + RI = e_{in}(t) \tag{5.113}$$

We can derive the transfer function by applying the D operator to Eq. (5.113) and forming the ratio of current to source voltage to obtain

$$\frac{I(t)}{e_{in}(t)} = \frac{1}{LD + R} \tag{5.114}$$

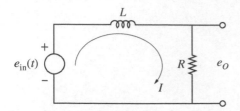

Figure 5.12 Series RL circuit for Example 5.16.

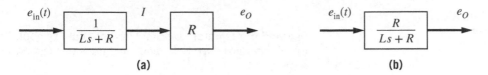

Figure 5.13 Block diagrams for Example 5.16: (a) two blocks in series and (b) single equivalent block.

Next, we substitute $s = D$ to obtain the transfer function $G(s)$

$$\frac{1}{Ls + R} = \frac{I(s)}{E_{in}(s)} \tag{5.115}$$

This transfer function has source voltage $e_{in}(t)$ as the input and current I as the output; however, we want resistor voltage $e_O = RI$ as the output variable. Therefore, we can multiply the transfer function output (current) by resistance R to obtain the desired system output e_O. Figure 5.13a shows this configuration in a block diagram with two blocks in series: the current signal from the transfer function block is sent to a gain block with magnitude R in order to produce the desired system output e_O. Figure 5.13b shows an equivalent block diagram for the RL circuit where the gain block R is simply factored into the RL circuit transfer function.

A second block diagram is drawn for this system using an integrator block instead of a transfer function block. The key to drawing a system block diagram using integrator blocks is to write the mathematical model in state-variable form and then send signal paths that represent each right-hand side of the state equations to separate integrators. Because this example involves a first-order system, we need only one integrator block. We begin by rewriting the modeling equation (5.113) as a state-variable equation

$$\dot{I} = \frac{1}{L}\left(e_{in}(t) - RI\right) \tag{5.116}$$

Equation (5.116) shows that the signal path that represents dI/dt can be constructed by computing the difference between the input voltage $e_{in}(t)$ and the resistor voltage e_O and multiplying this difference by the gain block $1/L$. Figure 5.14 shows the block diagram using this approach, where the summing junction is used to compute the voltage difference $e_{in}(t) - RI$. The output of the integrator is current I, which is multiplied by resistance R to produce the desired output. Note that the output voltage e_O is "fed back" to the summing junction. The reader should also note that all signals at a summing junction must have the same units, which are volts in this case.

Example 5.17

Consider the electromechanical DC motor system presented in Chapter 3 and Examples 5.6 and 5.7. Draw the complete block diagram of the DC motor using transfer functions to represent the armature circuit and mechanical rotor dynamics. The desired system output is angular velocity of the rotor, ω.

The mathematical model of the DC motor consists of two linear, coupled, first-order ODEs

$$L\dot{I} + RI = e_{in}(t) - K_b\omega \tag{5.117}$$

$$J\dot{\omega} + b\omega = K_m I - T_L \tag{5.118}$$

Source voltage

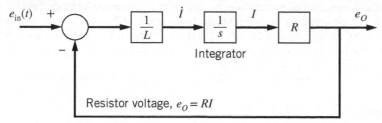

Figure 5.14 Block diagram for Example 5.16 using an integrator block.

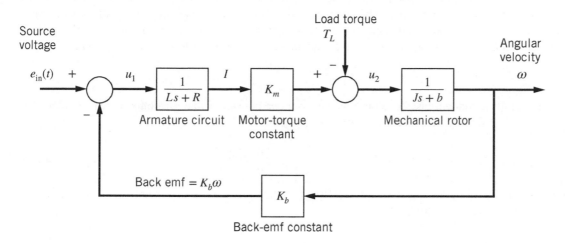

Figure 5.15 Block diagram of the DC motor (Example 5.17).

We choose to utilize the first-order model of the mechanical rotor (5.118) because the desired system output is angular velocity ω. The transfer function for the armature circuit can be developed by considering the right-hand side of Eq. (5.117) as input u_1

$$L\dot{I} + RI = u_1 \tag{5.119}$$

Therefore, the transfer function for the armature circuit is

$$\frac{1}{Ls + R} = \frac{I(s)}{U_1(s)} \tag{5.120}$$

where the input to the circuit transfer function is $u_1 = e_{in}(t) - K_b\omega$, or the "net voltage" comprised the source voltage $e_{in}(t)$ minus the "back-emf" voltage $K_b\omega$. Similarly, the mechanical rotor transfer function can be developed from Eq. (5.118) by defining the net torque input $u_2 = K_mI - T_L$

$$J\dot{\omega} + b\omega = u_2 \tag{5.121}$$

The mechanical system transfer function is

$$\frac{1}{Js + b} = \frac{\Omega(s)}{U_2(s)} \tag{5.122}$$

where $\mathscr{L}\{\omega(t)\} = \Omega(s)$ is the Laplace transform of angular velocity. The input to the mechanical transfer function is the net torque, which is motor torque K_mI minus the load torque T_L.

We are now ready to construct the overall block diagram of the DC motor. The armature circuit transfer function (5.120) will appear first in the diagram, as its output variable (current I) determines the motor torque K_mI. This torque is an *input* to the mechanical transfer function. Figure 5.15 shows the complete block diagram of the DC motor. Note that we have used two summing junctions to construct the input signals $u_1 = e_{in}(t) - K_b\omega$ and $u_2 = K_m I - T_L$. The first summing junction produces the "net voltage" input to the armature circuit, and the second summing junction produces the "net torque" input to the mechanical rotor. The output of the circuit transfer function (current I) is gained by the constant K_m to produce the motor torque. In a similar fashion, the output of the mechanical transfer function (angular velocity ω) is gained by the constant K_b to produce the "back-emf" voltage that is fed back to the source voltage. Note that the signals at each summing junction have the same units (volts in the first junction, N-m in the second junction).

5.8 STANDARD INPUT FUNCTIONS

The previous sections have presented various standard modeling forms for dynamic systems. In all cases, the dynamic system consists of a differential equation(s) with one or more input and output variables. The subsequent chapters emphasize obtaining the output or system response to a desired input function. System performance (such as response speed and damping attributes) is often characterized by the system's dynamic response to any number of *standard* input functions. We can think of these standard input functions as "test input signals" for evaluating the system's dynamic response. Many standard input functions have a basis in the realistic or expected input to a dynamic system.

Step Input

A *step input* function exhibits a sudden, instantaneous change from one constant value to another constant value. The *unit-step input* function $U(t)$ "steps up" from zero to unity at time $t = 0+$

$$\text{Unit-step input} \qquad U(t) = \begin{cases} 0 & \text{for} \quad t \leq 0 \\ 1 & \text{for} \quad t > 0 \end{cases} \tag{5.123}$$

We can represent a step input with magnitude A as $u(t) = AU(t)$. For example, a sudden constant 30-N force applied at time $t = 0$ would be expressed as $u(t) = 30U(t)$ N.

Ramp Input

A *ramp input* function increases linearly with time at a constant rate. The unit-ramp input function is $u(t) = t$. Clearly, the "slope" of the unit ramp is $du/dt = 1$. A general ramp input is $u(t) = at$ where a is the slope, which could be positive or negative.

Ramped Step Input

A *ramped step input* function exhibits a linear "ramp up" (or "ramp down") from zero to a constant (or step) value. Hence, the ramped step function is characterized by the slope a and step magnitude A

$$\text{Ramped step input} \qquad u(t) = \begin{cases} at & \text{for} \quad 0 \leq t \leq A/a \\ A & \text{for} \quad t > A/a \end{cases} \tag{5.124}$$

Clearly, the end of the ramp time is $t_1 = A/a$. In some cases, a ramped step input might be a more realistic representation of a step input as many physical inputs require a finite time interval to change to a new constant value.

Pulse Input

A *pulse input* consists of a constant (step) input that lasts for a finite duration, and then instantly steps to zero. Therefore, a pulse input with magnitude A can be described as

$$\text{Pulse input} \qquad u(t) = \begin{cases} 0 & \text{for} \quad t \leq 0 \\ A & \text{for} \quad 0 < t < t_1 \\ 0 & \text{for} \quad t \geq t_1 \end{cases} \tag{5.125}$$

In Eq. (5.125), the pulse input steps up to magnitude A at time $t = 0$, and then steps down to zero at time $t = t_1$.

Impulse Input

An *impulse function* is often used to represent an input with a constant magnitude that lasts for a very short duration. We can think of an impulse function as a pulse function where the pulse duration goes to zero in the limit. For example, consider a pulse input that consists of a 150-N force that lasts for 0.1 s as shown in Fig. 5.16a. From engineering mechanics, we know that the impulse of the force is its time integral, and therefore the area (or impulse) is 15 N-s. Figure 5.16b shows a pulse input with double the force (300 N) with half the duration (0.05 s) so that its area remains at 15 N-s. If we maintain the area as we shrink the pulse duration to zero, the magnitude approaches infinity and the input becomes an impulse function with a "strength" or "weight" of 15 N-s.

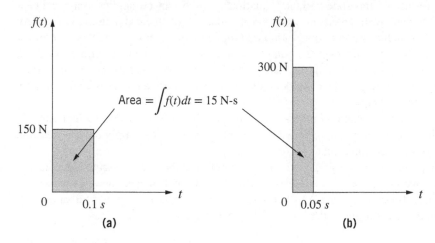

Figure 5.16 Pulse force inputs with an impulse of 15 N-s.

In the example shown in Fig. 5.16, the pulse function had an area of 15 N-s. If our original pulse function had an area of unity, we can develop the *unit-impulse function* by maintaining unity area and shrinking the pulse duration to zero. Mathematically, the unit impulse is represented by the Dirac delta function $\delta(t)$, which can be described by

$$\delta(t) = 0 \quad \text{for} \quad t \neq 0 \tag{5.126}$$

$$\int_{-\infty}^{\infty} \delta(t)dt = 1 \tag{5.127}$$

The impulse function $\delta(t)$ cannot exist in nature because of its discontinuous property, and applying an impulse input to a system can cause an instantaneous change in energy. An impulse input $A\delta(t)$ has "weight" or "strength" A, which might represent a very "skinny" pulse input with area A. For example, the impulsive force input shown in Fig. 5.16 would be represented mathematically as $f(t) = 15\delta(t)$ (in N), where the area (or weight) $A = 15$ N-s and the Dirac delta function $\delta(t)$ has units s^{-1}. The impulse input can be graphically represented by an arrow (with its weight value) placed at the appropriate point on the time axis.

Sinusoidal Input

A *sinusoidal input* function is a repeating, periodic input that can be represented by sine and/or cosine functions

$$\text{Sinusoidal input} \qquad u(t) = A \sin \omega t + B \cos \omega t \tag{5.128}$$

where A and B are the amplitudes and ω is the frequency in rad/s. If the sinusoidal input is zero at time $t = 0$, then we would use $u(t) = A \sin \omega t$. Sinusoidal inputs are used to represent periodic inputs such as oscillatory forces in mechanical systems and oscillatory voltage sources in electrical systems. Sometimes the frequency of the input signal is given in units of hertz (Hz) or cycles per second. Angular frequency ω is greater than frequency in hertz by a factor of 2π, for example $\omega = 2\pi$ rad/s = 1 Hz (or, one cycle per second). The system response to a sinusoidal input is called the *frequency response* and is studied in detail in Chapter 9.

SUMMARY

In this chapter, we introduced and discussed the standard forms for representing the mathematical models of physical systems. The standard forms include (1) state-variable equations, (2) the state-space representation (SSR), (3) input-output (I/O) equations, (4) transfer functions, and (5) block diagrams. Of course, the governing system dynamics (defined by the collection of ODEs) is not altered if we choose to represent the system in one or more different standard forms. As we see in the following chapters, each standard form has its own advantages and disadvantages when it comes to obtaining the system's response through numerical simulation or analytical techniques.

State-variable equations are a collection of n first-order differential equations, which may be linear or nonlinear. An SSR is a convenient vector-matrix format of the state-variable equations, which can be applied only to linear ODEs. As we see in Chapter 6, the SSR is very well-suited for computer simulations of linear dynamic systems. An I/O equation is a single nth-order ODE that includes one output variable and one or more input variables. A transfer function, by definition, is the ratio of the Laplace transform of the output variable to the Laplace transform of the input variable with zero initial conditions. In this volume, we emphasize using transfer functions to *represent* the dynamics of a single-input, single-output (SISO) system. In this chapter, we demonstrated how to derive the transfer function by applying the differential (or D) operator to the I/O equation.

We also presented the basic steps for the linearization process. This process allows us to develop a linear model that represents the system dynamics in the vicinity of a nominal operating point. Hence, we can apply the linearization process to nonlinear state-variable equations to develop a vector-matrix SSR.

Finally, we presented the block diagram, which is a graphical representation of a dynamic system. The dynamic variables are shown as signal paths, and operations such as addition, multiplication, and integration are represented as "blocks" with input and output variables. Block diagrams are the foundation of the extremely powerful simulation software Simulink. Numerical simulation using Simulink is the focus of the next chapter.

REFERENCE

1. Whitehead, J.C., "Rear Wheel Steering Dynamics Compared to Front Steering," *ASME Journal of Dynamic Systems, Measurement, and Control*, Vol. 112, Mar. 1990, pp. 88–93.

Chapter 6

Numerical Simulation of Dynamic Systems

6.1 INTRODUCTION

The first section of this book emphasized how to derive the mathematical models of an array of physical engineering systems. In Chapter 5, these mathematical models (i.e., the ordinary differential equations, or ODEs) were written in various "standard forms," including state-variable equations, state-space representation (SSR), input–output (I/O) equations, and transfer functions. It should be noted that the system dynamics do not change when we choose to represent a given mathematical model in a particular "standard form" such as SSR or a transfer function. Chapter 5 concluded with a discussion on block diagrams, which are graphical representations of interconnected systems with clearly defined input and output signals where the "blocks" denote specific I/O relationships.

Developing a mathematical model is always the first step in the analysis and design of dynamic systems. Determining the system's response to a specific input (i.e., solving the governing ODEs) is the second step, as the engineer is interested in characteristics such as response speed, maximum output, and the time to achieve a steady or constant output. Once the mathematical model has been developed, the engineer has two options for determining the system's response: (1) analytical methods or (2) numerical simulations using a digital computer. Analytical techniques involve solving the governing ODEs "by hand," which is feasible for linear, time-invariant (LTI) first- and second-order systems. Engineers should be able to ascertain the fundamental response characteristics of LTI first- and second-order systems by employing a few relatively simple analytical calculations. These analytical techniques are discussed in Chapter 7. However, when the system dynamics involve nonlinear terms, a numerical simulation is likely the only option for determining the system response. Furthermore, a numerical simulation is the best option for determining the response of a higher-order, real-world system that involves the interaction of multiple inputs and outputs.

Simulation is the process of obtaining the system's dynamic response by numerically solving the governing modeling equations; in other words, numerical integration of the model's differential equations. In this chapter, we introduce the MATLAB simulation software and present several examples that illustrate simulating dynamic systems. The two goals of this chapter are (1) to simulate LTI systems using built-in MATLAB commands and (2) to explain how to build and execute a simulation of a dynamic system using MATLAB's graphical software Simulink. Because the built-in MATLAB commands can be used only to simulate LTI systems (whereas Simulink can accommodate both linear and nonlinear models) Simulink is the primary focus of this chapter and is the principal simulation tool for the remainder of the book. We begin our discussion with simple, linear systems and end the chapter with more complex, integrated systems with interconnected linear and nonlinear components.

6.2 SYSTEM RESPONSE USING MATLAB COMMANDS

MATLAB has a suite of built-in commands for obtaining the response of a linear dynamic system due to its initial conditions and/or the system input function. It should be emphasized that these MATLAB commands can be used only for *linear* systems. Furthermore, these MATLAB commands are easy to use and allow the user to quickly obtain the dynamic response of linear systems that are represented by either a transfer function or an SSR. Appendix B presents a primer on using MATLAB and Table B.2 summarizes many useful MATLAB commands for simulating dynamic systems.

The four basic MATLAB simulation commands that we present are `step`, `impulse`, `lsim`, and `initial`. All four commands require that the user define the LTI system model using either the transfer function or an SSR format. For example, consider the second-order I/O equation

$$\ddot{y} + 3\dot{y} + 12y = 0.8u \tag{6.1}$$

Applying the D operator to replace the derivative terms, that is, $\dot{y} = Dy$ and $\ddot{y} = D^2 y$, Eq. (6.1) becomes

$$(D^2 + 3D + 12)y = 0.8u \tag{6.2}$$

Next, form the ratio of output to input and replace D with the Laplace variable s to derive the transfer function $G(s)$

$$G(s) = \frac{0.8}{s^2 + 3s + 12} = \frac{Y(s)}{U(s)} \tag{6.3}$$

The following MATLAB commands create the object `sys` that represents the transfer function in Eq. (6.3)

```
>> numG = 0.8;          % numerator of G(s) = 0.8/(s² + 3s + 12)
>> denG = [1 3 12];     % denominator of G(s) = 0.8/(s² + 3s + 12)
>> sys = tf(numG,denG)  % create LTI transfer function G(s)
```

Note that `denG` is a row vector of the denominator polynomial coefficients in descending powers of s. After hitting the return key, MATLAB displays `sys` as

```
Transfer function:
     0.8
- - - - - - - - - - - -
s^2 + 3 s + 12
```

so that the user can verify that he or she has properly defined the desired transfer function. Next, the user can obtain a plot of the unit-step response using the built-in MATLAB command `step` as follows

```
>> step(sys)
```

The `step` command simulates the response to the input $u(t) = U(t)$ and automatically plots the output $y(t)$ to the screen. In this case, the LTI system `sys` is transfer function $G(s)$.

In a similar fashion, the built-in MATLAB command `impulse` simulates the response to the unit-impulse input $u(t) = \delta(t)$:

```
>> impulse(sys)
```

As before, a plot of the response is automatically created. The user has the option to define a desired simulation time vector `t` and invoke left-hand-side arguments in order to create their own plot as shown below

```
>> t = 0:0.01:5;        % define time vector from 0 to 5 in steps Δt = 0.01 s
>> [y,t] = step(sys,t); % compute the unit-step response y(t) (no plot)
>> plot(t,y)            % plot the unit-step response y(t)
```

We can also use the `step` and `impulse` commands with the LTI system `sys` defined as a state-space representation. Using the LTI system presented by Eq. (6.1), we can define the following SSR for states $x_1 = y$ and $x_2 = \dot{y}$

$$\dot{\mathbf{x}} = \begin{bmatrix} 0 & 1 \\ -12 & -3 \end{bmatrix} \mathbf{x} + \begin{bmatrix} 0 \\ 0.8 \end{bmatrix} u \qquad (6.4)$$

$$y = \begin{bmatrix} 1 & 0 \end{bmatrix} \mathbf{x} + [0]u$$

Next, define the SSR matrices noting that MATLAB uses square brackets to denote matrices and vectors, where each row is separated by a semicolon (see Appendix B for a brief MATLAB tutorial):

```
>> A = [ 0 1 ; -12 -3 ];        % define state matrix A
>> B = [ 0 ; 0.8 ];             % define input matrix B
>> C = [ 1 0 ];                 % define output matrix C
>> D = 0;                       % define direct-link "matrix" D
```

We can create the SSR object sys using the ss command

```
>> sys = ss(A,B,C,D)
```

Finally, we can simulate and plot the unit-step or unit-impulse responses using the step or impulse commands with SSR object sys representing the system dynamics. Because transfer function Eq. (6.3) and SSR Eq. (6.4) both represent the system dynamics, that is, Eq. (6.1), we get identical simulation results whether we choose to define sys using tf(numG,denG) or ss(A,B,C,D).

The MATLAB command lsim ("linear simulation") allows the user to simulate the response of a linear system to an arbitrary user-defined input function. For example, suppose the desired input is a pulse with a magnitude of 20 that lasts for 5 s. To simulate the pulse response, we type the commands

```
>> t = 0:0.01:10;               % define time vector from 0 to 10 in steps Δt = 0.01 s
>> u(1:501) = 20;               % define pulse u(t) = 20 for 0 ≤ t ≤ 5 s
>> u(502:1001) = 0;             % define zero input u(t) = 0 for t > 5 s
>> [y,t] = lsim(sys,u,t);       % obtain system output y(t) for user-defined input u(t)
```

The reader should note that the first three commands define the time vector t and the pulse input vector u. Of course, the object sys must be defined and the user may choose either a transfer function or an SSR.

The built-in MATLAB commands step and impulse assume that the LTI system has zero initial conditions (of course, by definition systems represented by transfer functions have zero initial conditions). In many cases, we want to simulate systems that have nonzero initial conditions (hence the system *must* be created as a state-space model). We may include initial conditions using the MATLAB commands lsim and initial. The lsim command is modified by adding the initial state vector x0 to the right-hand-side argument list:

```
>> [y,t] = lsim(sys,u,t,x0);
```

As before, the user must define the input vector u and simulation time vector t. Of course, the initial conditions x0 can be used only when the object sys is defined by an SSR.

The command initial simulates the response of an LTI system to its initial conditions (with zero input). If initial conditions exist, the user must define the system using a state-space model. The usage of initial is similar to the step, impulse, and lsim commands:

```
>> [y,t] = initial(sys,x0,t);
```

Again, it should be emphasized that the initial command *cannot* be used with transfer-function models as (by definition) transfer functions are derived by assuming zero initial conditions.

In summary, the built-in MATLAB commands allow a user to quickly obtain the system response to standard inputs (step for a unit-step and impulse for a unit-impulse input, respectively), arbitrary input functions (lsim), and arbitrary initial conditions (initial, for SSR models only). These commands can be used only for *linear* systems that can be represented by a transfer function or state-space model. The following examples illustrate the use of these built-in MATLAB commands.

Example 6.1

Figure 6.1 shows the simple series RL circuit from Example 5.16. Use MATLAB to determine the current $I(t)$ and the resistor voltage $e_O(t)$ if the voltage input $e_{in}(t)$ is a 1-V step function applied at time $t = 0$. The current is initially zero, and $L = 0.1$ H and $R = 1.6 \, \Omega$.

The mathematical model of the RL circuit is

$$L\dot{I} + RI = e_{in}(t) \tag{6.5}$$

A transfer function is probably the easiest way to represent this simple first-order model. However, the reader should note that transfer functions assume zero initial conditions, which may not apply to all problems (fortunately, the initial current is zero in this case). Using the D operator to replace the derivative term, that is, $\dot{I} = DI$, Eq. (6.5) becomes

$$(LD + R)I = e_{in}(t) \tag{6.6}$$

Next, form the ratio of current to input voltage, $I(t)/e_{in}(t)$, replace D with the Laplace variable s, and substitute the numerical values for L and R to derive the transfer function

$$\frac{1}{0.1s + 1.6} = \frac{I(s)}{E_{in}(s)} \tag{6.7}$$

Because the voltage input $e_{in}(t)$ is a constant with unity magnitude, we can use the `step` command to obtain the unit-step response. The following MATLAB commands obtain the responses for current $I(t)$ and resistor voltage $e_O(t) = RI(t)$

```
>> sys = tf(1,[0.1 1.6]);    % create transfer function object sys
>> t = 0:0.001:0.5;          % define simulation time vector t
>> [I,t] = step(sys,t);      % obtain unit-step response for current I(t)
>> e_O = 1.6*I;              % compute resistor voltage e_O(t) = RI(t)
```

Plots of the desired variables can be created using MATLAB's `plot` command along with the desired plotting options; the commands for plotting current versus time are

```
>> plot(t,I)                 % plot current versus time
>> grid                      % add grid lines to plot
>> xlabel('Time, s')         % add x-axis label
>> ylabel('Current, A')      % add y-axis label
```

A similar set of commands plot resistor voltage $e_O(t)$. Figure 6.2 presents plots of current $I(t)$ and resistor voltage $e_O(t)$ versus time. Note that both responses exhibit an exponential rise from a zero initial condition to a constant value, which is a characteristic of a first-order system response that we investigate in Chapter 7.

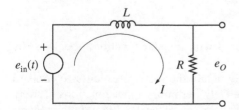

Figure 6.1 Series *RL* circuit for Example 6.1.

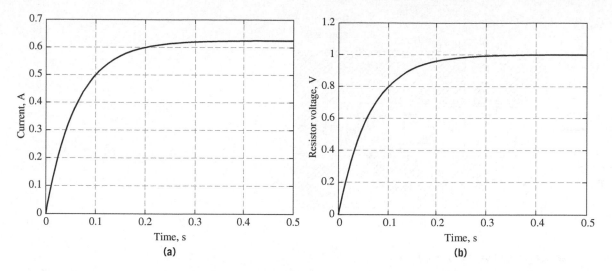

Figure 6.2 *RL* circuit unit-step response for Example 6.1: (a) current versus time and (b) resistor voltage e_O versus time.

Note that we can use the `lsim` command to obtain the unit-step response, as shown below:

```
>> t = 0:0.001:0.5;          % define simulation time vector t
>> u = ones(size(t));        % define unit-step input vector u(t) = U(t)
>> [I,t] = lsim(sys,u,t);    % obtain unit-step response for current I(t)
```

Here we must define the input `u`, which is a vector of unity values with the same dimension as simulation time vector `t`.

Example 6.2

Figure 6.3 shows a three-way spool valve used to meter flow in a hydraulic system, and Eq. (6.8) is the mathematical model of the valve. Use MATLAB to obtain the response of the system if the applied force $f(t)$ is a pulse function that steps up from zero to 12 N at time $t = 0.02$ s and steps down to zero at $t = 0.06$ s. The spool valve is initially at rest.

$$0.04\ddot{y} + 16\dot{y} + 7000y = f(t) \tag{6.8}$$

Equation (6.8) is the governing mathematical model of the spool valve, which consists of a single mass ($m = 0.04$ kg), a linear friction force (viscous friction coefficient $b = 16$ N-s/m), and a linear spring force (spring constant $k = 7000$ N/m). Variable $y(t)$ is the displacement of the spool valve (in m), and $f(t)$ is the force from an electromagnetic actuator that can push on the valve. We assume that there is no hydraulic fluid pressure imbalance on the valve, and that flow forces are neglected; hence, the actuator force $f(t)$ is the sole applied force acting on the valve mass. The system is initially at rest ($\dot{y}_0 = y_0 = 0$) at time $t = 0$, and the force steps from zero to 12 N at time $t = 0.02$ s.

We derive the system transfer function from the I/O equation (6.8) by using the D operator to replace the derivative terms, that is, $\ddot{y} = D^2 y$ and $\dot{y} = Dy$, which yields

$$(0.04D^2 + 16D + 7000)y = f(t) \tag{6.9}$$

Next, form the ratio of output to input, $y(t)/f(t)$, and replace D with the Laplace variable s to derive the transfer function

$$\frac{1}{0.04s^2 + 16s + 7000} = \frac{Y(s)}{F(s)} \tag{6.10}$$

Equation (6.10) is the transfer function that represents the spool-valve system. Because we have an arbitrary input (a pulse), we must use the `lsim` command. The following MATLAB commands will produce the pulse response:

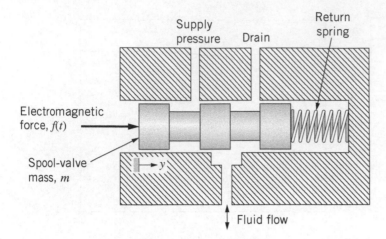

Figure 6.3 Three-way spool valve for Example 6.2.

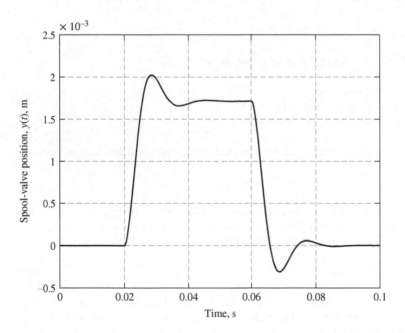

Figure 6.4 Spool-valve response to a 12-N pulse force (Example 6.2).

```
>> sys = tf(1,[0.04 16 7000]);      % create transfer function object sys
>> t = 0:0.0001:0.1;                % define simulation time vector 0 ≤ t ≤ 0.1 s
>> f(1:200) = 0;                    % define zero input for 0 ≤ t < 0.02 s
>> f(201:601) = 12;                 % define 12-N input force for 0.02 ≤ t ≤ 0.06 s
>> f(602:1001) = 0;                 % define zero input for t > 0.06 s
>> [y,t] = lsim(sys,f,t);           % obtain pulse response for valve position y(t)
```

MATLAB's plot command can be used to create the plot of $y(t)$ shown in Fig. 6.4. Note that valve position begins at zero (its initial condition), responds to the 12-N step force applied at $t = 0.02$ s, reaches a peak response of about 0.002 m (2 mm), and finally settles to a constant displacement of 0.0017 m (1.7 mm). At time $t = 0.06$ s the applied force is instantly stepped to zero and it is clear that the valve responds in a reversed but symmetric manner as it returns to the zero position.

6.3 BUILDING SIMULATIONS USING SIMULINK

MATLAB's Simulink is an extremely useful and powerful software tool for simulating dynamic systems and obtaining the dynamic response. Universally accepted, it is used in both academic settings and the engineering industry. Simulink is a graphical tool based on block diagrams composed of individual I/O blocks. A working outline of Simulink and the basic steps for building a simulation are presented in this section; Appendix C presents a more complete Simulink tutorial.

A good numerical simulation tool successfully incorporates the following desirable traits:

1. Ease of entering the mathematical models that represent the system dynamics
2. Ability to include standard and arbitrary inputs to the system
3. Ease of storing and plotting the desired output variables
4. Ability to include arbitrary initial conditions for the dynamic variables
5. Ease of adjusting the run-time and numerical integration parameters

Simulink uses a graphical user interface (GUI) that allows the user to browse and select I/O blocks from several libraries, such as the `Continuous` and `Math Operations` libraries. The `Continuous` library includes block icons for standard models such as the transfer function, SSR, and the integrator block for I/O equations. The `Math Operations` library includes useful blocks such as the gain and summing junction that can be used to build the desired block-diagram representation of the system dynamics. The user "drags and drops" the desired blocks onto a working template and connects the input and output ports in order to build a simulation. He or she may select a wide range of available input functions from the `Sources` library, such as the step function, signal generator, and sine wave. Hence, all blocks in the `Sources` library have output ports but do not have input ports. The signal paths that connect the various I/O blocks and input functions contain the time-history information of the signals, which can be routed to various data storage and plotting functions contained in the `Sinks` library. Therefore, all blocks in the `Sinks` library (such as the `Scope` and `To Workspace` blocks) have input ports but do not have output ports. The initial conditions of dynamic variables can be set by double-clicking (opening) the appropriate model block (such as the state-space or integrator block) and entering the desired values. Finally, the parameters that define the numerical integration process are contained in the `Model Configuration Parameters` option under the `Simulation` menu in the model workspace window. Here, the user can enter the simulation time, the type of numerical integration method (Euler integration, Runge–Kutta integration, etc.), and either the integration step size for fixed-step methods or the error tolerances that determine the variable integration step size.

To start the Simulink program, simply enter the following command in the MATLAB environment

```
>> simulink
```

that opens the Simulink library browser. Clicking on the "new model" icon in the upper left corner will create a new Simulink model, which begins as a blank template for building the simulation block diagram. The user may now access the Simulink libraries and add and connect the desired I/O blocks to create a simulation. Appendix C presents figures depicting the Simulink libraries, blocks, and associated windows used to set the simulation parameters, as well as additional details for constructing a simulation model.

Building a simulation using Simulink is best demonstrated by presenting examples. This section concludes with a simulation of the simple first-order block-diagram example from Chapter 5.

Example 6.3

Consider again the simple *RL* circuit from Example 6.1 and Fig. 6.1. Using Simulink, build a simulation of this system and determine the current $I(t)$ and the resistor voltage $e_O(t)$ if the voltage input $e_{in}(t)$ is a 2-V step function applied at time $t = 0$. The current is initially zero, and $L = 0.1$ H and $R = 1.6$ Ω.

We need to construct a block diagram of this system and a transfer function is probably the easiest way to represent this simple first-order model (the reader should note that this system has zero initial conditions and hence we may use a transfer function). The transfer function was derived in Example 6.1 and it is repeated below

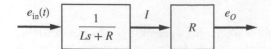

Figure 6.5 Block diagram for Example 6.3.

$$\frac{1}{Ls + R} = \frac{I(s)}{E_{in}(s)} \tag{6.11}$$

Figure 6.5 shows the I/O relationship between source voltage $e_{in}(t)$ and current I. Because we want $e_O = RI$ as the output variable, we multiply current I by resistance R as shown in Fig. 6.5. We construct our Simulink model using the block diagram in Fig. 6.5, which consists of a single transfer function for the RL circuit dynamics, Eq. (6.11), followed by a gain block (resistance R) that produces the desired output signal e_O.

After starting the Simulink program and opening a new model template, we open the `Continuous` library and "drag and drop" the transfer function icon (`Transfer Fcn`) onto the new model window. The `Gain` block (from the `Math Operations` library) is added to the right of the transfer function. Next, we open the `Sources` library and include the step function (`Step`) for the voltage source $e_{in}(t)$ and the clock icon (`Clock`) for the simulation time. We open the `Sinks` library and include the `To Workspace` icon after the gain block. Signal-path connections are made from the step function to the input port of the transfer function, from the transfer function's output port to the input port of the gain block, and from the gain block's output port to the `To Workspace` block. Finally, we include three additional `To Workspace` icons from the `Sinks` library and make the signal-path connections from both `Sources` (`Step` and `Clock`) to two `Sinks` (`To Workspace`) in order to store $e_{in}(t)$ and time t. Because we also want to store current $I(t)$, we send a signal to the third `To Workspace` block. We can define the variable for the stored signals by double-clicking the appropriate `To Workspace` block and changing the `Variable name` dialog box. For example, we can define the variable name `e_in` for the source voltage sent to the `To Workspace` block. It is important to save the format of the stored variable as an array so that it may be plotted from the MATLAB command line. To change the format double-click each `To Workspace` block and select `Array` in the `Save format` dialog box.

We set the desired system parameters (input voltage $e_{in} = 2\,\text{V}$, inductance $L = 0.1\,\text{H}$, and resistance $R = 1.6\,\Omega$) by double-clicking on the appropriate `Step`, `Transfer Fcn`, and `Gain` blocks and entering the numerical values. The transfer function block has two dialog boxes for the numerator and denominator coefficients, which are entered as row vectors in descending powers of s. Note that Simulink displays the numerical values in the `Transfer Fcn` and `Gain` blocks once they have been set. The fixed-step, fourth-order Runge–Kutta method `ode4` is selected as the numerical solver under the `Simulation > Model Configuration Parameters` menu, and the fixed time step is set at 10^{-3} s. Labels for blocks and signal paths may be added for clarity. Figure 6.6 shows the final Simulink model, which essentially matches the block diagram in Fig. 6.5.

It should be emphasized that Simulink uses the transfer function block in Fig. 6.6 to *represent* the I/O equation of the RL circuit dynamics, and even though the transfer function includes the complex Laplace variable s, Simulink determines the system response by using MATLAB's numerical integration algorithms and *not* by using Laplace transform theory. Hence, it is correct to use the time-domain labels, such as $e_O(t)$, on the signal paths and not the Laplace-transformed variables, such as $E_O(s)$.

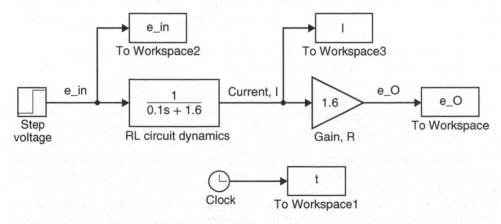

Figure 6.6 Simulink diagram for Example 6.3.

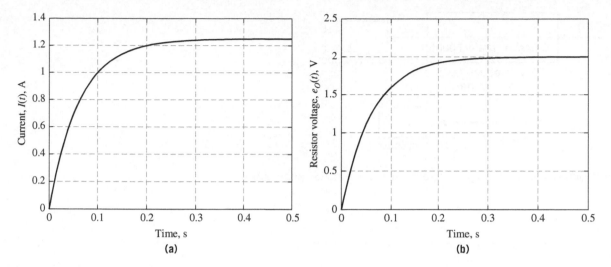

Figure 6.7 RL circuit response for Example 6.3: (a) current versus time and (b) resistor voltage e_O versus time.

After the Simulink model is constructed and all parameters are set, the simulation is executed by selecting `Simulation > Run` (or, by clicking the `Run` button). Plots of the stored variables `I` and `e_O` can be created using MATLAB's `plot` command. Figure 6.7 presents plots of current $I(t)$ and resistor voltage $e_O(t)$ versus time. Note that both step responses are identical to the *unit*-step responses (Fig. 6.2 from Example 6.1) if we apply a scaling factor of 2. This comparison makes sense for linear systems because the input here is a 2-V step whereas the source voltage in Example 6.1 was a unit-step input.

6.4   SIMULATING LINEAR SYSTEMS USING SIMULINK

In this section, we present three different methods for simulating a linear system with Simulink. These three methods are

1. Transfer functions
2. State-space representation
3. Integrator blocks for each state-variable equation

Clearly, we may use transfer functions or an SSR *only* if the mathematical model consists of linear ODEs. Example 6.3 demonstrated the transfer-function approach to simulating a simple first-order electrical system. The third method, integrating each state-variable equation using an integrator block, can be applied to both linear and nonlinear mathematical models. All three methods have their advantages and drawbacks, as we demonstrate and discuss in this section by presenting the multiple simulation approaches applied to a common linear dynamic system.

Example 6.4

Consider again the three-way spool-valve system described in Example 6.2 (Fig. 6.3). Simulate the system response using Simulink with a transfer-function representation. The applied force $f(t)$ is a step function with a magnitude of 12 N.

In Example 6.2, we derived the transfer function that relates spool-valve position $y(t)$ to actuator force $f(t)$:

$$\frac{1}{0.04s^2 + 16s + 7000} = \frac{Y(s)}{F(s)} \tag{6.12}$$

We construct our Simulink diagram by using the `Transfer Fcn` block from the `Continuous` library, the `Clock` and `Step` blocks from the `Sources` library, and the `To Workspace` blocks from the `Sinks` library. Double-clicking the

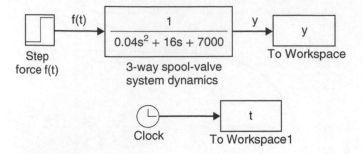

Figure 6.8 Simulink diagram for Example 6.4: transfer-function approach.

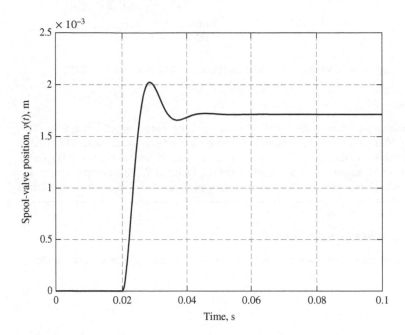

Figure 6.9 Spool-valve response to a 12-N step force (Example 6.4).

`Transfer Fcn` block opens a dialog box, where we can enter the numerator and denominator coefficients of the desired transfer function in Eq. (6.12). The parameters governing the force input are also set by double-clicking the `Step` block: step time is 0.02 s, initial value is 0 N, and final value is 12 N. Finally, we choose the fixed-step, Runge–Kutta numerical integration algorithm `ode4` (with step size of 10^{-4} s) under the `Simulation > Model Configuration Parameters` menu. Figure 6.8 presents the Simulink diagram of the spool valve using the transfer-function approach. Figure 6.9 presents the response of the spool-valve position $y(t)$ to the 12-N step input force. Note that $y(t)$ begins at zero (its initial condition), responds to the step force applied at $t = 0.02$ s, reaches a peak response of about 0.002 m (2 mm), and settles to its final constant value of 0.0017 m (1.7 mm). The step response shown in Fig. 6.9 is identical to the 12-N pulse response from Example 6.2 until $t = 0.06$ s at which time the pulse input steps to zero (see Fig. 6.4).

Example 6.5

Given the spool-valve system in Fig. 6.3 and Example 6.2, create and execute a simulation using a state-space model. The applied force $f(t)$ is a step function with a magnitude of 12 N.

We begin the derivation of an SSR from the second-order I/O equation (6.8) by defining the two state variables: $x_1 = y$ (valve position) and $x_2 = \dot{y}$ (valve velocity). Therefore, two state-variable equations are

$$\dot{x}_1 = \dot{y} \tag{6.13}$$

$$\dot{x}_2 = \ddot{y} = \frac{1}{0.04}\left(-16\dot{y} - 7000y + f(t)\right) \tag{6.14}$$

Substituting the state variables, $x_1 = y$ and $x_2 = \dot{y}$, and input $u = f(t)$ into Eqs. (6.13) and (6.14) yields

$$\dot{x}_1 = x_2 \tag{6.15}$$

$$\dot{x}_2 = -400x_2 - 175{,}000x_1 + 25u \tag{6.16}$$

which can be assembled into the matrix-vector state equation

$$\dot{\mathbf{x}} = \begin{bmatrix} \dot{x}_1 \\ \dot{x}_2 \end{bmatrix} = \begin{bmatrix} 0 & 1 \\ -175{,}000 & -400 \end{bmatrix} \begin{bmatrix} x_1 \\ x_2 \end{bmatrix} + \begin{bmatrix} 0 \\ 25 \end{bmatrix} u \tag{6.17}$$

Recall that the 2×2 square matrix in Eq. (6.17) is the state matrix $\mathbf{A}$, and the 2×1 column vector is the input matrix $\mathbf{B}$. In order to complete the SSR, we write the matrix-vector output equation, where the output (valve position) is the first state variable, or $y = x_1$

$$y = \begin{bmatrix} 1 & 0 \end{bmatrix} \begin{bmatrix} x_1 \\ x_2 \end{bmatrix} + [0]u \tag{6.18}$$

The 1×2 row vector in Eq. (6.18) is the output matrix $\mathbf{C}$, and the $\mathbf{D}$ "matrix" is null (zero).

We construct our Simulink diagram by using the `State-Space` block from the `Continuous` library, the `Clock` and `Step` blocks from the `Sources` library, and the `To Workspace` blocks from the `Sinks` library. Double-clicking the `State-Space` block opens a dialog box, where we can enter the appropriate numerical values for the $\mathbf{A}$, $\mathbf{B}$, $\mathbf{C}$, and $\mathbf{D}$ matrices as demonstrated below

```
A = [ 0 1 ; -175e3 -400 ]
B = [ 0 ; 25 ]
C = [ 1 0 ]
D = 0
```

Note that even though a matrix may be zero (such as the direct-link matrix $\mathbf{D}$ in this case) Simulink requires that all matrices be defined. The reader should carefully enter the matrices as Simulink will fail to run if all four matrices do not have the proper dimensions: $\mathbf{A}$ must be $n \times n$, $\mathbf{B}$ must be $n \times r$, $\mathbf{C}$ must be $m \times n$, and $\mathbf{D}$ must be $m \times r$ (in this example, $n = 2$, $r = 1$, and $m = 1$).

Simulink's `State-Space` block has a dialog box for the initial state vector, $\mathbf{x}(0)$. In this example, the initial states are $x_1(0) = y_0 = 0$ and $x_2(0) = \dot{y}_0 = 0$. Therefore, the initial state is entered as the 2×1 column vector `[ 0 ; 0 ]` in the `Initial conditions` dialog box. The ability to include initial conditions for all states is a distinct advantage of using the state-space approach in Simulink.

The remaining steps for building the Simulink diagram (`Step`, `Clock`, and `To Workspace` blocks) are identical to the previous example, and Fig. 6.10 presents the Simulink diagram of the spool valve using the state-space approach. Note that Simulink does not display the numerical values of the SSR matrices; the user must open the dialog box to see the respective values. Executing the simulation and plotting the output $y(t)$ (valve position) yields a plot identical to Fig. 6.9.

A final note is in order: if we wanted to observe the position *and* velocity of the valve, we could do so with the state-space method by redefining the output vector as both state variables, that is, $\mathbf{y} = \mathbf{x}$. Therefore, the $\mathbf{C}$ matrix is the 2×2 identity matrix and the $\mathbf{D}$ matrix is a 2×1 null vector:

```
C = [ 1 0 ; 0 1 ]
D = zeros(2,1)
```

When we run the simulation, the output `y` sent to the MATLAB workspace will contain two columns. The first column is the valve's position, and the second column is the valve's velocity data for the simulation time. This example demonstrates another advantage of using the state-space method: we can obtain the dynamic responses of all states. The transfer-function method only provides the dynamic response of a single output variable, which is valve displacement in this example.

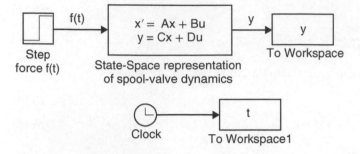

Figure 6.10 Simulink diagram for Example 6.5: state-space approach.

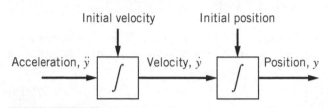

Figure 6.11 Block diagram of two successive integrals of acceleration.

Example 6.6

Given the three-way spool-valve system in Fig. 6.3 and Example 6.2, create and execute a simulation using the integrator-block approach. The applied force $f(t)$ is a 12-N step function.

The basic concept of the integrator-block approach is to simply "chain together" a series of n integrator blocks for each nth-order I/O equation. For this example, we have a single second-order I/O equation, and therefore the "core" of the simulation will be two successive integrals of acceleration, as shown in Fig. 6.11. Note that each integrator block has an initial condition (the integration constant) that is added to the time integral of the input signal.

In order to use the integrator block method, we start with an expression for the nth-order derivative term (acceleration in this case), which we obtain from Eq. (6.8)

$$\ddot{y} = \frac{1}{0.04}\left(-16\dot{y} - 7000y + f(t)\right) \tag{6.19}$$

Therefore, the left-hand side of the block diagram in Fig. 6.11 (acceleration) must be equal to the right-hand side of Eq. (6.19), which is the sum of the friction, stiffness, and applied forces divided by mass.

Figure 6.12 shows the Simulink diagram of the spool-valve system using the integrator-block approach. The simulation is initially constructed by connecting two `Integrator` blocks from the `Continuous` library. The friction force is produced by "picking off" the velocity signal (`y-dot`, the output of the first integrator) and multiplying by a `Gain` block that represents the viscous friction coefficient. The stiffness force is constructed in a similar manner using position y (the user should note that the triangular `Gain` block can be "flipped" in direction by highlighting the block and selecting

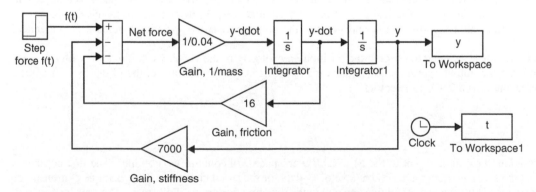

Figure 6.12 Simulink diagram for Example 6.6: integrator-block approach.

Rotate & Flip under the Diagram menu). All three forces are added together in a summing junction (Sum) from the Math Operations library (note the signs), and the net force is divided by mass (Gain block) to produce acceleration (y-ddot in Fig. 6.12). The initial conditions for each integrator ($\dot{y}_0$ and y_0) are set by double-clicking the respective Integrator block and entering the numerical values in the dialog box (both initial conditions are zero in this example).

Executing the Simulink diagram in Fig. 6.12 and plotting $y(t)$ produces a plot that is identical to Fig. 6.9. Note that we could "pick off" and send several signals from the Simulink diagram in Fig. 6.12 to the MATLAB workspace for plotting: net force, acceleration $\ddot{y}(t)$, velocity $\dot{y}(t)$, friction force, and spring force. Hence, the integrator-block approach may be the most versatile of the three simulation methods demonstrated by these examples.

We may summarize the characteristics of the three Simulink approaches based on the results of the previous examples:

1. The transfer-function approach is likely the most concise and easiest method of the three, because it is relatively easy to derive the transfer function(s) for linear systems. In addition, Simulink displays the numerator and denominator of the transfer function, which provides a means to check the construction of the simulation diagram. However, by definition, the transfer function assumes zero initial conditions. Therefore, a transfer function cannot be used when the system has nonzero initial conditions. Furthermore, each transfer function provides a single output variable, so some "internal" dynamic variables may be impossible to obtain. Note that in our previous example, position $y(t)$ is available by using the transfer-function approach, but velocity $\dot{y}(t)$ is not.

2. The state-space approach is concise (it represents an entire linear system with a single block), but it requires that the user derive the complete SSR matrices. However, it is flexible as it allows the user to set arbitrary initial conditions for all n states. In addition, the user can obtain the dynamic response of all n states by setting the output matrix **C** to the $n \times n$ identity matrix.

3. The integrator-block approach is likely the most flexible method, as it allows the user to set arbitrary initial conditions for all dynamic variables; plus, the user can store and plot any signal-path variable, which may aid in trouble-shooting complex simulations. However, for complex systems, the user may need to carefully plan the simulation diagram because it will involve multiple signal paths, summing junctions, and blocks, which may make it difficult to read.

6.5 SIMULATING NONLINEAR SYSTEMS

Next we focus on using Simulink to simulate systems that are governed by nonlinear mathematical models. As stated in the Introduction, numerical simulation methods are likely the only option available for obtaining the dynamic response of nonlinear systems, whereas analytical techniques (described in Chapter 7) can be used to obtain the solution of linear systems, such as the second-order mechanical system from the previous examples. It is important to remember that we cannot use transfer functions or an SSR when dealing with nonlinear systems; our only option is to use numerical integration of each nonlinear ODE, which we perform using Simulink and the integrator-block approach. We demonstrate nonlinear system simulations by revisiting the spool-valve system from Example 6.2 and the hydraulic tank system from Chapter 5.

Example 6.7

Consider again the spool-valve system in Fig. 6.3 and Example 6.2, but with the inclusion of Coulomb or dry friction along with viscous friction. Simulate the response to a 12-N step input.

Let us assume that the dry friction force F_{dry} has a magnitude of 0.4 N. Because the dry friction force always opposes the direction of motion, it can be modeled as $F_{dry}\,\text{sgn}(\dot{y})$, where the signum function "sgn" returns the sign of its input value, which is velocity $\dot{y}$. Adding the dry friction term $F_{dry}\,\text{sgn}(\dot{y})$ to the linear spool-valve equation (6.8) yields a *nonlinear* mathematical model of the spool valve

$$0.04\ddot{y} + 16\dot{y} + 0.4\,\text{sgn}(\dot{y}) + 7000y = f(t) \tag{6.20}$$

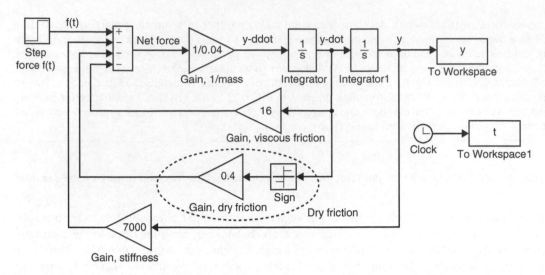

Figure 6.13 Simulink diagram for Example 6.7: nonlinear mechanical system with integrator-block approach.

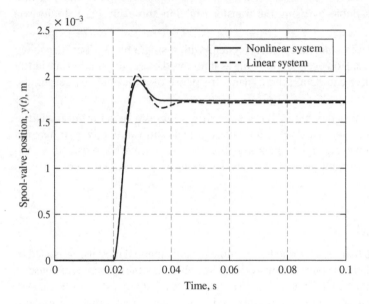

Figure 6.14 Spool-valve responses to a 12-N step force: nonlinear and linear system models (Example 6.7).

Because the mathematical model is nonlinear, we must use the integrator-block approach to obtain the system's response. Recall that in Example 6.6 we constructed our block diagram from the double integration of acceleration of the valve mass, and therefore we solve Eq. (6.20) for acceleration

$$\ddot{y} = \frac{1}{0.04}\left(-16\dot{y} - 0.4\,\mathrm{sgn}\,(\dot{y}) - 7000y + f(t)\right) \tag{6.21}$$

All bracketed terms in Eq. (6.21) are forces: viscous friction force, dry friction force, spring force, and the applied force $f(t)$. We can modify the linear system Simulink diagram shown in Fig. 6.12 and include an additional feedback path for the dry friction force. Figure 6.13 shows the Simulink diagram of the spool valve with the addition of the dry friction force. This force is created by sending velocity $\dot{y}$ to the signum (or `Sign`) function from the `Math Operations` library, followed by a `Gain` block that represents the dry force magnitude (0.4 N in this case). Note that the summing junction has been modified to accept four (force) signals with the appropriate signs in order to match Eq. (6.21).

Figure 6.14 presents the response of the nonlinear spool-valve system to the 12-N step input force applied at time $t = 0.02$ s. The solid line in Fig. 6.14 is the step response of the *nonlinear* system with dry friction, and the dashed line is

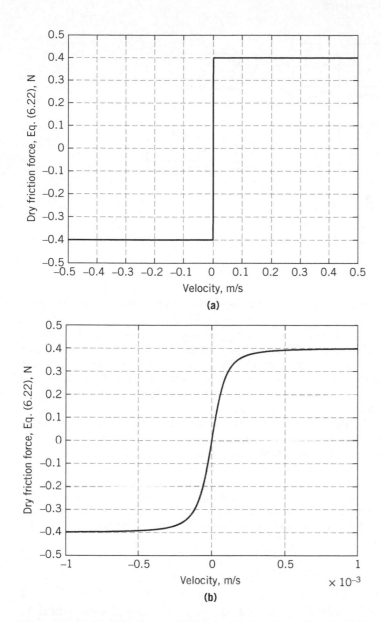

Figure 6.15 Continuous model for dry friction force, Eq. (6.22): (a) operating range for velocity and (b) "zoomed-in" view near zero velocity.

the step response of the linear system from Example 6.2 (no dry friction). Note that the inclusion of nonlinear dry friction slightly reduces the peak response of $y(t)$ as compared to the peak response of the linear model. In addition, the nonlinear response does not exhibit "undershoot" after the peak response, but rather reaches its constant value at about 0.035 s (the linear response reaches its constant value after 0.04 s).

Modeling Coulomb or dry friction using the signum function can at times cause simulation problems because of its discontinuity at zero velocity. The discontinuity often requires a small integration step size so that the force is accurately computed near zero velocity, which in turn slows down the simulation run time. Furthermore, the signum function can lead to "chatter" where the dry friction force rapidly switches signs between $\pm F_{\text{dry}}$ because of the velocity signal switching signs as it approaches zero (equilibrium). For these reasons, it may be useful to approximate the discontinuous dry friction $F_{\text{dry}}\, \text{sgn}(\dot{y})$ force with the following continuous function

$$F_{\text{DF}} = \frac{F_{\text{dry}}\dot{y}}{\sqrt{\dot{y}^2 + \varepsilon^2}} \tag{6.22}$$

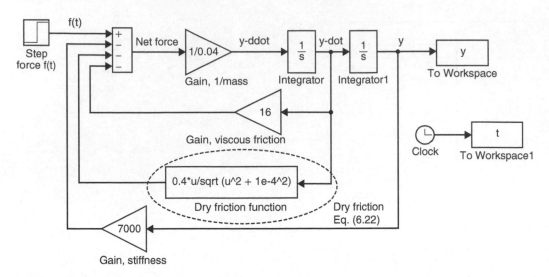

Figure 6.16 Simulink diagram for Example 6.7: nonlinear mechanical system with continuous function for dry friction.

where F_{DF} is the dry friction force, and ε is a constant with units of velocity. As parameter ε becomes "small," Eq. (6.22) provides a good approximation of the discontinuous function $F_{dry} \, \text{sgn}(\dot{y})$, and when ε is exactly zero, Eq. (6.22) is equivalent to the discontinuous dry friction model. Figure 6.15 shows the continuous dry friction equation (6.22) for the expected velocity range of the spool-valve mass (-0.5 to 0.5 m/s) with $\varepsilon = 10^{-4}$ m/s and $F_{dry} = 0.4$ N. From Fig. 6.15a, it appears that Eq. (6.22) models discontinuous switching between $\pm F_{dry}$ when velocity switches signs. However, the "zoomed-in" view near the origin as shown in Fig. 6.15b demonstrates that the function is indeed continuous.

We can modify the nonlinear Simulink diagram shown in Fig. 6.13 and use the continuous equation (6.22) to model dry friction force. Figure 6.16 shows the Simulink diagram of the spool valve with the continuous dry friction force model. In this case, we have defined the dry friction force with the `Fcn` (or function) block from the `User-Defined Functions` library. The `Fcn` block allows the user to define any functional output equation in terms of a single input (u). To do so, the user must double-click the `Fcn` block and enter the desired equation in the `Expression` dialog box. Figure 6.16 shows Eq. (6.22) in the function block, where u is the generic symbol that represents the input to the block (velocity, $\dot{y}$, in this case). Note that the constant ε is set to 10^{-4} m/s. Executing the Simulink diagram in Fig. 6.16 produces a response for valve position $y(t)$ that is essentially identical to the response shown in Fig. 6.14 for the nonlinear Simulink diagram that uses the discontinuous signum function for the dry friction force.

Example 6.8

Consider again the simple hydraulic tank system from Example 5.10, shown in Fig. 6.17. Water is the working fluid. In Example 5.10, we derived the governing nonlinear model, and then through linearization developed a linear model about a nominal point. Using Simulink, simulate the nonlinear and linear systems under the same operating conditions and compare the system responses.

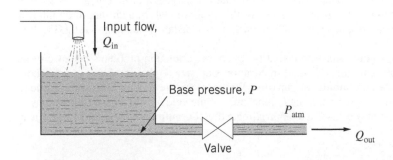

Figure 6.17 Hydraulic tank system for Example 6.8.

Nonlinear system simulation

Figure 6.17 shows a single tank with constant cross-sectional area A. It is being filled with water with input volumetric-flow rate Q_{in}. The output volumetric flow Q_{out} is modeled by turbulent flow through the valve to atmospheric pressure P_{atm}. The nonlinear modeling equation for the hydraulic tank system is

$$C\dot{P} = Q_{in} - K_T \sqrt{P - P_{atm}} \tag{6.23}$$

where C is the capacitance of the tank, P is the pressure at the base of the tank, and K_T is a turbulent flow coefficient.

   We develop the nonlinear simulation first, using the following numerical parameters: cross-sectional area of the tank $A = 1.962\,\text{m}^2$, water density $\rho = 1000\,\text{kg/m}^3$, turbulent flow coefficient $K_T = 4(10^{-4})\,\text{m}^{3.5}/\text{kg}^{0.5}$, and atmospheric pressure $P_{atm} = 1.0133(10^5)\,\text{N/m}^2$. Therefore, fluid capacitance is $C = A/\rho g = 2(10^{-4})\,\text{m}^4\text{-s}^2/\text{kg}$. The Simulink diagram consists of a single integrator block that integrates the nonlinear state-variable equation for tank pressure P, which is obtained from Eq. (6.23)

$$\dot{P} = \frac{1}{C}\left(Q_{in} - K_T \sqrt{P - P_{atm}}\right) \tag{6.24}$$

Figure 6.18 presents the Simulink diagram of the nonlinear tank model. The system input, Q_{in}, is represented by the `Constant` block from the `Sources` library. In this example, the variable name `Q_in` has been entered in the dialog box for the `Constant` block instead of a particular numerical value. In a similar fashion, a `Constant` block is used to define atmospheric pressure `P_atm`, which is subtracted from the tank pressure P to form the pressure difference. Therefore, the user must define the constants `Q_in` and `P_atm` in the MATLAB workspace before executing the Simulink model. The nonlinear term in Eq. (6.24), the square root of the pressure difference, is clearly seen in Fig. 6.18. The pressure difference is computed using a summing junction, and the square-root operation is performed using the `Sqrt` block from the `Math Operations` library (note that in previous versions of MATLAB the square-root function resides in the `Math Function` block). The remainder of the Simulink diagram includes a summing junction to compute the net volumetric flow and a gain ($1/C = 5000\,\text{Pa/m}^3$) to produce the time-derivative of pressure, $\dot{P}$. The reader should be able to see how the Simulink diagram in Fig. 6.18 matches the governing nonlinear model (Eq. 6.24).

   The user must set the parameters `Q_in`, `P_atm`, and `P0` (the initial tank pressure for the integrator block) in the MATLAB workspace before running the Simulink model in Fig. 6.18. One way to do this is to write a MATLAB program (or script) called an M-file, which is simply a collection of the required single-line MATLAB commands. MATLAB M-file 6.1 (called `run_tank_NL.m`) sets the required parameters and executes the Simulink model `tank_NL.mdl` (Fig. 6.18) using the `sim` command. This M-file also plots the tank pressure P. We present the nonlinear simulation response after discussing the linear simulation.

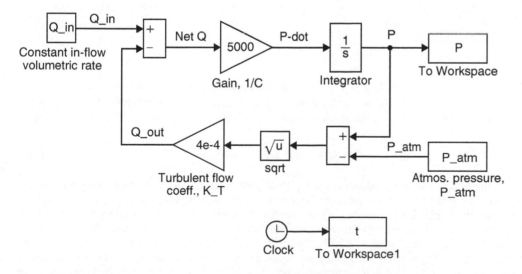

Figure 6.18 Simulink diagram for Example 6.8: nonlinear tank model.

MATLAB M-file 6.1

```
%
%   run_tank_NL.m
%
%   This M-file sets the parameters for
%   the hydraulic tank system, executes
%   the nonlinear Simulink model, and
%   plots the results
%

%   tank system parameters
Q_in = 0.052;        %   constant in-flow volumetric rate, m^3/s
P0 = 1.15e5;         %   initial tank base pressure, N/m^2
P_atm = 1.0133e5;    %   atmospheric pressure, N/m^2

%   execute nonlinear Simulink model
sim tank_NL

%   plot the tank pressure
plot(t,P)
grid
title('Nonlinear model: tank pressure vs. time')
xlabel('Time, s')
ylabel('Tank pressure, P(t), N/m^2')
```

Linear system simulation

The linear hydraulic model was developed in Example 5.10 by linearizing the nonlinear system Eq. (6.24) about a nominal (constant) input flow rate Q_{in}^* and the corresponding nominal (constant) pressure P^*. The resulting linear hydraulic model is

$$\delta \dot{P} = \left. \frac{\partial f}{\partial P} \right|_* \delta P + \left. \frac{\partial f}{\partial Q_{in}} \right|_* \delta Q_{in} \tag{6.25}$$

where $\delta P = P - P^*$ and $\delta Q_{in} = Q_{in} - Q_{in}^*$ are perturbations from their nominal conditions, and $\dot{P} = f(P, Q_{in})$ is the nonlinear ODE presented by Eq. (6.24). The first-order partial derivatives were determined in Example 5.10 and are repeated below

$$\left. \frac{\partial f}{\partial P} \right|_* = \frac{-K_T^2}{2CQ_{in}^*} \qquad \left. \frac{\partial f}{\partial Q_{in}} \right|_* = \frac{1}{C} \tag{6.26}$$

The nominal pressure P^* is

$$P^* = \frac{Q_{in}^{*2}}{K_T^2} + P_{atm} \tag{6.27}$$

which produces an equilibrium condition ($C\dot{P} = 0$) given a nominal volumetric flow Q_{in}^*. Using the numerical value for K_T and the nominal input flow rate $Q_{in}^* = 0.05 \, \text{m}^3/\text{s}$, the nominal pressure is $P^* = 1.16955(10^5) \, \text{N/m}^2$. Using the numerical values for C, K_T, and Q_{in}^* we can compute the first-order partial derivatives in Eq. (6.26). Therefore, the linear model is

$$\delta \dot{P} = -0.008 \, \delta P + 5000 \, \delta Q_{in} \tag{6.28}$$

We are now ready to construct the linear Simulink diagram using Eq. (6.28). Because Eq. (6.28) is linear, we could use the transfer-function approach; however, this approach would apply only to problems with zero initial conditions, or $\delta P(0) = P_0 - P^* = 0$. Instead, we will use the integrator-block approach for the linear model so that we may apply arbitrary initial conditions. Figure 6.19 presents the Simulink diagram for the linear tank model. Note that the core of the simulation is Eq. (6.28), which involves δQ_{in} as the input variable and δP as the dynamic variable. Therefore, the simulation computes the perturbation input flow rate $\delta Q_{in} = Q_{in} - Q_{in}^*$ by using a summing junction that compares two constant input signals. Because the solution to the linear model Eq. (6.28) is in terms of *perturbation* pressure δP, we must add the nominal pressure P^* in order to obtain a linearized approximation to tank pressure P as shown in Fig. 6.19.

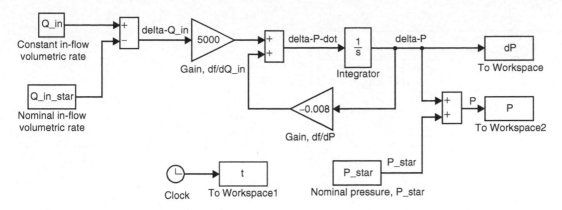

Figure 6.19 Simulink diagram for Example 6.8: linear tank model.

MATLAB M-file 6.2

```
%
%    run_tank_linear.m
%
%    This M-file sets the parameters for
%    the hydraulic tank system, executes
%    the linear Simulink model, and plots
%    the results
%

%  tank parameters
P_atm = 1.0133e5;    %  atmospheric pressure, N/m^2
K_T = 4e-4;          %  turbulent flow coefficient

%  nominal values (linearization point)
Q_in_star = 0.05;
P_star = Q_in_star^2/K_T^2 + P_atm;

%  tank system parameters
Q_in = 0.052;        %  constant in-flow volumetric rate, m^3/s
P0 = 1.15e5;         %  initial tank base pressure, N/m^2
dP0 = P0 - P_star;   %  initial perturbation pressure, N/m^2

%  execute linear Simulink model
sim tank_linear

%  plot the tank pressure
plot(t,P)
grid
title('Linear model: tank pressure vs. time')
xlabel('Time, s')
ylabel('Tank pressure, P(t), N/m^2')
```

As with the nonlinear model, the user must set the constant parameters in the MATLAB workspace before running the simulation. For the simulation in Fig. 6.19, the user must set the parameters Q_in, Q_in_star, P_star, and dP0 (the initial *perturbation* pressure for the integrator block). MATLAB M-file 6.2 (called run_tank_linear.m) sets the required parameters, executes the linear Simulink model tank_linear.mdl (Fig. 6.19), and plots the linear approximation for pressure $P(t)$.

We may now run both the nonlinear and linear models and compare the responses. As previously shown, the nominal input volumetric-flow rate is $Q_{in}^* = 0.05 \, \text{m}^3/\text{s}$, and the corresponding nominal pressure is $P^* = 1.16955(10^5) \, \text{N/m}^2$. The actual input flow rate is set to $Q_{in} = 0.052 \, \text{m}^3/\text{s}$, and the initial tank pressure is set at $P_0 = 1.15(10^5) \, \text{N/m}^2$, which is less than the nominal pressure. Therefore, the (constant) perturbation flow rate is $\delta Q_{in} = 0.002 \, \text{m}^3/\text{s}$ and the initial perturbation pressure is $\delta P_0 = -1955 \, \text{N/m}^2$. Figure 6.20 presents the tank pressure responses from the nonlinear and linear Simulink

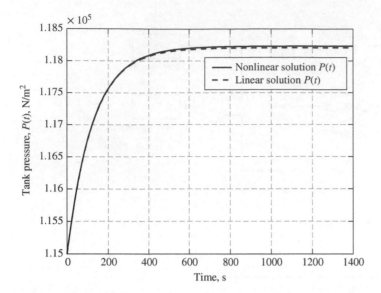

Figure 6.20 Pressure responses for nonlinear and linear hydraulic tank models (Example 6.8).

models. The linear solution accurately matches the true nonlinear solution for tank pressure and underestimates the final (steady) pressure by less than $25\,N/m^2$. The linear solution is accurate because the perturbations are relatively small: the flow-rate perturbation is within 4% of its nominal value, while the initial pressure perturbation is less than 2% of its nominal value.

6.6 BUILDING INTEGRATED SYSTEMS

Most physical, real-world engineering systems are mixed-discipline systems; that is, they are composed of interconnected mechanical, electrical, and fluid subsystems. A good example is a hydraulic or pneumatic actuator: a solenoid coil (an electrical subsystem) is used to actuate a valve (a mechanical subsystem), which meters fluid flow into a cylinder (a hydraulic or pneumatic subsystem), which provides a force to move a piston and load mass (a second mechanical subsystem). Each physical subsystem may be modeled by linear or nonlinear ODEs, and the output of one subsystem (fluid pressure, for example) will be the input to another subsystem (such as a mechanical piston actuator). It is not difficult to imagine that building a Simulink diagram of a complex, interconnected system may become unwieldy and intractable with a web of crossing signal paths.

Fortunately, there is a "clean" method for constructing complex systems by using interconnected subsystem blocks. Figure 6.21 presents a functional block diagram of an *integrated system* consisting of three subsystems, each with their own input and output variables. Note that in this general example, subsystem 1 has one input u and two output variables, y_1 (which is an input to subsystem 3) and y_2 (which is an input to subsystem 2). Subsystem 2 has one output variable y_3, which is an input to subsystem 3. Finally, subsystem 3 has two output variables, y_4 and y_5 (which is an input to subsystem 2). Each subsystem will have its own modeling equations, which may be linear or nonlinear, and its own set of "internal" state variables that are contained within the subsystem block (the individual system output variables will be functions of the state variables).

Simulink has two methods for constructing subsystem blocks: (1) using the `Subsystem` block from the `Ports & Subsystems` library and (2) grouping an existing Simulink diagram into a subsystem. For the first method, the user drags and drops the `Subsystem` icon from the `Ports & Subsystems` library to the workspace, double-clicks on the subsystem block, and then constructs the subsystem model using the typical Simulink blocks (integrators, gains, summing junctions, etc.). The default `Subsystem` block has one input and one output, and the user can add inputs and outputs by selecting the `In1` and `Out1` icons from the `Ports & Subsystems` library. The second method involves constructing the subsystem model first (such as Fig. 6.12) and then selecting the desired blocks and connecting signal paths with a bounding box (click and hold outside of the block diagram, drag the cursor across the diagram, release the mouse button). When the desired blocks and signal paths are selected, use the `Diagram` menu to select `Subsystem &`

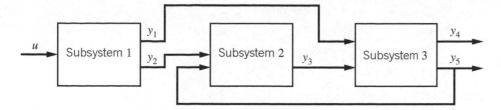

Figure 6.21 Functional block diagram of an integrated system.

`Model Reference > Create Subsystem from Selection` to construct a subsystem. The user can open the subsystem to see and edit the inner model by double-clicking the subsystem block.

Building an integrated system using Simulink is best demonstrated by an example. The following example presents the electromechanical solenoid previously described in Chapters 2, 3, and 5.

Example 6.9

Figure 6.22 shows the solenoid actuator–valve system described in Chapters 2, 3, and 5. Construct an integrated model using Simulink and determine the response of the solenoid coil current $I(t)$ and valve position $x(t)$ for a 10-V constant voltage input $e_{in}(t)$ applied at time $t = 0.05$ s. The system has zero stored energy at time $t = 0$.

Recall that the electromechanical system consists of a solenoid coil circuit and a valve mass constrained by a return spring. For this example, let us assume that the inductance L of the solenoid coil is constant and that the "back-emf" and force coefficient $K(= dL/dx)$ is also constant. Let us also assume that the return spring does not include a preload force and that the armature–valve mass m experiences viscous and dry friction effects. The complete mathematical model of the electromechanical system is

$$L\dot{I} + RI = e_{in}(t) - KI\dot{x} \tag{6.29}$$

$$m\ddot{x} + b\dot{x} + F_{dry}\,\mathrm{sgn}(\dot{x}) + kx = F_{em} \tag{6.30}$$

The electromagnetic force F_{em} is a nonlinear function of current, I

$$F_{em} = 0.5KI^2 \tag{6.31}$$

First, we can plan the structure of the integrated system by identifying the various input and output variables. The first-order electrical model, Eq. (6.29), involves a single state variable (current I), and two input variables, source voltage $e_{in}(t)$ and armature–valve velocity $\dot{x}$, for the back-emf term. The second-order mechanical model, Eq. (6.30), involves two state variables (position and velocity) and requires one input variable (force F_{em}). Because we want to determine the dynamic response of the valve position, x is an output variable of the mechanical model. Because velocity is needed as an input to the electrical model, it is the second output variable of the mechanical model. Figure 6.23 shows a functional block diagram of the integrated system, where the electrical and mechanical subsystem blocks represent the modeling equations (6.29) and (6.30), respectively.

Next, we construct each subsystem model, just as we constructed Simulink block diagrams in the previous examples. Figure 6.24 presents the electrical subsystem model, which matches the mathematical modeling equation (6.29). Note that the gain blocks do not contain numerical values but rather use variable names (R, L, and K) that are set to the respective numerical parameters in the MATLAB workspace before executing the simulation. The simulation diagram shows that the input voltage $e_{in}(t)$, resistor voltage RI, and back-emf voltage $KI\dot{x}$ are summed together (with the proper signs) and divided by inductance L to produce the time-rate of current, $\dot{I}$, which is integrated to produce current I. The electromagnetic force F_{em} is computed by squaring current I and multiplying by $K/2$; see Eq. (6.31). Note that current and force are saved to the workspace so that they may be plotted after the simulation is executed. The two inputs, $e_{in}(t)$ and $\dot{x}$, are designated by the In1 and In2 input port icons from the `Ports & Subsystems` library (they are relabeled as e_in and xdot), and the single output F_{em} is designated by the Out1 output port icon, also from the `Ports & Subsystems` library (relabeled as F_em). After the block diagram of the electrical system in Fig. 6.24 is completed, it is enclosed into a group and the corresponding subsystem block shown in Fig. 6.25 is created using the method previously described. Double-clicking the

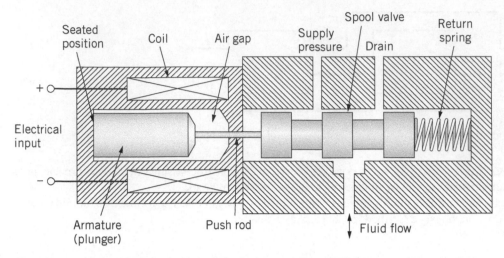

Figure 6.22 Solenoid actuator–valve system for Example 6.9.

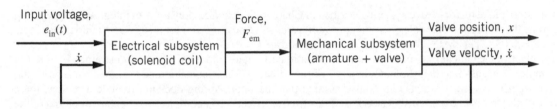

Figure 6.23 Functional block diagram for the solenoid actuator (Example 6.9).

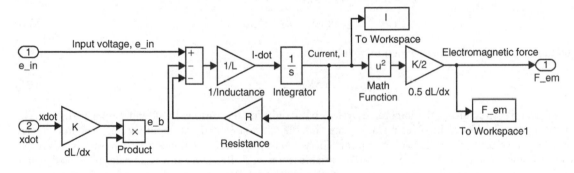

Figure 6.24 Simulink diagram for Example 6.9: electrical subsystem.

subsystem block in Fig. 6.25 will open the subsystem and display the detailed block diagram of the electrical model shown in Fig. 6.24.

The same procedure is applied to the mechanical modeling equation (6.30). Figure 6.26 presents the mechanical subsystem model, which also uses variables in the gain blocks (such as 1/m, b, k, and F_dry) instead of fixed numerical values. The electromagnetic force (the single input variable), friction forces, and return-spring force are summed together to create the net force. The net force is divided by mass m and then integrated (twice) to produce velocity and position, which are the two output variables of the subsystem. The signum (or Sign) function from the Math Operations library operates on velocity in order to compute the dry friction force. After the mechanical system in Fig. 6.26 is completed, the corresponding subsystem block shown in Fig. 6.27 is created. Double-clicking the subsystem block in Fig. 6.27 will open and display the detailed block diagram of the mechanical model shown in Fig. 6.26.

The integrated electromechanical system is constructed by properly connecting the electrical and mechanical Simulink subsystems (Figs. 6.25 and 6.27), and Fig. 6.28 presents the complete Simulink model (because the input and output ports have been labeled inside each subsystem these labels appear in Fig. 6.28). Note that the Simulink model in Fig. 6.28 matches

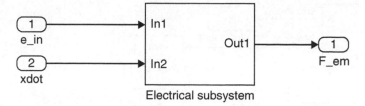

Figure 6.25 Simulink subsystem for Example 6.9: electrical subsystem.

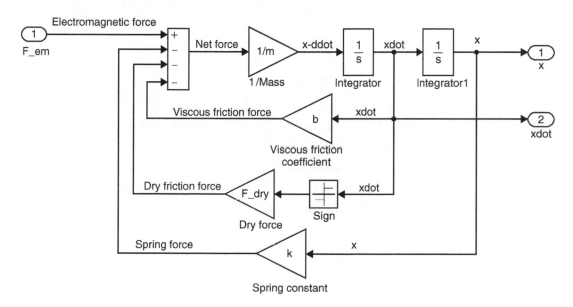

Figure 6.26 Simulink diagram for Example 6.9: mechanical subsystem.

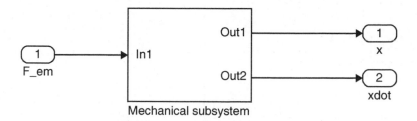

Figure 6.27 Simulink subsystem for Example 6.9: mechanical subsystem.

the functional block diagram shown in Fig. 6.23. A `Step` block is used to produce the 10-V input $e_{in}(t)$, and the input block is modified so that the step occurs at time $t = 0.05$ s.

We can now simulate the integrated system. Table 6.1 presents the numerical values of the system parameters for the solenoid actuator. All integrators in the subsystem models are initialized with values of zero. MATLAB M-file 6.3 sets the required parameters and executes the Simulink model `integrated_EMA.mdl` (Fig. 6.28) using the `sim` command (note that "EMA" denotes "electromagnetic actuator"). This M-file also plots the desired dynamic variables, current $I(t)$ and armature–valve position $x(t)$ (note that the M-file converts position to units of millimeters for the plot).

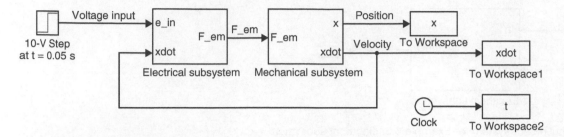

Figure 6.28 Simulink diagram for Example 6.9: integrated solenoid actuator system.

Table 6.1 System Parameters for the Solenoid Actuator (Example 6.9)

System Parameter	Value
Coil resistance, R	$3\,\Omega$
Coil inductance, L	$0.005\,H$
Force/back-emf constant, K	$6\,N/A^2$
Armature−valve mass, m	$0.03\,kg$
Viscous friction coefficient, b	$12\,N\text{-}s/m$
Spring constant, k	$6000\,N/m$
Dry friction force, F_{dry}	$0.5\,N$

MATLAB M-file 6.3

```
%
%   run_EMA.m
%
%   This M-file sets the modeling parameters for the
%   electromagnetic actuator (EMA), executes the
%   Simulink model, and plots the dynamic variables
%


% Solenoid electrical circuit parameters
e_in = 10;              % step input voltage
R = 3;                  % coil resistance, Ohms
L = 0.005;              % coil inductance, H
K = 6;                  % dL/dx (back-emf and force constant), N/A^2

% Mechanical parameters
m = 0.03;               % armature-valve mass, kg
b = 12;                 % viscous friction coefficient, N-s/m
k = 6000;               % return spring constant, N/m
F_dry = 0.5;            % dry friction force, N

% run Simulink model
sim integrated_EMA

% plots
figure(1)
plot(t,1e3*x)
grid
xlabel('Time, s')
ylabel('Armature-valve position, x(t), mm')


figure(2)
plot(t,I)
grid
xlabel('Time, s')
ylabel('Solenoid current, I(t), A')
```

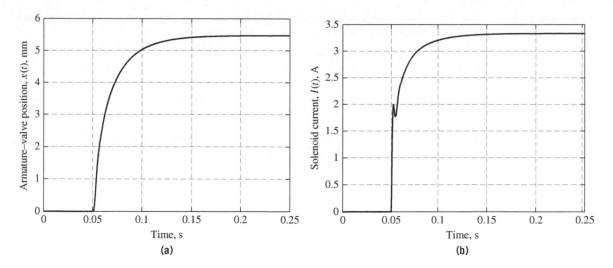

Figure 6.29 Response of the integrated solenoid actuator–valve system (Example 6.9): (a) armature–valve position and (b) solenoid current.

Figure 6.29a presents the armature–valve position $x(t)$ (in mm) for the 10-V step voltage input. Note that the valve rises from its seat at time $t = 0.05$ s when the voltage input is applied, and reaches a steady (constant) value of about 5.5 mm at time $t = 0.15$ s (i.e., 0.1 s after the voltage is applied). Figure 6.29b shows the current response $I(t)$ for the solenoid coil. Current $I(t)$ rises very quickly from zero when the voltage input is applied at $t = 0.05$ s and exhibits a brief peak at about 2 A, followed by a small decrease in current, and then a steady rise to a constant value of 3.333 A. The small decrease in current is due to the "back-emf" voltage term $e_b = KI\dot{x}$ that is induced by the large positive velocity of the armature as it is lifted off its seat. The back-emf voltage diminishes as the armature's velocity decreases. Hence, the current quickly begins to increase and rise to its steady value.

The purpose of this example is to demonstrate how to build integrated systems using subsystem blocks in Simulink. We revisit the solenoid actuator in Chapter 11, where we include the effects of nonlinear inductance $L(x)$ that depends on the armature position, a preloaded spring, and the wall-contact force.

SUMMARY

In this chapter, we have demonstrated how to use MATLAB to numerically simulate the dynamic response of a system. Two simulation methods were presented: the built-in MATLAB commands and the graphical software Simulink. We showed how to simulate dynamic systems with Simulink by using (1) transfer functions (2) a state-space representation (SSR) or (3) the integrator-block method. It is important for the reader to note that transfer functions and SSRs can be used only with linear models, while the integrator-block approach is likely the only option when faced with nonlinear models. Furthermore, transfer functions cannot be used for systems that involve nonzero initial conditions, while both the SSR and integrator-block approaches can handle initial conditions. We also demonstrated how to construct nonlinear models in Simulink and how to simplify the Simulink block diagram for complex systems by utilizing interconnected subsystem blocks. Chapter 11 presents engineering case studies in dynamic systems; these case studies rely heavily on Simulink simulations of complex, integrated systems.

REFERENCES

1. Whitehead, J.C., "Rear Wheel Steering Dynamics Compared to Front Wheel Steering," *ASME Journal of Dynamic Systems, Measurement, and Control*, Vol. 112, Mar. 1990, pp. 88–93.
2. O'Connor, D.N., Eppinger, S.D., Seering, W.P., and Wormley, D.N., "Active Control of a High-Speed Pantograph," *ASME Journal of Dynamic Systems, Measurement, and Control*, Vol. 119, Mar. 1997, pp. 1–4.
3. Dunne, J.F., "Dynamic Modelling and Control of Semifree-Piston Motion in a Rotary Diesel Generator Concept," *ASME Journal of Dynamic Systems, Measurement, and Control*, Vol. 132, Sept. 2010, pp. 051003/1–051003/12.

4. Genin, J., Ginsberg, J.H., Ting, E.C., "Longitude Train-Track Dynamics: A New Approach," *ASME Journal of Dynamic Systems, Measurement, and Control*, Vol. 96, Dec. 1974, pp. 466–469.

5. Wait, K.W., and Goldfarb, M., "Enhanced Performance and Stability in Pneumatic Servosystems with Supplemental Mechanical Damping," *ASME Journal of Dynamic Systems, Measurement, and Control*, Vol. 132, Jul. 2010, pp. 041012–1-8.

6. Pfafflin, J.R., "Space Heating Dynamics," *IEEE Transactions on Industry Applications*, Vol. IA-19, No. 5, 1983, pp. 844–847.

7. Langjord, H., and Johansen, T.A., "Dual-Mode Switched Control of an Electropneumatic Clutch Actuator," *IEEE/ASME Transactions on Mechatronics*, Vol. 15, No. 6, 2010, pp. 969–981.

Chapter 7

Analytical Solution of Linear Dynamic Systems

7.1 INTRODUCTION

So far, we have discussed modeling dynamic systems, the standard forms for representing system models, and obtaining the system's response using numerical simulation methods. In this chapter, we discuss how to obtain the system response using analytical techniques, that is, obtaining the solution of the governing ordinary differential equations (ODEs) "by hand."

The reader may wonder why it is important to obtain analytical solutions to differential equations when numerical packages such as the MATLAB and Simulink software are readily available. Dynamic systems and control engineers should know how to evaluate the response of first- and second-order systems by using "back-of-the-envelope" calculations as many real engineering systems can be adequately modeled by low-order, linear differential equations. Students of dynamic systems should be able to develop an understanding of how basic system parameters affect the system's response. For example, we see that varying the stiffness or damping parameters in a mechanical system affects the vibration frequency of the response and the time it takes the system to reach a steady-state response. Therefore, the goal of this chapter is not a full, rigorous treatment of linear differential equations and their solution; these topics are reserved for the prerequisite course in differential equations. Instead, the objective of this chapter is to develop an understanding of first- and second-order system responses for standard input functions such as the step and impulse. After completing this chapter, the reader should be able to predict the responses of first- and second-order systems by employing a few simple hand calculations.

7.2 ANALYTICAL SOLUTIONS TO LINEAR DIFFERENTIAL EQUATIONS

In this section, we present an overview of the solution of linear ODEs with constant coefficients. To begin, consider the general nth-order input–output (I/O) equation

$$a_n y^{(n)} + a_{n-1} y^{(n-1)} + \cdots + a_2 \ddot{y} + a_1 \dot{y} + a_0 y = b_m u^{(m)} + \cdots + b_1 \dot{u} + b_0 u \tag{7.1}$$

where $y^{(n)} \equiv d^n y/dt^n$ and $u^{(m)} \equiv d^m u/dt^m$. We can replace the right-hand side (the input terms) with a general "forcing function" $f(t)$ so that Eq. (7.1) becomes

$$a_n y^{(n)} + a_{n-1} y^{(n-1)} + \cdots + a_2 \ddot{y} + a_1 \dot{y} + a_0 y = f(t) \tag{7.2}$$

The complete or total solution to Eq. (7.2) has the general form

$$y(t) = y_H(t) + y_P(t) \tag{7.3}$$

where $y_H(t)$ is the called the *homogeneous solution* and $y_P(t)$ is called the *particular solution*. The homogeneous (or *complementary*) solution $y_H(t)$ is the solution to the differential equation (7.2) when the right-hand side (the input) is zero:

$$a_n y_H^{(n)} + a_{n-1} y_H^{(n-1)} + \cdots + a_2 \ddot{y}_H + a_1 \dot{y}_H + a_0 y_H = 0 \tag{7.4}$$

Equation (7.4) is the *homogeneous differential equation*. Euler noted that the homogeneous solution has the form $y_H(t) = ce^{rt}$, where c is a constant. Taking successive time derivative of this assumed solution form yields

$$\dot{y}_H(t) = rce^{rt}, \ \ddot{y}_H(t) = r^2 ce^{rt}, \ \dddot{y}_H(t) = r^3 ce^{rt}, \ \ldots \ y_H^{(n)}(t) = r^n ce^{rt}$$

After substituting these derivatives into Eq. (7.4), we obtain

$$(a_n r^n + a_{n-1} r^{n-1} + \cdots + a_2 r^2 + a_1 r + a_0)ce^{rt} = 0 \tag{7.5}$$

Because ce^{rt} cannot be zero for all time, the terms in the bracket in Eq. (7.5) must be zero. Therefore, we obtain the nth-order polynomial equation

$$a_n r^n + a_{n-1} r^{n-1} + \cdots + a_2 r^2 + a_1 r + a_0 = 0 \tag{7.6}$$

Equation (7.6) is called the *characteristic equation*, and its solution yields the *characteristic roots* $r_i, i = 1, 2, \ldots, n$. If all n roots are distinct or unique, then the homogeneous solution is

$$y_H(t) = c_1 e^{r_1 t} + c_2 e^{r_2 t} + c_3 e^{r_3 t} + \cdots + c_n e^{r_n t} \tag{7.7}$$

If we have two repeated roots (say $r_1 = r_2$), then the homogeneous solution is

$$y_H(t) = c_1 e^{r_1 t} + c_2 t e^{r_1 t} + c_3 e^{r_3 t} + \cdots + c_n e^{r_n t} \tag{7.8}$$

In either case, the n coefficients c_i are computed after the particular solution $y_P(t)$ is determined.

The particular solution $y_P(t)$ must satisfy the nonhomogeneous differential equation (7.2) and it can be found using the method of undetermined coefficients, where we assume a functional form for $y_P(t)$ that generally matches the forcing (or input) function $f(t)$ and its derivatives. For example, if the forcing function $f(t)$ is a constant, we can assume that the particular solution is also an undetermined constant. And, if the forcing function is $f(t) = \sin 4t$, we can assume that the particular solution is $y_P(t) = a \sin 4t + b \cos 4t$, with unknown amplitude coefficients a and b. The assumed solution form for $y_P(t)$ is substituted into the original differential equation (7.2) and the unknown coefficients are determined by equating the corresponding terms. After the particular solution is found, the unknown coefficients c_i for the homogeneous solution in Eq. (7.7) are determined by applying the known initial conditions of the output $y(t)$ and its derivatives at time $t = 0$:

$$y(0) = y_0, \ \dot{y}(0) = \dot{y}_0, \ \ddot{y}(0) = \ddot{y}_0, \ \ldots, \ y^{(n-1)}(0) = y_0^{(n-1)}$$

The following examples illustrate the general steps for solving low-order differential equations.

Example 7.1

Consider the first-order linear differential equation with initial condition $y(0) = 3$.

$$4\dot{y} + 8y = 6$$

Determine the complete solution $y(t)$.

First, we determine the characteristic equation by equating the terms between Eqs. (7.6) and (7.2). For this first-order system, we get a first-order characteristic equation

$$4r + 8 = 0$$

and the single characteristic root is $r = -8/4 = -2$. Therefore, the homogeneous solution has the form $y_H(t) = ce^{-2t}$. Next, we determine the particular solution. Because the forcing function (right-hand side) is $f(t) = 6$, we assume that the particular solution is also a constant, $y_P(t) = a$. Substituting $y_P(t) = a$ and $\dot{y}_P(t) = 0$ into the original differential equation yields $8a = 6$, and therefore the particular solution is $y_P(t) = a = 6/8 = 0.75$. Finally, the complete solution is the sum of the homogeneous and particular solutions, or

$$y(t) = y_H(t) + y_P(t) = ce^{-2t} + 0.75$$

The final step is to determine the coefficient c from the known initial condition, $y(0) = 3$:

$$y(0) = ce^0 + 0.75 = 3$$

which yields $c = 3 - 0.75 = 2.25$. Therefore, the complete solution of the differential equation is

$$y(t) = 2.25e^{-2t} + 0.75$$

Example 7.2

Determine the complete solution $y(t)$ of the following second-order linear differential equation

$$2\ddot{y} + 8\dot{y} + 6y = 10\sin 4t \tag{7.9}$$

with initial conditions $y(0) = 2$ and $\dot{y}(0) = -1$.

First, we write the second-order characteristic equation for this differential equation

$$2r^2 + 8r + 6 = 0$$

This characteristic equation can be factored as $2(r^2 + 4r + 3) = 2(r + 1)(r + 3) = 0$, and therefore the two roots are $r_1 = -1$ and $r_2 = -3$. Hence, the homogeneous solution has the form

$$y_H(t) = c_1 e^{-t} + c_2 e^{-3t} \tag{7.10}$$

Because the particular solution must have the same form as the forcing function $f(t) = 10\sin 4t$, we can select

$$y_P(t) = a\sin 4t + b\cos 4t \tag{7.11}$$

The successive derivatives of the particular solution are $\dot{y}_P = 4a\cos 4t - 4b\sin 4t$ and $\ddot{y}_P = -16a\sin 4t - 16b\cos 4t$. Substituting for $y_P(t)$ and its derivatives in the original differential equation (7.9) yields

$$2(-16a\sin 4t - 16b\cos 4t) + 8(4a\cos 4t - 4b\sin 4t) + 6(a\sin 4t + b\cos 4t) = 10\sin 4t$$

After equating the $\cos 4t$ and $\sin 4t$ terms from both sides of the equal sign, we obtain

$$\cos 4t \text{ terms: } -32b + 32a + 6b = 0$$
$$\sin 4t \text{ terms: } -32a - 32b + 6a = 10$$

Solving these two equations for the two unknowns yields $a = -0.1529$, $b = -0.1882$. The complete solution is the sum of the homogeneous solution Eq. (7.10) and the particular solution Eq. (7.11)

$$y(t) = c_1 e^{-t} + c_2 e^{-3t} - 0.1529\sin 4t - 0.1882\cos 4t \tag{7.12}$$

Applying the first initial condition $y(0) = 2$ to Eq. (7.12) yields

$$y(0) = c_1 e^0 + c_2 e^0 - 0.1882 = 2 \tag{7.13}$$

The first time derivative of Eq. (7.12) is

$$\dot{y}(t) = -c_1 e^{-t} - 3c_2 e^{-3t} - 0.6118 \cos 4t + 0.7529 \sin 4t \qquad (7.14)$$

Applying the second initial condition $\dot{y}(0) = -1$ to Eq. (7.14) yields

$$\dot{y}(0) = -c_1 e^0 - 3c_2 e^0 - 0.6118 = -1 \qquad (7.15)$$

We obtain the unknown coefficients from the simultaneous solution of Eqs. (7.13) and (7.15), and the result is $c_1 = 3.0882$ and $c_2 = -0.9000$. Finally, the complete solution is determined from Eq. (7.12)

$$y(t) = 3.0882e^{-t} - 0.9000e^{-3t} - 0.1529 \sin 4t - 0.1882 \cos 4t$$

As a final check, we see that the initial conditions

$$y(0) = 3.0882 - 0.9 - 0.1882 = 2$$

$$\dot{y}(0) = -3.0882 - (-3)(0.9) - (4)(0.1529) = -1$$

are satisfied as expected.

The Complete Response

As demonstrated by the two previous examples, the complete solution to the linear time-invariant (LTI) differential equation (7.2) comprises the homogeneous solution $y_H(t)$ plus the particular solution $y_P(t)$. We can call $y_H(t)$ the *natural* or *free response* as its form is obtained by solving the homogeneous differential equation that results from setting the forcing function $f(t)$ equal to zero. Therefore, the natural response depends on the system's "natural dynamics" that are revealed by the constant coefficients a_i in the governing differential equation (7.2). The reader can see that the a_i coefficients in the differential equation (7.2) directly lead to the characteristic equation (7.6), and the roots of the characteristic equation determine the form of the natural response $y_H(t)$. The particular solution $y_P(t)$ is also called the *forced response* because it depends on the form of the forcing function $f(t)$, or right-hand side of the differential equation (7.2). It should be emphasized that the unknown coefficients c_i of the natural response $y_H(t)$ are determined from the known initial conditions only after the forced response $y_P(t)$ has been determined.

Another way to categorize the complete response $y(t)$ is to break it into the *transient response* and *steady-state response*. The transient response can be defined as the part of the complete response that goes to zero as time t approaches infinity. The steady-state response is the part of the complete response that remains as t approaches infinity. Figure 7.1 shows a generic total response $y(t)$ that initially begins at zero, and exhibits an oscillatory transient response that eventually "dies out" and transitions to a constant steady-state response. The total or complete response is the sum of the transient and steady-state responses.

Characteristic Roots and the Transfer Function

The previous subsections have discussed the solution to LTI differential equations, and we have seen that the homogeneous or free response $y_H(t)$ depends on the roots of the characteristic equation (7.6). It is important for the reader to understand that the roots of the characteristic equation can also be determined from the corresponding system transfer function. Recall that the transfer function $G(s)$ is the ratio of two polynomials in the complex variable s, that is, $G(s) = b(s)/a(s)$. The values of s that make the denominator polynomial $a(s)$ equal zero are called the *poles* of the transfer function, and the poles are identical to the roots of the characteristic equation. The following example demonstrates that the characteristic roots are the same as the poles of the transfer function, and that both are easily derived from the I/O equation.

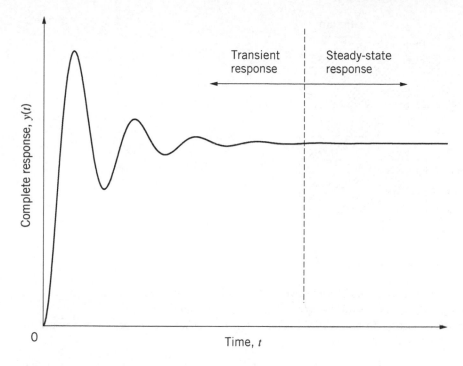

Figure 7.1 Transient and steady-state responses.

Example 7.3

Figure 7.2 shows a three-way spool valve used to meter flow in a hydraulic system. Given the following I/O equation for the valve, determine (a) the characteristic equation and characteristic roots and (b) the transfer function and poles of the transfer function.

$$0.04\ddot{y} + 16\dot{y} + 7000y = f(t) \tag{7.16}$$

Equation (7.16) is the governing mathematical model of the spool valve, which consists of a single mass, a linear viscous friction force, and a linear spring force. Variable $y(t)$ is the displacement of the spool valve, and $f(t)$ is the force from an electromagnetic actuator (i.e., solenoid) that can push on the valve. We assume that there is no hydraulic fluid pressure imbalance on the valve, and that flow forces are neglected; hence, the electromagnetic force $f(t)$ is the sole applied force acting on the valve mass.

We see that by comparing the general I/O equation (7.1) with its corresponding characteristic equation (7.6), the characteristic equation of the valve model (7.16) can be written by using the left-hand side a_i coefficients

$$0.04r^2 + 16r + 7000 = 0 \tag{7.17}$$

Equation (7.17) is the characteristic equation of the valve model. In summary, the characteristic equation of an I/O equation is an nth-order polynomial in r, where the nth derivative $y^{(n)}$ is replaced by r^n, $y^{(n-1)}$ is replaced by r^{n-1}, etc. The characteristic roots are determined by setting the characteristic equation to zero. We can use MATLAB's `roots` command to determine the roots of this second-order polynomial:

```
>> r = roots([0.04 16 7000])
```

where the row vector `[0.04 16 7000]` contains the a_i coefficients of the characteristic polynomial in descending order from the second- to zeroth-order terms. The two characteristic roots are

$$r_1 = -200 + j367.42 \qquad r_2 = -200 - j367.42$$

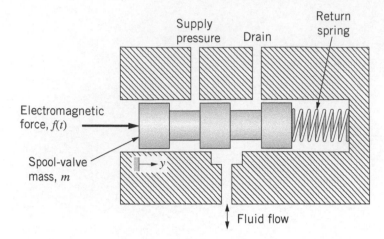

Figure 7.2 Three-way spool valve for Example 7.3.

These roots are called *complex conjugates* as they have the same real part (-200), but the imaginary parts ($j367.42$ and $-j367.42$) have equal magnitudes but opposite signs (we use $j = \sqrt{-1}$ as the imaginary number).

Next, we derive the system transfer function from the I/O equation (7.16) by using the D operator to replace the derivative terms, that is, $\ddot{y} = D^2 y$ and $\dot{y} = Dy$, which yields

$$(0.04D^2 + 16D + 7000)y = f(t) \tag{7.18}$$

Next, form the ratio of output to input, $y(t)/f(t)$, and replace D with the Laplace variable s to derive the transfer function $G(s)$

$$G(s) = \frac{1}{0.04s^2 + 16s + 7000} \tag{7.19}$$

Equation (7.19) is the transfer function that represents the spool-valve system. For this case, the numerator of $G(s)$ is 1, and the denominator of $G(s)$ is the polynomial $0.04s^2 + 16s + 7000$. The poles of the transfer function $G(s)$ are determined by setting the denominator polynomial to zero

$$0.04s^2 + 16s + 7000 = 0 \tag{7.20}$$

Equation (7.20) is identical to the characteristic equation (7.17), and hence the poles of the transfer function are

$$s_1 = -200 + j367.42 \qquad s_2 = -200 - j367.42$$

which are the same as the characteristic roots.

We can use MATLAB's `pole` command to determine the poles of the transfer function $G(s)$. First, we must construct the system transfer function using the `tf` command as follows:

```
>> numG = 1;                    % numerator of G(s) – see Eq. (7.19)
>> denG = [0.04 16 7000];       % denominator of G(s) – see Eq. (7.19)
>> sysG = tf(numG,denG)         % define sysG as transfer function G(s)
```

Note that the row vector `denG = [0.04 16 7000]` contains the coefficients of the denominator of transfer function $G(s)$ in descending powers of s. The MATLAB `pole` command below computes the poles of the transfer function defined as `sysG`.

```
>> p = pole(sysG)               % compute poles of G(s)
```

Executing these MATLAB line commands yields the vector of poles p with components $p_1 = -200 + j367.42$ and $p_2 = -200 - j367.42$, which is the same result previously obtained in this example.

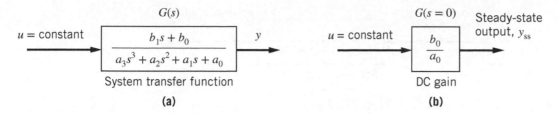

Figure 7.3 SISO system: (a) system transfer function and (b) system gain at steady state (DC gain).

DC Gain

The *DC gain* is a useful analysis technique for computing a system's steady-state response to a constant input. The name arises from circuit analysis, where "direct current" or DC implies a constant, nonoscillatory input, as opposed to "alternating current" or AC. The definition of the system DC gain is the steady-state gain to a constant input, for the case when the output reaches a constant steady-state value. The DC gain can be computed from the system's transfer function by setting the complex variable $s = 0$, which is a consequence of the final-value theorem in Laplace transform theory (see Section 8.2 for details). However, we can show the concept of the DC gain using the differential or D operator method instead of using Laplace transforms.

For example, consider the third-order, linear I/O equation

$$a_3 \ddot{y} + a_2 \ddot{y} + a_1 \dot{y} + a_0 y = b_1 \dot{u} + b_0 u \tag{7.21}$$

If the input $u(t)$ is a constant, and if the output $y(t)$ reaches a constant steady-state value, then all derivative terms of the input and output variables go to zero at steady state, that is, when $t \to \infty$. In the case of constant input ($\dot{u} = 0$) and constant output at steady state ($\dddot{y}(\infty) = \ddot{y}(\infty) = \dot{y}(\infty) = 0$), we can easily solve Eq. (7.21) for the steady-state output

$$y_{ss} = \frac{b_0}{a_0} u \tag{7.22}$$

The ratio b_0/a_0 in Eq. (7.22) acts as a constant gain at steady state, and thus it is the DC gain for this case.

We can show the same result using the D operator method, which we apply to the I/O equation (7.21) to obtain the ratio of output to input

$$\frac{y(t)}{u(t)} = \frac{b_1 D + b_0}{a_3 D^3 + a_2 D^2 + a_1 D + a_0} \tag{7.23}$$

If we set $D = 0$ in Eq. (7.23), that is, constant input and constant output, this expression yields the steady-state output $y_{ss} = b_0 u/a_0$, which is the same result as Eq. (7.22). Finally, we can derive the system transfer function from Eq. (7.23) by replacing D with s

$$G(s) = \frac{b_1 s + b_0}{a_3 s^3 + a_2 s^2 + a_1 s + a_0} \tag{7.24}$$

We obtain the DC gain by setting $s = 0$ in Eq. (7.24), which is the same result obtained by steady-state analysis of the I/O equation, or by use of the D operator method. Figure 7.3a shows the single-input, single-output (SISO) system Eq. (7.21) in a block-diagram format using transfer function $G(s)$. Figure 7.3b shows that when the input is constant, the DC gain can be used to compute the steady-state output y_{ss}.

Example 7.4

Figure 7.4 presents the block diagram of a simplified solenoid actuator–valve system (the spool valve is shown in Fig. 7.2). If the input voltage $e_{in}(t)$ is a constant 2 V, determine the steady-state electromagnetic force f_{ss} and steady-state valve position y_{ss}.

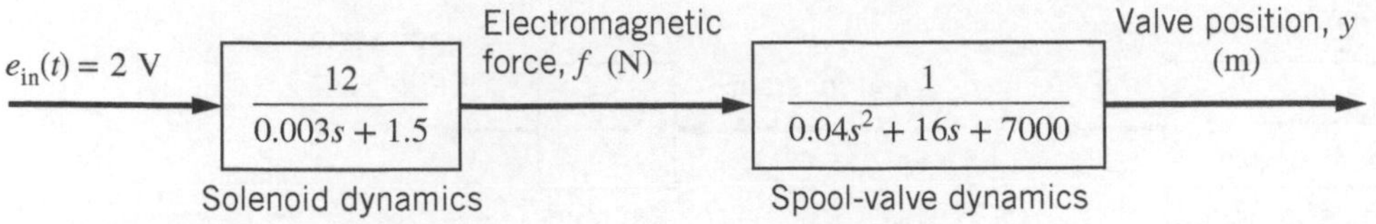

Figure 7.4 Solenoid actuator and spool valve for Example 7.4.

In Chapter 3, we developed a complex model of the solenoid actuator that involved a nonlinear relationship between current and electromagnetic force and armature motion and induced voltage (the "back emf"). In this example, we use a linear current–force relationship and neglect the motion-induced voltage, and hence the electromagnetic force can be modeled as the response of a first-order system (an RL coil circuit) to an applied voltage, as shown in Fig. 7.4. The spool valve is composed of a single mass with a return spring and viscous friction. Its transfer function in Fig. 7.4 is therefore second-order and linear. The solenoid is modeled by a first-order transfer function with source voltage $e_{in}(t)$ as the input and force f as the output, while the spool valve is modeled by a second-order transfer function with force f as the input and valve position y as the output.

The DC gain of the solenoid transfer function is obtained by setting $s = 0$, which results in a gain of $12/1.5 = 8$. Therefore, the steady-state force is $f_{ss} = 2 \text{ V} \cdot 8 = 16 \text{ N}$. The DC gain of the spool-valve transfer function is also obtained by setting $s = 0$, which results in a gain of $1/7000 = 1.4286(10^{-4})$. The steady-state valve position is computed using the steady-state force f_{ss} as the constant input to the spool-valve transfer function, which yields $y_{ss} = 16 \text{ N} \cdot 1.4286(10^{-4}) = 0.0023 \text{ m}$, or 2.3 mm.

The steady-state valve position can also be obtained by multiplying the solenoid and spool-valve transfer functions to obtain a single transfer function relating input voltage $e_{in}(t)$ to valve position y:

$$\frac{12}{(0.003s + 1.5)(0.04s^2 + 16s + 7000)} = \frac{Y(s)}{E_{in}(s)}$$

The DC gain of this total transfer function is $12/(1.5 \cdot 7000) = 0.001143$, and therefore the steady-state valve position is $y_{ss} = 2 \text{ V} \cdot 0.001143 = 0.002286 \text{ m}$, or 2.3 mm, as before.

7.3   FIRST-ORDER SYSTEM RESPONSE

Recall that many of the mathematical models we derived in Chapters 2–4 for physical engineering systems were first-order differential equations. Table 7.1 summarizes several examples of systems represented by first-order I/O equations. In the case of the electrical, pneumatic, and thermal systems in Table 7.1, we obtain a first-order model because each system has only one energy-storage element, namely, a single electrical capacitance, fluid capacitance, or thermal capacitance element, all denoted by the symbol C. Furthermore, the electrical, pneumatic, and thermal systems in Table 7.1 all contain a single resistance element (electrical resistor, fluid resistance, or thermal resistance), all denoted by symbol R. The mechanical rotor example in Table 7.1 can be written as a first-order system only if angular position θ does not appear in the modeling equation.

Note that all first-order models of the physical systems in Table 7.1 have the same standard form

$$\tau \dot{y} + y = bu \tag{7.25}$$

Table 7.1   Examples of Engineering Systems Modeled by a First-Order Input–Output Equation

Physical System	Input Variable	Output Variable	Modeling Equation
Mechanical rotor with friction	Torque, $T_{in}(t)$	Angular velocity, ω	$J\dot{\omega} + b\omega = T_{in}(t)$
Electrical series RC circuit	Voltage source, $e_{in}(t)$	Capacitor voltage, e_C	$RC\dot{e}_C + e_C = e_{in}(t)$
Pneumatic vessel with laminar valve	Pressure source, $P_{in}(t)$	Vessel pressure, P	$RC\dot{P} + P = P_{in}(t)$
Thermal chamber	Ambient temperature, $T_a(t)$	Chamber temperature, T	$RC\dot{T} + T = T_a(t)$

Table 7.2 Output Variable and Constant τ for the First-Order Systems of Table 7.1

Physical System	Output Variable	Constant τ
Mechanical rotor with friction	Angular velocity, ω	J/b
Electrical series RC circuit	Capacitor voltage, e_C	RC
Pneumatic vessel with laminar valve	Vessel pressure, P	RC
Thermal chamber	Chamber temperature, T	RC

where y is the dynamic or output variable of interest, u is the input variable, and b is the right-hand side coefficient. The constant τ in Eq. (7.25) must have units of time so that the dimensions of the first term on the left-hand side ($\tau \dot{y}$) match the dimensions of the second term (y). Table 7.2 defines the output variable and constant τ for every first-order model presented in Table 7.1. The constant τ is the product of the resistance and capacitance elements for the electrical, pneumatic, and thermal systems. We characterize the first-order system response in terms of the constant τ, so it is important for the reader to note that we can write *any* linear, first-order model in the "standard form" of Eq. (7.25).

First-Order Response with Zero Input

To begin the analysis of a first-order system response, we consider the case with zero input, which leads to the homogeneous differential equation

$$\tau \dot{y} + y = 0 \tag{7.26}$$

The characteristic equation can be obtained by inspection

$$\tau r + 1 = 0 \tag{7.27}$$

and therefore the single characteristic root is $r = -1/\tau$. The homogeneous solution is an exponential function

$$y_H(t) = ce^{rt} = ce^{-t/\tau} \tag{7.28}$$

For the case with zero input, there is no particular solution $y_P(t)$, so the homogeneous solution $y_H(t)$ is the total or complete solution. Furthermore, the constant c is determined by applying the initial condition at time $t = 0$, so $y(0) = ce^0 = c = y_0$. Hence, the total response of the first-order system for the case of zero input is

$$y(t) = y_0 e^{-t/\tau} \tag{7.29}$$

which is either an exponentially decaying or a divergent function, depending on the sign of the characteristic root, r. If root $r < 0$ (or, constant $\tau > 0$), then the solution $y(t)$ decays from its initial condition y_0 to zero as time $t \to \infty$ and the response is bounded or stable. If $r > 0$, then $y(t)$ diverges to infinity as time $t \to \infty$ and the response is unbounded or unstable. We see from Tables 7.1 and 7.2 that the constant τ is always positive for these physical systems, and hence the free response Eq. (7.29) will be an exponential decay to zero at steady state. The parameter τ is called the *time constant* for the first-order system.

Figure 7.5 presents the free or natural response of a general first-order system with zero input and initial condition y_0. Note that when time $t = \tau$, the free response has decayed to 36.8% of its initial value because $e^{-1} = 0.368$. At time $t = 4\tau$, the free response has decayed to less than 2% of its initial value as $e^{-4} = 0.018$, and therefore we can say that the first-order free response has essentially reached the zero steady-state value as shown in Fig. 7.5. Clearly, the time constant τ determines the response time of the first-order free response.

Step Response of a First-Order System

Next, we consider the response of a first-order system with a constant or step input with magnitude A; that is, $u(t) = AU(t)$. Therefore, our general first-order model Eq. (7.25) becomes

$$\tau \dot{y} + y = bA \tag{7.30}$$

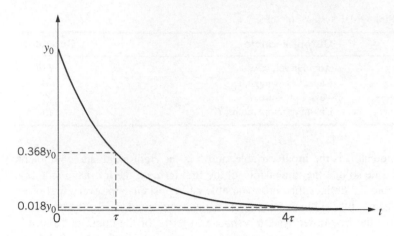

Figure 7.5 Free response of the first-order model $\tau\dot{y} + y = 0$.

The total response to the step input (that is, the *step response*) is

$$y(t) = y_H(t) + y_P(t) \tag{7.31}$$

where the homogeneous solution was found to be $y_H(t) = ce^{-t/\tau}$. The particular solution $y_P(t)$ must satisfy Eq. (7.30) with constant input bA, and we see that $y_P(t) = bA$ is the solution. The step response Eq. (7.31) is rewritten as

$$y(t) = ce^{-t/\tau} + bA \tag{7.32}$$

Because the homogeneous solution $ce^{-t/\tau}$ goes to zero as $t \to \infty$, the steady-state response is $y(\infty) = y_{ss} = bA$, which depends on the coefficient b and the magnitude of the step input A. Finally, we solve for the coefficient c by applying the initial condition $y(0) = y_0$ to Eq. (7.32), which yields

$$y(0) = ce^0 + y_{ss} = y_0 \tag{7.33}$$

and hence $c = y_0 - y_{ss}$. The step response is

$$y(t) = (y_0 - y_{ss})e^{-t/\tau} + y_{ss} \tag{7.34}$$

We see that the first term of the step response is the *transient response* as it dies out over time, and that the second term is the *steady-state response* because it remains as $t \to \infty$. We define t_S as the *settling time*, or the time it takes to reach steady state, and we can estimate settling time for a first-order system as four time constants, or $t_S = 4\tau$.

In many instances, we may simply wish to sketch the step response of a first-order system. The following steps will lead to an accurate step-response sketch:

1. Rewrite the first-order model in the "standard form" of Eq. (7.25) and compute the time constant, τ.
2. Estimate the settling time as four time constants, that is, $t_S = 4\tau$.
3. Compute the steady-state response for the step input $u(t) = AU(t)$, that is, $y_{ss} = bA$.
4. Sketch an exponential response from the known initial condition y_0 to the steady-state value y_{ss}. The response approximately reaches y_{ss} at $t = t_S$. The total response will "decay down" to steady state if $y_0 > y_{ss}$, or show an "exponential rise" to steady state if $y_{ss} > y_0$.

Example 7.5

Consider again the solenoid actuator dynamics from Example 7.4, as shown in Fig. 7.4. Sketch the step response of the first-order solenoid model for a step voltage input, $e_{in}(t) = 2$ V. The system is initially at rest at time $t = 0$, or $f(0) = 0$.

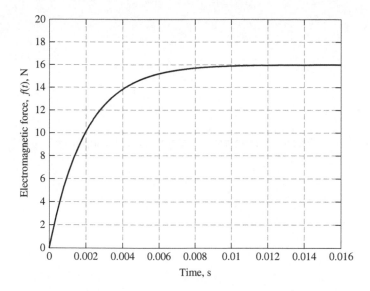

Figure 7.6 Step response of the solenoid actuator for Example 7.5.

We can derive the first-order differential equation of the solenoid model from the transfer function shown in Fig. 7.4, which relates voltage input $e_{in}(t)$ to electromagnetic force f, and the result is

$$0.003\dot{f} + 1.5f = 12e_{in}(t) \tag{7.35}$$

Following our procedure for sketching the first-order step response, we rewrite the first-order model in our standard form by dividing Eq. (7.35) by 1.5

$$0.002\dot{f} + f = 8e_{in}(t) \tag{7.36}$$

Therefore, the time constant is identified as $\tau = 0.002$ s. The solenoid will reach its steady-state force output in approximately four time constants, or $t_S = 4\tau = 0.008$ s. From Eq. (7.36), we see that the force reaches steady state for a 2-V step input when $\dot{f} = 0$, and therefore $f_{ss} = 8 \cdot 2$ V $= 16$ N. Finally, the step response $f(t)$ can be sketched by drawing an exponential rise from the given initial force $f(0) = 0$ to the steady-state force $f_{ss} = 16$ N at settling time $t_S = 0.008$ s. Figure 7.6 presents the step response of the solenoid actuator for a 2-V step input.

Pulse Response of a First-Order System

Recall that a pulse input consists of a constant input that lasts for a finite duration, and then instantly steps to zero. Therefore, a pulse input with magnitude P can be described as

$$u(t) = \begin{cases} 0 & \text{for} \quad t \leq 0 \\ P & \text{for} \ 0 < t < T \\ 0 & \text{for} \quad t \geq T \end{cases} \tag{7.37}$$

Figure 7.7 shows a first-order system with a pulse input of magnitude P, where we have used the standard form Eq. (7.25) for the first-order transfer function (because we are using a transfer function to depict the system, the initial output $y(0)$ is zero). Before we obtain the mathematical solution of the pulse response, we can use the previous results for the zero-input and step-input cases to characterize the output. If the pulse duration T is greater than the settling time of the first-order system, that is, $T > 4\tau$, then the initial part of the pulse response will simply "look" like a step response, and the output will initially exhibit an exponential rise to a steady value. When the pulse input is stepped to zero at $t = T$, the latter part of the pulse response (i.e., $t > T$) will "look" like the free response shown in Fig. 7.5, and the output will eventually decay to zero in four time constants.

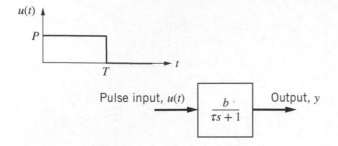

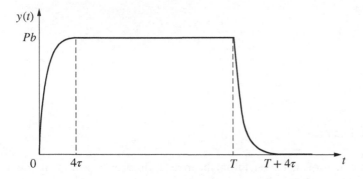

Figure 7.7 First-order system with pulse input.

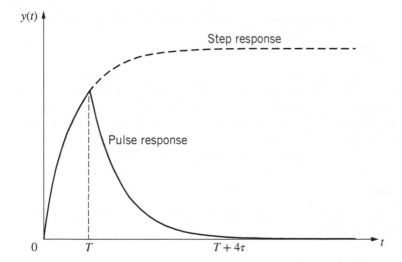

Figure 7.8 Pulse response of a first-order system where pulse time $T > 4\tau$.

Figure 7.9 Pulse response of a first-order system where pulse time $T < 4\tau$.

Figure 7.8 shows the pulse response of the first-order system where pulse time T is greater than the system's settling time t_S. Note that the output $y(t)$ exhibits an exponential rise from zero to a steady value, which it reaches in the settling time $t_S = 4\tau$. The steady-state output is the product of the pulse magnitude P and the DC gain of the transfer function in Fig. 7.7, which is b. At time $t = T$, the pulse is stepped to zero, and therefore the output $y(t)$ shows an exponential decay to zero, which it reaches at approximately $t = T + 4\tau$.

Next, we consider the case when the pulse duration is less than the system's settling time. Figure 7.9 shows the pulse response of the first-order system where $T < 4\tau$, along with the step response to a constant input with magnitude P. The pulse response initially exhibits an exponential rise from zero as the pulse input is applied, thus matching the step response. However, at time $t = T$ the pulse response begins its decay to zero because the pulse input has vanished.

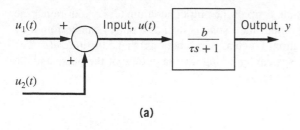

(a)

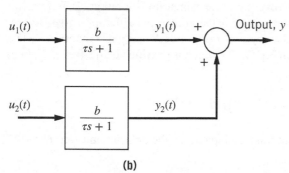

(b)

Figure 7.10 Equivalent simulation diagrams of a first-order system with two input functions.

The decay occurs before the response has reached its steady value because pulse time T is less than the system's settling time $t_S = 4\tau$. The pulse response for $t > T$ resembles the free response, and reaches zero in approximately four time constants after the pulse input goes to zero, or $t = T + 4\tau$ as shown in Fig. 7.9.

We can compute the pulse response by applying the *superposition property*, which states that the system response to two or more simultaneous input functions is equivalent to the sum of the individual responses to the individual inputs. As we noted in Chapter 1, linear systems obey the superposition property. Figure 7.10 shows two equivalent simulation diagrams demonstrating the superposition property for our standard first-order system with input $u(t) = u_1(t) + u_2(t)$. We can synthesize the pulse input described by Eq. (7.37) by adding a step function with magnitude P to a second step function with magnitude $-P$ and step time $t = T$. Mathematically, the two input functions are

$$u_1(t) = PU(t)$$
$$u_2(t) = -PU(t - T) \tag{7.38}$$

Note that $U(t-T)$ is a "delayed" unit-step function; that is, it is zero when $t \le T$, and unity when $t > T$. Using the simulation diagram shown in Fig. 7.10b, each output will be a step response, and therefore we can use the step-response Eq. (7.34) to write equations for each component of the pulse response

$$y_1(t) = Pb\left(1 - e^{-t/\tau}\right) \tag{7.39}$$

$$y_2(t) = \begin{cases} 0 & \text{for } 0 \le t \le T \\ -Pb\left(1 - e^{-(t-T)/\tau}\right) & \text{for } \quad t > T \end{cases} \tag{7.40}$$

The second output $y_2(t)$ must be zero for $t \le T$ as its input in Fig. 7.10b is a delayed step function. Another way to write Eq. (7.40) is to use the delayed unit-step function to replace the discontinuity at $t = T$

$$y_2(t) = -Pb\left(1 - e^{-(t-T)/\tau}\right)U(t - T) \tag{7.41}$$

The pulse response $y(t)$ is the sum of $y_1(t)$ and $y_2(t)$

$$y(t) = Pb\left(1 - e^{-t/\tau}\right) - Pb\left(1 - e^{-(t-T)/\tau}\right)U(t - T) \tag{7.42}$$

which is a mathematical representation of the pulse responses shown in Figs. 7.8 and 7.9.

The important feature of this pulse-response example is not the form of the output equation (7.42), but rather the understanding that the response of a linear system to an arbitrary input function $u(t)$ (such as a pulse) can be found by summing the individual responses to potentially simpler input functions (such as step functions) that constitute $u(t)$. The reader should be able to sketch the pulse response of a first-order system by noting the magnitude of the pulse and the relationship between the time constant τ and the pulse duration T.

Impulse Response of a First-Order System

Recall that an impulse input is a constant-magnitude input applied over an infinitesimal time duration. Therefore, we can obtain the impulse response of a system by evaluating the pulse response in the limit as pulse duration goes to zero. Consider again the first-order system shown in Fig. 7.7 with a pulse input of magnitude P and pulse duration T. The area or weight of the pulse is $A = PT$. The pulse response is given in Eq. (7.42), which is rewritten below (as y_{pulse}) with pulse magnitude $P = A/T$, or area/pulse duration

$$y_{\text{pulse}}(t) = \frac{Ab}{T}\left(1 - e^{-t/\tau}\right) - \frac{Ab}{T}\left(1 - e^{-(t-T)/\tau}\right)U(t-T) \tag{7.43}$$

If we allow pulse duration T to go to zero in the limit to produce an impulse input, then the delayed unit step $U(t-T)$ in Eq. (7.43) becomes $U(t)$, which leads to

$$\frac{Ab}{T}\left(1 - e^{-t/\tau}\right) - \frac{Ab}{T}\left(1 - e^{-(t-T)/\tau}\right) = \frac{Ab}{T}e^{-t/\tau}\left(e^{T/\tau} - 1\right) \tag{7.44}$$

The first-order system's response to impulse $A\delta(t)$ is obtained by taking the limit of Eq. (7.44) as T approaches zero

$$y_{\text{impulse}}(t) = \lim_{T \to 0} \frac{Ab}{T}e^{-t/\tau}\left(e^{T/\tau} - 1\right) \tag{7.45}$$

Applying l'Hopital's rule to evaluate the limit in Eq. (7.45), we obtain the impulse response

$$y_{\text{impulse}}(t) = \lim_{T \to 0} \frac{Ab}{\tau}e^{-t/\tau}e^{T/\tau} = \frac{Ab}{\tau}e^{-t/\tau} \tag{7.46}$$

Equation (7.46) shows that the impulse response of a first-order system exhibits a discontinuous step output at time $t = 0$, which then decays to zero in approximately four time constants. The magnitude of the impulse response at $t = 0$ is Ab/τ, which makes sense intuitively, as A is the weight or strength of the impulse input, b is the coefficient that multiplies the input (see Eq. (7.25) or Fig. 7.7), and τ is the system's time constant. A first-order system with a small τ has a very short transient response (i.e., a rapid response), which tends to increase the initial magnitude of the impulse response. The reader should be able to use the basic form of Eq. (7.46) to easily sketch the impulse response of a first-order system. All that is required is knowledge of the basic parameters of the first-order model (constants τ and b) and the strength of the impulse input, A.

Example 7.6

Figure 7.11 shows a series RL circuit with voltage source $e_{\text{in}}(t)$. Sketch the current $I(t)$ for an impulse voltage input, $e_{\text{in}}(t) = 0.08\delta(t - t_1)$ V. The system has zero energy at time $t = 0$, or $I(0) = 0$, and the impulse is applied at time $t_1 = 0.1$ s. The inductance is $L = 0.02$ H, and the resistance is $R = 1.2$ Ω.

We begin with the mathematical model of the RL circuit, which we determined in Chapter 3:

$$L\dot{I} + RI = e_{\text{in}}(t) \tag{7.47}$$

Because the circuit has only one energy-storage element (inductor L), we have a first-order system. We substitute the numerical values for R and L and divide by resistance R to write the first-order system in our standard form

$$0.0167\dot{I} + I = 0.8333e_{\text{in}}(t) \tag{7.48}$$

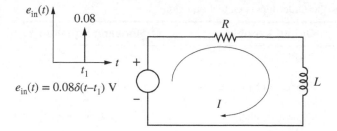

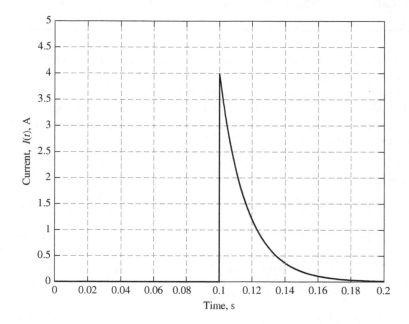

Figure 7.11 Electrical system with impulse input (Example 7.6).

Figure 7.12 Impulse response of electrical system (Example 7.6).

Therefore, the time constant is identified as $\tau = 0.0167$ s. The impulse response is found using Eq. (7.46), where A is the strength of the impulse (0.08 V-s) and b is the input coefficient ($1/R = 0.8333\,\Omega^{-1}$). The initial magnitude of the impulse response is $Ab/\tau = (0.08)(0.8333)/0.0167 = 4$ A. Note that the units are correct, as A/τ has units of (V-s)/s $=$ V, and b has units Ω^{-1}, which results in volts/ohms $=$ amperes. Hence, the impulse response for current shows a discontinuous step to 4 A at the instant the impulse is applied at $t_1 = 0.1$ s. Total energy of the electrical system is $LI^2/2$, and therefore the system exhibits an instantaneous jump in energy at $t = 0.1$ s because of the impulse input. The impulse response exponentially decays to zero in about four time constants after the impulse is applied, or $t = 0.1 + 4\tau = 0.1667$ s.

Figure 7.12 shows the impulse response, which the reader should be able to sketch from the information computed in this example. The strength of the impulse 0.08 V-s might be an idealized representation of an 80-V pulse input that is applied for pulse duration of 0.001 s (or, 1 ms).

7.4 SECOND-ORDER SYSTEM RESPONSE

Recall from our previous chapters on system modeling that a mechanical system with a single inertia element will be modeled by a single second-order differential equation, which arises from applying force $=$ mass $\times$ acceleration. An electrical system with two energy-storage elements (inductor or capacitor) will generally result in a second-order system model. If these systems have linear stiffness and friction laws (mechanical systems) or linear electrical element laws, the respective second-order models will be linear. Some complex, nonlinear, higher-order systems, such as aircraft pitch and yaw dynamics, have dominant modes of motion that can be approximated by linearized second-order models. Table 7.3 summarizes several examples of systems represented by second-order I/O equations.

Table 7.3   Examples of Engineering Systems Modeled by a Second-Order Input–Output Equation

Physical System	Input Variable	Output Variable	Modeling Equation
Mechanical mass with stiffness and friction	Force, $F_{in}(t)$	Position, x	$m\ddot{x} + b\dot{x} + kx = F_{in}(t)$
Electrical series RLC circuit	Voltage source, $e_{in}(t)$	Capacitor charge, q	$L\ddot{q} + R\dot{q} + 1/C\,q = e_{in}(t)$
Electrical series RLC circuit	Voltage source, $e_{in}(t)$	Capacitor voltage, e_C	$LC\ddot{e}_C + RC\dot{e}_C + e_C = e_{in}(t)$
Electrical parallel RLC circuit	Current source, $I_{in}(t)$	Magnetic flux linkage, λ	$C\ddot{\lambda} + 1/R\dot{\lambda} + 1/L\lambda = I_{in}(t)$

We can obtain the response of a linear second-order system using the methods outlined in the earlier sections of this chapter. However, just as in the case of first-order systems, we can predict the behavior of a second-order system based on a few key parameters. For first-order systems, the key parameter is a single time constant τ, which determines the rate of exponential rise or decay, and the magnitude of the initial response to an impulse input. For second-order systems, the behavior of the transient response depends on *two* parameters, and knowledge of these two parameters allows us to quickly sketch the second-order system response.

Second-Order Response with Zero Input

To begin the analysis of a second-order system response, we consider first the case with zero input, which leads to the homogeneous differential equation

$$\ddot{y} + a_1\dot{y} + a_0 y = 0 \tag{7.49}$$

Note that we can write the left-hand side of any second-order system with a unity coefficient for $\ddot{y}$ by performing simple division. The free response (for zero input) will depend on the roots of the characteristic equation, which can be obtained from Eq. (7.49) by inspection

$$r^2 + a_1 r + a_0 = 0 \tag{7.50}$$

The two characteristic roots depend on coefficients a_1 and a_0

$$r = \frac{-a_1 \pm \sqrt{a_1^2 - 4u_0}}{2} \tag{7.51}$$

We have four possible cases for the two roots, r_1 and r_2

1. Both roots are real numbers and distinct (the radicand of Eq. (7.51) is positive).
2. Both roots are real numbers and equal (the radicand of Eq. (7.51) is zero).
3. Both roots are complex conjugates (the radicand of Eq. (7.51) is negative).
4. Both roots are purely imaginary numbers (coefficient $a_1 = 0$ and $a_0 > 0$).

Case 1: two real, distinct roots

If the roots are *real* and *distinct* (not repeated), we have Case 1 and the homogeneous solution has the form

$$y_H(t) = c_1 e^{r_1 t} + c_2 e^{r_2 t} \tag{7.52}$$

Therefore, the free response consists of two exponential functions, where the two coefficients c_1 and c_2 are determined from the two initial conditions $y(0) = y_0$ and $\dot{y}(0) = \dot{y}_0$. The free response decays to zero as $t \to \infty$ only if both roots are negative. If *either* root is positive, the free response diverges to infinity as $t \to \infty$. If one root is zero (say, $r_1 = 0$) and the other root is negative, then the free response will involve a constant value c_1 that remains at steady state. Figure 7.13

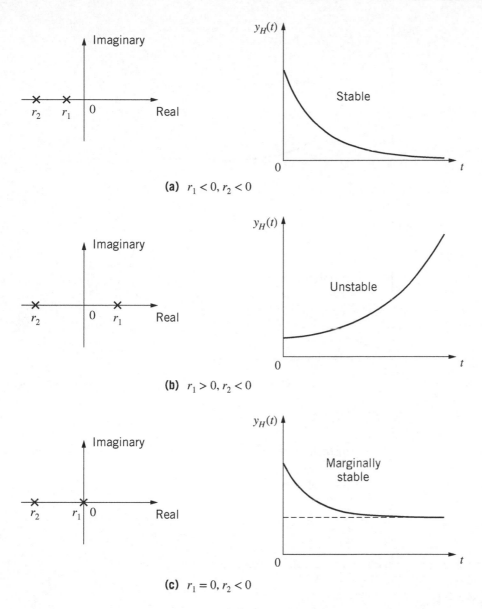

(a) $r_1 < 0, r_2 < 0$

(b) $r_1 > 0, r_2 < 0$

(c) $r_1 = 0, r_2 < 0$

Figure 7.13 Second-order system root locations and a typical corresponding free response: real and distinct roots (Case 1).

provides general examples of root locations and the corresponding free response for Case 1. The root locations are marked with an "×" in the *complex plane*, which has all real numbers on the *x*-axis, and all imaginary numbers on the *y*-axis. Because both roots for Case 1 are strictly real numbers, all root locations in Fig. 7.13 lie on the *x*-axis. The free response is said to be *stable* if it decays to zero (Fig. 7.13a). If one real root is positive, the free response is *unstable* (Fig. 7.13b), as it diverges to infinity. If one real root is zero, the free response is *marginally stable* (Fig. 7.13c), because the response remains bounded at a constant value but does not decay to zero as $t \to \infty$.

Case 2: two real, repeated roots

If the roots are real and equal ($r_1 = r_2$), the homogeneous solution has the form

$$y_H(t) = c_1 e^{r_1 t} + c_2 t e^{r_1 t} \tag{7.53}$$

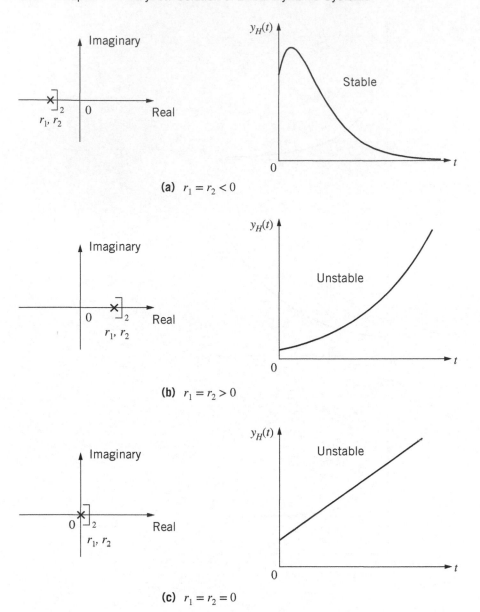

(a) $r_1 = r_2 < 0$

(b) $r_1 = r_2 > 0$

(c) $r_1 = r_2 = 0$

Figure 7.14 Second-order system root locations and a typical corresponding free response: real, repeated roots (Case 2).

Figure 7.14 shows three examples of the double-root locations in the complex plane and the corresponding free response. The "double-root" locations are marked with an "×" with bracket and subscript 2. The free response is stable and decays to zero only if the repeated roots are negative, as shown in Fig. 7.14a. If the double roots are positive, then the free response diverges to infinity as $t \to \infty$, as shown in Fig. 7.14b. If the double roots are zero, then the free response changes linearly with time (unstable) as shown in Fig. 7.14c. We see from Eq. (7.51) that double roots at $r_1 = 0$ occur only when both coefficients are zero ($a_1 = a_0 = 0$), in which case the second-order system is $\ddot{y} = 0$. Therefore, two successive integrations result in a free response $y_H(t) = c_1 + c_2 t$, where constants $c_1 = y_0$ and $c_2 = \dot{y}_0$ are the initial conditions.

Case 3: two complex conjugate roots

If the radicand in Eq. (7.51) is negative (and coefficient a_1 is not zero), then the two roots are *complex* and will be composed of a real part α and an imaginary part β

$$r_1 = \alpha + j\beta \quad \text{and} \quad r_2 = \alpha - j\beta \tag{7.54}$$

These roots are *complex conjugates* because they have the same real part (α) and the imaginary parts ($j\beta$ and $-j\beta$) have equal magnitudes but opposite signs. Therefore, complex conjugate roots exhibit symmetry about the real or x-axis when plotted in the complex plane. The homogeneous solution has the form

$$y_H(t) = c_1 e^{(\alpha+j\beta)t} + c_2 e^{(\alpha-j\beta)t} = e^{\alpha t}\left(c_1 e^{j\beta t} + c_2 e^{-j\beta t}\right) \tag{7.55}$$

Using Euler's formula, $e^{j\theta} = \cos\theta + j\sin\theta$ with $\theta = \beta t$, Eq. (7.55) becomes

$$y_H(t) = e^{\alpha t}[c_1(\cos\beta t + j\sin\beta t) + c_2(\cos\beta t - j\sin\beta t)]$$

$$= e^{\alpha t}[(c_1 + c_2)\cos\beta t + j(c_1 - c_2)\sin\beta t] \tag{7.56}$$

Because the response $y_H(t)$ must be a real number, the sine factor ($c_1 - c_2$) must be an imaginary number, while the cosine factor ($c_1 + c_2$) must be a real number. Therefore, the constants c_1 and c_2 are complex conjugates. Defining two new real constants $c_3 = c_1 + c_2$ and $c_4 = j(c_1 - c_2)$, we can rewrite Eq. (7.56) as

$$y_H(t) = e^{\alpha t}[c_3 \cos\beta t + c_4 \sin\beta t] \tag{7.57}$$

Another (perhaps simpler) form of the free response Eq. (7.57) is

$$y_H(t) = Ke^{\alpha t}\cos(\beta t + \phi) \tag{7.58}$$

where we have used the trigonometric identity for linear combinations of sine and cosine functions

$$A\cos\beta t + B\sin\beta t = \sqrt{A^2 + B^2}\cos\left(\beta t - \tan^{-1}\frac{B}{A}\right)$$

The reader should note that Eqs. (7.57) and (7.58) are equivalent, and both involve two unknown constants. We will use Eq. (7.58) for the free response of a second-order system with complex roots, where the two unknowns are amplitude K and phase angle ϕ, which can be determined from the two initial conditions y_0 and $\dot{y}_0$.

Figure 7.15 shows two examples of the complex conjugate root locations in the complex plane and the corresponding free response. The free response Eq. (7.58) is the product of the "amplitude envelope" $Ke^{\alpha t}$ (dashed curve in Fig. 7.15) and a sinusoidal function with frequency β (rad/s). When the complex roots have a *negative* real part (i.e., $\alpha < 0$), the amplitude function $Ke^{\alpha t}$ in Eq. (7.58) decays to zero, and therefore the system is stable as shown in Fig. 7.15a. We call the decaying, oscillating response a *damped sinusoid*. When the complex roots have a *positive* real part (i.e., $\alpha > 0$), the amplitude function $Ke^{\alpha t}$ diverges to infinity, and the free response is unstable, as shown in Fig. 7.15b.

Case 4: two imaginary roots

If the coefficients $a_1 = 0$ and $a_0 > 0$, then the radicand in Eq. (7.51) is negative, and the two roots are purely imaginary numbers (no real part, or $\alpha = 0$)

$$r_1 = j\beta \quad \text{and} \quad r_2 = -j\beta \tag{7.59}$$

The homogeneous solution has the form

$$y_H(t) = c_1 e^{j\beta t} + c_2 e^{-j\beta t}$$

$$= c_1(\cos\beta t + j\sin\beta t) + c_2(\cos\beta t - j\sin\beta t)$$

$$= (c_1 + c_2)\cos\beta t + j(c_1 - c_2)\sin\beta t \tag{7.60}$$

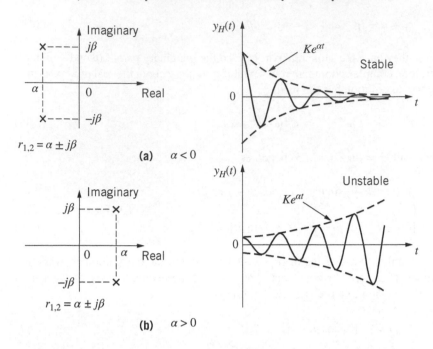

Figure 7.15 Second-order system root locations and a typical corresponding free response: complex roots (Case 3).

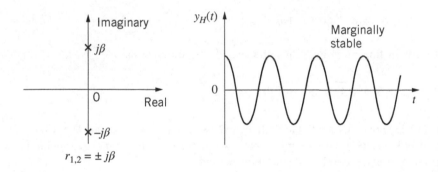

Figure 7.16 Second-order system root locations and a typical corresponding free response: imaginary roots (Case 4).

Constants c_1 and c_2 are complex conjugates (just as in Case 3), and therefore the free response can be written in the same form as Eq. (7.58), but without the exponential function

$$y_H(t) = K\cos(\beta t + \phi) \tag{7.61}$$

Figure 7.16 shows an example of the imaginary root locations in the complex plane and the corresponding free response. The free response is a purely harmonic (sinusoidal) function, which oscillates with constant amplitude. Therefore, the free response is marginally stable.

The consequences of the root locations for a second-order system can be summarized as follows:

- If the two roots are real, nonzero numbers, the free response will be the sum of two exponential functions. If both roots are negative, the free response decays to zero at steady state and the system is stable. If *either* root is positive, the free response diverges to infinity and the system is unstable.

- If one root is equal to zero, part of the free response will be a constant, and therefore the system is marginally stable if the other root is negative. If the two roots are *both* equal to zero, the free response changes at a linear rate with time (provided $\dot{y}_0 \neq 0$), and the system is unstable.

- If the two roots are complex numbers, they appear as complex conjugate pairs. The free response will oscillate at frequency β rad/s, where β is the imaginary part of the complex roots. If the real part of the two roots (α) is negative, the free response is a damped sinusoid that decays to zero at steady state (stable). If α is positive, the free response is a diverging sinusoid (unstable).

- If the two roots are purely imaginary numbers, the free response is a sinusoid function that oscillates at frequency β rad/s. The system is marginally stable because the amplitude remains constant.

Example 7.7

Several LTI second-order I/O equations are given below. Describe the free response for each I/O equation, and (if applicable) estimate the time to reach steady state and the number of periodic oscillations before the free (or transient) response "dies out" to zero.

(a) $2\ddot{y} + 22\dot{y} + 48y = 3u$

The nature of the free response $y_H(t)$ depends on the characteristic roots, so the starting point is the characteristic equation, which can be written from the inspection of the left-hand side coefficients of the I/O equation

$$2r^2 + 22r + 48 = 0$$

or

$$2(r^2 + 11r + 24) = 2(r + 3)(r + 8) = 0$$

Therefore, the characteristic roots are $r_1 = -3$ and $r_2 = -8$. Immediately, we know that the free response will be composed of two decaying exponential functions as both roots are negative real numbers, and the free response will resemble the response in Fig. 7.13a. The free response has the form

$$y_H(t) = c_1 e^{-3t} + c_2 e^{-8t}$$

where the constants c_1 and c_2 can be determined from knowledge of the two initial conditions y_0 and $\dot{y}_0$, which are not provided in this example. The first exponential function $e^{-3t} = e^{-t/0.3333}$ has a time constant $\tau_1 = 0.3333$ s, so it "dies out" to zero in approximately $4\tau_1 = 1.3333$ s. The second exponential function $e^{-8t} = e^{-t/0.125}$ has a time constant $\tau_2 = 0.125$ s, so it "dies out" to zero in approximately $4\tau_2 = 0.5$ s. Hence, the settling time of $y_H(t)$ is 1.3333 s, which is dictated by its "slowest" of the two exponential functions, in this case e^{-3t}. Obviously, the free response does not exhibit oscillations because it is composed of two exponential functions. The complete system response will depend on the input $u(t)$, but we know that the homogeneous or free response will go to zero in 1.3333 s.

(b) $\ddot{y} - 4\dot{y} + 40y = 3u$

The characteristic equation is
$$r^2 - 4r + 40 = 0$$

and the characteristic roots are $r_1 = 2 + j6$ and $r_2 = 2 - j6$, which are complex conjugate pairs. Immediately, we know that the free response will involve a sinusoidal function, where the frequency of oscillation will be the imaginary part, or 6 rad/s in this case. The "amplitude envelope" will involve an exponential function of the real part, or e^{2t}, which diverges to infinity. Therefore, this system is *unstable* because the real part of the roots is positive. We can write the form of the free response using Eq. (7.58) and the known roots for this example

$$y_H(t) = K e^{2t} \cos(6t + \phi)$$

where the amplitude K and phase angle ϕ can be determined from knowledge of the two initial conditions. Clearly, this free response oscillates and diverges to infinity, and resembles the response in Fig. 7.15b.

(c) $3\ddot{y} + 75y = 3u$

The characteristic equation is

$$3r^2 + 75 = 0$$

or

$$3(r^2 + 25) = 0$$

and the characteristic roots are $r_1 = j5$ and $r_2 = -j5$, which are conjugate imaginary numbers. Because the two roots are purely imaginary, we immediately know that the free response will be a sinusoidal function with constant amplitude that resembles the response in Fig. 7.16. Equation (7.61) gives us the form of the free response, where the frequency of oscillation is the imaginary part of the roots, or 5 rad/s in this case

$$y_H(t) = K\cos(5t + \phi)$$

where the amplitude K and phase angle ϕ can be determined from knowledge of the two initial conditions. This free response will oscillate at 5 rad/s with constant amplitude.

   (d) $\ddot{y} + 0.4\dot{y} + 18y = 3u$

The characteristic equation is

$$r^2 + 0.4r + 18 = 0$$

and the characteristic roots are $r_1 = -0.2 + j4.2379$ and $r_2 = -0.2 - j4.2379$, which are complex conjugate pairs. Immediately, we know that the free response will involve a sinusoidal function, where the frequency of oscillation will be the imaginary part, or 4.2379 rad/s in this case. The "amplitude envelope" will involve an exponential function of the real part, or $e^{-0.2t}$, which decays to zero at steady state. Therefore, the free (or transient) response is a *damped sinusoid* because the roots are complex, and the real part is negative. We can write the form of the free response using Eq. (7.58) and the known roots for this example

$$y_H(t) = Ke^{-0.2t}\cos(4.2379t + \phi)$$

where the amplitude K and phase angle ϕ can be determined from knowledge of the two initial conditions. Clearly, this free response oscillates and decays to zero, and resembles the response in Fig. 7.15a. The settling time is dictated by the real part of the complex roots, and the time constant for $e^{-0.2t} = e^{-t/5}$ is $\tau = 5$ s. Hence, the free response decays to zero in $4\tau = 20$ s. The number of oscillations in the free response can be determined by dividing the settling time by the period of one oscillation. Because the frequency of oscillation is $\omega = 2\pi/T_{period}$, the period is $T_{period} = (2\pi \text{ rad})/(4.2379 \text{ rad/s}) = 1.4826$ s. Therefore, the number of oscillations in the free response is 20 s/1.4826 s $\approx$ 13.5 cycles.

Damping Ratio and Undamped Natural Frequency

We have seen from the previous section and Example 7.7 that the form of the free (or transient) response of a second-order system is completely determined by the two characteristic roots. When the two roots are complex with a negative real part, the transient response is a damped sinusoid. Another way to categorize the second-order transient response is to use the *damping ratio*. Consider again the general LTI second-order I/O equation

$$\ddot{y} + a_1\dot{y} + a_0y = b_0u(t) \tag{7.62}$$

Its characteristic roots are determined by Eq. (7.51), which is repeated here

$$\text{Characteristic roots:} \quad r = \frac{-a_1 \pm \sqrt{a_1^2 - 4a_0}}{2} \tag{7.63}$$

Let's assume that coefficients a_1 and a_0 are positive. Note that if coefficient a_1 is too small so that $a_1^2 < 4a_0$, the roots will be complex conjugates because the radicand is negative. In this case, the transient response will be a damped sinusoid, and the system is said to be *underdamped*. Furthermore, if coefficient a_1 is large enough so that $a_1^2 > 4a_0$, the two roots will be real, distinct, and negative. In this case, the transient response will consist of two decaying exponential functions, and the system is said to be *overdamped*. The transitional condition is when $a_1 = 2\sqrt{a_0}$ and the two roots are real, negative, and equal. In this transitional case, the system is said to be *critically damped*. We can characterize these three cases by defining the *damping ratio* ζ as the ratio of coefficient a_1 to its *critical* value ($2\sqrt{a_0}$) when both roots become real and equal

$$\text{Damping ratio:} \quad \zeta = \frac{a_1}{2\sqrt{a_0}} \tag{7.64}$$

Therefore, we can classify the three cases in terms of the damping ratio ζ:

1. $\zeta > 1$: Overdamped system (see Fig. 7.13a)
2. $\zeta = 1$: Critically damped system (see Fig. 7.14a)
3. $0 < \zeta < 1$: Underdamped system (see Fig. 7.15a)

Another way to think of the damping ratio is to observe the mathematical model of a mechanical valve in a hydraulic fluid with a return spring, which might be modeled by the mass–spring–damper system:

$$m\ddot{x} + b\dot{x} + kx = f(t)$$

where x is the valve position from static equilibrium and $f(t)$ is the applied force. The roots of the mass-spring-damper system are

$$r = \frac{-b \pm \sqrt{b^2 - 4mk}}{2m}$$

Using Eq. (7.64) and coefficients $a_1 = b/m$ and $a_0 = k/m$, we see that the damping ratio is

$$\zeta = \frac{b}{2\sqrt{km}}$$

Therefore, for a mechanical system, damping ratio is the system's *actual* damping coefficient (viscous friction b) divided by the *critical* damping coefficient ($2\sqrt{km}$). This result makes intuitive sense: if the viscous friction coefficient b is very "small," very little friction is present, and the valve mass will exhibit oscillations during the transient response. Conversely, if b is very "large," then the friction force will inhibit oscillations, and the transient response will show an exponential decay to steady state.

Note that when coefficient $a_1 = 0$, the two roots are imaginary numbers and the damping ratio $\zeta = 0$. Consequently, the free response is an *undamped* sinusoid with constant amplitude. The frequency of oscillation for the case of no damping is called the *undamped natural frequency* ω_n, and it is the magnitude of the two imaginary roots, which can be computed from Eq. (7.63) with $a_1 = 0$

$$\text{Undamped natural frequency:} \quad \omega_n = \sqrt{a_0} \ (\text{rad/s}) \tag{7.65}$$

For the mass–spring–damper mechanical system, coefficient $a_0 = k/m$, and therefore the undamped natural frequency is $\omega_n = \sqrt{k/m}$ rad/s. This result shows that increasing the spring constant k for a frictionless mass-spring system with a fixed mass m will increase the vibration frequency, which again makes intuitive sense.

We can replace the coefficients a_0 and a_1 in the general second-order I/O equation (7.62) with damping ratio and undamped natural frequency: using Eq. (7.65) we see that coefficient $a_0 = \omega_n^2$, and Eq. (7.64) shows that $a_1 = 2\zeta\sqrt{a_0} = 2\zeta\omega_n$. Therefore, our general second-order I/O equation (7.62) can be written as

$$\ddot{y} + 2\zeta\omega_n\dot{y} + \omega_n^2 y = b_0 u(t) \tag{7.66}$$

Equation (7.66) is our "standard form" for an LTI second-order I/O equation. Recall that we are able to put any LTI *first*-order I/O equation into the standard form of Eq. (7.25) and consequently identify the time constant τ, which is the important parameter for a first-order system response. Likewise, we can put any LTI second-order I/O equation into the standard form of Eq. (7.66) and identify the damping ratio ζ and the undamped natural frequency ω_n, which are the two important parameters for an underdamped second-order system response.

When the system is undamped ($\zeta = 0$) or underdamped ($0 < \zeta < 1$), we can write the characteristic roots in terms of ζ and ω_n by using Eq. (7.63) with $a_1 = 2\zeta\omega_n$ and $a_0 = \omega_n^2$, and the resulting complex roots are

$$r = -\zeta\omega_n \pm j\omega_n\sqrt{1 - \zeta^2} \tag{7.67}$$

Hence the product $-\zeta\omega_n$ is the *real* part of the two complex roots, and $\omega_n\sqrt{1 - \zeta^2}$ is the *imaginary* part of the two complex roots. Recall that Eq. (7.58) describes the free response of an underdamped system (Case 3), where the real part of the roots determines the "exponential envelope" while the imaginary part determines the frequency of oscillation. Therefore, the free response of an underdamped system is

$$y_H(t) = Ke^{-\zeta\omega_n t}\cos(\omega_d t + \phi) \tag{7.68}$$

where $\omega_d = \omega_n\sqrt{1 - \zeta^2}$ is called the *damped frequency* (rad/s). It is important to note that an underdamped system oscillates at frequency ω_d when damping is present (i.e., $0 < \zeta < 1$), and that a second-order response oscillates at frequency ω_n *only* when there is no damping (i.e., $\zeta = 0$).

Figure 7.17 shows two complex conjugate roots in the complex plane. Note the right triangle formed by leg $A = \zeta\omega_n$ (magnitude of real part) and leg $B = \omega_d = \omega_n\sqrt{1 - \zeta^2}$ (magnitude of imaginary part). The distance from either root to the origin is the undamped natural frequency, as demonstrated by the length of the hypotenuse C

$$C = \sqrt{A^2 + B^2} = \sqrt{\zeta^2\omega_n^2 + \omega_n^2(1 - \zeta^2)} = \omega_n$$

The cosine of angle θ measured clockwise from the negative real axis to the radial line connecting the origin to the root is

$$\cos\theta = \frac{A}{C} = \frac{\zeta\omega_n}{\omega_n} = \zeta$$

These geometric results lead to the following statements regarding the loci of constant ω_n, ζ, and ω_d, and these statements are illustrated in Fig. 7.18.

1. Complex roots that lie on a semicircle have the *same* undamped natural frequency ω_n. Increasing the radius of the semicircle increases ω_n.

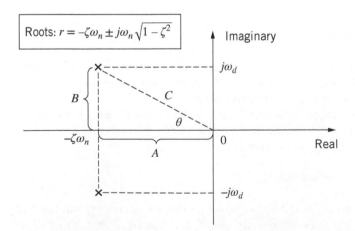

Figure 7.17 Complex root locations and computation of ζ and ω_n.

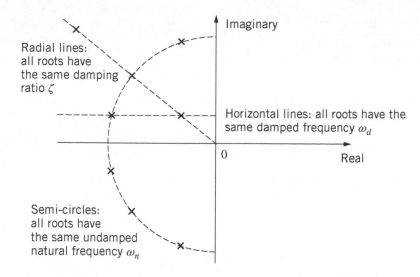

Figure 7.18 Complex root locations and loci of constant ζ, ω_n, and ω_d.

2. Complex roots that lie on a radial line from the origin have the *same* damping ratio ζ. As the radial line approaches the negative real axis, the angle θ becomes smaller and the damping ratio increases, where $\zeta \to 1$ as $\theta \to 0$. As the radial line approaches the imaginary axis, the angle θ becomes larger and damping ratio decreases, where $\zeta \to 0$ as $\theta \to 90°$.

3. Complex roots that lie on a horizontal line have the same *damped* frequency ω_d. As the horizontal line moves away from the real axis, ω_d increases.

Step Response of an Underdamped Second-Order System

A step input is often used as a "test input" in order to establish performance or design specifications for a second-order system, such as its damping level and settling time. In this subsection, we derive basic performance equations for an underdamped second-order system based on a constant (step) input.

To begin, consider the second-order LTI system that has been written in our "standard form" for an underdamped system

$$\ddot{y} + 2\zeta\omega_n\dot{y} + \omega_n^2 y = \omega_n^2 u(t) \tag{7.69}$$

We have assumed that the coefficient multiplying the input $u(t)$ is equal to the coefficient for y. Therefore, if the input is a constant $u(t) = A$, then the steady-state output is the same constant, or $y(\infty) = A$ (recall that at steady state $\ddot{y} = \dot{y} = 0$ for a stable system with a constant input). The complete system response will be the sum of the homogeneous and particular solutions, or $y(t) = y_H(t) + y_P(t)$. If the input is a unit-step function, then the particular solution $y_P(t)$ is also a constant with a value of one. Because we have assumed that the system is underdamped ($0 < \zeta < 1$), the homogeneous or transient response is a damped sinusoid represented by Eq. (7.57). The complete unit-step response is

$$y(t) = 1 + e^{\alpha t}[c_3 \cos \beta t + c_4 \sin \beta t] \tag{7.70}$$

where $\alpha = -\zeta\omega_n$ (real part of the complex roots) and $\beta = \omega_n\sqrt{1 - \zeta^2}$ (imaginary part of the complex roots). For now, let us leave the step response Eq. (7.70) in terms of α and β and substitute their respective values later.

We can obtain the constants c_3 and c_4 in Eq. (7.70) from the initial conditions, which we assume are zero: $y(0) = \dot{y}(0) = 0$. Setting $t = 0$ in Eq. (7.70) yields

$$y(0) = 1 + c_3 = 0$$

and therefore constant $c_3 = -1$. The time derivative of Eq. (7.70) is

$$\dot{y}(t) = \alpha e^{\alpha t} c_3 \cos \beta t - e^{\alpha t} \beta c_3 \sin \beta t + \alpha e^{\alpha t} c_4 \sin \beta t + e^{\alpha t} \beta c_4 \cos \beta t \tag{7.71}$$

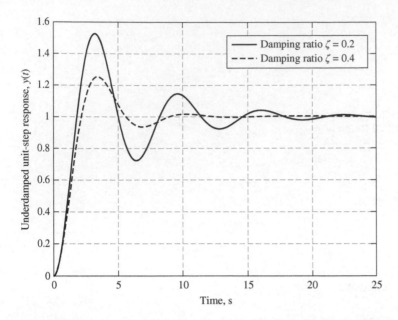

Figure 7.19 Unit-step responses for a second-order system with $\omega_n = 1$ rad/s.

Setting $t = 0$ in Eq. (7.71) yields

$$\dot{y}(0) = \alpha c_3 + \beta c_4 = 0$$

Substituting $c_3 = -1$, we find that constant $c_4 = \alpha / \beta$. Using Eq. (7.70), the complete unit-step response is

$$y(t) = 1 - e^{\alpha t} \left(\cos \beta t - \frac{\alpha}{\beta} \sin \beta t \right) \tag{7.72}$$

The "exponential envelope" depends on the real part of the roots, $\alpha = -\zeta \omega_n$, and the sinusoidal (damped) frequency depends on the imaginary part of the roots, $\beta = \omega_n \sqrt{1 - \zeta^2} = \omega_d$. The real and imaginary parts (α and β) are functions of damping ratio ζ and undamped natural frequency ω_n. Figure 7.19 shows the unit-step response Eq. (7.72) for two cases: both with $\omega_n = 1$ rad/s, and two damping ratios $\zeta - 0.2$ and 0.4. Clearly, the damping ratio ζ affects the peak value of the transient response.

Next, we present the performance equations for a step response. The *peak time* t_p is the time to reach the peak (maximum) output during the transient response. Figure 7.19 shows that the peak output occurs at one-half of the period of a damped sinusoidal cycle. The period of one cycle is $T_{\text{period}} = 2\pi / \omega_d$, where $\omega_d = \omega_n \sqrt{1 - \zeta^2}$ is the damped frequency. Therefore, the peak time is

$$t_p = \frac{\pi}{\omega_n \sqrt{1 - \zeta^2}} \tag{7.73}$$

The peak or maximum output y_{max} is obtained from the unit-step response Eq. (7.72) at peak time $t = \pi / \omega_d$, with the substitution $\beta = \omega_d$

$$y_{\text{max}} = y(t_p) = 1 - e^{\alpha \pi / \omega_d} \left(\cos \pi - \frac{\alpha}{\omega_d} \sin \pi \right) \tag{7.74}$$

Substituting for the real part of the roots ($\alpha = -\zeta \omega_n$) and the damped frequency $\omega_d = \omega_n \sqrt{1 - \zeta^2}$ in Eq. (7.74), we obtain the peak unit-step response

$$y_{\text{max}} = 1 + e^{-\zeta \pi / \sqrt{1 - \zeta^2}} \tag{7.75}$$

A more general expression is the *maximum overshoot*, M_{os}, which is defined as the difference between the peak (y_{max}) and steady-state (y_{ss}) values, normalized by the steady-state value

$$M_{os} = \frac{|y_{max} - y_{ss}|}{y_{ss}} \tag{7.76}$$

In our derivation of the peak output in Eq. (7.75), the steady-state output is unity. Therefore, an equation for maximum overshoot can be obtained by combining Eqs. (7.75) and (7.76) with $y_{ss} = 1$

$$M_{os} = e^{-\zeta\pi/\sqrt{1-\zeta^2}} \tag{7.77}$$

Now, the peak output for the general case can be written as

$$y_{max} = y_{ss}(1 + M_{os}) \tag{7.78}$$

Peak response and/or maximum overshoot depends only on damping ratio ζ and does not depend on the undamped natural frequency. The reader should note that when $M_{os} = 0.45$, the peak output overshoots its steady-state value by 45%.

The settling time, or time to reach steady state, can be estimated from the exponential envelope term $e^{\alpha t}$ in Eq. (7.72). When the envelope term $e^{\alpha t} = e^{-4}$, the transient response has essentially "died out" as $e^{-4} = 0.0183$, and hence the magnitude of the transient sinusoidal terms are less than 2% of the steady-state response. Settling time t_S for an underdamped system is therefore $t_S = -4/\alpha$, where $\alpha = -\zeta\omega_n$, or

$$t_S = \frac{4}{\zeta\omega_n} \tag{7.79}$$

Finally, the number of oscillations (cycles) N_{cycles} during the transient response can be estimated by dividing the settling time Eq. (7.79) by the damped period $T_{period} = 2\pi/\omega_d$. After substituting $\omega_d = \omega_n\sqrt{1-\zeta^2}$ and some simplification, we obtain

$$N_{cycles} = \frac{2\sqrt{1-\zeta^2}}{\pi\zeta} \tag{7.80}$$

Equation (7.80) shows that the number of cycles during the transient response is a function of damping ratio only.

Table 7.4 summarizes the equations for the important performance criteria that characterize an underdamped step response. Note that every underdamped response criteria in Table 7.4 depends solely on damping ratio ζ and undamped natural frequency ω_n. It should be heavily stressed that these performance equations apply *only* to *underdamped* systems where $0 < \zeta < 1$. Therefore, given any LTI second-order system, the first step in the analysis is to determine whether or not the system is underdamped by computing either the characteristic roots or the damping ratio ζ. If a system is critically damped ($\zeta = 1$) or overdamped ($\zeta > 1$), then the concepts of overshoot and period of oscillation are meaningless, as the transient response is composed of two exponential functions.

Table 7.4   Performance Equations for the Step Response of an Underdamped Second-Order System

Performance Criteria	Equation
Peak time, t_p	$t_p = \dfrac{\pi}{\omega_n\sqrt{1-\zeta^2}}$
Maximum overshoot, M_{os}	$M_{os} = e^{-\zeta\pi/\sqrt{1-\zeta^2}}$
Settling time to steady state, t_S	$t_S = \dfrac{4}{\zeta\omega_n}$
Period of oscillation, T_{period}	$T_{period} = \dfrac{2\pi}{\omega_n\sqrt{1-\zeta^2}}$
Number of cycles to steady state, N_{cycles}	$N_{cycles} = \dfrac{2\sqrt{1-\zeta^2}}{\pi\zeta}$

Example 7.8

Figure 7.20 shows a one degree-of-freedom (1-DOF) rotational mechanical system that possesses stiffness and friction elements. If the system is initially at rest, $\theta(0) = \dot{\theta}(0) = 0$, and the input torque is a step function $T_{in}(t) = 2.5U(t)$ N-m, sketch the angular response $\theta(t)$ of the mechanical system and note the important response criteria.

The system has inertia $J = 0.2$ kg-m^2, viscous friction coefficient $b = 1.6$ N-m-s/rad, and torsional spring constant $k = 65$ N-m/rad. Therefore, the mathematical model is

$$0.2\ddot{\theta} + 1.6\dot{\theta} + 65\theta = T_{in}(t) \tag{7.81}$$

First, we rewrite the modeling equation (7.81) in our standard form by dividing by moment of inertia $J = 0.2$ kg-m^2 in order to have a unity coefficient on angular acceleration

$$\ddot{\theta} + 8\dot{\theta} + 325\theta = 5T_{in}(t) \tag{7.82}$$

Comparing Eq. (7.82) to the standard form of a second-order system Eq. (7.66), we can compute the undamped natural frequency

$$\omega_n = \sqrt{325} = 18.0278 \text{ rad/s}$$

The damping ratio is computed from the term involving angular velocity

$$2\zeta\omega_n = 8 \quad \text{or} \quad \zeta = 0.2219$$

Because damping ratio $\zeta < 1$, we can use the performance equations in Table 7.4, which pertain only to an underdamped second-order system (the reader should note that if damping ratio is computed to be greater than or equal to unity, then the performance equations contained in Table 7.4 do not apply and cannot be used).

Using the two parameters ζ and ω_n, the important response characteristics are computed as follows:

Peak time: $t_p = \dfrac{\pi}{\omega_n\sqrt{1 - \zeta^2}} = 0.1787$

Maximum overshoot: $M_{os} = e^{-\zeta\pi/\sqrt{1-\zeta^2}} = 0.4893$ (48.93% overshoot)

Settling time: $t_S = \dfrac{4}{\zeta\omega_n} = 1.0$ s

Period of oscillation: $T_{period} = \dfrac{2\pi}{\omega_n\sqrt{1 - \zeta^2}} = 0.3574$ s

Number of cycles to steady state: $N_{cycles} = \dfrac{2\sqrt{1 - \zeta^2}}{\pi\zeta} = 2.8$ cycles

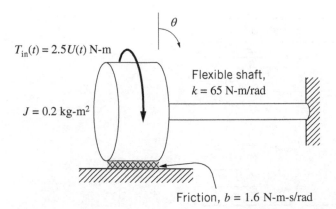

θ

$T_{in}(t) = 2.5U(t)$ N-m

Flexible shaft,
$k = 65$ N-m/rad

$J = 0.2$ kg-m^2

Friction, $b = 1.6$ N-m-s/rad

Figure 7.20 1-DOF rotational mechanical system for Example 7.8.

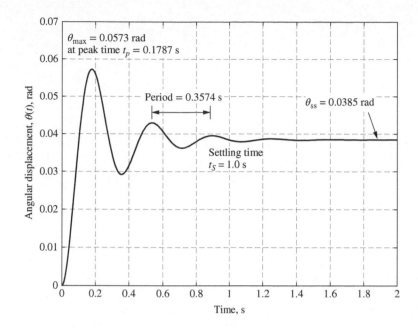

Figure 7.21 Step response of the underdamped, second-order mechanical system for Example 7.8.

The steady-state value is also needed for an accurate sketch of the step response. Using the modeling equation (7.81) with the steady-state condition $\ddot{\theta} = \dot{\theta} = 0$, the steady-state angular displacement is $\theta_{ss} = 2.5$ N-m/65 N-m/rad $= 0.0385$ rad (or 2.20°). The peak value of the transient response is

$$\theta_{max} = \theta_{ss}(1 + M_{os}) = 0.0573 \ \text{rad (or 3.28°)}$$

Figure 7.21 shows the step response of the rotational mechanical system that can be obtained by using the following MAT-LAB commands:

```
>> sys = tf(1,[0.2 1.6 65]);      % define LTI system, Eq. (7.81)
>> t = 0:0.001:2;                 % define time vector t
>> Tin = 2.5*ones(size(t));       % define input torque vector Tin(t)
>> [theta,t] = lsim(sys,Tin,t);   % obtain step response θ(t) using lsim
```

The important response characteristics are labeled on Fig. 7.21. The reader should be able to make an accurate sketch of the complete response from the performance criteria computed in this example.

Log Decrement and the Damping Ratio

In many mechanical system applications, characterizing the system's friction or damping is often a difficult task. The damping ratio ζ for an underdamped system can be estimated from the peak values of the transient response and the "amplitude envelope." Figure 7.22 shows the step and impulse responses of an underdamped second-order system. Expressions for the complete system responses shown in Fig. 7.22 are

$$\text{Step response:} \quad y(t) = K_1 e^{-\zeta \omega_n t} \cos(\omega_d t + \phi_1) + y_{ss} \tag{7.83}$$

$$\text{Impulse response:} \quad y(t) = K_2 e^{-\zeta \omega_n t} \cos(\omega_d t + \phi_2) \tag{7.84}$$

where constants $K_1, K_2, \phi_1,$ and ϕ_2 depend on the initial conditions and the magnitude of the input function. Note that the steady-state step response in Fig. 7.22a is $y_{ss} = 1$, while the steady-state impulse response in Fig. 7.22b is zero because the input is zero for $t > 0$. We can define the peak values relative to the steady-state value as

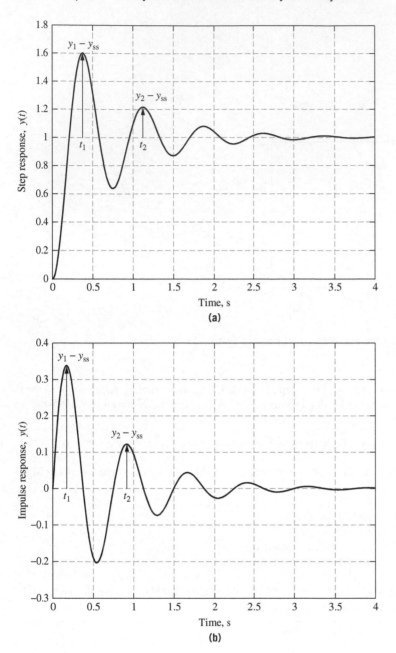

Figure 7.22 Underdamped second-order system response: (a) step response and (b) impulse response.

$$\text{First relative peak value:} \quad x_1 = y_1 - y_{ss}$$
$$\text{Second relative peak value:} \quad x_2 = y_2 - y_{ss}$$

where y_1 and y_2 are the first two peak values of the complete response. The relative peak values are denoted in Figs. 7.22a and 7.22b. If we compute the ratio of the relative peak values, x_1/x_2, for the step response, the sinusoidal term cancels because

$$K_1 \cos(\omega_d t_1 + \phi_1) = K_1 \cos(\omega_d t_2 + \phi_1)$$

where $t_2 = t_1 + T_{period}$. The same cancellation occurs if we use the ratio of relative peaks for the impulse response. Therefore, the ratio of relative peak values involves only the ratio of the exponential envelope terms

$$\frac{x_1}{x_2} = \frac{e^{-\zeta\omega_n t_1}}{e^{-\zeta\omega_n t_2}} = \frac{e^{-\zeta\omega_n t_1}}{e^{-\zeta\omega_n(t_1+T_{\text{period}})}} = \frac{1}{e^{-\zeta\omega_n T_{\text{period}}}} = e^{\zeta\omega_n T_{\text{period}}} \tag{7.85}$$

Taking the natural logarithm of the ratio of relative peaks yields

$$\ln\frac{x_1}{x_2} = \zeta\omega_n T_{\text{period}} \tag{7.86}$$

Substituting for the period of the underdamped response, $T_{\text{period}} = 2\pi/\omega_n\sqrt{1-\zeta^2}$, in Eq. (7.86) results in the cancellation of undamped natural frequency and yields

$$\ln\frac{x_1}{x_2} = \frac{2\pi\zeta}{\sqrt{1-\zeta^2}} \tag{7.87}$$

Finally, we can solve Eq. (7.87) for ζ to obtain an expression for damping ratio

$$\zeta = \frac{\delta}{\sqrt{4\pi^2 + \delta^2}} \tag{7.88}$$

where the symbol δ is the natural logarithm of the ratio of relative peaks, or *log decrement*

$$\text{Log decrement:} \quad \delta = \ln\frac{x_1}{x_2}$$

Hence, Eq. (7.88) can be used to estimate damping ratio ζ from the log decrement δ, which may be obtained from either a step or an impulse response. The reader should note that the peak values x_1 and x_2 must be computed relative to the respective steady-state value, as shown in Fig. 7.22. In addition, the reader should note that the log decrement method is difficult to apply to systems with "moderate to large" damping ratios, such as $0.4 < \zeta < 1$, because the second peak response is difficult to discern from the steady-state response. Therefore, this method has practical use only for lightly damped systems.

7.5 HIGHER-ORDER SYSTEMS

Once we understand the nature of first- and second-order system responses, we can develop a qualitative feel for the response of third- and higher-order systems. Determining the characteristic roots (or, equivalently, the poles of the transfer function) should always be our first step in analyzing a linear system, whether it is a first-, second-, or higher-order system. The roots will tell us whether or not the system is stable, and if the free response is composed of exponential, sinusoidal, or damped sinusoidal functions. For example, consider the third-order linear I/O equation

$$\dddot{y} + 2.5\ddot{y} + 38\dot{y} + 18.5y = u(t) \tag{7.89}$$

The characteristic equation is

$$r^3 + 2.5r^2 + 38r + 18.5 = 0 \tag{7.90}$$

which has three roots at $r_1 = -0.5$ and $r_{2,3} = -1 \pm j6$. Hence, the homogeneous or free response will be the sum of an exponential function (the real root at $r_1 = -0.5$) and a damped sinusoidal function (the complex conjugate roots at $r_{2,3} = -1 \pm j6$)

$$y_H(t) = c_1 e^{-0.5t} + c_2 e^{-t}\cos(6t + \phi) \tag{7.91}$$

Note that the first term in Eq. (7.91) is due to the real root r_1 and the second term is due to the complex conjugate roots r_2 and r_3. Part of the free response $y_H(t)$ is an exponential function $e^{-0.5t}$ that "dies out" in about 8 s (note that its time constant is $\tau = 2$ s), while the other component of $y_H(t)$ is a damped sinusoid that "dies out" in about 4 s and oscillates with a frequency of 6 rad/s. The three unknown constants c_1, c_2, and ϕ are determined from the three initial conditions $y(0)$, $\dot{y}(0)$, and $\ddot{y}(0)$, and the input $u(t)$ (the input also determines the particular or forced solution y_P).

This simple example demonstrates that the free response of third- or higher-order systems is simply composed of a sum of first- and second-order response functions. Our knowledge of the free response for first- and second-order systems thus allows us to obtain a qualitative feel for the free response of a higher-order system.

Example 7.9

Given the linear system shown in Fig. 7.23, determine the characteristics of the natural response $y_H(t)$ and estimate its time to reach steady state if the input $u(t)$ is a unit-step function.

The natural response is dictated by the characteristic roots, which are equivalent to the poles of the transfer function $G(s)$. The poles of $G(s)$ satisfy

$$s^5 + 24s^4 + 240s^3 + 1392s^2 + 4400s + 4800 = 0$$

which can be solved using the MATLAB `roots` command

```
>> roots([ 1 24 240 1392 4400 4800 ])
```

Using this command, we find that the five poles (or, characteristic roots r_i) are $r_1 = -2$, $r_2 = -6$, $r_3 = -10$, and $r_{4,5} = -3 \pm j5.5678$. Therefore, the natural response has the form

$$y_H(t) = c_1 e^{-2t} + c_2 e^{-6t} + c_3 e^{-10t} + c_4 e^{-3t} \cos(5.5678t + \phi)$$

Hence, the natural response is composed of three exponential functions (due to the three real roots), and a damped sinusoidal function (due to the two complex roots). The "fastest" part of the natural response is the exponential function e^{-10t} because it reaches steady state in about 0.4 s. We see that the "slowest" part of $y_H(t)$ is the exponential function e^{-2t} as it reaches steady state in about 2 s, and therefore the step response $y(t)$ will reach its steady-state value in about 2 s. The reader should see that the steady-state response to a unit-step input is $y_{ss} = 0.5125$ (use the system DC gain). Furthermore, because the natural response $y_H(t)$ dies out as time $t \to \infty$ it is also the transient response.

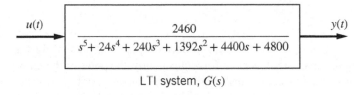

LTI system, $G(s)$

Figure 7.23 LTI system for Example 7.9.

7.6 STATE-SPACE REPRESENTATION AND EIGENVALUES

In Section 7.2, we presented an overview of the solution of linear ODEs with constant coefficients. Our solution process began with an assumed form for the homogenous solution, namely $y_H(t) = c e^{rt}$, which we applied to a homogeneous I/O equation. As a result, we obtained the characteristic equation (an nth-order polynomial in r), which leads to the characteristic roots. It is the characteristic roots that tell us the nature of the homogeneous response; that is, whether or not the response is stable or unstable, or the transient response is underdamped or overdamped. Recall that the *poles* of the system transfer function $G(s)$ are the values of s that cause its denominator to equal zero, and that the poles of $G(s)$ are identical to the characteristic roots. This fact should not be surprising as the system transfer function is derived from the governing I/O equation, and hence the denominator of $G(s)$ is the same nth-order polynomial that is the characteristic equation. The reader may wish to review Example 7.3 to see the connection between the characteristic equation and its roots and the poles of the system transfer function.

In this section, we present another approach for determining the characteristic equation that is based on the state-space representation (SSR). To begin, consider the linear homogeneous state equation

$$\dot{\mathbf{x}} = \mathbf{A}\mathbf{x} \tag{7.92}$$

where $\mathbf{x}$ is the $n \times 1$ state vector, and $\mathbf{A}$ is the $n \times n$ state matrix composed of constant coefficients. Following our previous method of solving linear differential equations, let us assume an exponential solution form for *each* state variable:

$$x_1(t) = c_1 e^{\lambda t}$$
$$x_2(t) = c_2 e^{\lambda t}$$
$$\vdots$$
$$x_n(t) = c_n e^{\lambda t} \tag{7.93}$$

where the constants c_i are generally different for each state-variable solution, but the exponential function $e^{\lambda t}$ is the same for each state solution. Equation (7.93) can be written in a compact vector format

$$\mathbf{x}(t) = \mathbf{c}e^{\lambda t} \tag{7.94}$$

where $\mathbf{c}$ is an $n \times 1$ column vector containing the constants $c_1, c_2, \ldots, c_n$. The time derivative of the assumed solution form, Eq. (7.94), is

$$\dot{\mathbf{x}}(t) = \lambda \mathbf{c}e^{\lambda t} \tag{7.95}$$

Equation (7.95) is equal to the state equation (7.92). Equating the right-hand side of the state equation (7.92) with Eq. (7.95) (along with the substitution $\mathbf{x} = \mathbf{c}e^{\lambda t}$) yields

$$\lambda \mathbf{c}e^{\lambda t} = \mathbf{A}\mathbf{c}e^{\lambda t} \tag{7.96}$$

Moving the right-hand side of Eq. (7.96) to the left-hand side and factoring out $\mathbf{c}e^{\lambda t}$ yields

$$[\lambda \mathbf{I} - \mathbf{A}]\mathbf{c}e^{\lambda t} = \mathbf{0} \tag{7.97}$$

Note that the bracket term in Eq. (7.97) must be an $n \times n$ matrix, and therefore the scalar λ is multiplied by the $n \times n$ identity matrix $\mathbf{I}$, which consists of ones on the main diagonal and zeros elsewhere. The right-hand side of Eq. (7.97) is an $n \times 1$ column vector of zeros. Except for the trivial solution $\mathbf{x} = \mathbf{c}e^{\lambda t} = \mathbf{0}$, the term $\mathbf{c}e^{\lambda t}$ is not zero for all values of time t, and therefore Eq. (7.97) is satisfied only if the determinant of the matrix in the brackets is zero, or

$$|\lambda \mathbf{I} - \mathbf{A}| = \det(\lambda \mathbf{I} - \mathbf{A}) = 0 \tag{7.98}$$

Expanding the determinant in Eq. (7.98) produces an nth-order polynomial in λ, which is best demonstrated by an example. Consider the third-order system matrix

$$\mathbf{A} = \begin{bmatrix} 0 & 1 & 0 \\ 0 & 0 & 1 \\ -12 & -8 & -3 \end{bmatrix} \tag{7.99}$$

The matrix $\lambda \mathbf{I} - \mathbf{A}$ is

$$\lambda \mathbf{I} - \mathbf{A} = \begin{bmatrix} \lambda & 0 & 0 \\ 0 & \lambda & 0 \\ 0 & 0 & \lambda \end{bmatrix} - \begin{bmatrix} 0 & 1 & 0 \\ 0 & 0 & 1 \\ -12 & -8 & -3 \end{bmatrix} = \begin{bmatrix} \lambda & -1 & 0 \\ 0 & \lambda & -1 \\ 12 & 8 & \lambda+3 \end{bmatrix} \tag{7.100}$$

The determinant of Eq. (7.100) is

$$\begin{vmatrix} \lambda & -1 & 0 \\ 0 & \lambda & -1 \\ 12 & 8 & \lambda+3 \end{vmatrix} = \lambda^3 + 3\lambda^2 + 8\lambda + 12 = 0 \tag{7.101}$$

Equation (7.101) is the system's characteristic equation, which in this case is a third-order polynomial in the parameter λ. The n values of λ for which the characteristic equation (7.101) is zero are called the *eigenvalues* of the system matrix **A**. They are used in the homogeneous state solution $\mathbf{x}(t) = \mathbf{c}e^{\lambda t}$ just as we used the characteristic roots r_i in the homogeneous response $y_H(t) = ce^{rt}$ shown in Eq. (7.7). Hence, the eigenvalues tell us the nature of the free or transient response of a dynamic system.

Knowledge of the characteristic roots allows one to understand the natural response of linear dynamic systems, and the characteristic roots are easily determined from the system's mathematical model, which may be represented as an I/O equation, transfer function, or SSR. This important result can be summarized as follows:

1. If the system is represented as an SISO I/O equation, then the n characteristic roots r_i can be obtained by solving the nth-order characteristic equation, which is readily determined from the coefficients in the I/O equation.

2. If the system is represented as a transfer function $G(s)$, then the n poles are the values of s that make the nth-order denominator polynomial of $G(s)$ equal zero. The poles of $G(s)$ are equivalent to the characteristic roots r_i.

3. If a state-space approach is used, then the n eigenvalues can be obtained from the determinant $|\lambda \mathbf{I} - \mathbf{A}| = 0$. The eigenvalues λ_i are equivalent to the characteristic roots r_i and the poles of the system transfer function.

We can use MATLAB to compute the eigenvalues of the system matrix **A** by using the command

```
>> eig(A)
```

The following example demonstrates the equivalence between the characteristic roots, the poles of the transfer function, and the eigenvalues of the system matrix **A**.

Example 7.10

Given the SISO I/O equation

$$2\ddot{y} + 8\dot{y} + 40y = 3u \tag{7.102}$$

determine (a) the characteristic roots, (b) the poles of the transfer function, and (c) the eigenvalues of the system matrix. First, let us write the second-order characteristic equation for this ODE

$$2r^2 + 8r + 40 = 0$$

or, dividing by 2

$$r^2 + 4r + 20 = 0 \tag{7.103}$$

MATLAB can be used to determine the roots

```
>> roots([ 1 4 20 ])      or      >> roots([ 2 8 40 ])
```

which yields the two characteristic roots, $r_{1,2} = -2 \pm j4$. Because the roots are complex, the system is underdamped. We can derive the transfer function by applying the D operator to the governing I/O equation (7.102), which yields

$$(2D^2 + 8D + 40)y = 3u(t)$$

Next, form the ratio of output to input, $y(t)/u(t)$, and replace D with the Laplace variable s to derive the transfer function $G(s)$

$$G(s) = \frac{3}{2s^2 + 8s + 40} = \frac{1.5}{s^2 + 4s + 20}$$

The poles of the transfer function $G(s)$ are determined by setting the denominator polynomial to zero

$$2s^2 + 8s + 40 = 0 \quad \text{or} \quad s^2 + 4s + 20 = 0 \tag{7.104}$$

which is identical to the characteristic equation (7.103). Hence, the poles of $G(s)$ are equivalent to the characteristic roots $r_{1,2} = -2 \pm j4$.

Finally, to find the eigenvalues we must obtain an SSR. We can select the two states as $x_1 = y$ and $x_2 = \dot{y}$, and consequently the two state-variable equations are

$$\dot{x}_1 = \dot{y} = x_2$$

$$\dot{x}_2 = \ddot{y} = \frac{1}{2}(3u - 8\dot{y} - 40y) = 1.5u - 4x_2 - 20x_1$$

which can be assembled in the matrix-vector format as the state equation

$$\dot{\mathbf{x}} = \begin{bmatrix} 0 & 1 \\ -20 & -4 \end{bmatrix} \mathbf{x} + \begin{bmatrix} 0 \\ 1.5 \end{bmatrix} u$$

The eigenvalues are computed from the determinant $|\lambda \mathbf{I} - \mathbf{A}| = 0$

$$|\lambda \mathbf{I} - \mathbf{A}| = \begin{vmatrix} \lambda & -1 \\ 20 & \lambda + 4 \end{vmatrix} = \lambda^2 + 4\lambda + 20 = 0 \tag{7.105}$$

This polynomial equation is identical to Eqs. (7.103) and (7.104), and hence it is the characteristic equation. The eigenvalues are $\lambda_{1,2} = -2 \pm j4$ and are equivalent to the characteristic roots and the poles of $G(s)$.

7.7 APPROXIMATE MODELS

Oftentimes during the preliminary design stage, it is beneficial to use a *simplified* or *approximate* system model instead of the complete (more complex) higher-order model. For example, control system analysis and design can be performed using approximate subsystem models without significant loss of accuracy. The subsequent simulation run times can be greatly reduced by replacing higher-order, nonlinear, complex models with simplified, linear, reduced-order models. In other cases it is difficult to develop an accurate mathematical model from fundamental physical laws because of complex geometry, uncertain parameters (such as friction), or nonlinear effects (such as flow forces in fluid systems). If I/O data is available from conducting experiments with a physical system, then it may be possible to develop a simple model by "fitting" the measurements with a first- or second-order response.

In many cases, a properly designed first- or second-order model can adequately represent the full complex model. Figure 7.24 shows a schematic block diagram where the complex or uncertain system is replaced by an approximate model. It is important for the reader to understand that (1) not every complex system can be accurately replaced by a simple approximate model and (2) the original I/O relationship must be maintained by the approximate model.

This brief discussion alludes to the very important topic of model accuracy versus model simplicity, which we discussed in Chapter 1. The engineer must always remember that a compromise exists between modeling effort and computational time and model accuracy. In some applications, leaving out the higher-order and/or nonlinear terms may

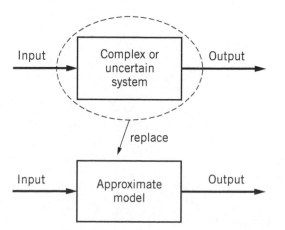

Figure 7.24 Replacing a complex or uncertain system with an approximate model.

not degrade the accuracy of the simplified model; in some cases including the complex dynamics is critical. Engineering experience is extremely important in making these decisions. The following example illustrates replacing a complex system with an approximate model.

Example 7.11

Figure 7.25 shows the measured step response of a solenoid actuator–valve system. This response has been experimentally obtained by measuring the valve displacement that results from a 38-mA current step input applied at $t = 1$ s [1]. Note that the valve response in Fig. 7.25 displays a very small amount of random "noise" or "chatter" because of imprecise measurements in the experimental setting. If possible, develop an approximate actuator–valve model from the experimental data.

   Note that the measured step response approximates an exponential rise to a constant steady-state displacement. Because the step response does not exhibit any overshoot, we can model the I/O relationship between current (input) and valve position (output) with a simple first-order linear ODE

$$\tau \dot{y} + y = au(t)$$

where y is the valve position (mm), and $u(t)$ is the current input (mA). Figure 7.25 shows that the steady-state position is ~0.44 mm for a step input $u = 38$ mA, and therefore coefficient $a = 0.44/38 = 0.0116$. Figure 7.25 also shows that the valve position reaches its steady-state value with settling time $t_S = 1.2$ s, and therefore the time constant of the approximate first-order model is $\tau = t_S/4 = 0.3$ s. Using these numerical values, the approximate first-order model of the actuator–valve system becomes

$$0.3\dot{y} + y = 0.0116u(t)$$

or, the first-order transfer function is

$$G(s) = \frac{0.0116}{0.3s + 1}$$

Figure 7.26 shows the first-order model that approximates the complex actuator–valve system. If we applied a 38-mA step input to the approximate model shown in Fig. 7.26, its response would look very much like the response from the experimental hydraulic test rig (Fig. 7.25). Figure 7.27 shows the first-order response of the approximate model plotted with the measured valve response. The approximate model accurately predicts the settling time and steady-state response, but it

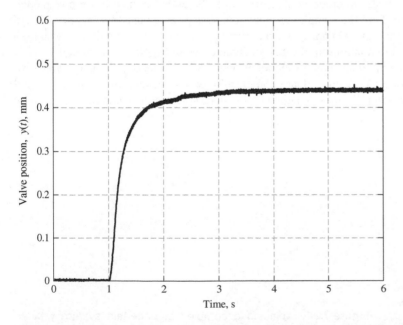

Figure 7.25 Measured step response of an actuator–valve system (Example 7.11).

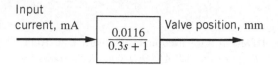

Input
current, mA → $\dfrac{0.0116}{0.3s + 1}$ → Valve position, mm

Figure 7.26 First-order model that approximates the actuator–valve system (Example 7.11).

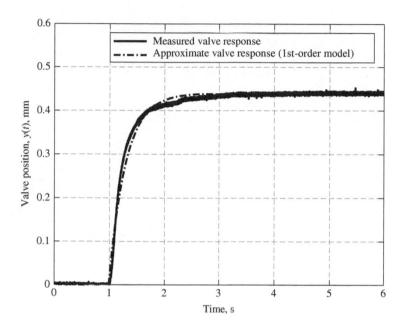

Figure 7.27 Step responses for the complex actuator–valve system and the first-order approximate model (Example 7.11).

exhibits some error with the measured valve position during the transient-response phase. An experienced engineer would need to determine if the first-order transfer function could be used to model the complex actuator–valve system based on factors such as accuracy requirements.

SUMMARY

This chapter has presented a brief review of the solution of linear ODEs. The complete solution of an ODE is composed of the homogeneous (free) and particular (forced) responses, where the free response is dictated by the characteristic roots, and the forced response depends on the nature of the input function. First- and second-order systems were studied in depth, and the reader should recall that the free or natural response of a first-order system is characterized by the time constant τ, while the free response of an underdamped second-order system depends on damping ratio ζ and undamped natural frequency ω_n. A major emphasis of this chapter is obtaining the *form* of the free (or transient) response rather than the complete mathematical solution including its constants of integration. In most cases, an accurate sketch of a first- or second-order system response can be produced with knowledge of the time constants or ζ and ω_n. The reader may want to review Table 7.4 for a summary of the step-response characteristics of an underdamped second-order system.

Perhaps the most important concept contained in this chapter is the calculation of the characteristic roots and their effect on the free or transient response. The characteristic roots are computed from the characteristic equation, which can be derived from the system's I/O equation, transfer function, or state-space model. We have seen that the poles of the transfer function $G(s)$ and the eigenvalues of the system matrix $\mathbf{A}$ are identical to the characteristic roots. The location of the roots in the complex plane determines the system's response speed and damping characteristics.

REFERENCE

1. Carpenter, R., and Fales, R., "Mixed Sensitivity H-infinity Control Design with Frequency Domain Uncertainty Modeling for a Pilot Operated Proportional Control Valve," *Proceedings of the 5th ASME Dynamic Systems and Control Conference*, Fort Lauderdale, FL, Oct. 2012, Vol. 1, pp. 733–741.

Chapter **8**

System Analysis Using Laplace Transforms

8.1 INTRODUCTION

In Chapters 5, 6, and 7, we presented the system's transfer function as a convenient means to represent and analyze the input–output relationship of a single-input, single-output (SISO) dynamic system. Furthermore, we derived the system transfer function in Chapter 5 by using the differential (or D) operator method without formally applying the Laplace transformation. Transfer functions were used in Chapters 5 and 6 to represent the system dynamics in the block-diagram format (e.g., Simulink models). In addition, transfer functions are used extensively in analyzing a system's frequency response (Chapter 9) and in closed-loop control systems (Chapter 10).

In this chapter, we present a brief overview of Laplace transform theory and its use in obtaining the response of dynamic systems that are modeled by linear, time-invariant (LTI) differential equations. Laplace transformation offers a systematic approach for solving an LTI differential equation by transforming its variables in time t to an algebraic equation in the domain of the complex Laplace variable s. Any existing initial conditions of the dynamic system are handled in a systematic manner using the Laplace transformation, and the system's time response is finally obtained by determining the inverse Laplace transform. The reader should recall that we obtained the analytical solution of an LTI differential equation in Chapter 7 by assuming a solution form in the time domain. While this approach is typically the first technique presented in a standard course in differential equations, Laplace transformations offer an alternate approach to solving LTI differential equations "by hand."

8.2 LAPLACE TRANSFORMATION

The *Laplace transform* converts the function $f(t)$ from the time domain to the domain of the complex variable s, and is defined by

$$\mathscr{L}\{f(t)\} = F(s) = \int_0^\infty f(t)e^{-st}dt \tag{8.1}$$

The Laplace transform variable $s = \sigma + j\omega$ is a complex variable where σ and ω are the real and imaginary parts, respectively. The operation defined by Eq. (8.1) can be stated as "the Laplace transform of $f(t)$ is the complex function $F(s)$." Typically, the uppercase letter is used for the Laplace transform of the corresponding time function (lowercase letter) and the parenthetical s indicates that the complex (Laplace) variable is the independent variable. The Laplace transform converts an ordinary differential equation (ODE) into an algebraic equation in s, which can be easily manipulated. The inverse operation

$$\mathscr{L}^{-1}\{F(s)\} = f(t) \tag{8.2}$$

converts the complex function $F(s)$ into the time function $f(t)$ and is known as the "inverse Laplace transform of $F(s)$."

Laplace Transforms of Common Time Functions

We can use Eq. (8.1) to compute the Laplace transforms of common time functions (such as exponential and sinusoidal functions) and construct a table of these "standard" Laplace transforms. Evaluating the Laplace transformation integral

(Eq. 8.1) requires the mathematical knowledge of an introductory calculus course. We demonstrate the Laplace transform operation (Eq. 8.1) with the following examples.

Example 8.1

Compute the Laplace transform of the exponential function $f(t) = Ae^{-at}$ for $t \geq 0$ where A and a are constants. The function $f(t) = 0$ for $t < 0$.

Using the definition of the Laplace transform (Eq. 8.1)

$$\mathcal{L}\{Ae^{-at}\} = \int_0^\infty Ae^{-at}e^{-st}dt$$

$$= A\int_0^\infty e^{-(a+s)t}dt$$

$$= A \left.\frac{e^{-(s+a)t}}{-(s+a)}\right|_{t=0}^{t=\infty} = \frac{Ae^{-\infty}}{-(s+a)} - \frac{Ae^0}{-(s+a)}$$

Hence, the Laplace transform of the exponential function $f(t) = Ae^{-at}$ is

$$F(s) = \frac{A}{s+a}$$

Example 8.2

Compute the Laplace transform of the step function $f(t) = A$ for $t > 0$ where A is a constant. The step function $f(t) = 0$ for $t \leq 0$.

Using the definition of the Laplace transform (Eq. 8.1)

$$\mathcal{L}\{A\} = \int_0^\infty Ae^{-st}dt$$

$$= A \left.\frac{e^{-st}}{-s}\right|_{t=0}^{t=\infty} = \frac{Ae^{-\infty}}{-s} - \frac{Ae^0}{-s}$$

Hence, the Laplace transform of the step function $f(t) = A$ is

$$F(s) = \frac{A}{s}$$

If $f(t)$ is a *unit*-step function, then $A = 1$ and $F(s) = 1/s$.

Example 8.3

Compute the Laplace transform of the sinusoidal function $f(t) = A\sin\omega t$ for $t \geq 0$ where A is the constant amplitude and ω is the constant frequency. The function $f(t) = 0$ for $t < 0$.

Using Eq. (8.1)

$$\mathcal{L}\{A\sin\omega t\} = \int_0^\infty A\sin\omega t e^{-st}dt$$

We can use Euler's formula to relate sinusoidal functions to a complex exponential function

$$e^{j\omega t} = \cos\omega t + j\sin\omega t$$

or, $\sin \omega t = (e^{j\omega t} - \cos \omega t)/j$. Using the complex conjugate of Euler's formula

$$e^{-j\omega t} = \cos \omega t - j \sin \omega t$$

we can substitute $\cos \omega t = e^{-j\omega t} + j \sin \omega t$ into the previous expression to yield

$$\sin \omega t = \frac{1}{2j}\left(e^{j\omega t} - e^{-j\omega t}\right)$$

Using this result the Laplace transform integral becomes

$$\mathscr{L}\{A \sin \omega t\} = \frac{A}{2j}\int_0^\infty \left(e^{j\omega t} - e^{-j\omega t}\right)e^{-st}dt$$

$$= \frac{A}{2j}\int_0^\infty \left(e^{-(s-j\omega)t} - e^{-(s+j\omega)t}\right)dt$$

$$= \frac{A}{2j}\left[\frac{e^{-(s-j\omega)t}}{-(s-j\omega)}\Big|_{t=0}^{t=\infty} + \frac{e^{-(s+j\omega)t}}{(s+j\omega)}\Big|_{t=0}^{t=\infty}\right]$$

$$= \frac{A}{2j}\left[\frac{1}{s-j\omega} - \frac{1}{s+j\omega}\right]$$

Multiplying both terms by the appropriate complex conjugate $(s+j\omega$ or $s-j\omega)$ will produce a common denominator and the result is

$$F(s) = \frac{A\omega}{s^2 + \omega^2}$$

which is the Laplace transform of the sine function $f(t) = A \sin \omega t$.

In summary, the examples show that computing the Laplace transform of a time function $f(t)$ requires integrating the product of $f(t)$ and e^{st} from $t = 0$ to $t = \infty$. Once we have computed the Laplace transform of a common time function, we do not need to recalculate it every time it is needed; instead we can consult a table of precomputed Laplace transforms. Table 8.1 summarizes the Laplace transforms of common time functions, including the results from Examples 8.1, 8.2, and 8.3.

Laplace Transform Using MATLAB

MATLAB's Symbolic Math Toolbox can be used to compute the Laplace transform of a given time function $f(t)$. To do so, the user must define the time function $f(t)$ as the symbolic object f and the corresponding Laplace transform is computed using the single-line command F = laplace(f). For example, the following MATLAB commands will determine the Laplace transform of $f(t) = 3 \sin 2t$

```
>> syms t              % define variable t (time) as a symbolic object
>> f = 3*sin(2*t)      % define function f(t) = 3 sin 2t as a symbolic object
>> F = laplace(f)      % compute the Laplace transform F(s)
>> pretty(F)           % display F in a format similar to typeset mathematics
```

The third and fourth commands will display $F(s)$ as a symbolic function of s. Typing all four line commands shown above results in

```
          6
        ------
          2
        s   + 4
```

which is easily identified as the answer for Example 8.3 with $A = 3$ and $\omega = 2$.

Table 8.1   Laplace Transforms of Common Time Functions

Number	Time Function, $f(t)$	Laplace Transform, $\mathscr{L}\{f(t)\} = F(s)$
1	Unit impulse, $\delta(t)$	1
2	Unit step, $U(t)$	$\dfrac{1}{s}$
3	A	$\dfrac{A}{s}$
4	Unit ramp, t	$\dfrac{1}{s^2}$
5	t^2	$\dfrac{2}{s^3}$
6	e^{-at}	$\dfrac{1}{s+a}$
7	te^{-at}	$\dfrac{1}{(s+a)^2}$
8	$\sin \omega t$	$\dfrac{\omega}{s^2+\omega^2}$
9	$\cos \omega t$	$\dfrac{s}{s^2+\omega^2}$
10	$e^{-at}\sin \omega t$	$\dfrac{\omega}{(s+a)^2+\omega^2}$
11	$e^{-at}\cos \omega t$	$\dfrac{s+a}{(s+a)^2+\omega^2}$

Poles and Zeros of Laplace Transforms

In general, a Laplace transform can be expressed using the form

$$F(s) = \frac{a(s+z_1)(s+z_2)\cdots(s+z_m)}{(s+p_1)(s+p_2)\cdots(s+p_n)}$$

where a is a constant. The values $s = -z_1, s = -z_2, \ldots, s = -z_m$ that make $F(s) = 0$ are called the *zeros* of the transform $F(s)$ while the values $s = -p_1, s = -p_2, \ldots, s = -p_n$ that make $F(s) = \infty$ (or, the denominator of $F(s)$ equals zero) are called the *poles* of $F(s)$.

As a first example, consider the following Laplace transform

$$F_1(s) = \frac{1}{s+3}$$

we see that the single pole is at $s = -3$ (this Laplace transform has no zeros). Table 8.1 (number 6) shows us that the time function corresponding to this Laplace transform is the exponential function $f_1(t) = e^{-3t}$. As a second example consider the Laplace transform

$$F_2(s) = \frac{s}{s^2+16}$$

This Laplace transform has a single zero at $s = 0$ (the origin) and two complex poles at $s = j4$ and $s = -j4$ (the reader should recall that by definition $j^2 = -1$ and, therefore, $(j4)^2 = (-j4)^2 = j^2 16 = -16$). The time function corresponding to this Laplace transform (see number 9 in Table 8.1) is the cosine function $f_2(t) = \cos 4t$.

These two simple examples demonstrate that the poles of the Laplace transform $F(s)$ determine the nature of the corresponding time function $f(t)$. In general, when the poles of $F(s)$ are real numbers that are distinct (not repeated), the corresponding time function $f(t)$ is composed of exponential functions (see number 6 in Table 8.1). When the poles of

$F(s)$ are imaginary numbers (which must appear in conjugate pairs), the corresponding $f(t)$ is composed of sinusoidal functions (see numbers 8 and 9 in Table 8.1). If a Laplace transform has n poles at $s = 0$ (origin), then the corresponding time function is composed of n polynomial functions, or $f(t) = \sum t^{k-1}$, $k = 1, 2, \ldots, n$ (see numbers 2, 4, and 5 in Table 8.1). Finally, if $F(s)$ has complex conjugate poles (real and imaginary parts), then $f(t)$ is the product of an exponential function and a sinusoidal functions (see numbers 10 and 11 in Table 8.1). We further discuss the connection between a Laplace transform's poles and the corresponding time response in Sections 8.3 and 8.4.

Laplace Transform Properties and Theorems

In this subsection, we present a number of useful properties and theorems associated with the Laplace transform and demonstrate some of these properties with a few examples. Throughout this section, let $\mathcal{L}\{f(t)\} = F(s)$ and a_i be a constant. Details and derivations of these Laplace transformation properties can be found in many standard textbooks on ordinary differential equations such as Reference 1.

Superposition (linearity)

a. $\quad \mathcal{L}\{af(t)\} = aF(s)$ $\hfill (8.3)$

b. $\quad \mathcal{L}\{a_1 f_1(t) + a_2 f_2(t)\} = a_1 F_1(s) + a_2 F_2(s)$ $\hfill (8.4)$

Differentiation

a. $\quad \mathcal{L}\{\dot{f}(t)\} = sF(s) - f(0)$ $\hfill (8.5)$

b. $\quad \mathcal{L}\{\ddot{f}(t)\} = s^2 F(s) - sf(0) - \dot{f}(0)$ $\hfill (8.6)$

c. $\quad \mathcal{L}\left\{ \dfrac{d^n}{dt^n} f(t) \right\} = s^n F(s) - s^{n-1} f(0) - s^{n-2} \dot{f}(0) - \cdots - f^{(n-1)}(0)$ $\hfill (8.7)$

In the above equations, $f(0)$ is the initial value of $f(t)$ evaluated at $t = 0, \dot{f}(0)$ is the initial value of the first time derivative, $f^{(n-1)}(0)$ is the initial value of the $n - 1$ time derivative, and so on. If all initial conditions are zero, then the Laplace transform of the nth time derivative of $f(t)$ is $s^n F(s)$. In general, differentiation in the time domain is equivalent to multiplication by s in the Laplace domain.

Integration

$$\mathcal{L}\left\{ \int_0^t f(\tau)d\tau \right\} = \frac{F(s)}{s} \hfill (8.8)$$

In general, integration in the time domain is equivalent to division by s in the Laplace domain.

Multiplication $f(t)$ by e^{-at} in the time domain

$$\mathcal{L}\{e^{-at}f(t)\} = F(s + a) \hfill (8.9)$$

Therefore, taking the Laplace transform of $e^{-at}f(t)$ is equal to the Laplace transform of $f(t)$ with complex variable s replaced by $s + a$. For example, note that the Laplace transforms of $e^{-at}\sin \omega t$ and $e^{-at}\cos \omega t$ in Table 8.1 (numbers 10 and 11) are simply the Laplace transforms of $\sin \omega t$ and $\cos \omega t$ with s replaced by $s + a$.

Final-value theorem

As the name implies, the final-value theorem relates the steady-state (final) value of the time function $f(\infty)$ with its Laplace transform $F(s)$. The final-value theorem is

$$f(\infty) = \lim_{t \to \infty} f(t) = \lim_{s \to 0} sF(s) \tag{8.10}$$

Therefore, we can use Eq. (8.10) and the Laplace transform $F(s)$ to compute the steady-state value of the corresponding time response $f(t)$. It is important to remember that the final-value theorem holds only for cases where the time function $f(t)$ reaches a steady-state (constant) value as time $t \to \infty$. In general, the time function $f(t)$ does *not* reach a single steady-state value if its Laplace transform $F(s)$ contains poles *on* the imaginary axis (with the exception of a *single* pole at $s = 0$) or in the right-half of the complex plane. As a simple example, consider the sinusoidal function $f(t) = \sin 3t$ that clearly does not reach a single, constant steady-state value as time $t \to \infty$. Table 8.1 shows that the Laplace transform of $\sin 3t$ is $F(s) = 3/(s^2 + 9)$, which has two poles (at $s = \pm j3$) that lie on the imaginary axis and hence the final-value theorem does not apply.

Initial-value theorem

The initial-value theorem relates the initial value of the time function $f(0+)$ with its Laplace transform $F(s)$

$$f(0+) = \lim_{t \to 0+} f(t) = \lim_{s \to \infty} sF(s) \tag{8.11}$$

Note that $f(0+)$ is the function evaluated at time $t = 0+$, which is a small incremental amount greater than zero. Unlike the final-value theorem, the initial-value theorem does not have restrictions on the poles of $F(s)$ and, therefore, it can be used to find $f(0+)$ for a sinusoidal function, for example.

Example 8.4

Compute the Laplace transform of the following time function

$$f(t) = 0.2 + 2e^{-3t} - 5e^{-6t} + 8\sin 4t$$

The superposition (linearity) property Eq. (8.4) shows that the complete Laplace transform $F(s)$ is the sum of the Laplace transforms of each individual time function. Using Table 8.1, each Laplace transform is

$$\mathscr{L}\{0.2\} = \frac{0.2}{s} \qquad \text{(number 3 in Table 8.1)}$$

$$\mathscr{L}\{2e^{-3t}\} = \frac{2}{s+3} \qquad \text{(number 6 in Table 8.1)}$$

$$\mathscr{L}\{-5e^{-6t}\} = \frac{-5}{s+6} \qquad \text{(number 6 in Table 8.1)}$$

$$\mathscr{L}\{8\sin 4t\} = 8\frac{4}{s^2 + 4^2} \qquad \text{(number 8 in Table 8.1)}$$

Therefore, the complete Laplace transform is the sum of these four transforms

$$F(s) = \frac{0.2}{s} + \frac{2}{s+3} - \frac{5}{s+6} + \frac{32}{s^2 + 16}$$

The reader should be able to compute the original time function $f(t)$ from the above Laplace transform $F(s)$ and Table 8.1.

Example 8.5

Calculate the final value $f(\infty)$ (if it exists) and initial value $f(0+)$ from the given Laplace transforms $F(s)$

(a)

$$F(s) = \frac{7(s+3)}{s^2 + 2s + 10}$$

First, we check the poles of the $F(s)$ to determine if we can use the final-value theorem. The poles of $F(s)$ are the roots of its denominator polynomial $s^2 + 2s + 10 = 0$, which yields the two complex poles $s = -1 \pm j3$. Because the complex poles lie in the left half of the complex plane (i.e., the real part of the two poles is negative), we can use the final-value theorem Eq. (8.10)

$$f(\infty) = \lim_{s \to 0} sF(s) = \lim_{s \to 0} \frac{7s(s+3)}{s^2 + 2s + 10} = \frac{(0)(3)}{10} = 0$$

Hence, the steady-state value of $f(t)$ is zero.
   We find the initial value by using Eq. (8.11)

$$f(0+) = \lim_{s \to \infty} sF(s) = \lim_{s \to \infty} \frac{7s(s+3)}{s^2 + 2s + 10} = \lim_{s \to \infty} \frac{7s^2}{s^2} = 7$$

Hence, the initial value of $f(0+)$ is 7.

(b)

$$F(s) = \frac{3(s^2 + 2)}{s(s+4)(s^2 + 4s + 5)}$$

First, we check the poles of the $F(s)$ to determine if we can use the final-value theorem. The four poles are $s = 0$, $s = -4$, and $s = -2 \pm j$. Because we have one pole at the origin and three poles strictly in the left-half plane, we can use the final-value theorem:

$$f(\infty) = \lim_{s \to 0} sF(s) = \lim_{s \to 0} \frac{3s(s^2 + 2)}{s(s+4)(s^2 + 4s + 5)} = \frac{(3)(2)}{(4)(5)} = 0.3$$

The initial value is

$$f(0+) = \lim_{s \to \infty} sF(s) = \lim_{s \to \infty} \frac{3s(s^2 + 2)}{s(s+4)(s^2 + 4s + 5)} = \lim_{s \to \infty} \frac{3s^3}{s^4} = 0$$

(c)

$$F(s) = \frac{3(s^3 + 2s^2 + 6s + 12)}{s(s+4)(s^2 + 25)}$$

The four poles of $F(s)$ are $s = 0$, $s = -4$, and $s = \pm j5$. Because the Laplace transform has two poles *on* the imaginary axis at $s = \pm j5$, the time function $f(t)$ will include the harmonic functions ($\sin 5t$ and $\cos 5t$) and hence there is no final value $f(\infty)$.
   The initial value is

$$f(0+) = \lim_{s \to \infty} sF(s) = \lim_{s \to \infty} \frac{3s(s^3 + 2s^2 + 6s + 12)}{s(s+4)(s^2 + 25)} = \lim_{s \to \infty} \frac{3s^4}{s^4} = 3$$

8.3   INVERSE LAPLACE TRANSFORMATION

We should restate that our overall goal is to use Laplace transforms to obtain the response of a dynamic system that is subjected to initial conditions and/or a known input function. Consequently, we begin our analysis with the system's mathematical model, which is an ordinary differential equation. The Laplace transform methods discussed thus far provide a systematic approach for solving the ODE and obtaining the dynamic response. This systematic approach can be summarized as

1. Take the Laplace transform of every term in the input–output (I/O) mathematical model (the ODE) and incorporate the initial conditions using the differentiation properties, Eqs. (8.5), (8.6), and (8.7).
2. Using the result from step 1, solve for the Laplace transform of the dynamic variable, $Y(s)$.
3. Obtain the system's dynamic response by taking the *inverse* Laplace transform, $y(t) = \mathscr{L}^{-1}\{Y(s)\}$.

This systematic approach is best illustrated by the following example.

Example 8.6

Given the following I/O equation (mathematical model)

$$2\ddot{y} + 10\dot{y} + 12y = u(t) \quad \text{with } y(0) = -1, \ \dot{y}(0) = 0.5$$

determine the dynamic response $y(t)$ if the input is a step function $u(t) = 4$ for $t > 0$.

We begin by taking the Laplace transform of each term on the left-hand side of the I/O equation and incorporating the respective initial conditions using Eqs. (8.5) and (8.6)

$$\mathscr{L}\{2\ddot{y}\} = 2\left(s^2 Y(s) - sy(0) - \dot{y}(0)\right) = 2s^2 Y(s) + 2s - 1$$

$$\mathscr{L}\{10\dot{y}\} = 10\left(sY(s) - y(0)\right) = 10sY(s) + 10$$

$$\mathscr{L}\{12y\} = 12Y(s)$$

Next, take the Laplace transform of the step-function input (right-hand side)

$$\mathscr{L}\{u(t)\} = \mathscr{L}\{4\} = \frac{4}{s}$$

Combining all transform results yields

$$2s^2 Y(s) + 2s - 1 + 10sY(s) + 10 + 12Y(s) = \frac{4}{s}$$

or,

$$(2s^2 + 10s + 12)Y(s) = \frac{4}{s} - 2s - 9 = \frac{4 - 2s^2 - 9s}{s} \tag{8.12}$$

Solving Eq. (8.12) for the Laplace transform $Y(s)$ we obtain

$$Y(s) = \frac{-2s^2 - 9s + 4}{2s(s^2 + 5s + 6)} = \frac{-2s^2 - 9s + 4}{2s(s + 2)(s + 3)} \tag{8.13}$$

This Laplace transform does not appear in Table 8.1. However, we can use a partial-fraction expansion to express $Y(s)$ as the sum of three fractions involving the three poles at $s = 0$, -2, and -3

$$Y(s) = \frac{-2s^2 - 9s + 4}{2s(s + 2)(s + 3)} = \frac{a_1}{s} + \frac{a_2}{s + 2} + \frac{a_3}{s + 3} \tag{8.14}$$

The corresponding constants are $a_1 = 1/3$, $a_2 = -7/2$, and $a_3 = 13/6$. Note that the inverse Laplace transform of each of the partial-fraction terms in Eq. (8.14) is easily found in Table 8.1: the first term is the Laplace transform of a constant, while the next two terms are the Laplace transforms of exponential functions. Therefore, taking the inverse Laplace transform of $Y(s)$ yields the dynamic response

$$y(t) = \frac{1}{3} - \frac{7}{2}e^{-2t} + \frac{13}{6}e^{-3t}$$

We can check this result by evaluating the initial conditions: $y(0) = (1/3) - (7/2) + (13/6) = -1$ as required. The first time derivative is $\dot{y}(t) = 7e^{-2t} - (13/2)e^{-3t}$ and, therefore, $\dot{y}(0) = 7 - (13/2) = 0.5$ as required.

Partial-Fraction Expansion Method

The critical step in Example 8.6 is computing the inverse Laplace transform of $Y(s)$ in order to determine the time-response function $y(t)$. In Example 8.6, the Laplace transform $Y(s)$ as expressed in Eq. (8.13) does not appear in Table 8.1; however, the Laplace transforms of the three partial-fraction terms in Eq. (8.14) are easily found in Table 8.1. The partial-fraction expansion method breaks a complicated Laplace transform (i.e., a ratio of two polynomials in s) into a sum of simpler terms where each denominator involves a pole (or poles) associated with $Y(s)$. Therefore, the inverse Laplace transforms of the simpler partial-fraction terms are easy to obtain because they appear in Table 8.1.

The partial-fraction expansion method can be generalized for the cases where the poles are distinct, repeated, or complex. Let us discuss each case briefly.

Partial-fraction expansion with distinct poles

When its poles are distinct, the Laplace transform $Y(s)$ can be expanded as

$$Y(s) = \frac{a_1}{s + p_1} + \frac{a_2}{s + p_2} + \cdots + \frac{a_n}{s + p_n} \tag{8.15}$$

where $s = -p_i$ are the distinct poles of $Y(s)$ and constants a_i are called the *residues* of $Y(s)$. We can evaluate the residue a_1 by multiplying both sides of Eq. (8.15) by $s + p_1$ and setting $s = -p_1$, or

$$a_1 = (s + p_1)Y(s)|_{s=-p_1}$$

The above equation can be generalized for all n residues

$$a_i = (s + p_i)Y(s)|_{s=-p_i} \quad i = 1, 2, \ldots, n \tag{8.16}$$

The MATLAB command `residue` computes the residues, poles, and "direct" terms of the partial-fraction expansion of $Y(s)$

```
>> [a,p,k] = residue(numY,denY)
```

where `a` is the vector of residues, `p` is the vector of poles corresponding to residues `a`, `k` is the vector of "direct" terms, `numY` is a row vector of the $Y(s)$ numerator coefficients, and `denY` is a row vector of the $Y(s)$ denominator coefficients. The "direct" term `k` exists only if the order of the numerator (`numY`) is greater than or equal to the order of the denominator (`denY`).

Example 8.7

Compute the inverse Laplace transform of

$$Y(s) = \frac{2s + 5}{s^2 + 8s + 12} = \frac{2s + 5}{(s + 2)(s + 6)}$$

Clearly, the two poles ($s = -2$, $s = -6$) are distinct. Therefore, the partial-fraction expansion is

$$Y(s) = \frac{2s + 5}{(s + 2)(s + 6)} = \frac{a_1}{s + 2} + \frac{a_2}{s + 6}$$

Using Eq. (8.16), the first residue is

$$a_1 = (s + 2)Y(s)|_{s=-2} = \frac{2s + 5}{s + 6}\bigg|_{s=-2} = \frac{1}{4} = 0.25$$

The second residue is

$$a_2 = (s + 6)Y(s)|_{s=-6} = \frac{2s + 5}{s + 2}\bigg|_{s=-6} = \frac{-7}{-4} = 1.75$$

We can verify the residue calculation using the MATLAB commands:

```
>> numY = [ 2 5 ];              % Y(s) numerator coefficients in descending powers of s
>> denY = [ 1 8 12 ];           % Y(s) denominator coefficients; descending powers of s
>> [a,p,k] = residue(numY,denY)
```

The result is a = [1.75 0.25] (residues), p = [-6 -2] (poles), and k = [] (null, no direct terms). Note that a(1) = 1.75 ($= a_2$) is the residue for pole p(1) = -6 and a(2) = 0.25 ($= a_1$) is the residue for pole p(2) = -2.
    Using the residues, the partial-fraction expansion is

$$Y(s) = \frac{0.25}{s + 2} + \frac{1.75}{s + 6}$$

The inverse Laplace transforms of both partial-fraction terms are exponential functions (see number 6 in Table 8.1). Therefore, the inverse Laplace transform of $Y(s)$ is

$$y(t) = 0.25e^{-2t} + 1.75e^{-6t}$$

Partial-fraction expansion with repeated poles

When the poles of the Laplace transform $Y(s)$ are repeated, the partial-fraction expansion (Eq. 8.15) is no longer valid. To show this, consider the following simple example with two repeated poles

$$Y(s) = \frac{2s + 8}{s^3 + 7s^2 + 16s + 12} = \frac{2s + 8}{(s + 2)(s + 2)(s + 3)} \tag{8.17}$$

This Laplace transform has two repeated poles at $s = -2$ and a single pole at $s = -3$. If we simply used (Eq. 8.15) for the partial-fraction expansion then the corresponding time function would be $y(t) = a_1e^{-2t} + a_2e^{-2t} + a_3e^{-3t}$, which is not correct as the first two terms could be written as Ae^{-2t} where $A = a_1 + a_2$. The correct partial-fraction expansion of Eq. (8.17) is

$$Y(s) = \frac{a_1}{(s + 2)^2} + \frac{a_2}{s + 2} + \frac{a_3}{s + 3} \tag{8.18}$$

The inverse Laplace transform can be obtained by using entries 6 and 7 from Table 8.1:

$$y(t) = a_1te^{-2t} + a_2e^{-2t} + a_3e^{-3t}$$

The difficulty lies with obtaining the residues a_i for the case with repeated poles. We do not show the derivation of the technique for obtaining the residues; instead, we present the resulting method (the reader can consult References 2 and 3 for the details of the derivation).

Using the Laplace transform (Eq. 8.17) as an example, the three residues for the partial-fraction expansion (Eq. 8.18) can be computed as

$$a_1 = (s+2)^2 Y(s)|_{s=-2} = \frac{2s+8}{s+3}\bigg|_{s=-2} = \frac{4}{1} = 4$$

$$a_2 = \frac{d}{ds}[(s+2)^2 Y(s)]|_{s=-2} = \frac{d}{ds}\left[\frac{2s+8}{s+3}\right]\bigg|_{s=-2} = \frac{2}{s+3} - \frac{2s+8}{(s+3)^2}\bigg|_{s=-2} = \frac{2}{1} - \frac{4}{1} = -2$$

$$a_3 = (s+3)Y(s)|_{s=-3} = \frac{2s+8}{(s+2)^2}\bigg|_{s=-3} = \frac{2}{1} = 2$$

Therefore, the partial-fraction expansion is

$$Y(s) = \frac{4}{(s+2)^2} + \frac{-2}{s+2} + \frac{2}{s+3}$$

and the inverse Laplace transform is

$$y(t) = 4te^{-2t} - 2e^{-2t} + 2e^{-3t}$$

This result can be verified using the following MATLAB commands

```
>> numY = [ 2 8 ];              % Y(s) numerator coefficients in Eq. (8.17)
>> denY = [ 1 7 16 12 ];        % Y(s) denominator coefficients in Eq. (8.17)
>> [b,p,k] = residue(numY,denY)
```

The result is b = [2 -2 4] (residues), p = [-3 -2 -2] (poles), and k = [] (null). The convention for MAT-LAB's residue command reverses the order of the partial-fraction expansion in Eq. (8.18), and, therefore, $b_1 = a_3$, $b_2 = a_2$, and $b_3 = a_1$.

Partial-fraction expansion with complex poles

When the poles of the Laplace transform $Y(s)$ are complex, the partial-fraction expansion method presented by Eq. (8.15) can be used. In this case, however, the residues a_i associated with Eq. (8.15) will themselves be complex coefficients, which can lead to cumbersome algebra with complex numbers. An alternative approach is to "complete the square" of the denominator of the Laplace transform so that it fits the format of entries 10 and 11 in Table 8.1. This method is best illustrated with an example.

Example 8.8

Compute the inverse Laplace transform of

$$Y(s) = \frac{2s+9}{s^2 + 6s + 25}$$

The two poles are computed by solving $s^2 + 6s + 25 = 0$, which yields the complex poles $s = -3 \pm j4$. Note that we can "complete the square" and rewrite the denominator of $Y(s)$ as the sum of two squared terms

$$Y(s) = \frac{2s+9}{(s+3)^2 + 4^2} \tag{8.19}$$

This form of the Laplace transform matches entries 10 and 11 in Table 8.1, which are the transforms of exponentially damped sine and cosine functions:

$$\mathcal{L}\{e^{-at}\sin\omega t\} = \frac{\omega}{(s+a)^2 + \omega^2}$$

$$\mathscr{L}\{e^{-at}\cos\omega t\} = \frac{s+a}{(s+a)^2 + \omega^2}$$

Comparing these transforms with Eq. (8.19), we see that $a = 3$ and $\omega = 4$. Next, we must rewrite Eq. (8.19) as the sum of two fractions that match the transformed damped sine and cosine functions:

$$Y(s) = \frac{2s+9}{(s+3)^2 + 4^2} = \frac{2(s+3)}{(s+3)^2 + 4^2} + \frac{(0.75)(4)}{(s+3)^2 + 4^2} \tag{8.20}$$

Now we can take the inverse Laplace transform of Eq. (8.20), which yields the two exponentially damped harmonic functions

$$y(t) = 2e^{-3t}\cos 4t + 0.75e^{-3t}\sin 4t$$

Inverse Laplace Transform Using MATLAB

Recall that in Section 8.2 we showed how MATLAB's Symbolic Math Toolbox can be used to compute the Laplace transform of a given time function. The Symbolic Math Toolbox can also compute the inverse Laplace transform $y(t) = \mathscr{L}^{-1}\{Y(s)\}$ using the single-line command y = ilaplace(Y). Here, the user must define the Laplace transform $Y(s)$ as a symbolic object. For example, the following MATLAB commands will determine the inverse Laplace transform of $Y(s)$ presented in Example 8.8:

```
>> syms s                          % define Laplace variable s
>> Y = (2*s + 9)/(s^2 + 6*s + 25)  % define Laplace transform Y(s)
>> y = ilaplace(Y)                 % find inverse Laplace transform
>> pretty(y)                       % display y in math typeset
```

The third and fourth commands will display $y(t)$ as a symbolic function of time t. Typing all four line commands shown above yields

```
        2 exp(-3 t) cos(4 t) + 3/4 exp(-3 t) sin(4 t)
```

which is the solution to Example 8.8.

8.4 ANALYSIS OF DYNAMIC SYSTEMS USING LAPLACE TRANSFORMS

As stated in the Introduction and restated in the previous section, the purpose of employing the Laplace transform method is to obtain the response of a dynamic system. More directly, Laplace transform techniques allow a systematic method for solving the LTI differential equations that comprise the mathematical model of a dynamic system. It is important to reiterate that Laplace transform methods can be applied only to dynamic systems that are modeled by LTI ODEs. The Laplace transform method offers an alternate technique (when compared to the methods of Chapter 7) for analytically obtaining the solution of linear ODEs.

We can group the approaches for obtaining the dynamic response using Laplace methods into two categories: (1) applying the Laplace transform to the system's time-domain I/O equation or (2) using the system's transfer function. The first approach was illustrated by Example 8.6 in the previous section where any existing initial conditions and/or system inputs were handled during the Laplace transformation of the ODE. Consequently, this approach yields the complete solution. The transfer-function approach yields only the system's response to the input as (by definition) the transfer function assumes zero initial conditions (therefore, the transfer-function approach can be viewed as a subset of the first approach). One advantage of the transfer-function approach is that subsystems (such as electrical and mechanical components) can be modeled by individual transfer functions, which can then be connected to form an integrated system (e.g., see Example 5.17 from Chapter 5 or Example 7.4 from Chapter 7). The following two subsections demonstrate the two approaches.

Laplace Transform of the Input–Output Equation

The basic steps for using the Laplace transform method to obtain the solution of the time-domain I/O equation are

1. Take the Laplace transform of every term in the I/O equation and include the initial conditions. This step will convert an ODE into an algebraic equation in the Laplace variable s.
2. Solve the algebraic expression from step 1 for the Laplace transform of the output $Y(s)$.
3. Take the inverse Laplace transform of $Y(s)$ to obtain the time response of the output $y(t)$.

The following examples illustrate this procedure.

Example 8.9

Figure 8.1 shows a single-disk mechanical system, where a motor provides torque $T_{in}(t)$ directly to the rotor inertia J. Obtain the angular velocity response of this mechanical system if the rotor is initially spinning at $\omega(0) = 120$ rad/s (1146 rpm) and the input torque is a step function of magnitude 15 N-m (applied at $t > 0$). The rotor moment of inertia is $J = 0.4$ kg-m^2 and the viscous friction coefficient is $b = 0.06$ N-m-s/rad.

A first-order mathematical model of the mechanical system was derived in Example 2.7 and is repeated below

$$J\dot{\omega} + b\omega = T_{in}(t) \quad \text{with} \quad \omega(0) = 120 \text{ rad/s} \tag{8.21}$$

We begin by taking the Laplace transform of each term on the left- and right-hand sides of the governing I/O equation (8.21) and incorporating the single initial condition:

$$\text{Left-hand side:} \quad \mathscr{L}\{J\dot{\omega} + b\omega\} = J\big(s\Omega(s) - \omega(0)\big) + b\Omega(s)$$

$$= 0.4(s\Omega(s) - 120) + 0.06\Omega(s)$$

$$\text{Right-hand side (step input):} \quad \mathscr{L}\{T_{in}(t)\} = \frac{15}{s}$$

where $\mathscr{L}\{\omega(t)\} = \Omega(s)$ is the Laplace transform of the angular velocity. Equating the left- and right-hand side terms yields

$$(0.4s + 0.06)\Omega(s) - 48 = \frac{15}{s}$$

or, solving for the Laplace transform $\Omega(s)$

$$\Omega(s) = \frac{48s + 15}{s(0.4s + 0.06)} = \frac{120s + 37.5}{s(s + 0.15)} \tag{8.22}$$

The two poles of $\Omega(s)$ are distinct and are located at $s = 0$ and $s = -0.15$, and the partial-fraction expansion of Eq. (8.22) is

$$\Omega(s) = \frac{120s + 37.5}{s(s + 0.15)} = \frac{a_1}{s} + \frac{a_2}{s + 0.15}$$

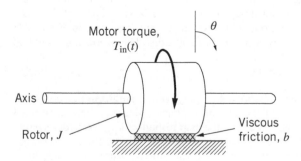

Motor torque,
$T_{in}(t)$

θ

Axis

Viscous
friction, b

Rotor, J

Figure 8.1 Single-disk mechanical system for Example 8.9.

The residues are

$$a_1 = s\Omega(s)|_{s=0} = \left.\frac{120s + 37.5}{s + 0.15}\right|_{s=0} = \frac{37.5}{0.15} = 250$$

$$a_2 = (s + 0.15)\Omega(s)|_{s=-0.15} = \left.\frac{120s + 37.5}{s}\right|_{s=-0.15} = \frac{19.5}{-0.15} = -130$$

Therefore, the Laplace transform $\Omega(s)$ in partial-fraction form is

$$\Omega(s) = \frac{250}{s} + \frac{-130}{s + 0.15}$$

and taking the inverse Laplace transform of $\Omega(s)$ yields the angular velocity response of the rotor

$$\omega(t) = 250 - 130e^{-0.15t} \text{ rad/s} \tag{8.23}$$

As a check, note that $\omega(0) = 250 - 130e^0 = 120$ rad/s as specified in the problem statement. The steady-state velocity response is $\omega(\infty) = 250 - 130e^{-\infty} = 250$ rad/s. Recall from Chapter 7 that the transient response of a first-order system "dies out" in about four time constants. For this problem, the time constant is $\tau = J/b = 6.667$ s and consequently the rotor reaches its steady-state velocity at about 26.67 s. The transient part of the complete response [Eq. (8.23)] is $-130e^{-0.15t}$ rad/s, which decays to a "small" value (< 2.4 rad/s, or less than 2% of its initial value) at time $t = 26.67$ s.

We can verify the initial and steady-state values by applying the initial- and final-value theorems to the Laplace transform, Eq. (8.22):

$$\text{Initial value:} \quad \omega(0+) = \lim_{s\to\infty} s\Omega(s) = \lim_{s\to\infty} \frac{120s^2 + 37.5s}{s^2 + 0.15s} = 120 \text{ rad/s}$$

$$\text{Final value:} \quad \omega(\infty) = \lim_{s\to 0} s\Omega(s) = \lim_{s\to 0} \frac{120s^2 + 37.5s}{s^2 + 0.15s} = \frac{37.5}{0.15} = 250 \text{ rad/s}$$

As a check, we can use the following MATLAB commands to compute the inverse Laplace transform of Eq. (8.22)

```
>> syms s                              % define Laplace variable s
>> W =(120*s + 37.5)/(s^2 + 0.15*s)    % define Ω(s) (capital W)
>> w = ilaplace(W)                     % inverse Laplace transform, ω(t)
>> pretty(w)                           % display w in math typeset
```

The result is

```
        -130 exp(-3/20 t) + 250
```

which matches Eq. (8.23), the angular velocity solution for $\omega(t)$.

As a final note, consider reworking this example using the methods of Chapter 7 where we directly solved the ODE. To begin, we rewrite the I/O equation (8.21) in the "standard form" of a first-order system

$$\frac{J}{b}\dot{\omega} + \omega = \frac{1}{b}T_{in}(t) \tag{8.24}$$

where $\tau = J/b$ is the time constant (in this case $\tau = 0.4/0.06 = 6.667$ s). The homogeneous (or transient) solution has the form $\omega_H(t) = ce^{-t/\tau}$, which is an exponential function that dies out in four time constants (or, $t_S = 4\tau = 26.67$ s). The particular (or forced) solution $\omega_P(t)$ is the steady-state response to the step input, which we can obtain directly from Eq. (8.24) or Eq. (8.21) with $\dot{\omega} = 0$, which yields $\omega_P(t) = T_{in}/b = (15 \text{ N-m})/(0.06 \text{ N-m-s/rad}) = 250$ rad/s. Therefore, the complete response is $\omega(t) = \omega_H(t) + \omega_P(t)$ or

$$\omega(t) = -130e^{-0.15t} + 250 \text{ rad/s} \tag{8.25}$$

where the constant c for the homogeneous response $\omega_H(t)$ has been obtained by applying the single initial condition $\omega(0) = 120$ rad/s. Equation (8.25) is identical to Eq. (8.23). If a sketch of the system response is desired, there is no need to obtain an explicit equation for the solution. Instead, the reader should note that the step response of a first-order system involves an exponential rise or decay from the initial condition to the steady-state value, and that the settling time is four time constants. Hence, all that is required to make an accurate sketch of the response is the time constant and steady-state value and both values can be quickly computed from the I/O equation. It is this author's opinion that direct analysis of the time-domain ODE is the preferred method of obtaining or characterizing the system's response. Laplace transform methods do provide a systematic approach for obtaining the complete system response but this author believes that the Laplace transform approach is less intuitive and unduly tedious.

Example 8.10

Figure 8.2 shows the 1-DOF rotational mechanical system studied in Example 7.8. Determine the dynamic response $\theta(t)$ if the disk is initially at rest (i.e., $\dot{\theta}(0) = 0$) with an initial angular position $\theta(0) = 0.1$ rad and the input torque is a step function $T_{\text{in}}(t) = 2.5U(t)$ N-m.

Using the numerical parameters from Example 7.8 (moment of inertia $J = 0.2$ kg-m^2, friction coefficient $b = 1.6$ N-m-s/rad, and torsional spring constant $k = 65$ N-m/rad) the mathematical model is

$$0.2\ddot{\theta} + 1.6\dot{\theta} + 65\theta = T_{\text{in}}(t) \tag{8.26}$$

Taking the Laplace transform of each term on the left- and right-hand sides of the I/O equation (8.26) yields

$$\text{Left-hand side:} \quad \mathscr{L}\{0.2\ddot{\theta}\} = 0.2\big(s^2\Theta(s) - s\theta(0) - \dot{\theta}(0)\big)$$

$$\mathscr{L}\{1.6\dot{\theta}\} = 1.6\big(s\Theta(s) - \theta(0)\big)$$

$$\mathscr{L}\{65\theta\} = 65\Theta(s)$$

$$\text{Right-hand side (step input):} \quad \mathscr{L}\{T_{\text{in}}(t)\} = \frac{2.5}{s}$$

After substituting initial conditions $\theta(0) = 0.1$ rad and $\dot{\theta}(0) = 0$ and collecting all left- and right-hand-side Laplace transforms, we obtain

$$(0.2s^2 + 1.6s + 65)\Theta(s) - 0.02s - 0.16 = \frac{2.5}{s} \tag{8.27}$$

or

$$(0.2s^2 + 1.6s + 65)\Theta(s) = \frac{0.02s^2 + 0.16s + 2.5}{s} \tag{8.28}$$

Solving Eq. (8.28) for the Laplace transform $\Theta(s)$ yields

$$\Theta(s) = \frac{0.02s^2 + 0.16s + 2.5}{s(0.2s^2 + 1.6s + 65)} \tag{8.29}$$

Dividing all terms by 0.2 yields

$$\Theta(s) = \frac{0.1s^2 + 0.8s + 12.5}{s(s^2 + 8s + 325)} \tag{8.30}$$

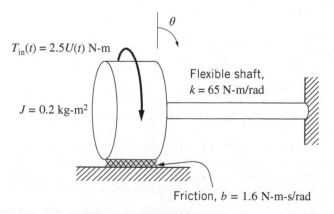

θ

$T_{\text{in}}(t) = 2.5U(t)$ N-m

Flexible shaft, $k = 65$ N-m/rad

$J = 0.2$ kg-m^2

Friction, $b = 1.6$ N-m-s/rad

Figure 8.2 1-DOF rotational mechanical system for Example 8.10.

The three poles of $\Theta(s)$ are $s = 0$ and $s = -4 \pm j17.5784$. Because two poles are complex conjugate pairs, we can "complete the square" and rewrite the corresponding second-order polynomial as

$$s^2 + 8s + 325 = (s + 4)^2 + 17.5784^2$$

Therefore, the partial-fraction expansion of Eq. (8.30) is

$$\Theta(s) = \frac{a_1}{s} + \frac{a_2(s + 4)}{(s + 4)^2 + 17.5784^2} + \frac{a_3(17.5784)}{(s + 4)^2 + 17.5784^2} \tag{8.31}$$

The residue for the single pole at the origin is

$$a_1 = s\Theta(s)|_{s=0} = \left.\frac{0.1s^2 + 0.8s + 12.5}{s^2 + 8s + 325}\right|_{s=0} = \frac{12.5}{325} = 0.0385$$

The residues for the complex poles are determined from Eqs. (8.30) and (8.31)

$$(s^2 + 8s + 325)\Theta(s)|_{s=-4+j17.5784} = \left.\frac{0.1s^2 + 0.8s + 12.5}{s}\right|_{s=-4+j17.5784}$$

$$= a_2(-4 + j17.5784 + 4) + a_3(17.5784) \tag{8.32}$$

The left-hand side of Eq. (8.32) is

$$\left.\frac{0.1s^2 + 0.8s + 12.5}{s}\right|_{s=-4+j17.5784} = \frac{0.1(-4 + j17.5784)^2 + 0.8(-4 + j17.5784) + 12.5}{-4 + j17.5784}$$

$$= \frac{-20}{-4 + j17.5784} = 0.2462 + j1.0817 \tag{8.33}$$

The right-hand side of Eq. (8.32) is

$$a_2(-4 + j17.5784 + 4) + a_3(17.5784) = 17.5784a_3 + j17.5784a_2 \tag{8.34}$$

Equating the real and imaginary parts of Eqs. (8.33) and (8.34) yields the residues

$$a_2 = \frac{1.0817}{17.5784} = 0.0615$$

$$a_3 = \frac{0.2462}{17.5784} = 0.0140$$

The numerical values of these residues could also be obtained by using Eqs. (8.30) and (8.31) evaluated at the conjugate pole $s = -4 - j17.5784$. Using the residues, the partial-fraction expansion (Eq. 8.31) becomes

$$\Theta(s) = \frac{0.0385}{s} + \frac{0.0615(s + 4)}{(s + 4)^2 + 17.5784^2} + \frac{0.0140(17.5784)}{(s + 4)^2 + 17.5784^2} \tag{8.35}$$

Taking the inverse Laplace transform (see numbers 3, 10, and 11 in Table 8.1) yields

$$\theta(t) = 0.0385 + 0.0615e^{-4t}\cos 17.5784t + 0.0140e^{-4t}\sin 17.5784t \text{ rad} \tag{8.36}$$

Figure 8.3 shows the step response equation (8.36). Equation (8.36) and Fig. 8.3 show that the dynamic response of the mechanical system is the sum of an exponentially damped sinusoidal function plus a constant (0.0385 rad). The exponentially damped sine and cosine functions are the transient response as they decay to zero in about 1 s. Note that at time $t = 0$ we have $\theta(0) = 0.0385 + 0.0615 = 0.1$ rad as required in this example (recall that the initial angular position was zero in Example 7.8). The steady-state response is $\theta(\infty) = 0.0385$ rad, which is identical to the steady-state response in Example 7.8.

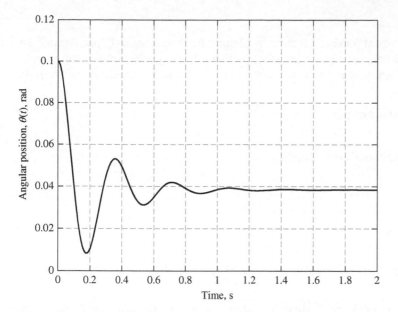

Figure 8.3 Step response of the rotational mechanical system (Example 8.10).

We can verify this result using the following MATLAB commands:

```
>> syms s                            % define Laplace variable s
>> numTh = 0.1*s^2 + 0.8*s + 12.5;   % define numerator of Θ(s)
>> denTh = s*(s^2 + 8*s + 325);      % define denominator of Θ(s)
>> Th = numTh/denTh;                 % define Θ(s)
>> th = ilaplace(Th);                % inverse Laplace transform, θ(t)
>> pretty(th)                        % display th in math typeset
```

The result is

```
                   1/2           16        1/2                              1/2
 4/65 exp(-4 t) cos(309    t) + ----- 309      exp(-4 t) sin(309      t) + 1/26
                                20085
```

Expressing the parameters using decimal representations the above solution becomes

```
0.0615 exp(-4t) cos(17.5784 t) + 0.0140 exp(-4t) sin(17.5784 t) + 0.0385
```

which is identical to the solution $\theta(t)$ presented by Eq. (8.36) and Fig. 8.3.

While the Laplace transformation offers a systematic approach to obtaining the dynamic response, it is this author's opinion that direct analysis of the system's I/O differential equation (as demonstrated in Chapter 7) provides a more intuitive approach. For example, the characteristic equation of this rotational mechanical system can be easily determined from the I/O equation (8.26)

$$0.2r^2 + 1.6r + 65 = 0 \tag{8.37}$$

where the characteristic roots are $r = -4 \pm j17.5784$. The characteristic roots match the two complex poles of the Laplace transform $\Theta(s)$ shown in Eq. (8.29). These complex roots are associated with the system's transient response (the exponentially damped sinusoidal functions). Furthermore, the important transient-response characteristics (such as settling time and period of oscillation) can be easily determined for an underdamped second-order system by using the damping ratio ζ and undamped natural frequency ω_n (see Example 7.8). The single real pole of $\Theta(s)$ at $s = 0$ corresponds to the system input (the step function) and its contribution to the Laplace transform $\Theta(s)$ is the steady-state response (0.0385 rad in this case), which can also be obtained directly from the system's I/O equation. The reader may wish to review Example 7.8 and compare the solution method to this example.

Transfer-Function Analysis

Recall that we derived an expression for the system transfer function in Chapter 5 without using Laplace-transform theory (see Section 5.6). Let us now present the definition of the transfer function using Laplace methods: the *transfer function G(s)* is defined as the ratio of the Laplace transform of the output $Y(s)$ to the Laplace transform of the input $U(s)$ with zero initial conditions

$$\text{Transfer function:}\quad G(s) = \frac{Y(s)}{U(s)}$$

As a quick example, consider the I/O equation presented in Example 5.14

$$\ddot{y} + 8\dot{y} + 10y = 4\dot{u} \tag{8.38}$$

Taking the Laplace transform of Eq. (8.38) yields

$$s^2 Y(s) - sy(0) - \dot{y}(0) + 8\big(sY(s) - y(0)\big) + 10Y(s) = 4\big(sU(s) - u(0)\big) \tag{8.39}$$

Because the definition of the transfer function requires *zero* initial conditions, or $\dot{y}(0) = y(0) = u(0) = 0$, Eq. (8.39) becomes

$$(s^2 + 8s + 10)Y(s) = 4sU(s) \tag{8.40}$$

Forming the ratio of transformed output $Y(s)$ to input $U(s)$ yields the transfer function

$$G(s) = \frac{Y(s)}{U(s)} = \frac{4s}{s^2 + 8s + 10} \tag{8.41}$$

Once we derive the transfer function, we can represent the system dynamics using the block diagram shown in Fig. 8.4. Note that the Laplace transform of the output is

$$Y(s) = G(s)U(s) \tag{8.42}$$

which can be derived from Eq. (8.41) or Fig. 8.4. The importance and usefulness of transfer-function analysis is encapsulated by noting the progression from the system I/O equation (8.38) to its transfer function Eq. (8.41) and ultimately a representation of the system output (Eq. 8.42 and Fig. 8.4).

The basic steps for using the transfer-function method to analytically obtain a system's dynamic response are

1. Derive the system transfer function $G(s)$ from the mathematical model (I/O equation).
2. Multiply the transfer function $G(s)$ by the Laplace transform of the given input function, $U(s)$, to obtain the Laplace transform of the output $Y(s)$.
3. Take the inverse Laplace transform of $Y(s)$ to obtain the time response of the output $y(t)$.

It is important for the reader to remember that the transfer-function approach can be used only for LTI systems with *zero* initial conditions. The following examples illustrate transfer-function analysis.

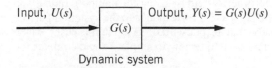

Input, $U(s)$ $G(s)$ Output, $Y(s) = G(s)U(s)$

Dynamic system

Figure 8.4 Transfer-function representation of a dynamic system.

Example 8.11

Figure 8.5 shows the series RL circuit with voltage source $e_{in}(t)$ from Example 7.6. Obtain the current response $I(t)$ for an impulse voltage input, $e_{in}(t) = 0.08\delta(t)$ V. The system has zero energy at time $t = 0$, or $I(0) = 0$, and the inductance and resistance values are $L = 0.02$ H and $R = 1.2$ Ω, respectively.

The mathematical model of the RL circuit was derived in Chapter 3 and used in Example 7.6

$$L\dot{I} + RI = e_{in}(t) \tag{8.43}$$

Taking the Laplace transform (with zero initial conditions) yields

$$(Ls + R)I(s) = E_{in}(s)$$

Forming the output/input ratio gives us the system transfer function

$$G(s) = \frac{I(s)}{E_{in}(s)} = \frac{1}{Ls + R} = \frac{1}{0.02s + 1.2} \tag{8.44}$$

Equation (8.44) is valid for any input. The Laplace transform of the current is

$$I(s) = G(s)E_{in}(s) = \frac{1}{0.02s + 1.2}E_{in}(s) \tag{8.45}$$

Equation (8.45) is also valid for any voltage input. For this problem the voltage input is an impulse, or $e_{in}(t) = 0.08\delta(t)$ V. Consulting entry 1 in Table 8.1, we see that $\mathscr{L}\{0.08\delta(t)\} = 0.08 = E_{in}(s)$. Hence, using the Laplace transform of the voltage impulse and Eq. (8.45) the Laplace transform of the current is

$$I(s) = \frac{0.08}{0.02s + 1.2} \tag{8.46}$$

or, dividing all terms by 0.02 we obtain

$$I(s) = \frac{4}{s + 60} \tag{8.47}$$

Clearly, the inverse Laplace transform of the current Eq. (8.47) is an exponential function (see number 6 in Table 8.1). Therefore, the dynamic response to the impulsive voltage input is

$$I(t) = 4e^{-60t} \text{ A} \tag{8.48}$$

Consequently, the current response shows an instantaneous jump from 0 to 4 A at time $t = 0+$ when the impulsive voltage input is applied to the electrical system. The time constant of this system is $\tau = L/R = 0.02/1.2 = 0.0167$ s and, therefore, the current decays to zero at the settling time $t_S = 4\tau = 0.0667$ s. This solution is identical to the current response obtained in Example 7.6 (note that in Example 7.6 the impulse is applied at time $t = 0.1$ s instead of time $t = 0$ as in this case).

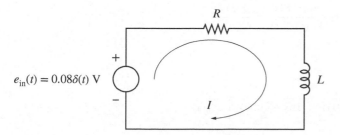

Figure 8.5 Electrical system with impulse input (Example 8.11).

Example 8.12

Repeat Example 8.11 and obtain the current response $I(t)$ of the series RL circuit with a step voltage input $e_{in}(t) = 2.4U(t)$ V.

Because we have the same system dynamics as Example 8.11 we can use Eq. (8.45) to express the Laplace transform of the current

$$I(s) = G(s)E_{in}(s) = \frac{1}{0.02s + 1.2}E_{in}(s)$$

The above equation is valid for any voltage input. For this problem, the voltage input is a step function, $e_{in}(t) = 2.4U(t)$ V, and consequently the Laplace transform of the input is $E_{in}(s) = 2.4/s$. Hence, the Laplace transform of the current is

$$I(s) = \frac{2.4}{s(0.02s + 1.2)}$$

Dividing all terms by 0.02 and expanding in partial fractions we obtain

$$I(s) = \frac{120}{s(s + 60)} = \frac{a_1}{s} + \frac{a_2}{s + 60} \tag{8.49}$$

The residues of Eq. (8.49) are $a_1 = 2$ and $a_2 = -2$. Taking the inverse Laplace transform of Eq. (8.49) we see that the current response $I(t)$ consists of a constant and an exponential function

$$I(t) = 2\left(1 - e^{-60t}\right) \text{ A} \tag{8.50}$$

Consequently, the current response shows an exponential rise from zero (at time $t = 0$) to a steady-state value of 2 A with a settling time of 0.0667 s. We could have obtained this result by direct analysis of the I/O equation (8.43) to see that at steady state the current reaches a constant value (hence $\dot{I} = 0$), which can be computed from Ohm's law: $e_{in}(t)/R = (2.4 \text{ V})/(1.2 \text{ }\Omega) = 2$ A. The settling time is four time constants, where $\tau = 0.0167$ s.

Example 8.13

Figure 8.6 presents the block diagram of the simplified solenoid actuator–valve system from Example 7.4. If the input voltage $e_{in}(t)$ is a constant 2 V (step input for $t > 0$), determine the response of the valve position $y(t)$. The system is in equilibrium (i.e., zero initial conditions) at time $t = 0$.

The two transfer functions in Fig. 8.6 are

$$\text{Solenoid:} \quad G_1(s) = \frac{12}{0.003s + 1.5} = \frac{F(s)}{E_{in}(s)}$$

$$\text{Spool valve:} \quad G_2(s) = \frac{1}{0.04s^2 + 16s + 7000} = \frac{Y(s)}{F(s)}$$

The *overall* system transfer function $G(s)$ relating the valve position y (output) to the voltage $e_{in}(t)$ (input) can be obtained by multiplying the solenoid and spool-valve transfer functions

$$G(s) = G_1(s)G_2(s) = \frac{F(s)}{E_{in}(s)}\frac{Y(s)}{F(s)} = \frac{Y(s)}{E_{in}(s)}$$

Multiplying $G_1(s)$ and $G_2(s)$ yields

$$G(s) = \frac{12}{(0.003s + 1.5)(0.04s^2 + 16s + 7000)} = \frac{Y(s)}{E_{in}(s)}$$

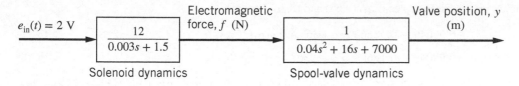

Figure 8.6 Solenoid actuator and spool valve for Example 8.13.

or, equivalently

$$G(s) = \frac{100,000}{(s+500)(s^2 + 400s + 175,000)} = \frac{Y(s)}{E_{\text{in}}(s)} \tag{8.51}$$

The Laplace transform of the position (output) is $Y(s) = G(s)E_{\text{in}}(s)$, or

$$Y(s) = \frac{100,000}{(s+500)(s^2 + 400s + 175,000)} E_{\text{in}}(s) \tag{8.52}$$

Equations (8.51) and (8.52) are valid for any voltage input $e_{\text{in}}(t)$. For a 2-V step input, the Laplace transform of the input is $E_{\text{in}}(s) = 2/s$ and Eq. (8.52) becomes

$$Y(s) = \frac{200,000}{s(s+500)(s^2 + 400s + 175,000)} \tag{8.53}$$

The four poles of $Y(s)$ are located at $s = 0$, $s = -500$, and $s = -200 \pm j367.42$. Note that we can "complete the square" and rewrite the second-order polynomial associated with the two complex poles as

$$s^2 + 400s + 175,000 = (s+200)^2 + 367.42^2$$

Therefore, the partial-fraction expansion of Eq. (8.53) is

$$Y(s) = \frac{a_1}{s} + \frac{a_2}{s+500} + \frac{a_3(s+200)}{(s+200)^2 + 367.42^2} + \frac{a_4(367.42)}{(s+200)^2 + 367.42^2} \tag{8.54}$$

The two residues associated with the real poles are

$$a_1 = sY(s)|_{s=0} = \frac{200,000}{(s+500)\left(s^2 + 400s + 175,000\right)}\Bigg|_{s=0} = \frac{200,000}{500 \times 175,000} = 0.002286$$

$$a_2 = (s+500)Y(s)|_{s=-500} = \frac{200,000}{s\left(s^2 + 400s + 175,000\right)}\Bigg|_{s=-500} = \frac{200,000}{-500 \times 225,000} = -0.001778$$

The third and fourth residues are determined from the two equations evaluated at the two complex conjugate poles

$$(s^2 + 400s + 175,000)Y(s)|_{s=-200+j367.42} = \frac{200,000}{s(s+500)}\Bigg|_{s=-200+j367.42}$$

$$= a_3(-200 + j367.42 + 200) + a_4(367.42) \tag{8.55}$$

and

$$(s^2 + 400s + 175,000)Y(s)|_{s=-200-j367.42} = \frac{200,000}{s(s+500)}\Bigg|_{s=-200-j367.42}$$

$$= a_3(-200 - j367.42 + 200) + a_4(367.42) \tag{8.56}$$

After substituting the complex values for the poles, Eqs. (8.55) and (8.56) become

$$-0.99049 - j0.18663 = a_3(j367.42) + a_4(367.42) \tag{8.57}$$

$$-0.99049 + j0.18663 = a_3(-j367.42) + a_4(367.42) \tag{8.58}$$

Equations (8.57) and (8.58) yield $a_3 = -5.0795(10^{-4})$ and $a_4 = -0.002696$. Therefore, the partial-fraction expansion (Eq. 8.54) becomes

$$Y(s) = \frac{0.002286}{s} + \frac{-0.001778}{s + 500} + \frac{-5.0795(10^{-4})(s + 200)}{(s + 200)^2 + 367.42^2} + \frac{-0.002696(367.42)}{(s + 200)^2 + 367.42^2} \tag{8.59}$$

The inverse Laplace transform of Eq. (8.59) consists of a constant, exponential function, and two exponentially damped sinusoidal terms (entries 10 and 11 in Table 8.1). Rounding to four decimal places, the spool-valve response is

$$y(t) = 0.0023 - 0.0018e^{-500t} - 0.0005e^{-200t}\cos 367.42t - 0.0027e^{-200t}\sin 367.42t \text{ m} \tag{8.60}$$

This solution can be verified by applying MATLAB's `ilaplace` command to the Laplace transform Eq. (8.53).

As a check, note that at time $t = 0$, $y(0) = 0.0023 - 0.0018 - 0.0005 = 0$ as expected (i.e., zero initial conditions). Figure 8.7 shows the 2-V step response computed using Eq. (8.60). Note that the slope of the response at time $t = 0$ is zero (i.e., $\dot{y}(0) = 0$) as specified in the problem statement. The steady-state value is clearly $y(\infty) = 0.0023$ m (or 2.3 mm), which is the residue a_1 because the other three terms involve an exponential decay to zero at steady state. We can obtain the steady-state valve position by applying the final-value theorem to Eq. (8.53), which yields $200{,}000/(500 \times 175{,}000) = 0.0023$. In Example 7.4, we computed the DC gain of the transfer function (i.e., evaluating $G(s)$ with $s = 0$) to find the same steady-state response to the 2-V step input. The reader should note that the DC gain method is an application of the final-value theorem when the input is a constant.

Finally, the reader should note that the step response in Fig. 8.7 can be quickly sketched by using the methods presented in Chapter 7: determine the transient-response characteristics from the roots (poles) of the solenoid's first-order transfer function $G_1(s)$ (i.e., compute time constant τ) and the spool valve's second-order transfer function $G_2(s)$ (i.e., compute ζ and ω_n). From these values, we can estimate settling time, percent overshoot, period of oscillation, and so on (see Table 7.4). The steady-state value of the step response is easily determined from the DC gain of the product of the two transfer functions $G_1(s)$ and $G_2(t)$.

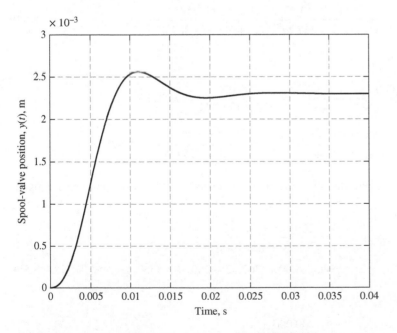

Figure 8.7 Step response of spool valve (Example 8.13).

SUMMARY

This chapter has presented Laplace transform methods for determining the response of dynamic systems. It is important to note that Laplace transform techniques can be applied only to systems represented by LTI differential equations. The Laplace transform method offers a systematic approach to obtaining the complete response by converting a differential equation in the time domain to an algebraic equation in terms of the complex Laplace variable s. Incorporating initial conditions is automatically handled by the Laplace transformation and the transformation of "standard" input functions (e.g., step, impulse, sinusoid, etc.) can be determined by consulting a table of Laplace transforms of common time functions. Determining the system response ultimately requires computing the inverse Laplace transform; this final (and possibly tedious) step may involve the partial-fraction expansion method.

When a system has zero initial conditions, the Laplace transform of the output is $Y(s) = G(s)U(s)$ where $G(s)$ is the system transfer function and $U(s)$ is the transform of the system input. With enough practice, the transfer function $G(s)$ can be derived by inspection from the system's I/O equation. The poles of $G(s)$ are identical to the characteristic roots studied in Chapter 7 and their location in the complex plane determines the system's response speed and damping characteristics.

REFERENCES

1. Creese, T.M., and Haralick, R.M., *Differential Equations for Engineers*, McGraw-Hill, New York, 1978, pp. 194–223, 342–349.
2. Ogata, K., *System Dynamics*, 4th ed., Pearson Prentice Hall, Upper Saddle River, NJ, 2004, pp. 32–34.
3. Close, C.M., Frederick, D.K., and Newell, J.C., *Modeling and Analysis of Dynamic Systems*, 3rd ed., Wiley, New York, 2002, pp. 222–224.

Chapter 9

Frequency-Response Analysis

9.1 INTRODUCTION

Chapter 7 presented analytical methods for determining the system response to inputs such as the step, pulse, and impulse functions. This chapter deals with determining the system's response to oscillating or harmonic input functions, where the forcing function is either $u(t) = U_0 \sin \omega t$ or $u(t) = U_0 \cos \omega t$ and the input has amplitude U_0 and frequency ω (rad/s). We show that the steady-state system response is also a sinusoidal function with the form $y(t) = Y_0 \sin(\omega t + \phi)$ if the input is a sine function. Hence, the steady-state output (defined as the *frequency response*) is also a sinusoidal function with the same frequency as the input. The frequency response differs from the input in its amplitude Y_0 and phase angle ϕ. We show that both of these parameters can be determined by using the system transfer function $G(s)$ that relates output $y(t)$ to input $u(t)$.

The objective of this chapter is to understand the frequency-response characteristics of first- and second-order systems as well as complex higher-order systems. By employing a graphical representation of the frequency-response parameters (the Bode diagram) we can quickly and easily determine the system's frequency response as well as identify phenomena such as resonance. In addition, this chapter introduces the topic of vibrations in mechanical systems.

9.2 FREQUENCY RESPONSE

The objective of this section is to develop the general solution for a linear time-invariant (LTI) system that is being driven by a sinusoidal (or oscillating) input. Figure 9.1 shows the LTI system with the harmonic input $u(t) = U_0 \sin \omega t$, where U_0 is the magnitude (or amplitude) of the input sine function and ω is the input frequency in rad/s (the reader should note that U_0 is an amplitude and not the unit-step function $U(t)$). Note that the harmonic input $u(t)$ could also be a cosine function; however, the sine function may be more realistic as it begins at zero at time $t = 0$ as opposed to the cosine function. The reader should recall that any LTI system that is modeled by an input–output (I/O) differential equation can be represented by the corresponding transfer function $G(s)$.

Recall that in Chapter 7 we showed that the complete (or total) solution of a linear differential equation has the general form

$$y(t) = y_H(t) + y_P(t) \tag{9.1}$$

where $y_H(t)$ and $y_P(t)$ are the homogeneous and particular solutions, respectively. In general, the form of the homogeneous (or natural) solution $y_H(t)$ depends on the characteristic roots of the I/O equation (or, the poles of the transfer function) while the form of the particular solution $y_P(t)$ depends on the nature of the input $u(t)$. Furthermore, if all characteristic roots have negative real parts (i.e., they lie in the left-half of the complex plane) then the homogeneous response $y_H(t)$ will "die out" at steady state. As a quick example, consider the following third-order LTI system

$$\dddot{y} + 8\ddot{y} + 37\dot{y} + 50y = u(t) \tag{9.2}$$

or, expressed as a transfer function

$$G(s) = \frac{1}{s^3 + 8s^2 + 37s + 50} = \frac{Y(s)}{U(s)} \tag{9.3}$$

The corresponding characteristic equation is

$$r^3 + 8r^2 + 37r + 50 = 0 \tag{9.4}$$

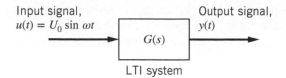

Figure 9.1 Linear time-invariant (LTI) system with a sinusoidal input.

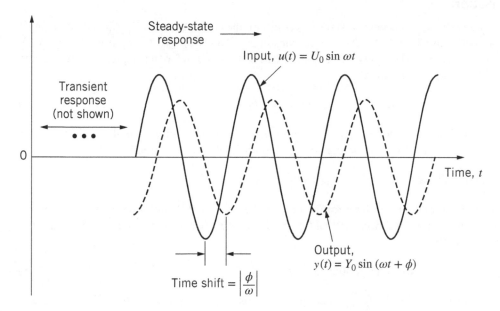

Figure 9.2 Frequency response: input and output sinusoidal functions.

and the three characteristic roots are $r_1 = -2$ and $r_{2,3} = -3 \pm j4$. We know from Chapter 7 that the general form of the homogeneous solution is

$$y_H(t) = c_1 e^{-2t} + c_2 e^{-3t} \sin 4t + c_3 e^{-3t} \cos 4t \tag{9.5}$$

Clearly, this homogenous solution "dies out" at steady state because the two exponential functions e^{-2t} and e^{-3t} decay to zero as $t \to \infty$. Only the particular solution $y_P(t)$ remains at steady state. Recall from Chapter 7 that the particular solution exhibits the same functional form as the input $u(t)$. Consequently, we expect the steady-state response of Eq. (9.2) to a *constant* (step) input to also be a constant (note that the DC gain of $G(s)$ is 1/50, and hence the steady-state output is the magnitude of the step input divided by 50). This intuitive reasoning tells us that the steady-state response of Eq. (9.2) when the input is a sinusoidal function (such as $u(t) = U_0 \sin \omega t$) will also be a sinusoidal function with the same frequency ω.

The previous discussion leads us to the following definition: the *frequency response* is the steady-state response of a system driven by a sinusoidal input. We show that if the sinusoidal input is $u(t) = U_0 \sin \omega t$ (as shown in Fig. 9.1) then the frequency response is $y(t) = Y_0 \sin(\omega t + \phi)$ where Y_0 is the magnitude (amplitude) of the output sinusoid and ϕ is the phase angle difference between the input and output sinusoidal functions. Figure 9.2 presents a general schematic of the frequency response of linear system $G(s)$ driven by the sinusoidal input $u(t) = U_0 \sin \omega t$. The transient response (not shown in Fig. 9.2) eventually dies out leaving the steady-state sinusoidal output $y(t)$, that is, the frequency response. The frequency response $y(t)$ has an amplitude Y_0 and the same frequency (or period) as the input sinusoid $u(t)$. When the output/input amplitude ratio $Y_0/U_0 < 1$, the output has been attenuated (as shown in Fig. 9.2), and when $Y_0/U_0 > 1$ the output has been amplified relative to the input signal. The following subsections demonstrate that the ratio Y_0/U_0 depends on the system transfer function and the input frequency ω. Figure 9.2 also shows the time shift between the output and input sinusoidal functions; this time shift is equal to the phase angle difference ϕ (rad) divided by the common frequency ω (rad/s). Hence, if $\phi = 0$ then the peaks and valleys of the input and output sinusoids are aligned (i.e., $y(t)$ and $u(t)$ are said to be "in phase"). Conversely, if $\phi = \pi$ rad then the peaks of the input are aligned with the valleys of the output (i.e., $y(t)$ and $u(t)$ are said to be "180° out of phase"). We show that the phase difference ϕ also depends on the system transfer function $G(s)$ and the input frequency ω. Consequently, it is important for the reader to note that the frequency response $y(t) = Y_0 \sin(\omega t + \phi)$ is *completely* determined by the amplitude ratio Y_0/U_0 and phase angle ϕ. This fundamental characteristic of the frequency response cannot be overemphasized and is mathematically derived in the next subsection.

Sinusoidal Transfer Function

In the previous subsection, we stated that the frequency response depends on the amplitude ratio Y_0/U_0 and phase angle ϕ and that these parameters are determined solely by the system transfer function $G(s)$ and the input frequency ω. To demonstrate this fact, we introduce the sinusoidal transfer function. Recall that in Section 5.6 in Chapter 5 we considered the response of the following third-order I/O equation

$$a_3\,\dddot{y} + a_2\ddot{y} + a_1\dot{y} + a_0 y = b_1\dot{u} + b_0 u \tag{9.6}$$

The input is a real-valued exponential function, $u(t) = U(s)e^{st}$, where $s = \sigma + j\omega$ is a complex variable (with real part σ and imaginary part ω) and $U(s)$ is a complex function. In general, the exponential input function is

$$u(t) = U(s)e^{st} = U(s)e^{\sigma t}e^{j\omega t} = U(s)e^{\sigma t}(\cos \omega t + j\sin \omega t) \tag{9.7}$$

where the last substitution comes from applying Euler's formula $e^{j\theta} = \cos\theta + j\sin\theta$. If the input $u(t)$ is a harmonic (sinusoidal) function, then there is no exponential decay $e^{\sigma t}$ and the real part of s is zero, or $\sigma = 0$. Rewriting Eq. (9.7) for a sinusoidal input with $s = j\omega$ we obtain

$$u(t) = U(j\omega)(\cos \omega t + j\sin \omega t) \tag{9.8}$$

Because the input $u(t)$ is a real function, $U(j\omega)$ is the complex conjugate of $\cos \omega t + j\sin \omega t$. Recall from Chapter 7 that if the input is $u(t) = U(j\omega)e^{j\omega t}$ the particular solution will also be an exponential function, $y(t) = Y(j\omega)e^{j\omega t}$, where $Y(j\omega)$ is a complex function. Next, substituting $u(t) = U(j\omega)e^{j\omega t}$ and $y(t) = Y(j\omega)e^{j\omega t}$ into the system I/O equation (9.6) and noting that their time derivatives are

$$\dot{u}(t) = j\omega U(j\omega)e^{j\omega t}, \ \ \dot{y}(t) = j\omega Y(j\omega)e^{j\omega t}, \ \ \ddot{y}(t) = (j\omega)^2 Y(j\omega)e^{j\omega t}, \text{etc.}$$

we see that the common complex functions $U(j\omega)$ and $Y(j\omega)$ can be factored out of all of the right- and left-hand-side terms. Consequently, the I/O equation (9.6) becomes

$$\left(a_3(j\omega)^3 + a_2(j\omega)^2 + a_1 j\omega + a_0\right)Y(j\omega)e^{j\omega t} = \left(b_1 j\omega + b_0\right)U(j\omega)e^{j\omega t}$$

Finally, forming the ratio of output/input yields

$$\frac{Y(j\omega)}{U(j\omega)} = \frac{b_1 j\omega + b_0}{a_3(j\omega)^3 + a_2(j\omega)^2 + a_1 j\omega + a_0} = G(j\omega) \tag{9.9}$$

The complex function $G(j\omega)$ is the *sinusoidal transfer function*. Note that the transfer function of the third-order I/O system (9.6) is

$$G(s) = \frac{Y(s)}{U(s)} = \frac{b_1 s + b_0}{a_3 s^3 + a_2 s^2 + a_1 s + a_0} \tag{9.10}$$

Comparing Eqs. (9.9) and (9.10), we see that the sinusoidal transfer function $G(j\omega)$ is simply the transfer function $G(s)$ with s replaced by $j\omega$, or $s = j\omega$. Figure 9.3 shows the frequency response in a block diagram format where the steady-state output sinusoid is

$$y(t) = G(j\omega)U(j\omega)e^{j\omega t} \tag{9.11}$$

In general, $G(j\omega)$, $U(j\omega)$, and $e^{j\omega t}$ are complex functions of the input frequency ω. Although we have not yet derived a simple expression for the frequency response, Eq. (9.11) shows that it depends on the sinusoidal transfer function $G(j\omega)$.

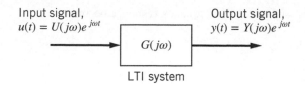

Input signal,
$u(t) = U(j\omega)e^{j\omega t}$

Output signal,
$y(t) = Y(j\omega)e^{j\omega t}$

$G(j\omega)$

LTI system

Figure 9.3 Sinusoidal transfer function and frequency response.

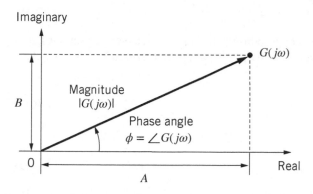

Figure 9.4 Magnitude and phase of the sinusoidal transfer function $G(j\omega)$.

Derivation of the Frequency Response

Repeating Eq. (9.11), the frequency response of the LTI system in Fig. 9.1 or Fig. 9.3 is

$$y_{ss}(t) = G(j\omega)U(j\omega)e^{j\omega t} \tag{9.12}$$

where the subscript ss denotes "steady state." The sinusoidal transfer function $G(j\omega)$ is a complex function of frequency ω and (generally) consists of real and imaginary parts. Figure 9.4 shows the sinusoidal transfer function $G(j\omega)$ as a point in the complex plane with real and imaginary components. The reader should recall that the complex plane shown in Fig. 9.4 is essentially a Cartesian coordinate frame where the horizontal axis consists of real numbers and the vertical axis consists of imaginary numbers. Therefore, we can represent the complex value $G(j\omega)$ using either Cartesian or polar coordinates. Referring to Fig. 9.4, the complex value of $G(j\omega)$ is

Cartesian form: $\quad G(j\omega) = A + jB$

Polar form: $\quad G(j\omega) = |G(j\omega)|e^{j\phi} = |G(j\omega)|(\cos\phi + j\sin\phi)$

where the magnitude (or absolute value) and phase (or argument) of $G(j\omega)$ are

$$\text{Magnitude:} \quad |G(j\omega)| = \sqrt{A^2 + B^2} \tag{9.13}$$

$$\text{Phase angle:} \quad \phi = \angle G(j\omega) = \tan^{-1}\left(\frac{B}{A}\right) \tag{9.14}$$

Computing the magnitude and phase angle of a complex number is relatively easy in MATLAB using the abs (absolute value) and `angle` commands. As a quick example, consider the complex number $X = 4 + j2$ and the following MATLAB commands:

```
>> X = 4 + j*2;          % define the complex number X = 4 + j2
>> magX = abs(X)         % compute the magnitude (absolute value) of X
>> phaseX = angle(X)     % compute the phase angle of X (rad)
```

After executing these commands, we obtain

$$\texttt{magX = 4.4721 and phaseX = 0.4636 (or, 26.57°)}$$

Next, we can replace $G(j\omega)$ in Eq. (9.12) with its polar form and hence the frequency response becomes

$$y_{ss}(t) = |G(j\omega)|e^{j\phi}U(j\omega)e^{j\omega t} = |G(j\omega)|U(j\omega)e^{j(\omega t+\phi)} \tag{9.15}$$

Expanding the exponential function using Euler's formula yields

$$y_{ss}(t) = |G(j\omega)|U(j\omega)\left[\cos(\omega t + \phi) + j\sin(\omega t + \phi)\right] \tag{9.16}$$

Recall that $U(j\omega)$ is a complex function of ω and that the frequency response $y_{ss}(t)$ is a real-valued function of time. If the input is a *sine* function, $u(t) = U_0 \sin \omega t$, then we use the imaginary part of the bracketed term in Eq. (9.16) and the steady-state output is also a sine function:

$$y_{ss}(t) = |G(j\omega)|U_0 \sin(\omega t + \phi) \tag{9.17}$$

If the input is a *cosine* function, $u(t) = U_0 \cos \omega t$, then we use the real part in Eq. (9.16)

$$y_{ss}(t) = |G(j\omega)|U_0 \cos(\omega t + \phi) \tag{9.18}$$

Let us summarize the key points of the frequency response:

1. Equation (9.17) is the frequency response of the LTI system shown in Fig. 9.1 when the input is a sine function, or $u(t) = U_0 \sin \omega t$. Equation (9.18) is the frequency response of the same LTI system when the input is a cosine function $u(t) = U_0 \cos \omega t$. The frequency response $y_{ss}(t)$ is a sinusoidal function with the same frequency ω (or, period) as the input $u(t)$.

2. In either case (sine or cosine input), the frequency response is completely determined by the *magnitude* and *phase* of the sinusoidal transfer function $G(j\omega)$.

3. The frequency-response equation (9.17) or (9.18) is valid *only* if the transient response "dies out" at steady state. In other words, the poles of the transfer function $G(s)$ must lie in the left-half of the complex plane.

We demonstrate the frequency response with the following examples.

Example 9.1

Figure 9.5 shows the series RL circuit from Example 7.6 and Example 8.11. Obtain the frequency response for the current $I(t)$ for a sinusoidal voltage input, $e_{in}(t) = 2 \sin 50t$ V. The system has zero energy at time $t = 0$, or $I(0) = 0$, and the inductance and resistance values are $L = 0.02$ H and $R = 1.2$ Ω, respectively.

The mathematical model of the RL circuit is

$$L\dot{I} + RI = e_{in}(t) \tag{9.19}$$

Hence the transfer function relating current (output) to the voltage source (input) is

$$G(s) = \frac{I(s)}{E_{in}(s)} = \frac{1}{Ls + R} = \frac{1}{0.02s + 1.2} \tag{9.20}$$

The reader should note that the system's I/O equation (9.19) or transfer function (9.20) remains unchanged whether the input is an impulse (as in Example 7.6) or a sinusoid as in this example.

Because the input voltage is a sine function, Eq. (9.17) provides the frequency response for the current at steady state. Furthermore, the sinusoidal voltage input is $e_{in}(t) = 2 \sin 50t$ V, which tells us that the input amplitude is 2 V and the input frequency is 50 rad/s. Using Eq. (9.17) with $U_0 = 2$ V and $\omega = 50$ rad/s yields

$$I_{ss}(t) = |G(j\omega)|2 \sin(50t + \phi) \tag{9.21}$$

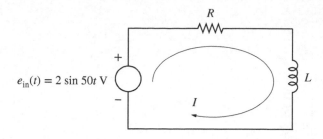

Figure 9.5 Electrical system with sinusoidal input (Example 9.1).

Therefore, we only need to compute the magnitude and phase of the sinusoidal transfer function for the input frequency $\omega = 50$ rad/s. We begin by writing the sinusoidal transfer function by using Eq. (9.20) with $s = j\omega$

$$G(j\omega) = \frac{1}{0.02j\omega + 1.2} \tag{9.22}$$

The magnitude of the complex function $G(j\omega)$ is computed by dividing the magnitude of the numerator by the magnitude of the denominator, or

$$|G(j\omega)| = \frac{\sqrt{1^2 + 0^2}}{\sqrt{1.2^2 + (0.02\omega)^2}} \tag{9.23}$$

Note that the magnitudes of the numerator and denominator terms are computing using Eq. (9.13) and the respective real and imaginary parts. Substituting the input frequency $\omega = 50$ rad/s into Eq. (9.23), we find that the magnitude is $|G(j\omega)| = 0.6402$.

The phase angle of $G(j\omega)$ is computed by subtracting the phase angle of the denominator from the phase angle of the numerator, or

$$\phi = \angle G(j\omega) = \angle(1 + j0) - \angle(1.2 + j0.02\omega) \tag{9.24}$$

Using Eq. (9.14) to compute each angle (or argument) we obtain

$$\phi = \angle G(j\omega) = \tan^{-1}\left(\frac{0}{1}\right) - \tan^{-1}\left(\frac{0.02\omega}{1.2}\right) \tag{9.25}$$

Substituting frequency $\omega = 50$ rad/s into Eq. (9.25), the phase angle is $\phi = -0.6947$ rad (or, $\phi = -39.81°$). Finally, substituting the magnitude and phase of $G(j\omega)$ into Eq. (9.21) the frequency response is

$$I_{ss}(t) = 1.2804 \sin(50t - 0.6947) \text{ A} \tag{9.26}$$

Equation (9.26) is the frequency response of the current of the RL circuit driven by a sinusoidal voltage input. At steady state, the current $I(t)$ oscillates at the same frequency as the voltage input (50 rad/s) and with an amplitude of 1.2804 A. A phase difference of $\phi = -0.6947$ rad exists between the output and input, and because the phase angle is negative, the current (output) "lags" behind the input (voltage source). The time shift Δt between input and output sinusoids can be computed from the phase lag and the frequency:

$$\Delta t = \left|\frac{\phi}{\omega}\right| = \left|\frac{-0.6947 \text{ rad}}{50 \text{ rad/s}}\right| = 0.0139 \text{ s} \tag{9.27}$$

Figure 9.6 shows the frequency response (9.26) plotted on the same graph with the voltage input $e_{in}(t) = 2 \sin 50t$ V. The voltage input (left y axis) and steady-state current (right y axis) are both sinusoidal signals with the same frequency, and the time lag (or phase difference) between the two signals is apparent.

We can verify our result by computing the magnitude and phase angle of $G(j\omega)$ using the following MATLAB commands:

```
>> w = 50;                          % define input frequency, ω = 50 rad/s
>> Gj50 = 1/(1.2 + 0.02*j*w);       % define sinusoidal transfer function G(j50)
```

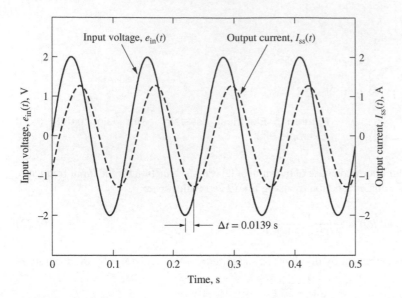

Figure 9.6 Frequency response of RL circuit (Example 9.1).

```
>> magGj50 = abs(Gj50)              % compute the magnitude of G(j50)
>> phaseGj50 = angle(Gj50)          % compute the phase angle of G(j50) (rad)
```

After executing these commands we obtain

$$\text{magGj50} = 0.6402 \quad \text{and} \quad \text{phaseGj50} = -0.6947$$

which matches our previous results.

Example 9.2

Use Simulink to simulate the RL circuit in Example 9.1 and plot the voltage input $e_{in}(t)$ and current response $I(t)$ on the same graph.

Because the initial conditions are zero, we can use a transfer function to represent the system dynamics in Simulink. Figure 9.7 shows the Simulink block diagram of this very simple system. The Simulink model is constructed by connecting the `Sine Wave` from the `Sources` library to the `Transfer Fcn` block, which has been edited to match $G(s)$ given in Eq. (9.20). The desired voltage input signal is created by editing the `Sine Wave` block and setting the `Amplitude` and `Frequency` dialog boxes to 2 (V) and 50 (rad/s), respectively.

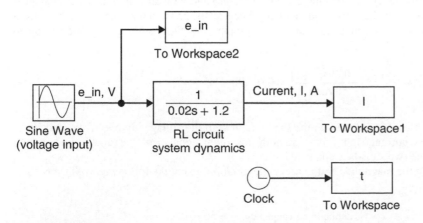

Figure 9.7 Simulink model of the RL circuit (Example 9.2).

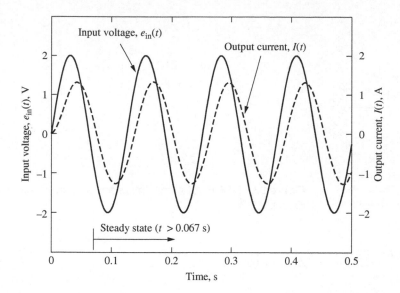

Figure 9.8 RL circuit response to sinusoidal voltage input (Example 9.2).

Figure 9.8 shows the sinusoidal input voltage and the resulting current response from executing the Simulink model. The reader should note that the current begins at zero as prescribed in the problem statement. Because the RL circuit is a first-order LTI system we can compute its time constant τ by dividing all terms in the transfer function (9.20) by R to yield

$$G(s) = \frac{1/R}{(L/R)s + 1} = \frac{0.8333}{\tau s + 1} \tag{9.28}$$

Therefore, the time constant is $\tau = L/R = 0.0167$ s and the first-order system reaches steady state in four time constants, or $4\tau = 0.067$ s as denoted in Fig. 9.8. If we compare the current responses presented in Fig. 9.8 (the complete response) and Fig. 9.6 (the steady-state or frequency response) we see that the only difference is the transient response (i.e., $0 \leq t < 0.067$ s). After time $t > 0.067$ s both figures show the same steady-state sinusoidal responses for current.

Example 9.3

Figure 9.9 shows the one-degree-of-freedom (1-DOF) rotational mechanical system from Example 7.8. If the system is initially at rest, $\theta(0) = \dot{\theta}(0) = 0$, and the input torque is a sine function $T_{in}(t) = 1.5 \sin 18t$ N-m, compute the frequency response $\theta_{ss}(t)$ of the mechanical system using analytical and numerical methods.

Using the parameters in Fig. 9.9 the mathematical model of the rotational mechanical system is

$$0.2\ddot{\theta} + 1.6\dot{\theta} + 65\theta = T_{in}(t) \tag{9.29}$$

Therefore, the transfer function is

$$G(s) = \frac{1}{0.2s^2 + 1.6s + 65} = \frac{\Theta(s)}{T_{in}(s)} \tag{9.30}$$

Using Eq. (9.17) with input amplitude $U_0 = 1.5$ N-m and frequency $\omega = 18$ rad/s the frequency response of the mechanical system is

$$\theta_{ss}(t) = |G(j\omega)|1.5 \sin(18t + \phi) \tag{9.31}$$

Consequently, we need to evaluate the magnitude and phase of the sinusoidal transfer function $G(j\omega)$ at frequency $\omega = 18$ rad/s. Substituting $s = j\omega$ in the transfer function (9.30) yields

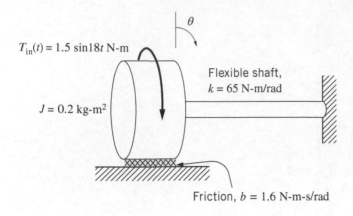

$T_{in}(t) = 1.5 \sin 18t$ N-m

Flexible shaft, $k = 65$ N-m/rad

$J = 0.2$ kg-m^2

Friction, $b = 1.6$ N-m-s/rad

Figure 9.9 1-DOF rotational mechanical system for Example 9.3.

$$G(j\omega) = \frac{1}{0.2(j\omega)^2 + 1.6(j\omega) + 65} \qquad (9.32)$$

Substituting $j^2 = -1$ into Eq. (9.32), we obtain

$$G(j\omega) = \frac{1}{65 - 0.2\omega^2 + j1.6\omega}$$

or, evaluating $G(j\omega)$ at input frequency $\omega = 18$ rad/s yields

$$G(j18) = \frac{1}{0.2 + j28.8}$$

The magnitude of $G(j18)$ is

$$|G(j18)| = \frac{\sqrt{1^2 + 0^2}}{\sqrt{0.2^2 + 28.8^2}} = 0.0347$$

The phase angle of $G(j18)$ is the phase of its numerator minus the phase of its denominator

$$\phi = \angle G(j18) = \angle(1 + j0) - \angle(0.2 + j28.8)$$
$$= \tan^{-1}\left(\frac{0}{1}\right) - \tan^{-1}\left(\frac{28.8}{0.2}\right) = 0 - 1.5639 \text{ rad}$$

Finally, substituting the magnitude and phase of $G(j\omega)$ into Eq. (9.31) gives us the frequency response

$$\theta_{ss}(t) = 0.0521 \sin(18t - 1.5639) \text{ rad} \qquad (9.33)$$

Equation (9.33) is the frequency response of the rotational mechanical system. The amplitude of the angular position response is 0.0521 rad or 2.99°. The phase lag between the input and output sine waves is 1.5639 rad or 89.6°.

Figure 9.10 shows the Simulink model of the 1-DOF rotational mechanical system that consists of the single transfer function (9.30) with a sinusoidal input (Sine Wave block from the Simulink Sources library). The Amplitude and Frequency in the Sine Wave dialog box are set to 1.5 (N-m) and 18 (rad/s), respectively. Figure 9.11 shows the response of the rotational mechanical system to the sinusoidal torque input. The angular position $\theta(t)$ exhibits a transient response as the amplitude increases and eventually reaches its steady-state value of about 0.052 rad (the settling time for this system is 1 s; see Example 7.8). At steady state, the input and output sinusoids clearly have the same frequency (period). Note that the frequency response $\theta_{ss}(t)$ in Fig. 9.11 displays a one-quarter-cycle shift (lag) with respect to the input (i.e., when the input $T_{in}(t)$ crosses zero the output $\theta(t)$ is at its maximum or minimum value). This one-quarter-cycle lag relates to a (negative) phase angle of $-2\pi/4 = -1.5708$ rad ($-90°$), which closely matches the phase angle of $G(j18)$, that is, $\phi = -1.5639$ rad.

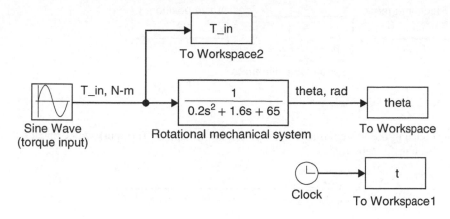

Figure 9.10 Simulink model of the rotational mechanical system (Example 9.3).

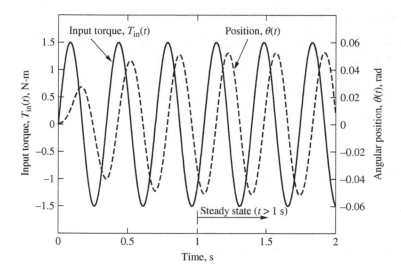

Figure 9.11 Rotational mechanical system response to a sinusoidal torque input (Example 9.3).

Example 9.4

Figure 9.12 presents the block diagram of the simplified solenoid actuator–valve system from Example 7.4. If the input voltage $e_{in}(t)$ is a sine wave with a magnitude of 6 V and frequency of 10 Hz, determine the frequency response of the spool valve using analytical and numerical methods. The system has zero initial conditions at time $t = 0$.

The input voltage signal $e_{in}(t)$ has frequency $f = 10$ Hz or 10 cycles per second. Therefore, the frequency in radians per second is $\omega = 2\pi f = 62.8319$ rad/s and the input voltage signal is

$$e_{in}(t) = 6 \sin 62.8319t \text{ V} \tag{9.34}$$

The two transfer functions in Fig. 9.12 are

$$\text{Solenoid:} \quad G_1(s) = \frac{12}{0.003s + 1.5} = \frac{F_{em}(s)}{E_{in}(s)}$$

$$\text{Spool valve:} \quad G_2(s) = \frac{1}{0.04s^2 + 16s + 7000} = \frac{Y(s)}{F_{em}(s)}$$

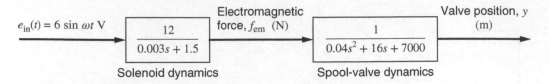

Figure 9.12 Solenoid actuator and spool valve for Example 9.4.

The *overall* system transfer function $G(s)$ relating the valve position y (output) to the voltage $e_{in}(t)$ (input) can be obtained by multiplying the solenoid and spool-valve transfer functions

$$G(s) = G_1(s)G_2(s) = \frac{F_{em}(s)}{E_{in}(s)} \frac{Y(s)}{F_{em}(s)} = \frac{Y(s)}{E_{in}(s)} = \frac{12}{(0.003s + 1.5)(0.04s^2 + 16s + 7000)}$$

Expanding the third-order denominator of $G(s)$, we obtain

$$G(s) = \frac{12}{0.00012s^3 + 0.108s^2 + 45s + 10{,}500} \tag{9.35}$$

Because the input is a sine signal the frequency response $y_{ss}(t)$ will also be a sine wave. Using Eq. (9.17) with $U_0 = 6$ V and $\omega = 62.8319$ rad/s yields the frequency response of the spool valve

$$y_{ss}(t) = |G(j\omega)|6 \sin(62.8319t + \phi) \tag{9.36}$$

As with the previous examples, we need to compute the magnitude and phase of the sinusoidal transfer function for the given input frequency. The sinusoidal transfer function is computed by using Eq. (9.35) with $s = j\omega$

$$G(j\omega) = \frac{12}{0.00012(j\omega)^3 + 0.108(j\omega)^2 + 45(j\omega) + 10{,}500} \tag{9.37}$$

Noting that $j^3 = -j$ and $j^2 = -1$, Eq. (9.37) becomes

$$G(j\omega) = \frac{12}{(10{,}500 - 0.108\omega^2) + j(-0.00012\omega^3 + 45\omega)} \tag{9.38}$$

Substituting the input frequency $\omega = 62.8319$ rad/s in Eq. (9.38), we obtain

$$G(j62.8319) = \frac{12}{10{,}073.63 + j2797.67} \tag{9.39}$$

The magnitude of $G(j62.8319)$ is

$$|G(j62.8319)| = \frac{\sqrt{12^2 + 0^2}}{\sqrt{10{,}073.63^2 + 2797.67^2}} = 0.001148$$

The phase angle of $G(j\omega)$ is computed by subtracting the phase angle of the denominator from the phase angle of the numerator, or

$$\phi = \angle G(j62.8319) = \angle(12 + j0) - \angle(10{,}073.63 + j2797.67)$$

$$= \tan^{-1}\left(\frac{0}{12}\right) - \tan^{-1}\left(\frac{2797.67}{10{,}073.63}\right) = 0 - 0.2709 \text{ rad}$$

Finally, substituting the magnitude and phase angle of $G(j\omega)$ into Eq. (9.36) the frequency response becomes

$$y_{ss}(t) = 0.006887 \sin(62.8319t - 0.2709) \text{ m} \tag{9.40}$$

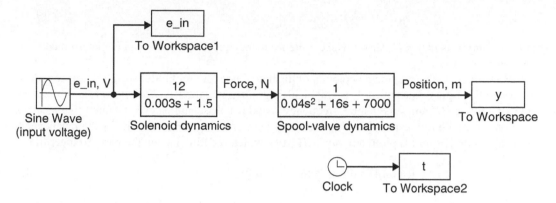

Figure 9.13 Simulink model of actuator–valve system (Example 9.4).

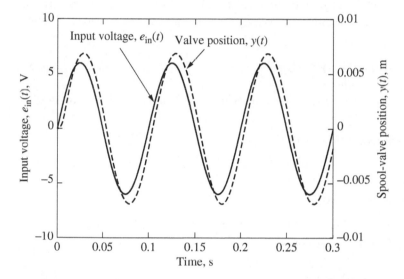

Figure 9.14 Spool-valve response to a sinusoidal voltage input (Example 9.4).

Equation (9.40) is the frequency response of the spool-valve position for a 10-Hz voltage input with a magnitude of 6 V. The steady-state amplitude of the valve is 0.0069 m (or 6.9 mm) and the phase lag is 0.2709 rad (or 15.52°).

Figure 9.13 shows the Simulink block diagram of the actuator–valve system where the transfer functions $G_1(s)$ and $G_2(s)$ are readily apparent (i.e., see Fig. 9.12). The sinusoidal voltage input is created by using the `Sine Wave` block (`Sources` library) with amplitude of 6 (V) and frequency of 62.8319 (rad/s). Figure 9.14 shows the complete response of the valve position $y(t)$ for the sinusoidal input. The first-order solenoid has a time constant $\tau = 0.003/1.5 = 0.002$ s and, therefore, its contribution to the transient response dies out in about 0.008 s (i.e., four time constants). The damping ratio and undamped natural frequency of the spool valve are $\zeta = 0.4781$ and $\omega_n = 418.33$ rad/s, respectively. Consequently, the transient-response contribution from the second-order valve dynamics dies out at time $t_S = 4/(\zeta\omega_n) = 0.02$ s, which is much sooner than one period of the input voltage signal (0.1 s for $f = 10$ Hz). Figure 9.14 shows that the amplitude of the steady-state valve position $y_{ss}(t)$ is about 0.0069 m, which verifies Eq. (9.40). Furthermore, Fig. 9.14 shows very little phase lag between the input and output sinusoids. The time shift is $\Delta t = |\phi/\omega| = 0.0043$ s, which matches the time lag between the input and output "peaks" or "valleys" in Fig. 9.14.

9.3 BODE DIAGRAMS

If the reader reviews the previous section and examples, he or she will see that the frequency response of an LTI system is *completely* determined by the magnitude and phase angle of the sinusoidal transfer function $G(j\omega)$. To emphasize this point, we repeat the frequency-response equation (9.17)

$$y_{ss}(t) = |G(j\omega)|U_0 \sin(\omega t + \phi) \tag{9.41}$$

where the (known) sinusoidal input is $u(t) = U_0 \sin \omega t$. Hence, the frequency response (9.41) can be determined if we can compute $|G(j\omega)|$ and $\phi = \angle G(j\omega)$.

In the 1930s, H.W. Bode developed a graphical depiction of the magnitude (or amplitude ratio) $|G(j\omega)|$ and phase angle ϕ plotted as a function of the input frequency ω. This graphical diagram (now called the *Bode plot* or *Bode diagram*) consists of two plots: (1) magnitude $|G(j\omega)|$ versus input frequency ω and (2) phase angle ϕ versus input frequency ω. The magnitude is plotted on a logarithmic scale and both plots share a common logarithmic scale for the independent variable (frequency, ω). Magnitude $|G(j\omega)|$ is plotted in *decibels* (dB), which is defined using the base-10 logarithm

$$|G(j\omega)| \text{ in dB} \equiv 20\log_{10}|G(j\omega)| \tag{9.42}$$

Let us denote the absolute-value magnitude as $|G(j\omega)|$ and its corresponding value in decibels as $|G(j\omega)|_{dB}$. As a quick example, consider the magnitude $|G(j\omega)| = 0.16$. Using Eq. (9.42) the corresponding magnitude in decibels is $|G(j\omega)|_{dB} = 20\log_{10}(0.16) = -36.65$ dB. Hence, while magnitude (or absolute value) $|G(j\omega)|$ is always positive, its corresponding magnitude in decibels may be positive or negative. Converting the magnitude in decibels to an absolute-value magnitude requires the inverse of Eq. (9.42):

$$|G(j\omega)| = 10^{|G(j\omega)|_{dB}/20} \tag{9.43}$$

We can summarize a few key properties linking the absolute-value magnitude $|G(j\omega)|$ with the magnitude in decibels, $|G(j\omega)|_{dB}$:

1. $20\log_{10}(1) = 0$ dB; therefore, unity amplitude ratio = 0 dB
2. If $|G(j\omega)| > 1$, $|G(j\omega)|_{dB} > 0$
3. If $|G(j\omega)| < 1$, $|G(j\omega)|_{dB} < 0$
4. Very small $|G(j\omega)|$ results in large negative $|G(j\omega)|_{dB}$

As we already know how to compute the magnitude and phase of $G(j\omega)$, we can illustrate the Bode diagram with the following simple example. Consider the simple first-order transfer function

$$G(s) = \frac{6}{s+4} \tag{9.44}$$

Replacing s with $j\omega$ the sinusoidal transfer function is

$$G(j\omega) = \frac{6}{j\omega + 4} \tag{9.45}$$

Using Eqs. (9.13) and (9.14) the magnitude and phase of the sinusoidal transfer function are

$$\text{Magnitude:} \quad |G(j\omega)| = \frac{\sqrt{6^2 + 0^2}}{\sqrt{4^2 + \omega^2}} \tag{9.46}$$

$$\text{Phase:} \quad \phi = \angle G(j\omega) = \tan^{-1}\left(\frac{0}{6}\right) - \tan^{-1}\left(\frac{\omega}{4}\right) \tag{9.47}$$

We can use Eqs. (9.46) and (9.47) to compute the magnitude and phase for a wide range of input frequencies. Table 9.1 summarizes these two key frequency response parameters for frequencies ranging from $\omega = 0.1$ rad/s (or, "low frequency" with a period of 62.8 s) to $\omega = 100$ rad/s (or, "high frequency" with a period of 0.06 s). Note that the corresponding magnitude in decibels using Eq. (9.42) is also presented in Table 9.1, and that the phase angle has been converted from radians to degrees. At very low frequency (say, as $\omega \to 0$), the magnitude approaches the DC gain of transfer function $G(s)$ (i.e., $6/4 = 1.5$, or $20\log_{10}(1.5) = 3.522$ dB) and the phase approaches zero. At very high frequency ($\omega \to \infty$), the magnitude approaches zero (or, $20\log_{10}(0+) \to -\infty$ dB) and the phase approaches $-90°$.

Table 9.1 Magnitude and Phase of First-Order Transfer Function $G(s) = 6/(s+4)$

Input Frequency, ω (rad/s)	Magnitude $\|G(j\omega)\|$	Magnitude $\|G(j\omega)\|_{dB}$ (dB)	Phase $\angle G(j\omega)$
0.1	1.4995	3.52	−1.43°
0.5	1.4884	3.45	−7.13°
1	1.4552	3.26	−14.04°
2	1.3416	2.55	−26.57°
4	1.0607	0.51	−45.00°
10	0.5571	−5.08	−68.20°
20	0.2942	−10.63	−78.69°
50	0.1196	−18.44	−85.43°
100	0.0600	−24.44	−87.70°

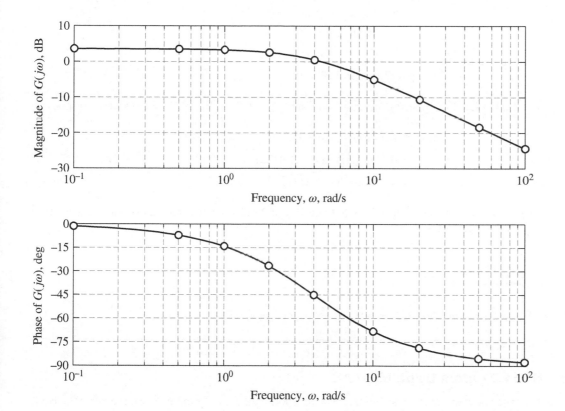

Figure 9.15 Bode diagram of first-order transfer function $G(s) = 6/(s+4)$ with data points from Table 9.1.

Figure 9.15 shows the Bode diagram for the first-order transfer function (9.44). The nine values of magnitude and phase from Table 9.1 are shown as discrete points in Fig. 9.15. Figure 9.15 shows that the Bode diagram consists of the two graphs: the top plot is magnitude (in decibels) versus frequency (rad/s) and the bottom plot is phase (degrees) versus frequency. The common frequency axis (i.e., the independent variable) is plotted on a logarithmic scale so that a wide range of input frequency can be shown.

Once we have constructed the Bode diagram, it can be used to efficiently compute the frequency response. The basic steps are summarized below:

1. Given the input sinusoid $u(t) = U_0 \sin \omega t$, read the magnitude $|G(j\omega)|_{dB}$ and phase ϕ (degrees) directly from the Bode diagram for the known input frequency ω.

2. Convert the magnitude $|G(j\omega)|_{dB}$ from decibels to an absolute-value magnitude using Eq. (9.43).

3. Convert the phase ϕ from degrees to radians.

4. Using the amplitudes U_0 and $|G(j\omega)|$ and phase ϕ compute the frequency response $y_{ss}(t) = |G(j\omega)|U_0 \sin(\omega t + \phi)$.

The following example illustrates how to utilize the Bode diagram.

Example 9.5

Figure 9.16 shows the block diagram of an LTI system where the input is a sine wave. Use the Bode diagram in Fig. 9.15 to compute the frequency response of this system.

We can use the Bode diagram in Fig. 9.15 because this diagram corresponds to the system transfer function in Fig. 9.16. Reading Fig. 9.15 with input frequency $\omega = 7$ rad/s, we obtain the following magnitude and phase:

$$|G(j7)|_{dB} = -2.5 \text{ dB}$$
$$\phi = \angle G(j7) = -60°$$

Therefore, the absolute-value magnitude is

$$|G(j7)| = 10^{-2.5/20} = 0.75$$

and the phase angle is $\phi = -1.0472$ rad. Finally, using the input amplitude $U_0 = 4.5$ and Eq. (9.17) the frequency response is

$$y_{ss}(t) = |G(j\omega)|U_0 \sin(\omega t + \phi)$$
$$= (0.75)(4.5) \sin(7t - 1.0472)$$
$$= 3.375 \sin(7t - 1.0472)$$

In summary, the frequency response has an amplitude of 3.375, frequency of 7 rad/s (same as the input), and phase angle of -1.0472 rad ($-60°$) relative to the input sinusoid.

Input,
$u(t) = 4.5\sin 7t$

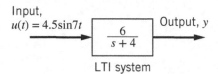

Output, y

LTI system

Figure 9.16 LTI system for Example 9.5.

Constructing the Bode Diagram Using MATLAB

The previous example illustrates how relatively easy the frequency response can be determined if we are given the Bode diagram. In Example 9.5, we used the Bode diagram in Fig. 9.15 to determine the frequency response for an input frequency $\omega = 7$ rad/s; however, we could have computed the frequency response for any input frequency between 0.1 and 100 rad/s as this is the range shown in Fig. 9.15.

Several textbooks such as References 1 and 2 present rules for constructing the approximate Bode diagram from linear asymptotes for the low- and high-frequency ranges (this calculation is presented by Problem 9.9 at the end of this chapter). Although these approximate methods can offer insight into how magnitude and phase vary with frequency, it is this author's opinion that it is more important for the system engineer to know how to *use* the Bode diagram than it is to know how to construct an approximate Bode diagram. This opinion is reinforced by the fact that the exact Bode diagram can be easily constructed by a single MATLAB command. To illustrate, let's construct the Bode diagram for the transfer function used in Table 9.1 and Fig. 9.15

$$G(s) = \frac{6}{s+4} \tag{9.48}$$

The required MATLAB commands are

```
>> sysG = tf(6,[1 4])          % create the system, transfer function G(s)
>> bode(sysG)                  % create and draw the Bode diagram for G(s)
```

The bode command draws the Bode diagram to the screen where magnitude is in decibels, phase is in degrees, and the frequency (rad/s) is plotted on a logarithmic scale.

The bode command can be modified to compute the magnitude and phase of the sinusoidal transfer function $G(j\omega)$ for a desired frequency ω by using the following format

```
>> w = 7;                      % set desired input frequency ω = 7 rad/s
>> [mag,phase] = bode(sysG,w)  % compute magnitude and phase (degrees)
```

No plot of the Bode diagram is drawn to the screen. The magnitude mag is the absolute value of $G(j\omega)$. If the magnitude in decibels is desired, the additional MATLAB command is required:

```
>> magdB = 20*log10(mag)       % magnitude of G(jω) in decibels
```

For LTI systems with multiple DOF (such as multiple-mass mechanical systems), it is probably easier to use bode with the system defined as a state-space representation (SSR). The MATLAB commands for a Bode diagram using an SSR are

```
>> A = [ ... ]                 % create state matrix A
>> B = [ ... ]                 % create input matrix B
>> C = [ ... ]                 % create output matrix C
>> D = [ ... ]                 % create direct-link matrix D
>> sys = ss(A,B,C,D)           % build system using SSR
>> bode(sys)                   % plot Bode diagram(s) for SSR
```

Of course, the user must fill in the appropriate matrices for the desired SSR. The number of Bode diagrams plotted by MATLAB corresponds to the number of inputs and outputs. The user can also compute the magnitude and phase for a desired frequency ω by using the command

```
>> w = 7;                      % set desired input frequency ω = 7 rad/s
>> [mag,phase] = bode(sys,w)   % compute magnitude and phase (degrees)
```

where sys corresponds to the desired SSR determined by matrices **A**, **B**, **C**, and **D**. For example, if an SSR has one input $u(t)$ and two outputs $y_1(t)$ and $y_2(t)$, then the above command will return *two* magnitudes and phase angles because the desired SSR is essentially defining two transfer functions: $G_1(s) = Y_1(s)/U(s)$ and $G_2(s) = Y_2(s)/U(s)$.

Bode Diagram of First-Order Systems

As previously mentioned, we focus on using the Bode diagram rather than the rules associated with sketching the approximate Bode diagram. Furthermore, MATLAB's bode command allows us to draw the exact Bode diagram. Therefore, we present examples of the Bode diagram for first-order systems and summarize their characteristics.

To begin, consider the first-order transfer expressed in our "standard form"

$$G(s) = \frac{K}{\tau s + 1} \tag{9.49}$$

where K is the DC gain of the transfer function (i.e., the value of $G(s)$ with $s = 0$) and τ is the time constant. Clearly, any first-order transfer function (with a constant numerator term) can be written in the standard form of Eq. (9.49). Let's first observe how the Bode diagram changes with the DC gain K. Figure 9.17 shows the Bode diagrams for transfer function (9.49) with time constant $\tau = 0.2$ s and three values of the DC gain K. Clearly, changing the DC gain K shifts the magnitude plot up or down but does *not* change the phase plot (all three gain settings produce the same phase plot). All magnitude plots begin with a "flat" asymptote for low frequencies followed by a linearly decreasing asymptote at high frequencies. The flat (low frequency) and sloped (high frequency) asymptotes are shown as dashed lines on the magnitude plot for $K = 1$ and these asymptotes intersect at the frequency $\omega = 1/\tau = 5$ rad/s. The frequency ω_c where

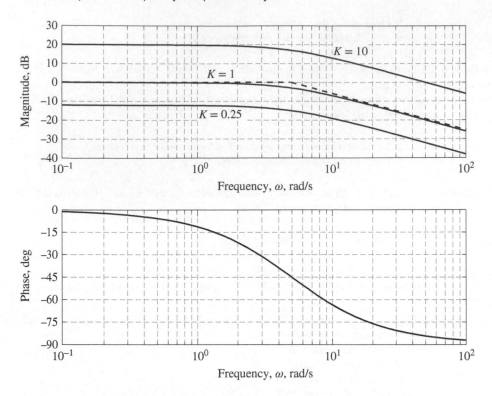

Figure 9.17 Bode diagrams of first-order system $K/(0.2s + 1)$.

the low- and high-frequency asymptotes intersect is called the *corner* or *break frequency* and is always equal to $1/\tau$ regardless of the DC gain K. Figure 9.17 shows that the three magnitude plots are identical except for a constant offset along the vertical axis. Recall that the DC gain is computed by evaluating the transfer function with $s = 0$. Therefore, because the sinusoidal transfer function is calculated by setting $s = j\omega$, the DC gain K corresponds to the magnitude at very low frequencies (i.e., $\omega \to 0$). Consequently, we can compute the magnitude (in dB) of the low-frequency asymptote for the three cases:

$$K = 10: \quad |G(j0)|_{dB} = 20\log_{10}(10) = 20 \text{ dB}$$

$$K = 1: \quad |G(j0)|_{dB} = 20\log_{10}(1) = 0 \text{ dB}$$

$$K = 0.25: \quad |G(j0)|_{dB} = 20\log_{10}(0.25) = -12.04 \text{ dB}$$

These values match the low-frequency asymptotes shown in Fig. 9.17. It is important for the reader to keep in mind that for a first-order system, a 0-dB low-frequency asymptote corresponds to a transfer function with a *unity* DC gain. If the low-frequency asymptote exhibits a negative decibel value, then the DC gain is less than one. In summary, varying the DC gain K will shift the magnitude plot up or down from the 0-dB line by $20\log_{10}(K)$ decibels but will have no effect on the phase plot. For a first-order system, the phase angle ϕ always begins at $0°$ at low frequencies and asymptotically approaches $-90°$ at very high frequencies.

Next, consider the "standard form" first-order transfer function (9.49) with a unity DC gain ($K = 1$) and different time constants. Figure 9.18 shows the Bode diagrams for transfer function (9.49) with $K = 1$ and three time constants τ. Clearly, all three magnitude plots have the same 0-dB low-frequency asymptote because the DC gain is fixed at $K = 1$. Changing the time constant τ changes the corner frequency ($\omega_c = 1/\tau$ rad/s) where the low- and high-frequency asymptotes intersect. The three corner frequencies on Fig. 9.18 are

$$\tau = 0.05 \text{ s}: \quad \omega_c = 20 \text{ rad/s}$$

$$\tau = 0.2 \text{ s}: \quad \omega_c = 5 \text{ rad/s}$$

$$\tau = 1 \text{ s}: \quad \omega_c = 1 \text{ rad/s}$$

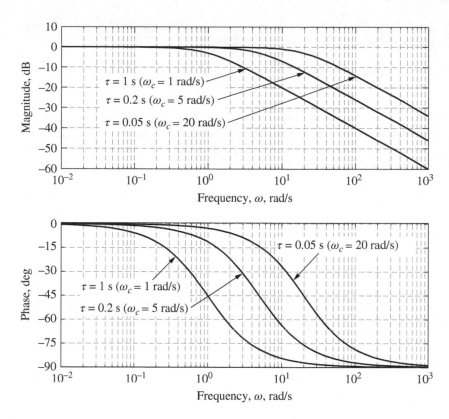

Figure 9.18 Bode diagrams of first-order system $1/(\tau s + 1)$.

It is also clear from Fig. 9.18 that the phase plot is shifted left or right as the time constant (or corner frequency) is varied. In all cases, the phase angle starts at 0° at low frequencies and asymptotically approaches −90° at high frequencies. The reader should note that the phase angle is −45° at the respective corner frequency. In other words, half of the total possible phase lag has occurred when the input frequency matches the corner frequency.

As a final note, we observe that the slope of the high-frequency asymptote remains unchanged despite variations in the DC gain K or time constant τ. Both magnitude plots in Figs. 9.17 and 9.18 show that the high-frequency asymptote drops 20 dB when the input frequency changes by a factor of 10 (a 10-fold factor in frequency is a "decade"). For example, consider the magnitude plot in Fig. 9.18 with $\tau = 1$ s: when $\omega = 10$ rad/s, the magnitude is −20 dB and when $\omega = 100$ rad/s, the magnitude has dropped to −40 dB. This characteristic of the high-frequency asymptote for first-order systems is proven in Problem 9.9 at the end of the chapter.

On the basis of the noted observations of Figs. 9.17 and 9.18, we can summarize the basic characteristics of the Bode diagram for a first-order transfer function in the standard form $G(s) = K/(\tau s + 1)$:

1. A flat low-frequency asymptote exists with a magnitude of $20\log_{10}(K)$ dB.
2. A high-frequency asymptote exists with a slope of −20 dB/decade.
3. The low- and high-frequency asymptotes intersect at the corner frequency $\omega_c = 1/\tau$ rad/s.
4. The phase angle ϕ starts at 0° for low frequencies and asymptotically approaches −90° at high frequencies.
5. The phase angle $\phi = -45°$ when the input frequency is the corner frequency ω_c.

The reader should be able to quickly compute the corner frequency and low-frequency asymptote from the first-order transfer function. For example, given the first-order system

$$G(s) = \frac{4}{s+8} = \frac{K}{\tau s + 1}$$

we see that the time constant is $\tau = 1/8 = 0.125$ s and the DC gain is $K = 4/8 = 0.5$. Hence, the low-frequency asymptote is $20\log_{10}(0.5) = -6.02$ dB and the corner frequency is $\omega_c = 1/\tau = 8$ rad/s.

Example 9.6

Figure 9.19 shows an RC circuit that is driven by sinusoidal voltage source $e_{in}(t) = 2.4\sin\omega t$ V. If the capacitance is $C = 0.003$ F and the resistance is $R = 4\,\Omega$, use the Bode diagram to determine the input frequency ω where the amplitude of the steady-state output voltage $e_C(t)$ is 1.2 V (or, the output/input amplitude ratio is one-half).

　　The mathematical model of the RC circuit can be determined by applying Kirchhoff's voltage law to the single loop and the resulting I/O equation is

$$RC\dot{e}_C + e_C = e_{in}(t)$$

Hence, the transfer function relating output voltage e_C to input voltage $e_{in}(t)$ is

$$G(s) = \frac{E_C(s)}{E_{in}(s)} = \frac{1}{RCs + 1}$$

Clearly, the DC gain is unity and the time constant is $\tau = RC = 0.012$ s. We know that the magnitude plot of the Bode diagram will remain near 0 dB for frequencies less than the corner frequency $\omega_c = 1/\tau = 83.33$ rad/s. At frequencies higher than ω_c, the magnitude plot will drop off at the rate of -20 dB/decade.

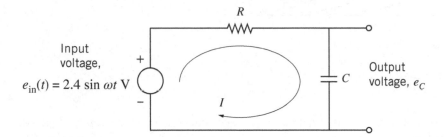

Figure 9.19 RC circuit for Example 9.6.

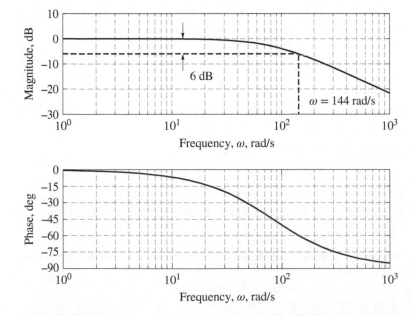

Figure 9.20 Bode diagram for RC circuit (Example 9.6).

Figure 9.20 shows the Bode diagram for the RC circuit and verifies our previous estimates for the magnitude plot. Note that the magnitude remains very close to 0 dB until the input frequency reaches about 40 rad/s; hence, the amplitude of the output voltage signal will essentially match the amplitude of the input voltage signal for $\omega < 40$ rad/s. Because we want to find the frequency where the output/input amplitude ratio is reduced to one-half, we must find the magnitude in decibels that corresponds to $|G(j\omega)| = 0.5$ in order to use the Bode diagram. Therefore,

$$|G(j\omega)|_{dB} = 20 \log_{10}(0.5) = -6.02 \text{ dB}$$

The 6-dB drop from the low-frequency asymptote is shown in Fig. 9.20 as a dashed line that intersects the magnitude plot at a frequency of about 144 rad/s. Hence, the amplitude of the output voltage $e_C(t)$ will be reduced to one-half of the input voltage amplitude when the input frequency is $\omega = 144$ rad/s (or, 22.9 Hz).

The RC circuit in Fig. 9.19 is a simple example of a *low-pass filter*, which is an electronic circuit that allows signals with "low" frequencies to "pass" without reduction in amplitude. Signals with frequencies higher than a prescribed "cutoff frequency" are attenuated (i.e., the output amplitude is reduced). In this example, the values of resistance R and capacitance C have been selected so that the output/input amplitude ratio is greater than 0.5 for input frequencies $\omega < 144$ rad/s (note that the amplitude ratio is essentially unity for $\omega < 40$ rad/s). We discuss and define the cutoff frequency in a later section.

Example 9.7

Consider again the RC circuit in Fig. 9.19 with capacitance $C = 0.003$ F and resistance $R = 4\ \Omega$. Suppose the input voltage is the sum of two sinusoidal signals

$$e_{in}(t) = e_S(t) + e_N(t)$$

where $e_S(t) = 2.4 \sin 10t$ V is the *desired* input signal and $e_N(t) = 0.2 \sin 800t$ V is an *undesirable* high-frequency "noise" signal. Simulate the response of the RC circuit to the noisy input and discuss the performance of the low-pass filter.

Figure 9.21 shows the Simulink model of the RC circuit (or, low-pass filter). Note that the noisy voltage input $e_{in}(t)$ is created by summing two `Sine Wave` sources where the input signal has an amplitude of 2.4 V and frequency $\omega = 10$ rad/s (1.6 Hz) and the noise signal has an amplitude of 0.2 V and frequency $\omega = 800$ rad/s (127.3 Hz). Figure 9.22 shows the desired input signal $e_S(t)$ and the noisy input signal $e_{in}(t)$ on the same graph. Note that the long-wavelength "clean" signal $e_S(t)$ is difficult to see in Fig. 9.22a as it is surrounded by the high-frequency noise; Fig. 9.22b shows an expanded view of the input signal near $t = 0.8$ s where both the low- and high-frequency signals are observable. The noisy signal $e_{in}(t)$ shown in Fig. 9.22 is the input to the low-pass filter transfer function shown in Fig. 9.21. Figure 9.23 shows the "clean" input signal $e_S(t)$ and the output voltage $e_C(t)$ from the low-pass filter. Clearly, the low-pass filter has performed as intended: the RC circuit has filtered or removed the high-frequency noise $e_N(t)$ from the input voltage $e_{in}(t)$ such that the filter output $e_C(t)$ nearly matches the "clean" 2.4-V input signal $e_S(t)$.

In summary, the low-pass filter has "passed" the low-frequency signal $e_S(t)$ with no attenuation and little phase lag. The filter performance can be quantified using its Bode diagram shown in Fig. 9.20: note that the frequency of the desired

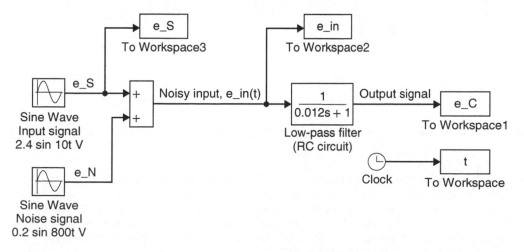

Figure 9.21 Simulink model of RC circuit or low-pass filter (Example 9.7).

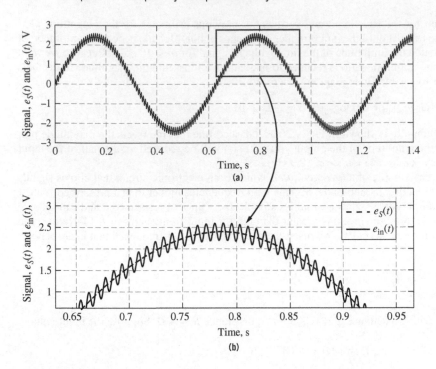

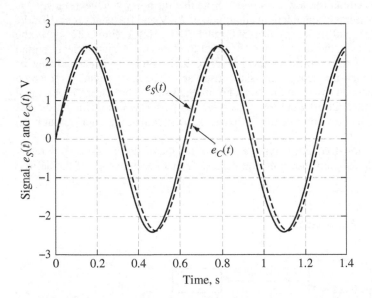

Figure 9.22 Desired input signal $e_S(t)$ and noisy signal $e_{in}(t)$ (Example 9.7).

Figure 9.23 Desired input signal $e_S(t)$ and filtered output signal $e_C(t)$ (Example 9.7).

signal $e_S(t)$ ($\omega = 10$ rad/s) is well below the filter's corner frequency $\omega_c = 83.33$ rad/s. Therefore, this low-frequency signal is passed with no change in amplitude because the filter's DC gain is unity (or, 0 dB). Figure 9.20 also shows that the phase angle is about $-7°$ for a frequency of 10 rad/s and consequently there is very little phase lag between $e_C(t)$ and $e_S(t)$ in Fig. 9.23. Finally, the Bode diagram in Fig. 9.20 shows that the magnitude is nearly -20 dB for the high-frequency noise signal ($\omega = 800$ rad/s). As an absolute-value magnitude -20 dB is 0.1, and, therefore, the low-pass filter has reduced the amplitude of the noise component $e_N(t)$ from 0.2 V to $0.2 \times 0.1 = 0.02$ V.

Bode Diagram of Second-Order Systems

Next, we present examples of the Bode diagram for underdamped second-order systems and summarize their characteristics. To begin, consider the second-order transfer expressed in our "standard form"

$$G(s) = \frac{K\omega_n^2}{s^2 + 2\zeta\omega_n s + \omega_n^2} \tag{9.50}$$

where ζ and ω_n are the damping ratio and undamped natural frequency, respectively. The numerator of $G(s)$ is a constant that is expressed in a form so that K is the DC gain of the transfer function (i.e., the value of $G(s)$ with $s = 0$). For now, let's set the DC gain to unity ($K = 1$) and the damping ratio and undamped natural frequency to $\zeta = 0.2$ and $\omega_n = 10$ rad/s, respectively. With these values, Eq. (9.50) becomes

$$G(s) = \frac{100}{s^2 + 4s + 100} \tag{9.51}$$

Figure 9.24 shows the Bode diagram for the second-order transfer function (9.51). As with the first-order Bode diagram, the magnitude plot exhibits a flat low-frequency asymptote (with a magnitude of 0 dB because the DC gain is $K = 1$). However, the second-order system displays an increase in magnitude (amplitude ratio) when input frequency is near the undamped natural frequency ($\omega_n = 10$ rad/s in this case). The peak magnitude in Fig. 9.24 is about 8 dB and occurs at a frequency slightly less than $\omega_n = 10$ rad/s. Therefore, when this second-order system is driven at an input frequency near 10 rad/s, the output/input amplitude ratio is about $10^{8/20} = 2.5$. Figure 9.24 also shows that the magnitude $|G(j\omega)|_{dB}$ decreases at a rate of -40 dB/decade when the input frequency is greater than $\omega_n = 10$ rad/s (the low- and high-frequency asymptotes are shown as dashed lines on Fig. 9.24). The low- and high-frequency asymptotes intersect at the corner frequency $\omega_c = \omega_n = 10$ rad/s for this case. Figure 9.24 also shows that the phase angle shifts from 0° (low frequency) to $-90°$ at the corner frequency and asymptotically approaches $-180°$ at high frequencies.

Figure 9.25 shows the Bode diagram for second-order system (9.50) with $K = 1$, $\omega_n = 10$ rad/s, and various values for damping ratio ζ. It is clear that the peak magnitude decreases as damping ratio is increased; for the two

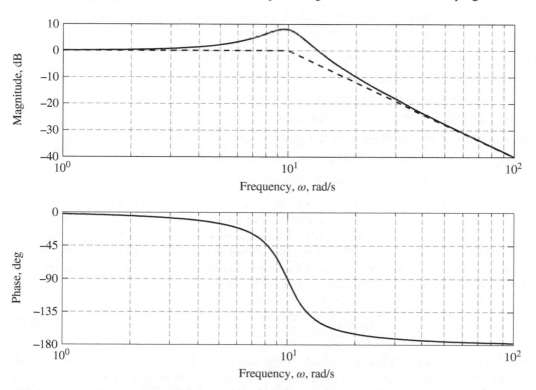

Figure 9.24 Bode diagram of second-order system $100/(s^2 + 4s + 100)$.

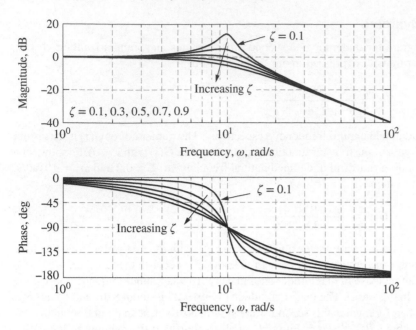

Figure 9.25 Bode diagram of second-order system $100/(s^2 + 20\zeta s + 100)$.

highest damping ratios ($\zeta = 0.7$ and 0.9) the peak magnitude disappears altogether and the entire magnitude plot is less than $0\,\text{dB}$. Figure 9.25 also shows that the phase plot exhibits a sharper shift from $0°$ to $-180°$ when the damping ratio is small (note that all phase plots pass through $-90°$ when input frequency is equal to the corner frequency $\omega_c = \omega_n$).

On the basis of the noted observations of Figs. 9.24 and 9.25, we can summarize the basic characteristics of the Bode diagram for a second-order transfer function in the standard form given by Eq. (9.50):

1. A flat low-frequency asymptote exists with a magnitude of $20\log_{10}(K)\,\text{dB}$.
2. A high-frequency asymptote exists with a slope of $-40\,\text{dB/decade}$.
3. The low- and high-frequency asymptotes intersect at the corner frequency $\omega_c = \omega_n$ rad/s.
4. The peak magnitude increases as damping ratio ζ is decreased.
5. The phase angle ϕ starts at $0°$ for low frequencies and asymptotically approaches $-180°$ at high frequencies.
6. The phase angle shows a sharper transition from $0°$ to $-180°$ as damping ratio ζ is decreased.
7. The phase angle $\phi = -90°$ when the input frequency is the corner frequency ω_c.

The above summary tells us that the Bode diagram of a second-order system is essentially the "double" of the Bode diagram of a first-order system; that is, the high-frequency asymptote slope is doubled and the total phase shift is doubled. A key difference, however, between first- and second-order Bode diagrams is that the shapes of the magnitude and phase plots for a second-order system are greatly affected by the damping ratio ζ. Lightly damped systems exhibit a peak magnitude or *resonant peak* at a particular input frequency that is less than the corner frequency ω_n. The frequency at which the maximum magnitude occurs is called the *resonant frequency* and is

$$\omega_r = \omega_n\sqrt{1 - 2\zeta^2} \tag{9.52}$$

The derivation of Eq. (9.52) can be found in Reference 1. As damping ratio $\zeta \to 0$, Eq. (9.52) tells us that the resonant frequency $\omega_r \to \omega_n$, which can be observed in Fig. 9.25. Because the radicand in Eq. (9.52) is less than or equal to one, the resonant frequency will always be less than the corner frequency ω_n. Finally, note that if damping ratio $\zeta > 0.7071$ a resonant peak does not exist and there is no resonant frequency.

Example 9.8

Consider again the 1-DOF rotational mechanical system presented in Fig. 9.9 (Example 9.3). Compute the resonant frequency ω_r and use the Bode diagram to determine the frequency response for a sinusoidal input torque $T_{in}(t) = 1.5 \sin 18t$ N-m.

Recall that the transfer function of the rotational mechanical system is

$$G(s) = \frac{1}{0.2s^2 + 1.6s + 65} = \frac{\Theta(s)}{T_{in}(s)} \tag{9.53}$$

We can express Eq. (9.53) in the standard form for a second-order system

$$G(s) = \frac{5}{s^2 + 8s + 325}$$

Clearly, the DC gain is $5/325 = 0.0154$, or -36.3 dB. It is important to note that the DC gain is a small value because the system transfer function relates the torque input to the angular displacement, and hence a constant 1 N-m torque input will result in a 0.0154 rad angular deflection at steady state. We see that the undamped natural frequency is $\omega_n = \sqrt{325} = 18.0278$ rad/s and the damping ratio is $\zeta = 0.2219$. Using Eq. (9.52) the resonant frequency is

$$\omega_r = \omega_n \sqrt{1 - 2\zeta^2} = 17.1172 \text{ rad/s}$$

The frequency response of the mechanical system with an input amplitude of 1.5 N-m and input frequency $\omega = 18$ rad/s is

$$\theta_{ss}(t) = |G(j18)|1.5 \sin(18t + \phi) \tag{9.54}$$

We can use MATLAB to construct the Bode diagram of the second-order mechanical system, and the resulting diagram is shown in Fig. 9.26. Note that the low-frequency asymptote is about -36 dB and the resonant peak occurs at a frequency of about 17 rad/s, which match our previous calculations. The magnitude and phase at the input frequency $\omega = 18$ rad/s are shown on the respective plots in Fig. 9.26, and their exact values can be computed using the MATLAB bode command:

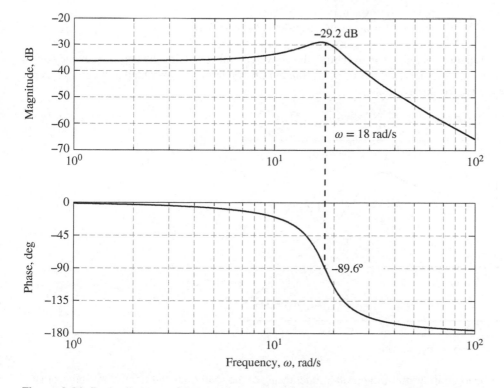

Figure 9.26 Bode diagram of the rotational mechanical system (Example 9.8).

```
>> sysG = tf(5,[ 1 8 325 ])          % create transfer function G(s)
>> w = 18;                           % desired input frequency, rad/s
>> [mag,phase] = bode(sysG,w)        % compute exact |G(j18)| and ∠G(j18)
>> magdB = 20*log10(mag)             % compute magnitude in decibels
```

Converting the Bode-plot magnitude of $-29.2\,\text{dB}$ to an absolute-value magnitude and phase angle to radians, we obtain $|G(j18)| = 0.0347$ and $\phi = -1.5639$ rad, respectively. Hence, the frequency response (9.54) becomes

$$\theta_{ss}(t) = 0.0521 \sin(18t - 1.5639) \text{ rad}$$

which is the same result obtained in Example 9.3. In summary, the mechanical system in this example is essentially being driven at its undamped natural frequency ω_n, which is slightly higher than its resonant frequency of 17.12 rad/s. The frequency response and resonant frequency are relatively easy to compute if an accurate Bode diagram is available.

Bandwidth

The *cutoff frequency* ω_B is defined as the maximum input frequency at which the output of the system will follow an input sinusoid in a satisfactory manner. Typically, "satisfactory tracking" includes all frequencies up to the cutoff frequency where the amplitude ratio is attenuated (reduced) from its low-frequency value by a factor of 0.7071 or $1/\sqrt{2}$. This 70.7% factor arises from electrical circuit theory and represents a one-half power loss because power is typically a squared quantity. In decibels, attenuation by a factor of 0.7071 is equal to a 3-dB drop from the low-frequency magnitude and, therefore, the cutoff frequency can be read directly from the Bode magnitude plot. The frequency range $0 \leq \omega \leq \omega_B$ where the magnitude of $G(j\omega)$ remains within 3 dB of the DC gain is called the *bandwidth* of the system.

Figure 9.27 shows the magnitude plot (dB) from the Bode diagram for the rotational mechanical system from Example 9.8. The low-frequency magnitude (DC gain) is -36.3 dB and is labeled on Fig. 9.27. The bandwidth is determined by the cutoff frequency ω_B where the magnitude plot has dropped to -39.3 dB (or, 3 dB below the DC gain value). We can estimate the cutoff frequency by reading the Bode diagram or we can use the MATLAB command `bandwidth`:

```
>> sysG = tf(5,[ 1 8 325 ])          % create transfer function G(s)
>> wB = bandwidth(sysG)              % compute the bandwidth of system G(s)
```

MATLAB returns bandwidth wB in units of rad/s, and executing the above commands yields the cutoff frequency $\omega_B = 27.023$ rad/s. This cutoff frequency is also depicted in Fig. 9.27. Bandwidth is sometimes presented in hertz (cycles per second), which is $\omega_B/(2\pi) = 4.3$ Hz for this example.

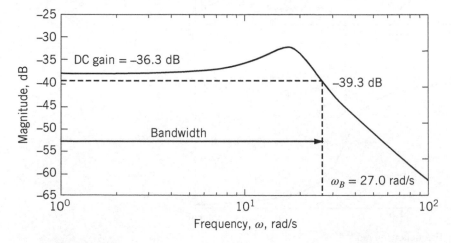

Figure 9.27 Magnitude plot showing bandwidth and cutoff frequency ω_B of the rotational mechanical system from Example 9.8.

In general, high bandwidth corresponds to a fast system response because the output can satisfactorily track a high-frequency input. For example, we might require an electromechanical actuator (solenoid) to possess a very high bandwidth so that it can quickly position a hydraulic valve and improve the overall system performance. However, we do not want the solenoid and valve to respond to undesirable high-frequency noise and, therefore, we might choose to send the voltage input signal through a low-pass filter in order to eliminate the high-frequency components.

Example 9.9

Figure 9.28 presents the spool-valve dynamics from Example 9.4, where force f_{em} (from the solenoid) is the input and valve position y is the output. Show the resonant frequency (if applicable) and bandwidth on the Bode diagram and present these frequencies in hertz.

The spool-valve transfer function is

$$G_2(s) = \frac{1}{0.04s^2 + 16s + 7000} = \frac{Y(s)}{F_{em}(s)}$$

Figure 9.29 shows the Bode diagram of the second-order spool-valve system. The DC gain of $G_2(s)$ is $1/7000 = 1.429(10^{-4})$ (or, -77 dB) and the corner frequency is $\omega_c = \sqrt{7000/0.04} = 418.3$ rad/s. From Fig. 9.29, we see that the magnitude drops 3 dB from its DC gain at a cutoff frequency of about $\omega_B = 542$ rad/s (86.3 Hz), hence the bandwidth is greater than the corner frequency. Because the undamped natural frequency and damping ratio of the spool valve are $\omega_n = 418.3$ rad/s and $\zeta = 0.478$, respectively, the resonant frequency of the spool valve is $\omega_r = \omega_n\sqrt{1 - 2\zeta^2} = 308.2$ rad/s or 49.1 Hz. The magnitude of the resonant peak is fairly small (due to the moderate amount of damping) as demonstrated by the magnitude plot in Fig. 9.29.

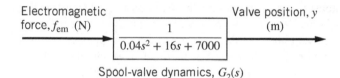

Electromagnetic force, f_{em} (N) → $\dfrac{1}{0.04s^2 + 16s + 7000}$ → Valve position, y (m)

Spool-valve dynamics, $G_2(s)$

Figure 9.28 Spool valve for Example 9.9.

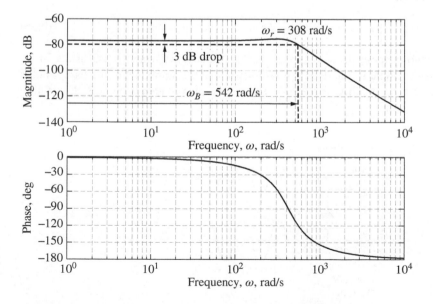

Figure 9.29 Bode diagram of spool valve $G_2(s)$ (Example 9.9).

9.4 VIBRATIONS

Vibrations in mechanical systems are generally undesirable phenomena as they transmit unwanted motion or forces, or generate noise. Suspensions systems consisting of stiffness and damping elements are often used to reduce the transmission of oscillatory motion or forces. Common examples include automobile suspension systems for suppressing road vibrations to its passengers, pneumatic "air springs" for reducing vibrations in industrial applications, and rubber mounts for suppressing vibrations in unbalanced rotating machines such as pumps and motors. Vibration analysis is an expansive subject that involves various topics such as computing the mode shapes by solving the eigenvalue problem. Many engineering programs require or offer a separate course in vibrations. In this section, we briefly present a few basic concepts that pertain to vibration analysis.

Vibration Isolation

In many industrial settings, it is important to reduce the transmission of vibrations from the environment to a sensitive machine, or, conversely, the transmission of vibrations from an unbalanced rotating machine to its environment. This task is called *vibration isolation* and it is often performed by a mounting or suspension system that consists of stiffness elements for support and friction elements for energy dissipation. For example, sensitive instruments for microgravity experiments on board the International Space Station use vibration mounts to isolate the instruments from steady-state environmental vibrations caused by pumps and fans, as well as transient vibrations caused by astronaut activity and thruster firings [3]. Figure 9.30a shows a schematic diagram of an instrument with rubber vibration mounts and Fig. 9.30b shows the vibration isolation system depicted as a simple mass–spring–damper mechanical system where the total stiffness and damping of the mounts are lumped into spring and damper elements k and b, respectively. Figure 9.30 shows that the objective of the vibration isolation system is to reduce the transmission of periodic base motion (x_b) to the instrument (mass m).

Transmissibility is a principal metric for vibration reduction, and for the system in Fig. 9.30 it is defined as the ratio of the amplitude of the transmitted displacement (output) to the amplitude of the input base displacement. For mechanical systems that experience a sinusoidal input *force* rather than an input base vibration (such as an unbalanced rotating machine), transmissibility is defined as the ratio of the amplitude of the transmitted force to the base (output) to the amplitude of the input force. Using the vibration isolation example shown in Fig. 9.30 with base vibrations, let us assume that the base motion is a sinusoidal function: $x_b(t) = U_0 \sin \omega t$ m. The steady-state response of the instrument mass (i.e., the frequency response) is determined by Eq. (9.17)

$$x_{ss}(t) = |G(j\omega)|U_0 \sin(\omega t + \phi) \text{ m} \tag{9.55}$$

where $G(s) = X(s)/X_b(s)$ is the transfer function relating the output (mass displacement x) to the input (base displacement x_b). Because the amplitude of the mass displacement in Eq. (9.55) is $|G(j\omega)|U_0$ and the amplitude of the base input is U_0, the output-to-input amplitude ratio (or transmissibility) is $|G(j\omega)|$. Hence, the magnitude plot of a Bode diagram essentially tells us the transmissibility of a system.

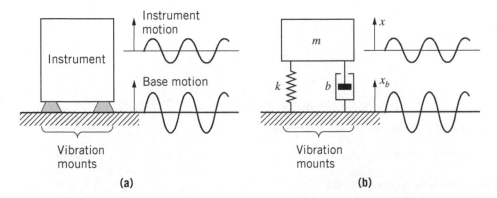

Figure 9.30 Vibration isolation system: (a) instrument with vibration mounts and (b) representative mechanical system.

For example, let us plot the transmissibility of the simple vibration isolation system shown in Fig. 9.30b. Applying Newton's laws to a free-body diagram of the mass–spring–damper system, we can derive the following mathematical model

$$m\ddot{x} + b(\dot{x} - \dot{x}_b) + k(x - x_b) = 0 \tag{9.56}$$

which can be rewritten with the input terms (base motion) on the right-hand-side

$$m\ddot{x} + b\dot{x} + kx = b\dot{x}_b + kx_b \tag{9.57}$$

Using either the D-operator or Laplace-transform methods we obtain the transfer function of the vibration isolation system:

$$G(s) = \frac{X(s)}{X_b(s)} = \frac{bs + k}{ms^2 + bs + k} = \frac{(b/m)s + (k/m)}{s^2 + (b/m)s + (k/m)} \tag{9.58}$$

The sinusoidal transfer function is obtained by substituting $s = j\omega$ into Eq. (9.58)

$$G(j\omega) = \frac{(b/m)j\omega + (k/m)}{-\omega^2 + (b/m)j\omega + (k/m)} \tag{9.59}$$

Because the vibration isolator is a mass–spring–damper system, we can substitute the standard second-order system parameters $\omega_n^2 = k/m$ and $2\zeta\omega_n = b/m$ into Eq. (9.59)

$$G(j\omega) = \frac{\omega_n^2 + j2\zeta\omega_n\omega}{\omega_n^2 - \omega^2 + j2\zeta\omega_n\omega} \tag{9.60}$$

Dividing all terms in Eq. (9.60) by ω_n^2, we obtain

$$G(j\omega) = \frac{1 + j(2\zeta\omega/\omega_n)}{1 - (\omega^2/\omega_n^2) + j(2\zeta\omega/\omega_n)} \tag{9.61}$$

We can simplify Eq. (9.61) by substituting the nondimensional parameter $\beta = \omega/\omega_n$

$$G(j\omega) = \frac{1 + j2\zeta\beta}{1 - \beta^2 + j2\zeta\beta} \tag{9.62}$$

It is important to note that parameter β is the ratio of the input frequency ω (a property of the base motion) and the undamped natural frequency ω_n (a function of the mount stiffness k and instrument mass m). Finally, the transmissibility is the magnitude of the sinusoidal transfer function and hence it can be computed from the real and imaginary parts of Eq. (9.62)

$$\text{Transmissibility:} \quad |G(j\omega)| = \frac{\sqrt{1^2 + (2\zeta\beta)^2}}{\sqrt{(1 - \beta^2)^2 + (2\zeta\beta)^2}} \tag{9.63}$$

Figure 9.31 shows transmissibility for the 1-DOF vibration isolation system depicted in Fig. 9.30. Transmissibility (9.63) is computed for the input frequency ratio range $0 < \beta < 3$ for four damping ratios: $\zeta = 0.1, 0.4, 0.7$, and 1. Analysis of Fig. 9.31 allows us to summarize the transmissibility characteristics for a 1-DOF mechanical vibration isolator:

1. When input frequency ratio $\beta = \omega/\omega_n$ is small, the transmissibility is unity (due to the unity DC gain of the transfer function) regardless of the damping ratio ζ. Consequently, a 1-cm amplitude base vibration will produce a 1-cm amplitude instrument vibration. Small β is due to a small input (base) frequency ω and/or a very large natural frequency ω_n of the isolation mounts (i.e., very stiff mounts).

2. Transmissibility shows a peak greater than one at the value of β that corresponds to the resonant frequency, that is, $\beta = \omega_r/\omega_n$. The magnitude of the resonant peak increases as damping ratio is decreased. When damping is very small, the transmissibility peak occurs near $\beta = 1$ because the resonant frequency is nearly equal to the undamped natural frequency ω_n. Note that unlike the standard second-order system in Eq. (9.50) and its

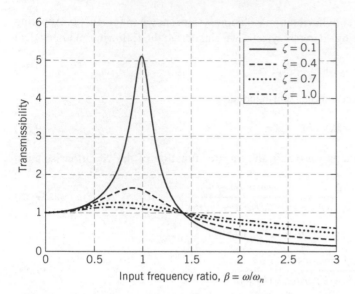

Figure 9.31 Transmissibility for a 1-DOF vibration isolation system.

corresponding Bode diagram shown in Fig. 9.25, transmissibility exhibits a resonant peak for all damping ratios. This difference is due to the first-order numerator of the vibration isolator transfer function (9.58).

3. When $\beta = \sqrt{2} = 1.4142$, transmissibility is unity for all damping ratios. Therefore, the excitation base amplitude is transmitted to the instrument without gain or loss if the input frequency is $\omega = \omega_n \sqrt{2}$.

4. When input frequency ratio $\beta < \sqrt{2}$, increasing ζ decreases transmissibility and improves vibration isolation (however, transmissibility is always greater than unity). Conversely, when input frequency ratio $\beta > \sqrt{2}$, increasing ζ *increases* transmissibility and degrades vibration isolation.

5. Transmitted vibrations are attenuated (i.e., transmissibility is less than one) only when $\beta > \sqrt{2}$.

Figure 9.31 and the above summary can help engineers select the best vibration isolation system for their particular application. For minimum transmitted motion, choose vibration isolation mounts with an undamped natural frequency (i.e., stiffness) much smaller than the expected input vibration frequency ω (therefore, $\beta = \omega/\omega_n$ is large). If possible, select mounts with light damping for minimum transmissibility (see Fig. 9.31 for $\beta > \sqrt{2}$). Many times the disturbing input frequency may not be a single value but instead may be in a known range and, therefore, the vibration isolator design must consider a range of β. Finally, the system engineer may need to balance transmissibility requirements with other design constraints such as shock response and settling time of the transient response.

Vibrations in Multiple-DOF Systems

It is important to reiterate that the transmissibility equation (9.63), Fig. 9.31, and the summaries regarding transmissibility characteristics apply only to a 1-DOF vibration isolation system that can be accurately modeled as a *single* mass−spring−damper system as shown in Fig. 9.30. In addition, transmissibility can be determined from the magnitude Bode diagram (of course, the Bode magnitude in decibels must be converted into an absolute-value magnitude in order to compute transmissibility).

In this subsection, we discuss vibrations in mechanical systems with multiple DOF. Modal analysis using matrix methods is a classical approach for analyzing vibrations in multiple-DOF systems and this technique relies on solving the eigenvalue and eigenvector problems in order to obtain the natural frequencies and mode shapes (e.g., see Reference 4). We analyze vibrations in multiple-DOF systems by using the Bode diagram as illustrated by the following example.

Example 9.10

Figure 9.32 presents a schematic diagram of an optical disk drive system that uses a laser to read data stored on a compact disk such as a DVD (this system has been analyzed in Problems 2.26 and 5.35). Determine the vibration frequencies for an impulsive input due to a sudden displacement of the frame and analyze the frequency response of the pick-up head (PUH) using the Bode diagram.

The cart motor in Fig. 9.32 translates the PUH along the radial direction of the spinning disk in order to properly position the focused laser (see References 5 and 6 for additional details). Figure 9.33 shows the optical disk drive as a simplified two-mass, 2-DOF mechanical system. The PUH mass m_1 is connected to the cart with stiffness k_1 and friction coefficient b_1, while the chassis and cart are lumped into mass m_2. A series of rubber mounts connect the chassis mass m_2 to the frame and these mounts have lumped stiffness k_2 and friction coefficient b_2. The mounts are used to suppress transmitted vibrations from the motion of the frame. The absolute displacements (as measured from the static equilibrium position) of the PUH and chassis/cart are x_1 and x_2, respectively. The system input is the absolute displacement of the frame $x_{in}(t)$.

The mathematical model of the 2-DOF mechanical system in Fig. 9.33 was derived in Chapter 2 (Problem 2.26) and is presented below:

PUH mass:
$$m_1\ddot{x}_1 + b_1(\dot{x}_1 - \dot{x}_2) + k_1(x_1 - x_2) = 0$$

Chassis/cart mass:
$$m_2\ddot{x}_2 + b_1(\dot{x}_2 - \dot{x}_1) + b_2\dot{x}_2 + k_1(x_2 - x_1) + k_2x_2 = b_2\dot{x}_{in}(t) + k_2x_{in}(t)$$

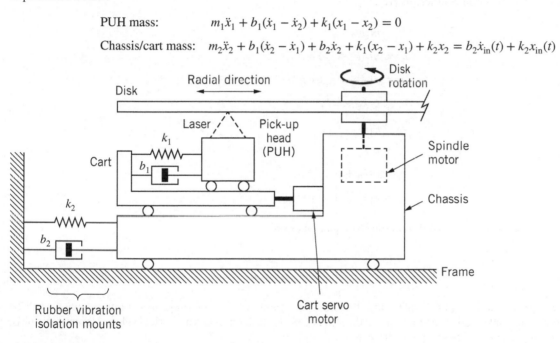

Figure 9.32 Optical disk drive system for Example 9.10.

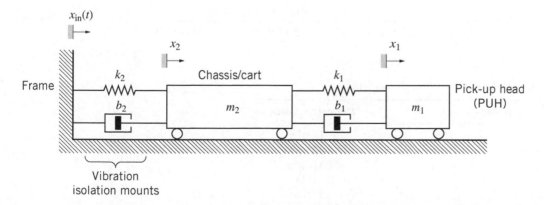

Figure 9.33 2-DOF mechanical model of the optical disk drive system (Example 9.10).

We can analyze the response of the PUH mass by using the transfer function that relates PUH displacement x_1 to frame displacement $x_{in}(t)$, or $G(s) = X_1(s)/X_{in}(s)$. However, the mathematical model consists of two coupled second-order ODEs and, therefore, we must manipulate both modeling equations in order to develop an I/O equation with PUH displacement x_1 as the single output and frame displacement $x_{in}(t)$ as the single input (see Problem 5.35). The first step is to apply the differential or D operator to the mathematical model to yield

PUH: $\qquad (m_1 D^2 + b_1 D + k_1)x_1 = (b_1 D + k_1)x_2$ $\qquad\qquad\qquad$ (9.64)

Chassis/cart: $\qquad (m_2 D^2 + (b_1 + b_2)D + k_1 + k_2)x_2 = (b_1 D + k_1)x_1 + (b_2 D + k_2)x_{in}(t)$ $\qquad$ (9.65)

Next, we can solve Eq. (9.65) for the chassis/cart position x_2 and substitute the result into Eq. (9.64), which will yield an equation in terms of x_1 and $x_{in}(t)$. After some algebra, the following I/O equation is obtained by noting that $Dx_1 = \dot{x}_1$, $D^2 x_1 = \ddot{x}_1$, etc.

$$a_4 x_1^{(4)} + a_3 \dddot{x}_1 + a_2 \ddot{x}_1 + a_1 \dot{x}_1 + a_0 x_1 = c_2 \ddot{x}_{in}(t) + c_1 \dot{x}_{in}(t) + c_0 x_{in}(t) \qquad (9.66)$$

where the left- and right-hand-side coefficients are

$$a_4 = m_1 m_2$$
$$a_3 = m_1(b_1 + b_2) + m_2 b_1$$
$$a_2 = m_1(k_1 + k_2) + m_2 k_1 + b_1 b_2$$
$$a_1 = b_1 k_2 + b_2 k_1$$
$$a_0 = k_1 k_2$$
$$c_2 = b_1 b_2$$
$$c_1 = b_1 k_2 + b_2 k_1 = a_1$$
$$c_0 = k_1 k_2 = a_0$$

Hence, the transfer function can be derived from the I/O equation (9.66)

$$G(s) = \frac{X_1(s)}{X_{in}(s)} = \frac{c_2 s^2 + c_1 s + c_0}{a_4 s^4 + a_3 s^3 + a_2 s^2 + a_1 s + a_0} \qquad (9.67)$$

Note that because $c_0 = a_0$ the DC gain is unity and, therefore, the amplitude of low-frequency frame vibrations will be transmitted without gain or attenuation to vibrations in the PUH mass. Table 9.2 presents numerical values for mass, stiffness, and friction that are representative of an optical drive system [6].

Using the values in Table 9.2, the transfer function (9.67) becomes

$$G(s) = \frac{2512.3 s^2 + 5.8145(10^6)s + 1.5806(10^9)}{s^4 + 100.484 s^3 + 1.1880(10^5)s^2 + 5.8145(10^6)s + 1.5806(10^9)} \qquad (9.68)$$

Table 9.2 Parameters for the Optical Disk Drive System [6]

Optical Disk Drive System Parameter	Numerical Value
PUH mass, m_1	$6(10^{-4})$ kg
PUH suspension stiffness, k_1	60 N/m
PUH suspension friction, b_1	0.03 N-s/m
Chassis/cart mass, m_2	0.124 kg
Chassis suspension stiffness, k_2	1960 N/m
Chassis suspension friction, b_2	6.23 N-s/m

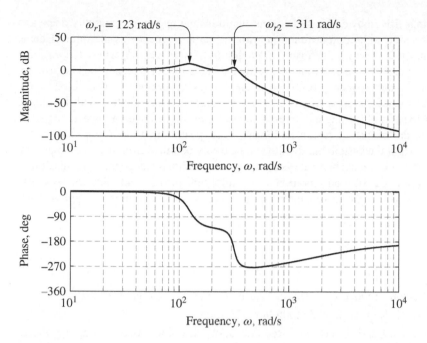

Figure 9.34 Bode diagram of the 2-DOF optical disk drive system (Example 9.10).

The characteristic roots of the I/O equation (9.66) or the poles of $G(s)$ are both determined by computing the roots of the fourth-order polynomial

$$s^4 + 100.484s^3 + 1.1880(10^5)s^2 + 5.8145(10^6)s + 1.5806(10^9) = 0 \tag{9.69}$$

The four roots (or poles) are two complex conjugate pairs:

$$s_{1,2} = -24.9587 \pm j122.8638 \quad \text{and} \quad s_{3,4} = -25.2832 \pm j316.1022$$

Consequently, the transient response of the PUH mass for an impulsive input $x_{in}(t)$ will be two exponentially damped sinusoidal functions. The two frequencies of the transient response are the imaginary parts of the two complex roots: 122.86 rad/s (or 19.6 Hz) and 316.10 rad/s (or 50.3 Hz). The PUH mass m_1 will vibrate at these two frequencies after an impulsive frame displacement. These transient vibrations will die out in approximately $t_S = 0.16$ s because both roots have real parts equal to -25 and $e^{-25t_S} = e^{-4}$.

Figure 9.34 shows the Bode diagram of the transfer function Eq. (9.68) where PUH displacement is the system output and frame displacement is the input. Note that at low input frequencies the magnitude is 0 dB as expected because of the unity DC gain of the transfer function. Hence, the transmissibility is one for low-frequency frame vibrations. As expected, the Bode diagram exhibits two resonant peaks because we have a two-mass (or 2-DOF) mechanical system. The first resonant frequency $\omega_{r1} = 123$ rad/s approximately matches the first damped frequency of the natural (transient) response. The first resonant peak magnitude is 10.15 dB or a transmissibility of 3.217 because $|G(j\omega_{r1})| = 10^{10.15/20} = 3.217$. Figure 9.34 shows that the second resonant frequency is $\omega_{r2} = 311$ rad/s (49.5 Hz), which is slightly less than the second damped frequency. The second resonant peak magnitude is about 4.53 dB or a transmissibility of about 1.685. For input frequencies greater than 341 rad/s (54.3 Hz), the magnitude in decibels is less than zero and hence the transmissibility is less than one. For example, if the input frequency is $\omega = 394$ rad/s (62.7 Hz) the magnitude is -10 dB and the transmissibility is 0.316; furthermore, if the input frequency is 550 rad/s (87.5 Hz) the magnitude is -25 dB and the transmissibility is 0.056. From this analysis, we can conclude that the vibration isolation mounts will provide good performance (at least 68% vibration reduction) for frame vibrations greater than 394 rad/s or 62.7 Hz.

SUMMARY

This chapter discussed the analysis of dynamic systems that are driven by a sinusoidal input function. In particular, we focused on the *frequency response* of an LTI dynamic system, which is defined as the steady-state response to a sinusoid input function. When a damped LTI system is driven by a sinusoidal input, the transient response eventually decays to zero and all that remains at steady state is a sinusoidal output function with the same frequency as the input. Hence, given the input $u(t) = U_0 \sin \omega t$ the frequency response has the form $y_{ss}(t) = |G(j\omega)|U_0 \sin(\omega t + \phi)$ where $|G(j\omega)|$ and ϕ are the magnitude and phase angle of the sinusoidal transfer function evaluated at the input frequency ω. The primary concept to be gleaned from this chapter is that the frequency response of an LTI system is *completely* defined by the magnitude and phase angle of the sinusoidal transfer function $G(j\omega)$, that is, the system transfer function with s replaced by $j\omega$. Bode diagrams present magnitude (dB) and phase angle (degrees) versus input frequency on a logarithmic scale and, therefore, facilitate the computation of the frequency response. Furthermore, peak output/input amplitude ratios (and the corresponding resonant frequencies) and the system bandwidth can be easily obtained from the Bode diagram. Finally, we presented a brief discussion of vibrations in mechanical systems. This discussion included the transmissibility metric, which is the ratio of the amplitude of the transmitted vibration to the amplitude of the input vibration. Because this output/input amplitude ratio is the magnitude $|G(j\omega)|$, we can determine transmissibility from the magnitude plot of the Bode diagram.

REFERENCES

1. Ogata, K., *System Dynamics*, 4th ed., Pearson Prentice Hall, Upper Saddle River, NJ, 2004, pp. 609–619.
2. Palm, W.J., *System Dynamics*, McGraw-Hill, New York, NY, 2005, pp. 553–568.
3. Whorton, M.S., Eldridge, J.T., Ferebee, R.C., Lassiter, J.O., and Redmon, J.W., "Damping Mechanisms for Microgravity Vibration Isolation," NASA TM-1998-206953, Jan., 1998.
4. Vu, H.V, and Esfandiari, R.S., *Dynamic Systems Modeling and Analysis*, McGraw-Hill, New York, 1997, pp. 401–437.
5. Liu, J.-J., and Yang, Y.-P., "Disk Wobble Control in Optical Disk Drives," *ASME Journal of Dynamic Systems, Measurement, and Control*, Vol. 127, Sept. 2005, pp. 508–514.
6. Yu, Z., "Anti-Shock Control for Optical Storage Drives," Ph.D. dissertation, Eindhoven University of Technology, The Netherlands, June 2011.
7. Wait, K.W., and Goldfarb, M., "Enhanced Performance and Stability in Pneumatic Servosystems with Supplemental Mechanical Damping," *ASME Journal of Dynamic Systems, Measurement, and Control*, Vol. 132, July 2010, pp. 041012–1-8.

Chapter 10

Introduction to Control Systems

10.1 INTRODUCTION

The previous eight chapters have dealt with two major topics: (1) developing models for physical engineering systems and (2) analyzing the system response to a known input function. In every case we have investigated thus far, the system's input has been independent of its output. In this chapter, we introduce the concept of a *feedback control system* where the system's output is measured and fed back in order to influence the system input. Knowledge of the system's response via feedback allows the control input to be shaped in order to improve the response characteristics. We show that certain control-logic schemes or "controllers" can be used to reduce the response time, improve the damping characteristics, and enhance the ability of the system output to match or track a desired reference signal.

The objective of this chapter is to introduce the reader to the analysis and design of feedback control systems. Because the focus of this textbook is modeling and analysis of dynamic systems, we only provide an introductory treatment of feedback control. We emphasize the proportional-integral-derivative (PID) control scheme (and its variants) as it is the most popular controller used in industrial settings. Controller design and stability analysis are presented using two graphical techniques: the root-locus method and the Bode diagram. Finally, we end this chapter with a brief discussion on the practical issues associated with implementing control schemes in the digital domain.

10.2 FEEDBACK CONTROL SYSTEMS

So far we have only considered the response of *open-loop* dynamic systems. For example, Fig. 10.1 shows the simplified solenoid actuator−valve system studied in Chapters 7 and 9 (see Examples 7.4 and 9.4). Figure 10.1 is an example of an *open-loop* system where the overall system input, voltage signal $e_{in}(t)$, does not depend on the system output, spool-valve position $y(t)$. Suppose we want to position and hold the valve at $y = 0.003$ m (3 mm). Because the DC gain of the complete system in Fig. 10.1 is $(12/1.5) \cdot (1/7000) = 0.001143$ m/V, a step voltage input $e_{in}(t) = 2.625$ V will produce the desired 3-mm steady-state valve position (recall that the steady-state response is the product of the constant input and the DC gain). However, the reader should note that the linear transfer functions in Fig. 10.1 may not *exactly* represent the true dynamics of the solenoid or spool valve owing to modeling uncertainties in viscous friction, solenoid inductance, and so on. Hence, our calculated step voltage $e_{in}(t) = 2.625$ V may not exactly produce the desired 3-mm valve position. Furthermore, the open-loop system in Fig. 10.1 has no way to automatically adjust the voltage input to position the valve at precisely 3 mm. In order to automatically correct the voltage signal, it must become a function of the valve-position error. Consequently, we must measure the valve position $y(t)$ and feed this information back to the solenoid actuator in order to "close the loop" and therefore create a *closed-loop* control system.

Figure 10.2 shows a general closed-loop feedback control system. The block labeled *plant* denotes the physical system to be controlled and is usually represented by one or more transfer functions or a state-space representation (SSR). Using our system in Fig. 10.1 as an example, the spool-valve transfer function would represent the plant or system to be controlled. The block labeled *sensor* in Fig. 10.2 denotes the physical measuring device that allows feedback information. For example, a linear variable differential transformer (LVDT) is an electromechanical device that measures translational displacement. The sensor block in Fig. 10.2 could either be a transfer function if the physical device exhibits a dynamic response or a simple gain if the sensor output is proportional to its input. The *controller* block in Fig. 10.2 denotes the control logic (or control rules) and the physical actuating

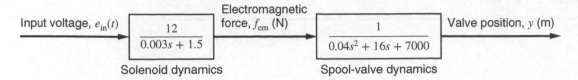

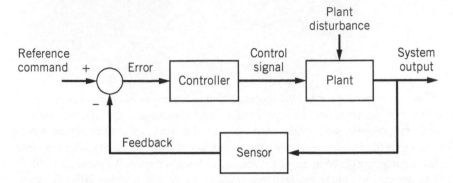

Figure 10.1 Open-loop system: solenoid actuator and spool valve.

Figure 10.2 General closed-loop feedback control system.

device that controls the plant and is usually represented by one or more transfer functions. The solenoid in Fig. 10.1 would be the physical actuating device because it drives the plant (spool valve, in this case). The control logic that determines the voltage input to the solenoid actuator is also part of the controller block. For real-world control systems, the control logic typically resides in a computer or a microprocessor. The input to the controller block is usually an error signal, which is the difference between the reference command (desired output) and the actual system output (feedback signal) as measured by the sensor. The output of the controller is the control signal that drives the plant and (if designed properly) ultimately produces a system output that matches the desired reference command (i.e., a zero error signal). Finally, the plant may be subjected to disturbance inputs from the operating environment such as winds, vibrations, etc.

Cruise control for an automobile is a good illustration of the closed-loop feedback structure shown in Fig. 10.2. The driver manually brings the automobile to the desired speed and presses a button to engage the cruise control, which therefore sets the reference command (desired speed). A speed sensor measures the actual speed of the automobile and this feedback signal is compared to the reference command to form a velocity error signal. The control-logic rules reside in a small computer onboard the automobile and the controller "rules" use the velocity error to determine an electronic throttle signal. The control signal (throttle signal) is the input to the engine, which is part of the plant. Because throttle signal is the plant input and speed is the plant output, the plant block in Fig. 10.2 would include models of the engine dynamics, powertrain, road terrain, and aerodynamic drag. The presence of uphill or downhill roads or winds would be examples of disturbance inputs to the plant.

The purpose of an automobile cruise control system is to automatically maintain a desired speed. We can list general performance requirements that dictate the design of feedback control systems:

1. **Stability margins:** the closed-loop system must demonstrate stable operation where the system output remains bounded for all bounded reference commands.

2. **Speed of response:** for example, the control system must quickly respond to a new reference command.

3. **Good damping characteristics:** for example, a good controller design for a cruise-control system should result in very low overshoot as the automobile's actual speed approaches the desired reference speed.

4. **Little or no steady-state error (tracking):** for example, a good cruise-control design would produce very small velocity error (good "tracking") at steady state.

5. Disturbance rejection: the closed-loop system should compensate for disturbance inputs and demonstrate adequate performance (i.e., good response speed, tracking, etc.).

Control System Transfer Functions

Figure 10.3 shows a simple closed-loop control system, which is essentially the same as Fig. 10.2 except that the controller, plant, and sensor blocks are represented by the single-input, single-output (SISO) transfer functions $G_C(s)$, $G_P(s)$, and $H(s)$, respectively. Note also that the respective path signals are labeled as both time- and Laplace-domain functions. The reader should recall that we labeled the path signals as time functions in the block diagrams discussed in Chapter 5 and the Simulink diagrams presented in Chapter 6. We choose to switch back and forth between labeling signals as either time- or Laplace-domain functions when it is convenient to do so. For example, we use time-domain signals for Simulink diagrams and Laplace-domain signals when we use and manipulate transfer functions.

The following transfer functions are now defined using Fig. 10.3. The *forward transfer function* is $G(s) = G_C(s)G_P(s)$. In other words, the forward transfer function contains all transfer functions in the forward path (controller and plant), and therefore relates the error signal $e(t)$ to the system output $y(t)$. The sensor transfer function $H(s)$ is the *feedback transfer function*. The *open-loop transfer function* is $G(s)H(s)$. Hence, the open-loop transfer function is the product of all transfer functions in the forward and feedback paths.

Figure 10.4a shows the closed-loop system in Fig. 10.3, with $G(s)$ replacing the controller and plant transfer functions in the forward loop. Let us develop a transfer function that relates the system output $y(t)$ to the reference input $r(t)$. To begin, we see that system output (in the Laplace domain) is the product of the forward transfer $G(s)$ and the error signal $E(s)$

$$Y(s) = G(s)E(s) = G(s)[R(s) - B(s)] \tag{10.1}$$

However, Fig. 10.4a shows that the feedback-path signal is $B(s) = H(s)Y(s)$ and therefore Eq. (10.1) becomes

$$Y(s) = G(s)[R(s) - H(s)Y(s)] \tag{10.2}$$

Moving the terms involving the output to the left-hand side, we obtain

$$Y(s)[1 + G(s)H(s)] = G(s)R(s) \tag{10.3}$$

Finally, solving Eq. (10.3) for the ratio of system output $Y(s)$ to reference input $R(s)$ yields

$$T(s) = \frac{Y(s)}{R(s)} = \frac{G(s)}{1 + G(s)H(s)} \tag{10.4}$$

Equation (10.4) is an extremely important result in analyzing closed-loop systems. The transfer function $T(s)$ in Eq. (10.4) is the *closed-loop transfer function* and it relates the overall system output $y(t)$ to the overall system input $r(t)$. Consequently, we can replace the closed-loop system shown in Fig. 10.4a by a single transfer function as shown in Fig. 10.4b. It should be emphasized that the two systems shown in Fig. 10.4 are equivalent. Once we derive the closed-loop transfer function, we can evaluate the *closed-loop* response characteristics by computing the poles of $T(s)$. In other words, computing the roots of the denominator polynomial (i.e., $1 + G(s)H(s) = 0$) determines the closed-loop poles

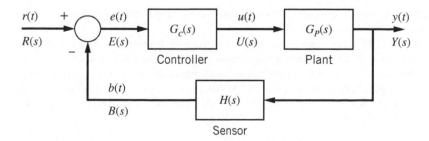

Figure 10.3 Closed-loop feedback control system.

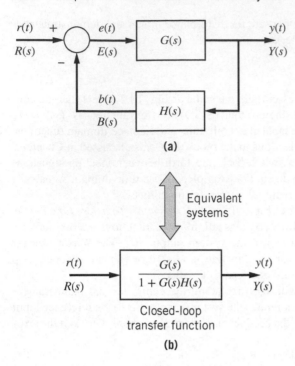

(a)

Equivalent
systems

$$\frac{G(s)}{1 + G(s)H(s)}$$

Closed-loop
transfer function

(b)

Figure 10.4 Closed-loop systems: (a) system with forward and
feedback paths and (b) equivalent closed-loop system.

and the time constants, damping ratio, natural frequency, etc., associated with the closed-loop system. We can apply
the various methods developed in Chapters 7–9 to analyze the closed-loop system's response to step, impulse, and
sinusoidal inputs.

MATLAB can compute the closed-loop transfer function using the `feedback` command. The user must define
the forward and feedback transfer functions $G(s)$ and $H(s)$ (see Fig. 10.4a), respectively:

```
>> sysG = tf(numG,denG)        % create forward transfer function G(s)
>> sysH = tf(numH,denH)        % create feedback transfer function H(s)
>> sysT = feedback(sysG,sysH)  % compute closed-loop transfer function T(s)
```

Of course, the user must define the numerator and denominator polynomials for $G(s)$ and $H(s)$, such as `numG` and `denG`.
The `feedback` command assumes a negative feedback sign at the summing junction, as shown in Fig. 10.4a.

Example 10.1

Figure 10.5 shows the block diagram of the DC motor from Example 5.17. Derive the closed-loop transfer function $T(s)$ that
relates angular velocity $\omega(t)$ to source voltage $e_{in}(t)$.

The DC motor consists of an armature circuit (RL circuit) and a mechanical rotor with friction b and inertia J. The
feedback signal in Fig. 10.5 is the "back-emf" voltage e_b that is induced by the angular velocity of the rotor windings. Hence,
the DC motor shown in Fig. 10.5 is *not* a feedback control system; rather, it is an electromechanical system that possesses
natural voltage feedback due to Faraday's laws of induction. However, the DC motor shown in Fig. 10.5 is a closed-loop
system and therefore we can determine its closed-loop transfer function using Eq. (10.4), where the forward transfer function
$G(s)$ is of the product of the three blocks in the forward path

$$G(s) = \frac{K_m}{(Ls + R)(Js + b)}$$

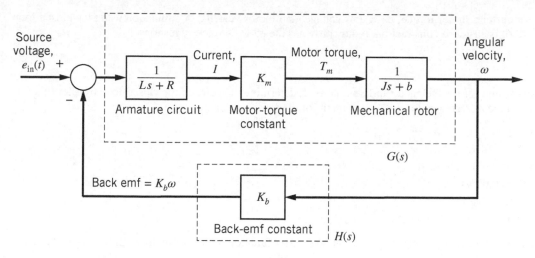

Figure 10.5 Block diagram of a DC motor (Example 10.1).

The feedback transfer function $H(s)$ is the back-emf gain K_b. Using Eq. (10.4) and the definitions of $G(s)$ and $H(s)$, the closed-loop transfer function is

$$T(s) = \frac{G(s)}{1 + G(s)H(s)} = \frac{\dfrac{K_m}{(Ls+R)(Js+b)}}{1 + \dfrac{K_m K_b}{(Ls+R)(Js+b)}} \tag{10.5}$$

Multiplying the numerator and denominator in Eq. (10.5) by $(Ls+R)(Js+b)$ in order to clear fractions we obtain

$$T(s) = \frac{K_m}{(Ls+R)(Js+b) + K_m K_b} \tag{10.6}$$

Equation (10.6) is the closed-loop transfer function of the DC motor. We can use MATLAB's `feedback` command to verify this result:

```
>> sysRL = tf(1,[L R])              % create RL circuit transfer function
>> sysMech = tf(1,[J b])            % create mechanical rotor transfer function
>> sysG = Km*sysRL*sysMech          % create forward transfer function G(s)
>> sysH = Kb                        % create feedback transfer function H(s) (gain)
>> sysT = feedback(sysG,sysH)       % compute closed-loop transfer function T(s)
```

Of course, the user must enter numerical values for the system parameters L, R, J, b, K_m, and K_b.

Given numerical values of the motor parameters, we could simulate the step response of the DC motor with Simulink using either the block diagram shown in Fig. 10.5 or the closed-loop transfer function shown in Fig. 10.6. Both simulations have advantages and disadvantages. Creating Fig. 10.5 using Simulink takes more effort but it provides complete system

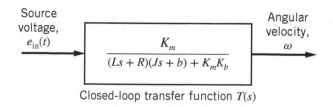

Closed-loop transfer function $T(s)$

Figure 10.6 Closed-loop transfer function representation of the DC motor (Example 10.1).

response information for current, motor torque, back-emf voltage, and motor velocity. A simulation using the closed-loop transfer function (Fig. 10.6) is easier to construct but it only provides the motor's velocity response.

Example 10.2

Using the following parameters for the DC motor in Example 10.1, determine the poles of the closed-loop transfer function and describe the angular velocity response to an 8-V step input for the source voltage.

Inductance, $L = 1.5$ mH
Resistance, $R = 0.5$ Ω
Motor torque constant, $K_m = 0.05$ N-m/A
Back-emf constant, $K_b = 0.05$ V-s/rad
Motor inertia, $J = 2.5(10^{-4})$ kg-m^2
Motor friction coefficient, $b = 10^{-4}$ N-m-s/rad

Equation (10.6) in Example 10.1 presents the closed-loop transfer function of the DC motor with source voltage $e_{in}(t)$ as the input and angular velocity ω as the output

$$T(s) = \frac{K_m}{(Ls + R)(Js + b) + K_m K_b} = \frac{K_m}{LJs^2 + (Lb + RJ)s + Rb + K_m K_b}$$

Using the numerical parameters for the DC motor, $T(s)$ becomes

$$T(s) = \frac{0.05}{3.75(10^{-7})s^2 + 1.2515(10^{-4})s + 0.00255} = \frac{1.3333(10^5)}{s^2 + 333.73s + 6800}$$

This result can be verified using the MATLAB command `feedback` as shown in Example 10.1.

The transient response of the DC motor depends on its characteristic roots or poles of the closed-loop transfer function $T(s)$. We compute the roots or poles by setting the denominator of $T(s)$ to zero

$$s^2 + 333.73s + 6800 = 0$$

The closed-loop roots or poles are two real, negative values: $s_1 = -21.80$ and $s_2 = -311.93$. Therefore, the transient response of the DC motor does not exhibit any oscillations (it is overdamped), but rather consists of two exponential functions $c_1 e^{-21.80t}$ and $c_2 e^{-311.93t}$, which both "die out" by time $t = 0.184$ s (note that the *slowest* root $s_1 = -21.80$ has a time constant of $\tau_1 = 1/21.80 = 0.0459$ s and hence the settling time is $4\tau_1 = 0.184$ s).

The steady-state motor speed can be computed from the product of the DC gain of the closed-loop transfer function $T(s)$ and the 8-V step input. Because the DC gain of $T(s)$ is $1.3333(10^5)/6800 = 19.608$, the steady-state angular velocity is (8 V)(19.608) = 156.86 rad/s (about 1498 revolutions per minute or rpm).

In summary, this example shows that the transient and steady-state response characteristics of a closed-loop system can be gleaned from knowledge of the poles and DC gain of the closed-loop transfer function; hence, we can use the techniques that were developed in Chapter 7. We can apply the analysis methods from Chapter 7 because both open- and closed-loop transfer functions are SISO systems.

Example 10.3

Repeat Examples 10.1 and 10.2 for the case where coil inductance is neglected and compare the results from the simplified model with the full DC motor model from Examples 10.1 and 10.2.

If we set inductance $L = 0$ in the DC motor block diagram (see Fig. 10.5), we see that motor current I changes instantaneously according to Ohm's law:

$$I = \frac{e_{in}(t) - e_b}{R}$$

where e_b is the back emf. Consequently, the forward transfer function becomes

$$G(s) = \frac{K_m}{R(Js + b)}$$

and the closed-loop transfer function is

$$T(s) = \frac{G(s)}{1 + G(s)H(s)} = \frac{\dfrac{K_m}{R(Js + b)}}{1 + \dfrac{K_m K_b}{R(Js + b)}}$$

or, simplifying

$$T(s) = \frac{K_m}{R(Js + b) + K_m K_b} \tag{10.7}$$

The reduced-order closed-loop system (10.7) is equivalent to the full (second-order) closed-loop system (10.6) with inductance $L = 0$. Note that Eq. (10.7) is a first-order system because the RL circuit dynamics have been removed. Using the numerical parameters for the DC motor, the reduced-order closed-loop transfer function is

$$T(s) = \frac{0.05}{1.25(10^{-4})s + 0.00255} = \frac{400}{s + 20.4} \tag{10.8}$$

Note that the DC gain is $400/20.4 = 19.608$, which matches the DC gain from the second-order closed-loop system where inductance was included. The single time constant for the reduced-order closed-loop system (10.8) is $\tau = 1/20.4 = 0.049$ s, which shows a very close match with the slowest time constant from the second-order model in Example 10.2. Hence, this reduced-order closed-loop DC motor model reaches steady state in the settling time $4\tau = 0.196$ s, which is very similar to the second-order model.

As a final note, we see that while inductance L can be neglected without too much loss of accuracy, we cannot neglect the back-emf constant K_b. Recall that the steady-state angular velocity for an 8-V step source voltage input is $\omega_{ss} = 156.86$ rad/s. The steady-state back-emf voltage for this speed is $e_b = K_b \omega_{ss} = 7.84$ V, which is nearly as large as the source voltage.

10.3   FEEDBACK CONTROLLERS

We stated earlier that the *controller* block in Figs. 10.2 and 10.3 consists of the control logic or "rules" required to turn feedback information into a control signal command that is sent to the plant. In this section, we briefly discuss and demonstrate the attributes of the following "standard" types of controllers for feedback systems:

1. On−off (or relay) controllers
2. Proportional (P) controllers
3. Proportional-integral (PI) controllers
4. Proportional-derivative (PD) controllers
5. Proportional-integral-derivative (PID) controllers

On−Off Controllers

As the name implies, on−off (or relay) controllers use the feedback error signal to determine an "on" or "off" command for the control signal. For example, a thermostat either turns the furnace on or off based on the difference between the desired temperature setting and the sensed room temperature. While on−off controllers use very simple control rules, they are nonlinear devices; hence, feedback control systems that use on−off controllers are nonlinear systems that cannot be analyzed using the methods we develop in this chapter. For this reason, on−off controllers are not emphasized in this chapter. The following example demonstrates the basic operating concept of an on−off controller.

Example 10.4

Figure 10.7 shows an electrical circuit known as a *buck converter*, which is used to step the source voltage $e_{in}(t)$ "down" to a lower desired output voltage (see Problem 3.27). The buck converter uses a switch to repeatedly connect and disconnect the voltage supply $e_{in}(t)$ from the remainder of the circuit until the output (capacitor) voltage e_C is equal to a desired reference voltage e_{ref}. Simulate the closed-loop system if the desired reference voltage is $e_{ref} = 12$ V and the supply voltage is constant at $e_{in} = 28$ V. At time $t = 0$, the system has zero stored energy.

The mathematical model of the buck converter circuit was developed in Problem 3.27 and an SSR was obtained in Problem 5.24. When the switch is in position "1," the voltage supply is connected to the circuit:

$$\text{Capacitor:} \quad RC\dot{e}_C = RI_L - e_C$$

$$\text{Inductor:} \quad L\dot{I}_L = e_{in} - e_C$$

Consequently, the state equation is

$$\dot{\mathbf{x}} = \begin{bmatrix} -1/(RC) & 1/C \\ -1/L & 0 \end{bmatrix} \mathbf{x} + \begin{bmatrix} 0 \\ 1/L \end{bmatrix} u \tag{10.9}$$

where the states are $x_1 = e_C$ and $x_2 = I_L$ and u is the input voltage to the circuit. The switching action is an on−off controller that is dictated by the simple equation

$$u(t) = \begin{cases} e_{in} & \text{if } e_{ref} - e_C \geq 0 \\ 0 & \text{if } e_{ref} - e_C < 0 \end{cases} \tag{10.10}$$

Figure 10.8 shows a diagram of the closed-loop system using the on−off controller described by Eq. (10.10). Capacitor voltage e_C is fed back and compared to the reference voltage e_{ref} to form the voltage error. The relay block in Fig. 10.8 denotes the on−off switching: the x-axis of the relay block is the input, while the y-axis is the output. Note that when the voltage error is *positive* or *zero* (i.e., $e_{ref} - e_C \geq 0$), Eq. (10.10) and the relay symbol in Fig. 10.8 indicate that the switch is in position "1" and the supply voltage e_{in} is connected to the energy-storage elements. When voltage error is negative, the switch is in position "2" and the supply voltage is disconnected from the circuit (i.e., $u = 0$). Figure 10.8 shows that

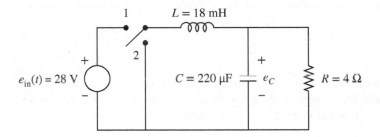

Figure 10.7 Buck converter electrical circuit (Example 10.4).

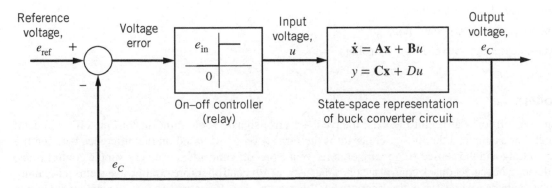

Figure 10.8 Buck converter circuit using an on−off controller (Example 10.4).

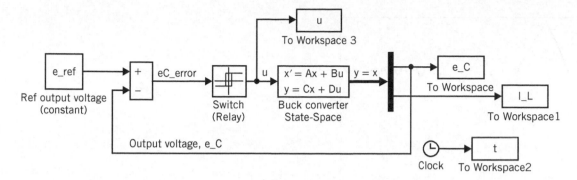

Figure 10.9 Simulink diagram of the buck converter circuit (Example 10.4).

an SSR is used to denote the plant or input–output relationship of the buck converter system dynamics where the two state variables are capacitor voltage e_C and inductor current I_L.

Figure 10.9 shows a Simulink model that represents the closed-loop system (Fig. 10.8). The `Relay` block is found in the `Discontinuities` Simulink library and the "on" and "off" output values are set to e_{in} (28 V) and zero, respectively. The state-equation matrices **A** and **B** are appropriately defined using Eq. (10.9); however, the output matrix **C** is set to an identity matrix so that both states (capacitor voltage and inductor current) can be plotted. Figures 10.10 and 10.11 show the closed-loop responses for capacitor voltage $e_C(t)$ and inductor current $I_L(t)$. We see that the on–off controller drives capacitor voltage to the desired reference voltage (12 V) in about 0.007 s (or 7 ms). Inductor current (Fig. 10.11) exhibits a sharp rise from zero and "jagged" chatter about a steady-state value of 3 A. The jagged current response can be explained by observing the dynamic equation for the inductor voltage:

$$L\dot{I}_L = u - e_C$$

and therefore the time-rate $\dot{I}_L$ is nearly constant and equal to $(28 - 12\,\mathrm{V})/18$ mH when the switch is in position 1 ($u = 28\,\mathrm{V}$) and $e_C \approx 12$ V. When the switch is in position 2 ($u = 0$) and $e_C \approx 12$ V, the slope $\dot{I}_L$ is $-12\,\mathrm{V}/18$ mH. Figure 10.12 shows the input voltage $u(t)$ to the circuit as dictated by the on–off controller equation (10.10). Clearly, the input voltage switches between 28 V (supply voltage e_{in}) and zero as the capacitor voltage $e_C(t)$ crosses the 12-V reference voltage as seen in Fig. 10.10. Current $I_L(t)$ switches from positive to negative slope as the input voltage switches from 28 to 0 V. Figure 10.12 also shows that the input voltage exhibits "chatter" or very rapid switching between 28 V (on) and 0 V (off) in order to maintain the desired 12-V capacitor voltage. If the switch remained on, then the capacitor would eventually reach 28 V; and if the switch remained off, then the resistor would eventually dissipate all energy stored in the electrical system. This high-frequency input "chatter" can be an undesirable consequence of using on–off controllers.

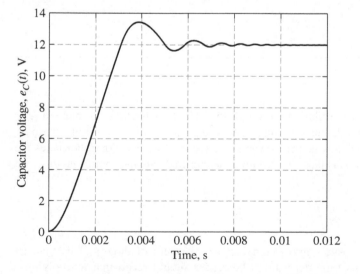

Figure 10.10 Capacitor voltage response $e_C(t)$ (Example 10.4).

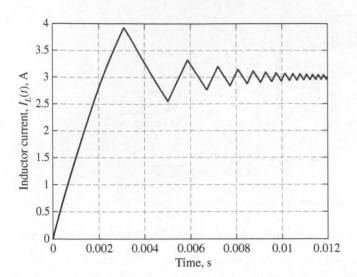

Figure 10.11 Inductor current response $I_L(t)$ (Example 10.4).

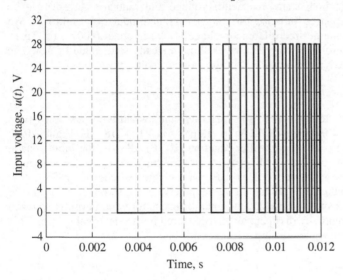

Figure 10.12 Input voltage $u(t)$ (Example 10.4).

PID Controllers

It is likely that PID controllers are the most common controllers used in industry. Their applications (to name a few) include motion control, hydraulic control systems, chemical processing plants, and guidance and control of aerospace vehicles. Figure 10.13 shows a PID controller in a feedback control system. We see that the PID controller takes the feedback error signal $e(t)$ and produces three control signals that are summed together to create the control input signal $u(t)$ to the plant. The PID control logic is

$$u(t) = K_P e(t) + K_I \int e(t) dt + K_D \dot{e}(t) \tag{10.11}$$

Equation (10.11) and Fig. 10.13 show that the composite control signal $u(t)$ is the sum of three signals that are proportional to the error $e(t)$, its time integral, and its time derivative (the reader should recall that multiplying a signal by $1/s$ is equivalent to integration and multiplying a signal by s is equivalent to differentiation). The three

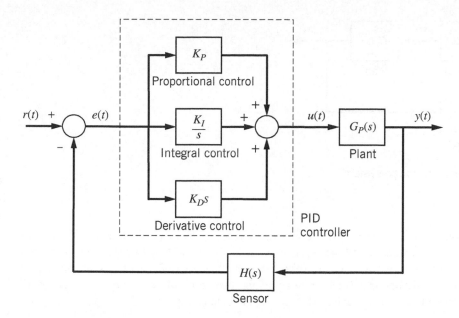

Figure 10.13 PID controller in a closed-loop system.

proportionality constants ("gains") in Eq. (10.11) and Fig. 10.13 are called the *proportional gain K_P*, the *integral gain K_I*, and the *derivative gain K_D*. Adjusting each individual gain changes the emphasis of the PID controller. In general, the effect or objective of each term of the PID controller can be summarized as listed:

1. Proportional control term, $K_P e(t)$: The control signal is proportional to the instantaneous error. Increasing the P-gain K_P will tend to speed up the system response. The proportional control term has a diminished effect as the feedback error goes to zero (i.e., good tracking at steady state).

2. Integral control term, $K_I \int e(t)dt$: The control signal is proportional to the summation (integral) of all past error signals and therefore the integral control term will be nonzero even when the feedback error goes to zero. For this reason, integral control is used to reduce the steady-state tracking error.

3. Derivative control term, $K_D \dot{e}(t)$: The control signal is proportional to the instantaneous derivative of the error signal. Hence, the derivative control signal "anticipates" the system response because it is based on the derivative or time rate of the error signal. In general, increasing the D-gain K_D reduces overshoot and adds damping to the closed-loop system.

In many cases a subset of the PID controller is used. For example, if a particular system has sufficient damping, we may not need the derivative control term and consequently we set the D-gain K_D to zero. As another example, some plants may include a "natural integrator" in their system dynamics and therefore may not require an additional integral control term for a good steady-state response, in which case we can set $K_I = 0$. In these cases we may choose to utilize a controller with only one or two terms, such as a P-controller, PI controller, or PD controller. We illustrate the various attributes of the PID controller and its variations in the example problems in this section and in the sections to follow.

We can derive the PID controller transfer function by expressing the control logic equation (10.11) using the Laplace variable s

$$U(s) = K_P E(s) + K_I \frac{E(s)}{s} + K_D s E(s) \tag{10.12}$$

The PID controller transfer function $G_C(s) = U(s)/E(s)$ is

$$G_C(s) = \frac{K_P s + K_I + K_D s^2}{s} \tag{10.13}$$

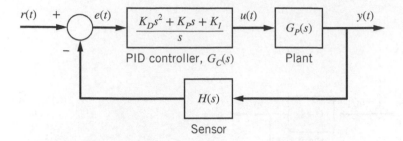

Figure 10.14 PID controller $G_C(s)$ in a closed-loop system.

Figure 10.14 shows a closed-loop control system with a PID controller represented as $G_C(s)$. The reader should see that the PID controllers shown in Figs. 10.13 and 10.14 are equivalent. It is also important to note that inserting a PID controller adds two zeros and one pole to the forward transfer function (the zeros of the PID controller $G_C(s)$ are the values of s that cause $G_C(s) = 0$). Equation (10.13) shows that the two zeros of $G_C(s)$ are computed from $K_D s^2 + K_P s + K_I = 0$ and therefore depend on the three gains while the single pole is $s = 0$.

Example 10.5

Figure 10.15 shows a closed-loop system for controlling the angular velocity of a DC motor. Investigate and compare the closed-loop speed responses using proportional and proportional-integral controllers.

Before we analyze the closed-loop system shown in Fig. 10.15, we note some of its features. First, the DC motor (plant) dynamics are modeled by the first-order system derived in Example 10.3, where coil inductance L has been ignored. The input to the DC motor transfer function is armature voltage $e_{in}(t)$ and the output is angular velocity ω. Figure 10.15 shows a "unity-feedback" system, where $H(s) = 1$ and the overall system output (angular velocity in rad/s) is fed back and compared directly to the reference velocity command (also in rad/s). It is important to note that signals that are compared at a summing junction must have the same units. Before microprocessors became commonplace in control systems, the angular velocity would typically be measured by a tachometer that converts the speed in rad/s to a voltage signal. This feedback voltage signal would be compared to a reference voltage signal that is proportional to the reference angular velocity ω_{ref}. An amplifier gain would be present in the controller $G_C(s)$ in order to boost the low-voltage error signal and create armature voltage $e_{in}(t)$. Because the same speed-to-voltage conversion gain (V-s/rad) would be applied to ω and ω_{ref}, we can factor out this common gain and assume a unity-feedback structure. With the advent of microprocessors and digital sensors, we can assume that the measured and commanded variables can have any units (such as rad/s, in this case). As long as we have common units at the feedback summing junction, we can usually utilize a unity-feedback system as shown in Fig. 10.15.

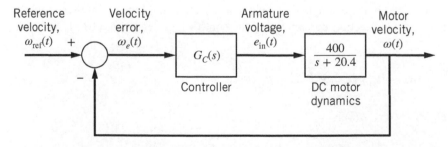

Figure 10.15 Closed-loop control system for DC motor (Example 10.5).

If we designate the DC motor dynamics as the plant transfer function $G_P(s)$, the closed-loop transfer function for *any* controller is

$$T(s) = \frac{G(s)}{1 + G(s)} = \frac{G_C(s)G_P(s)}{1 + G_C(s)G_P(s)} \tag{10.14}$$

(again, note that we have unity feedback so $H(s) = 1$). First, we consider a proportional (P) controller, and therefore the controller transfer function is simply a constant gain. Substituting $G_C(s) = K_P$ for the controller and $G_P(s) = 400/(s + 20.4)$ for the plant, the closed-loop transfer function (10.14) becomes

$$T(s) = \frac{K_P\dfrac{400}{s + 20.4}}{1 + K_P\dfrac{400}{s + 20.4}} = \frac{400K_P}{s + 20.4 + 400K_P} \tag{10.15}$$

Equation (10.15) relates the closed-loop speed response of the motor (ω) to the speed command (ω_{ref}). If the speed command ω_{ref} is a constant (step) input, then the steady-state motor speed can be determined from the DC gain:

$$\text{DC gain:} \quad T(s = 0) = \frac{400K_P}{20.4 + 400K_P} \tag{10.16}$$

For a step speed command, the steady-state motor speed is

$$\omega_{\text{ss}} = \frac{400K_P}{20.4 + 400K_P}\omega_{\text{ref}}$$

Hence, for a P-gain, $K_P = 0.5$ V-s/rad, the DC gain is $200/220.4 = 0.9074$, and therefore the steady-state motor speed ω_{ss} is 90.74% of the commanded reference speed ω_{ref}. Clearly, the steady-state motor speed exhibits better tracking (i.e., smaller velocity error) as the P-gain is increased. However, the P-controller cannot provide perfect steady-state tracking because a zero velocity error would produce a zero armature voltage (i.e., $e_{\text{in}}(t) = K_P\omega_e$) and the motor cannot maintain a constant angular velocity without an input torque (or, input voltage) to offset the friction torque.

Equation (10.15) shows that the time constant of the first-order closed-loop system is

$$\tau = \frac{1}{20.4 + 400K_P}$$

Increasing the P-gain K_P will reduce the time constant and hence reduce the settling time ($t_S = 4\tau$). Using the gain $K_P = 0.5$ V-s/rad, we see that the time constant is $\tau = 0.0045$ s and the settling time is 0.018 s. Figure 10.16 shows the closed-loop angular velocity response of the DC motor with reference command $\omega_{\text{ref}} = 50$ rad/s (477.5 rpm) and three proportional gains $K_P = 1$, 0.5, and 0.2 V-s/rad (the motor is starting from rest). As we increase K_P, the motor displays a faster response and achieves a steady-state speed that is closer to the 50-rad/s reference command. This trend might tempt the engineer to select an extremely large control gain for a very fast response with a small tracking error. However, a large gain will produce a large initial armature voltage $e_{\text{in}}(t)$. Initially, the speed error is 50 rad/s because the motor is starting from rest and the speed command is $\omega_{\text{ref}} = 50$ rad/s. Consequently, the initial armature voltage is 50 V if the controller gain is set at $K_P = 1$ V-s/rad. This input voltage would likely exceed the maximum allowable voltage for a small DC motor.

Figure 10.16 shows that a proportional controller alone cannot reduce the steady-state speed error of the motor. Furthermore, we need a controller that can provide an armature voltage signal (i.e., motor torque) even when the speed error $\omega_e(t)$ is driven to zero at steady state. A PI controller is the logical choice. The PI controller transfer function is

$$G_C(s) = K_P + \frac{K_I}{s} = \frac{K_Ps + K_I}{s} \tag{10.17}$$

The PI controller $G_C(s)$ has one zero at $s = -K_I/K_P$ and one pole at $s = 0$ (i.e., the integrator). Our closed-loop system with PI control is

$$T(s) = \frac{\left(\dfrac{K_Ps + K_I}{s}\right)\left(\dfrac{400}{s + 20.4}\right)}{1 + \left(\dfrac{K_Ps + K_I}{s}\right)\left(\dfrac{400}{s + 20.4}\right)} = \frac{400(K_Ps + K_I)}{s^2 + (20.4 + 400K_P)s + 400K_I} \tag{10.18}$$

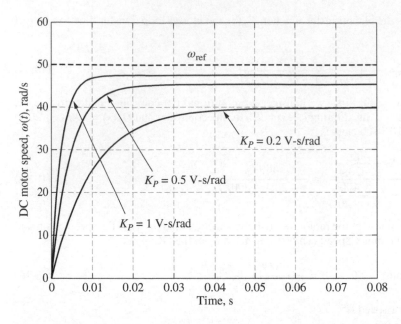

Figure 10.16 Closed-loop response of DC motor with P-controller (Example 10.5).

Equation (10.18) immediately tells us that the closed-loop DC gain is unity for *any* nonzero value of the integral gain K_I. In addition, the closed-loop transfer function now has a first-order numerator and second-order denominator due to the PI controller. A PI controller has two free gains for selection (K_P and K_I) and we will demonstrate the PI controller using the previous P-gains ($K_P = 1$, 0.5, and 0.2 V-s/rad). For example, when $K_P = 1$ V-s/rad, Eq. (10.18) shows that the closed-loop characteristic equation is

$$\text{P-gain } K_P = 1 \text{ V-s/rad:} \quad s^2 + 420.4s + 400K_I = 0$$

If the I-gain (K_I) is too large, the second-order characteristic equation will become underdamped and oscillations will occur. If K_I is too small, the characteristic roots will be real and negative, but they will be near the origin and hence the response will be slow. Figure 10.17 shows the closed-loop motor speed with three

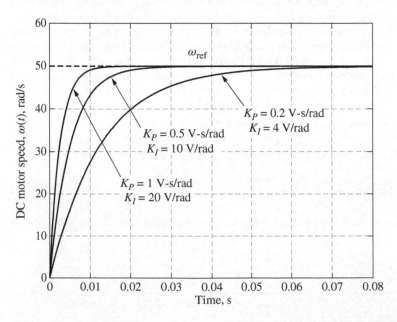

Figure 10.17 Closed-loop response of DC motor with PI controller (Example 10.5).

PI controller gain settings. All three motor responses reach the desired 50-rad/s reference speed at steady state because of the addition of the integral control loop. If we compare the responses in Figs. 10.16 and 10.17 for a particular K_P gain, we see that adding integral control has slightly slowed down the response. Again, if our DC motor has an armature voltage limit (say, a maximum of 12 V), then the PI controller with gains $K_P = 0.2$ V-s/rad and $K_I = 4$ V/rad is the only feasible choice for the three PI control options shown in Fig. 10.17.

In summary, the simple proportional (P) controller cannot provide good steady-state tracking for a step reference speed command to the DC motor. Adding an integral control loop (PI controller) eliminates the steady-state error and the motor speed eventually tracks the reference input. Increasing the P-gain speeds up the closed-loop response, but large values of K_P will lead to excessive control inputs (armature voltage $e_{in}(t)$, in this case).

Example 10.6

Figure 10.18 shows the closed-loop position control for a simple mechanical system modeled by a mass–damper system (mass $m = 1$ kg and viscous friction $b = 0.3$ N-s/m). Investigate and compare the use of proportional and proportional-derivative controllers for a reference step position command $x_{ref}(t) = 0.1$ m.

The controller $G_C(s)$ in Fig. 10.18 uses the feedback position error $x_e(t)$ to generate the voltage input $e_{in}(t)$ to an actuator device, which in turn produces the force $f(t)$ that is applied directly to the mechanical mass–damper system. We have neglected the actuator dynamics (they are very fast compared to the mechanical system) and therefore modeled the actuator as a simple gain K_A with units N/V. Figure 10.18 might represent the position control of a machine tool in a manufacturing process.

Using the block diagram in Fig. 10.18, the closed-loop transfer function for *any* controller is

$$T(s) = \frac{K_A G_C(s) G_P(s)}{1 + K_A G_C(s) G_P(s)} \tag{10.19}$$

where $G_P(s)$ is the mechanical system transfer function. Using the fixed actuator gain $K_A = 2$ N/V and a P-controller ($G_C(s) = K_P$), the closed-loop transfer function becomes

$$T(s) = \frac{\dfrac{2K_P}{s^2 + 0.3s}}{1 + \dfrac{2K_P}{s^2 + 0.3s}} = \frac{2K_P}{s^2 + 0.3s + 2K_P} \tag{10.20}$$

Immediately we see that the DC gain of $T(s)$ is always unity for any nonzero control gain K_P and therefore a P-controller will provide perfect steady-state tracking for a step position input. However, the transient response of the system using proportional control is very poor. To see this, note the closed-loop characteristic equation or denominator of $T(s)$

$$s^2 + 0.3s + 2K_P = 0 \tag{10.21}$$

Adjusting the P-gain K_P only changes the zeroth-order coefficient. If we use a "small" control gain $K_P = 0.01$ V/m, then the characteristic equation becomes $s^2 + 0.3s + 0.02 = 0$ and the closed-loop poles are $s_1 = -0.1$ and $s_2 = -0.2$. Consequently, the closed-loop response will be very slow (the settling time will be about 40 s). For $K_P > 0.01125$, the closed-loop poles become complex, with a real part that is always equal to -0.15 and hence the closed-loop transient response will *always*

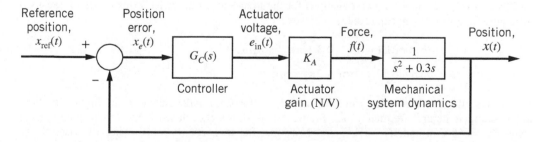

Figure 10.18 Closed-loop position control of a mechanical system (Example 10.6).

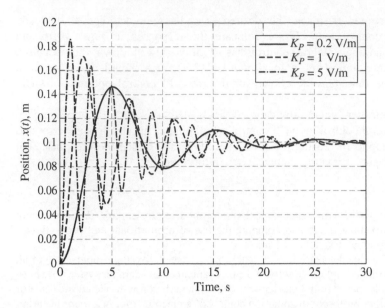

Figure 10.19 Closed-loop position response with P-controller (Example 10.6).

die out at settling time $t_S = 26.67$ s because of the exponential envelope term $e^{-0.15t}$. Increasing gain K_P will increase the undamped natural frequency (ω_n) and reduce the damping ratio (ζ), which increases the number of oscillations and the maximum overshoot, respectively (recall that the standard second-order underdamped model is $s^2 + 2\zeta\omega_n s + \omega_n^2$). Figure 10.19 shows the closed-loop step response ($x_{ref} = 0.1$ m) for three proportional controllers: $K_P = 0.2$, 1, and 5 V/m. All three responses reach the desired steady-state value of 0.1 m in about 26.7 s (as predicted), and the higher control gains result in higher frequency oscillations and a decrease in damping.

A proportional-derivative controller is needed to add damping to the closed-loop system and improve the transient response. The PD controller transfer function is

$$G_C(s) = K_P + K_D s \tag{10.22}$$

which has one zero at $s = -K_P/K_D$. After including the PD controller (again with fixed actuator gain $K_A = 2\,\text{N/V}$), the closed-loop transfer function (10.19) becomes

$$T(s) = \frac{\dfrac{2(K_P + K_D s)}{s^2 + 0.3s}}{1 + \dfrac{2(K_P + K_D s)}{s^2 + 0.3s}} = \frac{2(K_P + K_D s)}{s^2 + (0.3 + 2K_D)s + 2K_P} \tag{10.23}$$

Equation (10.23) shows that the resulting closed-loop system using PD control has a unity DC gain (i.e., $2K_P/2K_P$) for any nonzero gain K_P and hence it will also perfectly track a step input at steady state. Using the standard second-order underdamped model $s^2 + 2\zeta\omega_n s + \omega_n^2$, we can write two equations for the first- and zeroth-order terms of the closed-loop characteristic equation, or denominator of $T(s)$ in Eq. (10.23):

$$\text{First-order term:} \quad 0.3 + 2K_D = 2\zeta\omega_n$$

$$\text{Zeroth-order term:} \quad 2K_P = \omega_n^2$$

Therefore, the freedom to adjust the PD controller gains K_P and K_D allows for independent manipulation of both the damping ratio ζ and undamped natural frequency ω_n. Figure 10.20 shows the closed-loop step response using a PD controller with gains $K_P = 16$ V/m and $K_D = 4$ V-s/m. Note that the transient response using PD control has greatly improved: the overshoot is relative small, and the response reaches the desired steady-state position in less than 1 s.

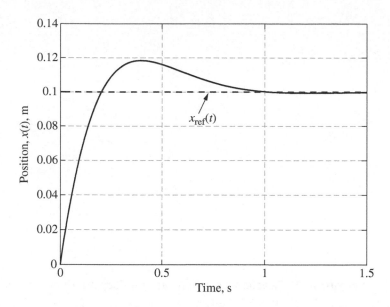

Figure 10.20 Closed-loop position response with PD controller (Example 10.6).

In summary, this example has demonstrated how adding a derivative control signal can improve the transient response of the closed-loop system by increasing the system damping. This particular problem did not require an integral control signal for improving the steady-state response due to the nature of the plant system dynamics.

PID Tuning Rules

The previous examples demonstrate the basic attributes of the PID controller: increasing the proportional gain K_P tends to speed up the response (reduce the peak time) but may increase the overshoot; increasing the integral gain K_I tends to reduce the steady-state error but may slow down the response; increasing the derivative gain K_D tends to reduce the overshoot and the settling time. It is clear that implementing a PID controller requires selecting three independent gains in order to achieve a good balance in the closed-loop performance criteria such as response time, overshoot, settling time, and steady-state error. In the early 1940s, Ziegler and Nichols [1] developed two methods for selecting "good" PID gains. These "PID tuning rules" were based on heuristic trials conducted by Ziegler and Nichols and they provide control engineers with a good starting point for selecting the PID gains that provide satisfactory closed-loop performance.

In the first method, Ziegler and Nichols noted that the open-loop step response of many dynamic systems exhibits an "S-shaped" curve with no overshoot. Figure 10.21 shows the characteristics of a general S-shaped open-loop response, which Ziegler and Nichols called the *reaction curve*. In practice, the reaction curve could be obtained experimentally by applying a step input and measuring the system output in an open-loop manner. The key parameters of the reaction curve are the delay time T_d and slope R shown in Fig. 10.21. Both parameters are obtained by drawing a line tangent to the inflection point of the S-curve (see Fig. 10.21), where the reaction curve has the steepest slope R. Ziegler and Nichols used these two parameters to develop PID gains that produced a closed-loop response that exhibited a one-quarter decay ratio, meaning that the transient response decays to one-quarter its peak value in one period of oscillation. Table 10.1 presents the Ziegler–Nichols rules for selecting the PID gains using the reaction-curve parameters delay time T_d and slope R. Note that Ziegler and Nichols present gain-tuning rules for P, PI, and PID controllers.

The second PID tuning method developed by Ziegler and Nichols relies on obtaining a *marginally* stable closed-loop response with a high gain setting. In this technique, the P-controller gain is continually increased until the closed-loop response transitions from damped sinusoidal oscillations to sustained oscillations with constant amplitude. Hence, the closed-loop system is marginally stable and on the brink of instability. Ziegler and Nichols called the P-gain setting that results in sustained oscillations the "ultimate gain" K_U. The period of the sustained oscillations is P_U (the "ultimate period") and is used in the PID tuning rules developed by Ziegler and Nichols. Table 10.2 presents the Ziegler–Nichols

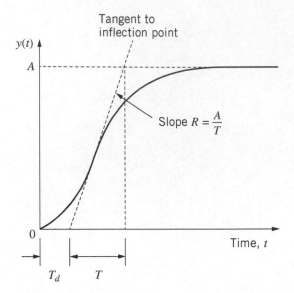

Figure 10.21 Reaction curve from an open-loop step input.

Table 10.1 Ziegler–Nichols Tuning Rules: Reaction-Curve Method

Controller Type	Gains
P	$K_P = \dfrac{1}{RT_d}$
PI	$K_P = \dfrac{0.9}{RT_d} \quad K_I = \dfrac{0.27}{RT_d^2}$
PID	$K_P = \dfrac{1.2}{RT_d} \quad K_I = \dfrac{0.6}{RT_d^2} \quad K_D = \dfrac{0.6}{R}$

Table 10.2 Ziegler–Nichols Tuning Rules: Ultimate Gain Method

Controller Type	Gains
P	$K_P = 0.5K_U$
PI	$K_P = 0.45K_U \quad K_I = \dfrac{0.54K_U}{P_U}$
PID	$K_P = 0.6K_U \quad K_I = \dfrac{1.2K_U}{P_U} \quad K_D = 0.075K_U P_U$

tuning rules using the "ultimate-gain method" and we see that the gains rely on just two parameters: ultimate gain K_U and ultimate period P_U. Note that the ultimate period P_U can be obtained from closed-loop experiments conducted with the physical system.

In summary, the Ziegler–Nichols tuning rules allow the control engineer to obtain a good starting point for selecting the PID gains. The final PID design may be obtained by further experimental or simulation trials with adjustments in a particular gain in order to improve a facet of the transient response (e.g., the derivative gain could be increased to further reduce overshoot if necessary). It should be noted that the Ziegler–Nichols tuning methods do not necessarily apply to all physical systems. Not all plants exhibit an S-shaped response to a step input and not all plants can be driven to sustained oscillations by increasing the proportional gain.

Example 10.7

Figure 10.22 shows a closed-loop system for controlling the pH balance in a chemical-processing system. The pH level of a solution in a stirred reaction tank is measured by a pH meter and fed back to form the pH error. The controller $G_C(s)$ uses the pH error to determine the acid/alkaline mixture ratio $u(t)$ of the input flow stream to the tank (if $u > 0$ the input flow is alkaline, if $u < 0$ the input flow is acidic). Use the Ziegler–Nichols tuning rules to design a PID controller that provides a good closed-loop response for a step reference pH command $r(t) = 9$ (alkaline) if the solution in the tank is initially neutral (pH = 7).

Reaction-curve method

Although the exact system dynamics of the reaction tank are not provided in Fig. 10.22, we assume that the physical system is available in order to derive the reaction curve by experimentation. Figure 10.23 shows the S-shaped reaction curve that results from a positive (alkaline) flow rate step input $u(t)$ applied to the open-loop plant. We can estimate the delay time as

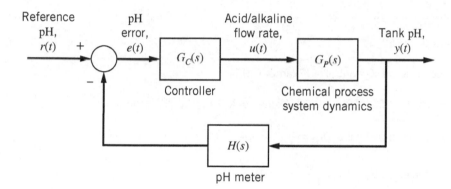

Figure 10.22 Closed-loop pH control of a chemical-processing system (Example 10.7).

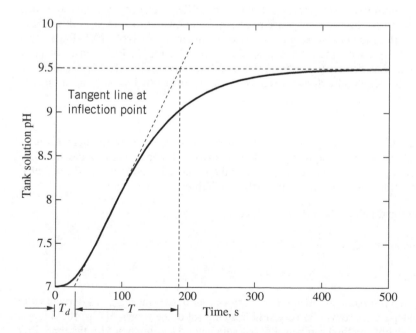

Figure 10.23 Reaction-curve response for a step input (Example 10.7).

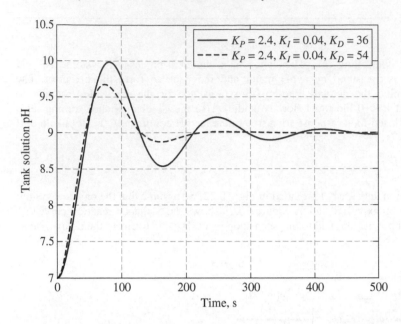

Figure 10.24 Closed-loop step response using a PID controller (Example 10.7).

$T_d = 30$ s and the maximum slope as $R = 2.5/T = 2.5/150 = 0.0167$. Using Table 10.1, the PID gains are

$$\text{Proportional gain: } K_P = \frac{1.2}{RT_d} = 2.4$$

$$\text{Integral gain: } K_I = \frac{0.6}{RT_d^2} = 0.04$$

$$\text{Derivative gain: } K_D = \frac{0.6}{R} = 36$$

Figure 10.24 shows the closed-loop step response with $r(t) = 9$ (reference pH) with these PID controller gains. The solid-line response shows that the PID controller has reduced the transient response to roughly one-quarter of its peak value (pH = 10) after one cycle, which is the design goal of the Ziegler–Nichols tuning rules summarized in Table 10.1. Recall that the Ziegler–Nichols rules provide the control engineer with a feasible starting point for selecting the PID gains. The peak overshoot can be reduced by increasing the D-gain as demonstrated by the dashed-line curve in Fig. 10.24, where K_D is increased by 50%. The closed-loop response with $K_D = 54$ exhibits a reduced peak response (pH = 9.7) and a quicker response with a settling time less than 200 s.

Ultimate-gain method

A simple P-controller is used in Fig. 10.22 and the closed-loop step response is obtained with increasing gain K_P until sustained oscillations are observed. Figure 10.25 shows the marginally stable closed-loop response that is achieved when gain K_P reaches the ultimate gain setting of $K_U = 4.134$. The ultimate period of the sustained oscillations is $P_U = 129$ s. Using the Ziegler–Nichols tuning rules for the ultimate-gain method summarized in Table 10.2, we obtain

$$\text{Proportional gain: } K_P = 0.6K_U = 2.48$$

$$\text{Integral gain: } K_I = \frac{1.2K_U}{P_U} = 0.0385$$

$$\text{Derivative gain: } K_D = 0.075K_U P_U = 40$$

We see that these three PID gains produced by the ultimate-gain method are very similar to the starting gains derived from the reaction-curve method. Hence, the closed-loop response will be very similar to the solid line shown in Fig. 10.24.

This example shows that either Ziegler–Nichols method can be used to obtain a good starting point for the three PID gains, after which the control engineer can observe the closed-loop response and subsequently adjust the gains to improve

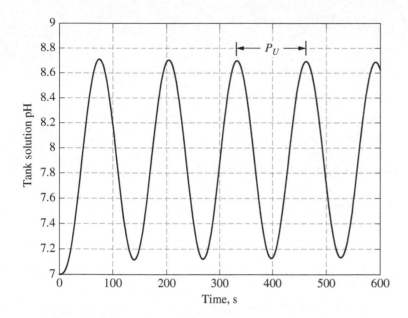

Figure 10.25 Closed-loop step response with ultimate gain K_U (Example 10.7).

the response. However, it should be emphasized that some systems (plants) will *not* exhibit an S-shaped reaction curve to a step input or sustained oscillations with a high-gain P-controller. Consequently, the Ziegler–Nichols tuning rules may not necessarily apply to a particular plant.

10.4 STEADY-STATE ACCURACY

In Section 10.2, we stated that good steady-state tracking is a desirable attribute of most feedback control systems. In addition, the previous discussion of PID controllers demonstrates that inserting an integral control signal generally improves the steady-state accuracy, or the steady-state error between the commanded reference signal and the system response. We present a systematic method for determining the steady-state accuracy of a linear closed-loop system when the reference signal is a step, ramp, and parabolic input function.

Figure 10.26 presents a unity-feedback (i.e., $H(s) = 1$) system where the forward transfer function is the product of the controller and plant transfer functions, or $G(s) = G_C(s)G_P(s)$. The *tracking error* in the time domain is $e(t)$ (or, $E(s)$ in the Laplace domain) and it is the difference between the reference command and the system output:

$$e(t) = r(t) - y(t) \quad \text{or} \quad E(s) = R(s) - Y(s) \tag{10.24}$$

If we divide Eq. (10.24) by the Laplace transform of the reference signal, we obtain

$$\frac{E(s)}{R(s)} = 1 - \frac{Y(s)}{R(s)} \tag{10.25}$$

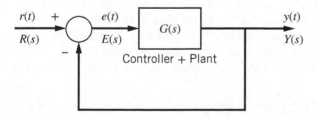

Figure 10.26 Unity-feedback closed-loop system.

Substituting the closed-loop transfer function $Y(s)/R(s) = G(s)/[1 + G(s)]$ yields

$$\frac{E(s)}{R(s)} = 1 - \frac{G(s)}{1 + G(s)} = \frac{1 + G(s)}{1 + G(s)} - \frac{G(s)}{1 + G(s)} \tag{10.26}$$

Finally, the transfer function relating tracking error to the reference command is

$$\frac{E(s)}{R(s)} = \frac{1}{1 + G(s)} \tag{10.27}$$

Equation (10.27) can be used to compute the Laplace transform of the tracking error

$$E(s) = \frac{1}{1 + G(s)} R(s) \tag{10.28}$$

Recall that the final-value theorem can be used to compute the final (or steady-state) tracking error

$$e(\infty) = \lim_{t \to \infty} e(t) = \lim_{s \to 0} sE(s) \tag{10.29}$$

Or, using Eq. (10.28)

$$e(\infty) = \lim_{s \to 0} \frac{sR(s)}{1 + G(s)} \tag{10.30}$$

Equation (10.30) shows that the steady-state tracking error depends on (1) the nature of the input, $R(s)$, and (2) the forward transfer function $G(s)$.

System type is a designation of the forward transfer function $G(s)$ that is useful for analyzing the steady-state error. It represents the number of "free integrators" that exist in the forward transfer function. To show this, let us write the forward transfer function using the following form:

$$G(s) = \frac{F(s)}{s^N Q(s)} \tag{10.31}$$

where $F(s)$ and $Q(s)$ are polynomials in s that do not contain a zero at $s = 0$. The index N is called the *system type* and it is the number of "free integrators" ($1/s$ terms) that can be divided out of $G(s)$. As a quick example, consider the forward transfer function:

$$G(s) = \frac{6(s + 2)}{s^2 + 14s + 36} = \frac{F(s)}{s^N Q(s)}$$

Hence, $N = 0$ and $F(s)$ and $Q(s)$ are the numerator and denominator polynomials of $G(s)$. Because $N = 0$, this forward transfer function $G(s)$ is called a "type 0 system." As a second example, consider the forward transfer function:

$$G(s) = \frac{6(2s + 4)}{s^2 + 8s} = \frac{12s + 24}{s(s + 8)} = \frac{F(s)}{s^N Q(s)}$$

Clearly, $N = 1$ and this is a "type 1 system." In other words, we can "divide out" a free integrator ($1/s$) from $G(s)$ in this example. The reader should recall that the forward transfer function $G(s)$ is the product of the controller and plant transfer functions, and therefore a PI or PID controller will introduce a "free integrator" into $G(s)$. As we shall see the system type N is a purely notational convenience that allows us to categorize the steady-state tracking error.

Now, let us compute the steady-state tracking error defined by Eq. (10.30) for different "standard" reference signals. The step input is the "least demanding" reference signal in terms of tracking requirements because it is a static input. If the reference signal is a unit-step function, $r(t) = 1 = U(t)$, then its Laplace transform is $\mathscr{L}\{r(t)\} = R(s) = 1/s$ and Eq. (10.30) becomes

$$\text{Unit-step input:} \quad e(\infty) = \lim_{s \to 0} \frac{s(1/s)}{1 + G(s)} = \frac{1}{1 + K_{sp}} \tag{10.32}$$

where K_{sp} is called the *static position error constant*. Equation (10.32) shows that the constant K_{sp} is equal to the DC gain of the forward transfer function, or $G(s = 0)$. Next, observe how the system type, Eq. (10.31), affects K_{sp}: if $N = 0$ (type 0 system), then the DC gain $G(s = 0)$ is *finite* and hence K_{sp} is a finite number. However, if $N \geq 1$ (type 1 or higher), then computing the DC gain $G(s = 0)$ results in dividing by zero and $K_{sp} = \infty$. Consequently, the steady-state error for a unit-step input is $e(\infty) = 1/\infty = 0$. In other words, the closed-loop system output can perfectly track a step input at steady state if we have a type 1 (or higher) system. It is important to note that the single "free integrator" may reside naturally in the plant transfer function $G_P(s)$ or it can be inserted into the forward path by using a PI or PID controller.

The unit-ramp input is a more demanding reference signal compared to the unit-step input as it is linearly increasing with time and thus more difficult to track at steady state. The unit ramp is $r(t) = t$ and its Laplace transform is $\mathscr{L}\{r(t)\} = R(s) = 1/s^2$, and hence Eq. (10.30) becomes

$$\text{Unit-ramp input:} \quad e(\infty) = \lim_{s \to 0} \frac{s(1/s^2)}{1 + G(s)} = \lim_{s \to 0} \frac{1/s}{1 + G(s)}$$

$$= \lim_{s \to 0} \frac{1}{s + sG(s)} = \frac{1}{K_{sv}} \tag{10.33}$$

where $K_{sv} = \lim_{s \to 0} sG(s)$ is called the *static velocity error constant*. As before, the system type N determines whether K_{sv} is zero, finite, or infinite. For a type 0 system (no free integrators), we have $K_{sv} = \lim_{s \to 0} sG(s) = 0$ because there is no $1/s$ term in $G(s)$ to cancel with the multiplicative s term. Because $K_{sv} = 0$, the steady-state tracking error is $e(\infty) = 1/K_{sv} = \infty$. Hence, the closed-loop response of a type 0 system *diverges* from the ramp input as time goes on, which results in a continually growing error signal. For a type 1 system ($N = 1$), we have one free integrator in $G(s)$ that cancels with the multiplicative s term and hence $K_{sv} = \lim_{s \to 0} sG(s)$ is a finite number. Therefore, the closed-loop response of a type 1 system will exhibit a constant offset from the reference ramp input at steady state. If we have a system that is type 2 (or higher), then $K_{sv} = \lim_{s \to 0} sG(s) = \infty$ and Eq. (10.33) shows that the steady-state error is zero. Hence, a system with two free integrators in the forward path can perfectly track a ramp input at steady state.

Finally, let us investigate the unit-parabolic input, $r(t) = t^2/2$, where the reference input signal increases at a quadratic rate with time. Clearly, a parabolic input is more difficult to track than a linear ramp input. The Laplace transform of the unit-parabolic input is $\mathscr{L}\{r(t)\} = R(s) = 1/s^3$ and hence Eq. (10.30) becomes

$$\text{Unit-parabolic input:} \quad e(\infty) = \lim_{s \to 0} \frac{s(1/s^3)}{1 + G(s)} = \lim_{s \to 0} \frac{1/s^2}{1 + G(s)}$$

$$= \lim_{s \to 0} \frac{1}{s^2 + s^2 G(s)} = \frac{1}{K_{sa}} \tag{10.34}$$

where $K_{sa} = \lim_{s \to 0} s^2 G(s)$ is called the *static acceleration error constant*. In order for K_{sa} to be nonzero, we must have a system that is type 2 or higher in order to cancel the multiplicative s^2 term. Consequently, $K_{sa} = 0$ for type 0 and type 1 systems, and the steady-state tracking error diverges to infinity. If $N = 2$ (type 2), then K_{sa} is finite and the tracking error exhibits a steady-state offset. For systems with *three* or more free integrators in the forward path ($N \geq 3$), the closed-loop response will perfectly track the parabolic input at steady state.

Table 10.3 summarizes the relationships between system type, steady-state tracking error $e(\infty)$, and the various error constants that we derived in the previous discussion. The table exhibits a sort of symmetry where finite tracking errors exist when the order of the input function (expressed as a polynomial in time t) matches the system type. For example, a unit ramp $r(t) = t$ is a first-order time polynomial and a type 1 system exhibits a finite steady-state tracking error. If the order of the input time polynomial is greater than the system type N, then the closed-loop output diverges from the reference and the steady-state error is infinity.

Table 10.3 and the previous analyses demonstrate that adding integrators to forward transfer function reduces or eliminates the steady-state tracking error. A review of Example 10.5 shows that the plant (speed response of a DC motor) is type 0 and therefore a finite speed error exists when the reference input is a step function and proportional (P) control is employed. When an integral control signal is added (PI control), the forward transfer function becomes type 1 and the

Table 10.3   Steady-State Errors for Closed-Loop Control Systems with Unity Feedback

System Type, N	Unit-Step Input $r(t) = 1$	Unit-Ramp Input $r(t) = t$	Unit-Parabola Input $r(t) = t^2/2$
0	$\dfrac{1}{1 + K_{sp}}$	∞	∞
1	0	$\dfrac{1}{K_{sv}}$	∞
2	0	0	$\dfrac{1}{K_{sa}}$

Static position error constant: $K_{sp} = \lim\limits_{s \to 0} G(s)$.
Static velocity error constant: $K_{sv} = \lim\limits_{s \to 0} sG(s)$.
Static acceleration error constant: $K_{sa} = \lim\limits_{s \to 0} s^2 G(s)$.

steady-state tracking error is zero. In contrast, Example 10.6 shows that the plant transfer function (mechanical system) is type 1 and consequently the closed-loop position response will perfectly track a step reference position input at steady state with either a P- or PD-controller (see Figs. 10.19 and 10.20). In this case where the plant is type 1 and the reference input is a step function, there is no need to add an integral control signal to improve steady-state accuracy.

The control engineer may be tempted to simply add one (or more) integrators to the controller in order to reduce (or eliminate) the steady-state error. However, adding integrators in the forward path adds lag to the closed-loop system, thus slowing down the response. In some cases, adding integrators can cause stability problems.

A final note is in order regarding Table 10.3: the tabulated steady-state errors are for *unit* input functions $r(t)$ such as the unit step, unit ramp, and unit parabola. Hence, the corresponding Laplace transforms of these unit inputs are $R(s) = 1/s$, $1/s^2$, and $1/s^3$, respectively. Because $R(s)$ has a unity numerator for a unit input function, the steady-state error equation (10.30) shows that the finite errors (the diagonal values in Table 10.3) all have unity numerators. However, if $r(t)$ is a non-unit input, then the numerator of the finite value in Table 10.3 will match the coefficient of the input function. For example, if the input is a ramp function $r(t) = 0.2t$, then steady-state error of a type 1 system is $e(\infty) = 0.2/K_{sv}$.

Example 10.8

Consider again the closed-loop position-control system from Example 10.6 and shown in Fig. 10.27. Determine the steady-state position errors for proportional and proportional-integral controllers. Let the controller gains be $K_P = 0.04$ V/m (P and PI control) and $K_I = 0.002$ V/m-s and the reference position input be $x_{ref}(t) = 0.01t$ m (ramp function). The fixed actuator gain is $K_A = 2$ N/V.

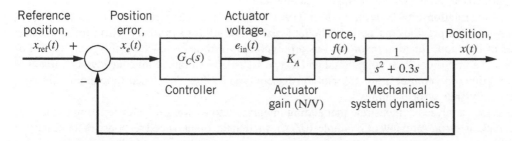

Figure 10.27 Closed-loop position control of a mechanical system (Example 10.8).

Proportional controller

Using proportional controller transfer function $G_C(s) = K_P = 0.04$ V/m, the forward transfer function of the system in Fig. 10.27 is

$$G(s) = \frac{2K_P}{s^2 + 0.3s} = \frac{0.08}{s(s + 0.3)}$$

Forward transfer function $G(s)$ is type 1 because we can factor out one free integrator. The reference input, $r(t) = 0.01t$ m, is a ramp function with slope of 0.01 m/s. Table 10.3 shows that the steady-state position error for a type 1 system with a ramp input is

$$x_e(\infty) = \frac{0.01}{K_{sv}}$$

where the static velocity error constant is

$$K_{sv} = \lim_{s \to 0} sG(s) = \lim_{s \to 0} \frac{0.08}{s + 0.3} = \frac{0.08}{0.3} = 0.2667$$

The steady-state position error for the ramp input is $x_e(\infty) = 0.01/K_{sv} = 0.0375$ m (or 3.75 cm). Figure 10.28 shows the closed-loop position response $x(t)$ to the ramp input $x_{ref}(t) = 0.01t$ m with a P-controller and gain $K_P = 0.04$ V/m. At steady state, the closed-loop position response $x(t)$ clearly exhibits an offset error of 0.0375 m with respect to the reference ramp input.

Proportional-integral controller

The PI controller transfer function is

$$G_C(s) = \frac{K_P s + K_I}{s}$$

The forward transfer function with $K_P = 0.04$ V/m and $K_I = 0.002$ V/m-s is

$$G(s) = \frac{2(K_P s + K_I)}{s^2(s + 0.3)} = \frac{0.08s + 0.004}{s^2(s + 0.3)}$$

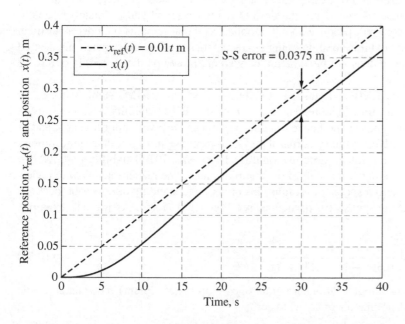

Figure 10.28 Closed-loop position response to ramp input with P-controller (Example 10.8).

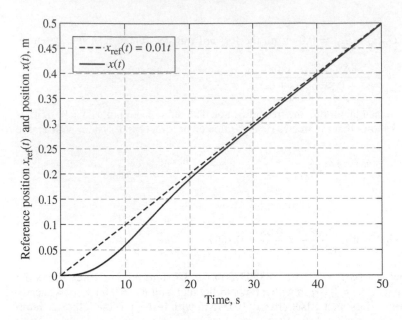

Figure 10.29 Closed-loop position response to ramp input with PI controller (Example 10.8).

Clearly, the forward transfer function $G(s)$ is type 2 because the PI controller has one free integrator and the plant contains one free integrator. Table 10.3 tells us that for a ramp input and a type 2 system, the steady-state error is zero. The values of the K_P and K_I gains do not affect the steady-state error, but they do affect the settling time and damping of the transient response. Figure 10.29 shows that the closed-loop response $x(t)$ using the PI controller eventually tracks the ramp input $x_{\text{ref}}(t)$ at steady state with zero offset error.

10.5 CLOSED-LOOP STABILITY

Stability is an essential attribute of a closed-loop control system. We expect that a stable closed-loop system remains "under control" during all normal modes of operation. For example, operating a stable closed-loop cruise-control system will never result in an automobile speed that diverges from the desired reference speed and ultimately becomes unbounded in time. A stable cruise-control system will produce an automobile speed that remains bounded over time. The closed-loop speed response may or may not exhibit good damping characteristics or ultimately match the desired reference speed, but as long as the speed response does not become unbounded or "blow up" the system is said to be stable.

We use the *bounded-input, bounded-output* (BIBO) definition of stability: a system is BIBO stable if for every bounded input the output remains bounded for all time. The reader should note that BIBO stability does not impose any specific performance criteria on the system response and hence a stable system can have very poor transient and/or steady-state responses. All that is required for BIBO stability is that the system response does not diverge to infinity when the reference input is a bounded function. For linear, time-invariant (LTI) systems, BIBO stability requires that *all* characteristic roots (or poles or eigenvalues) lie in the left-half of the complex plane. The reader may want to review Sections 7.3 and 7.4 (first- and second-order system response) to solidify his or her understanding of how root locations in the complex plane correspond to the system response (in particular, see Figs. 7.13–7.16). As a quick example, consider the transfer function of an LTI system

$$G_P(s) = \frac{1}{0.5s^3 + 4s^2 + 23s + 34} = \frac{Y(s)}{U(s)} \tag{10.35}$$

This transfer function could represent the input–output relationship of an open-loop system (plant). The characteristic roots (or poles) are determined by setting the denominator of $G_P(s)$ to zero:

Characteristic roots: $0.5s^3 + 4s^2 + 23s + 34 = 0$

The three roots are $s_1 = -2$ and $s_{2,3} = -3 \pm j5$. Because all three roots have negative real parts, they lie in the left-half of the complex plane. For this example the homogeneous or free response has the general form

$$y_H(t) = c_1 e^{-2t} + e^{-3t}[c_2 \cos 5t + c_3 \sin 5t]$$

Clearly, the free response eventually "dies out" owing to the exponential functions e^{-2t} and e^{-3t}, which depend on the real parts of the characteristic roots. The particular (or forced) response is bounded if the input is bounded. Hence, the LTI system $G_P(s)$ is BIBO stable. If even one characteristic root had a positive real part, then the corresponding exponential function would diverge to infinity over time (an unbounded output) and the system would be unstable.

This simple example shows that the stability of an LTI closed-loop control system can be determined by computing the characteristic roots or poles of the closed-loop transfer function. The MATLAB `roots` command can be used to quickly calculate the roots of the characteristic equation. Suppose the system $G_P(s)$ defined by Eq. (10.35) is the plant in a closed-loop, unity-feedback system with a proportional controller (K_P). The closed-loop transfer function is

$$T(s) = \frac{K_P G_P(s)}{1 + K_P G_P(s)} = \frac{K_P}{0.5s^3 + 4s^2 + 23s + 34 + K_P} \tag{10.36}$$

The following MATLAB commands will compute the closed-loop poles (closed-loop roots) for a proportional gain $K_P = 30$

```
>> Kp = 30;                          % proportional gain setting
>> denT = [ 0.5 4 23 (34+Kp) ];      % denominator of closed-loop T(s)
>> CLpoles = roots(denT)             % compute closed-loop poles
```

Executing the above MATLAB commands yields the closed-loop poles

```
CLpoles =
-1.8714 + 5.1540i
-1.8714 - 5.1540i
-4.2573
```

Because all three closed-loop poles have negative real parts, the closed-loop system is BIBO stable.

Before the advent of fast root-solving algorithms such as MATLAB's `roots` command, the *Routh–Hurwitz stability criterion* was used to determine stability. The Routh–Hurwitz stability criterion is a systematic procedure that involves constructing the so-called Routh array, where the first two rows consist of the coefficients of the characteristic equation. Subsequent rows of the Routh array are computed by determinant-like calculations using the previous two rows. If all elements of the first column of the Routh array have the same sign, then all characteristic roots have negative real parts; that is, the system is stable. The number of characteristic roots with *positive* real parts (unstable roots) is equal to the number of sign changes in the first-column elements. The Routh–Hurwitz stability criterion essentially provides a "yes or no" check on stability and does not provide information on the root locations. Clearly, a stability check can be determined by numerically computing the n closed-loop roots (e.g., MATLAB's `roots` command). One advantage of the Routh–Hurwitz stability criterion method is that it can be used to analytically determine the range of a parameter (such as a control gain) for closed-loop stability. However, as we show in the following example, MATLAB's `roots` command may also be used with several trial values of a gain to determine the range for stability. We do not present or utilize the Routh–Hurwitz criterion for stability analysis. Instead, we will rely on MATLAB's `roots` command or graphical techniques such as the root-locus method and the Bode diagram to analyze stability. The interested reader can consult References 2–5 for details regarding the Routh–Hurwitz stability criterion.

Example 10.9

Figure 10.30 shows a closed-loop system with proportional control and plant dynamics represented by $G_P(s)$. Use MATLAB to determine the closed-loop stability for control gains in the range $0 < K_P \le 250$.

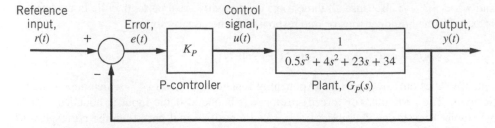

Figure 10.30 Closed-loop control system (Example 10.9).

Table 10.4 Stability Analysis of the Closed-Loop System in Fig. 10.30 (Example 10.9)

Control Gain, K_P	Closed-Loop Poles	Stability Status
1	$s_1 = -2.0774$, $s_{2,3} = -2.9613 \pm j4.9927$	Stable
50	$s_1 = -5.3009$, $s_{2,3} = -1.3495 \pm j5.4655$	Stable
100	$s_1 = -6.9377$, $s_{2,3} = -0.5312 \pm j6.1925$	Stable
150	$s_1 = -8.0$, $s_{2,3} = \pm j6.7823$	Marginally stable
200	$s_1 = -8.8091$, $s_{2,3} = 0.4045 \pm j7.2776$	Unstable
250	$s_1 = -9.4734$, $s_{2,3} = 0.7367 \pm j7.7081$	Unstable

Note that the plant $G_P(s)$ is identical to the transfer function (10.35) and the example from the previous discussion on stability. In order to check the stability of the closed-loop system, we must calculate the closed-loop transfer function:

$$T(s) = \frac{K_P G_P(s)}{1 + K_P G_P(s)} = \frac{K_P}{0.5s^3 + 4s^2 + 23s + 34 + K_P} = \frac{Y(s)}{R(s)}$$

The following MATLAB commands will compute the closed-loop poles of $T(s)$ for the gain $K_P = 1$

```
>> Kp = 1;                        % proportional gain setting
>> denT = [ 0.5 4 23 (34+Kp) ];   % denominator of closed-loop T(s)
>> CLpoles = roots(denT)          % compute closed-loop poles
```

We can repeat these commands for different control gains. Table 10.4 presents the three closed-loop poles for six discrete gains in the desired range: $K_P = 1, 50, 100, 150, 200,$ and 250. When the gain is small ($K_P = 1$), the closed-loop poles are close to the poles of the stable open-loop plant $G_P(s)$ ($s_1 = -2$ and $s_{2,3} = -3 \pm j5$) and hence the closed-loop system is stable. Table 10.4 shows that as K_P increases, the real pole becomes more negative and the real part of the complex conjugate poles increases. When the control gain is $K_P = 150$, the complex poles are $s_{2,3} = \pm j6.7823$ and therefore lie on the imaginary axis. Thus, the real part of the complex poles has become zero at this gain setting and the closed-loop system is *marginally stable*: the closed-loop response will exhibit sustained oscillations at a frequency of $\omega = 6.7823$ rad/s. When K_P is equal to 200 and 250, the two complex poles have positive real parts and therefore the closed-loop system is unstable. If we used a finer grid of discrete gains, we would find that the closed-loop system is stable for all gains $0 < K_P < 150$ and unstable for $K_P > 150$.

10.6 ROOT-LOCUS METHOD

Chapter 7 and the previous sections of this chapter have illustrated that the locations of the characteristic roots (or poles) strongly influence the transient-response performance criteria such as settling time, overshoot, and oscillation frequency. Furthermore, system stability is dictated by the root locations in the complex plane: if all characteristic roots have negative real parts, then the system is BIBO stable. It is clear that knowledge of the characteristic root locations is

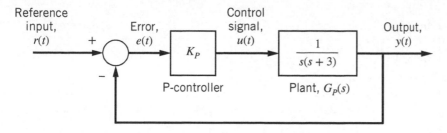

Figure 10.31 Closed-loop control system.

important to the control engineer. Changing the control gains and/or adding a dynamic controller (such as a PI or PD controller) will vary the root locations and therefore adjust the transient response. Large changes in the root locations will have dramatic effects on the transient response characteristics and may lead to instability.

In the late 1940s, W. R. Evans developed a method for computing the closed-loop roots based on knowledge of the open-loop transfer function. This technique, called the *root-locus method*, is a graphical method that determines how the closed-loop root locations change as a single parameter (usually a gain) is varied. As we shall see, the root-locus method is a valuable tool for determining if desirable transient-response characteristics (settling time, overshoot) can be achieved by adjusting a single control gain. In cases where simple gain adjustment is inadequate, the root-locus method can also be used to design dynamic controllers (such as PI and PD controllers) that can be introduced in order to improve the closed-loop response and stability margins.

Before we describe the theoretical basis of the root-locus method, it is useful to demonstrate how the closed-loop roots (poles) change as a single parameter is varied (the reader may want to review Example 10.9, which also illustrates how the closed-loop poles change with gain K_P). Figure 10.31 shows a very simple closed-loop system with a type-1 plant $G_P(s)$ and proportional controller (the plant has the same basic structure as the mechanical system shown in Examples 10.6 and 10.8). The closed-loop transfer function is

$$T(s) = \frac{K_P G_P(s)}{1 + K_P G_P(s)} = \frac{K_P}{s^2 + 3s + K_P}$$

The two closed-loop poles are determined by finding the roots of the characteristic equation

$$\text{Characteristic equation:} \quad s^2 + 3s + K_P = 0 \tag{10.37}$$

Now, let us compute the roots of the characteristic equation (10.37) (i.e., poles of the closed-loop transfer function) for the gains $K_P = 0.001, 1, 2.25, 4, 10,$ and 100. Table 10.5 shows the characteristic roots for these six gain settings. Note that when the control gain is small ($K_P = 0.001$), the closed-loop poles are very close to the open-loop poles located at $s = 0$ and $s = -3$. The closed-loop roots are negative real numbers for the gain range $0 < K_P \le 2.25$. When the gain is exactly $K_P = 2.25$, the characteristic equation has two repeated roots at $s = -1.5$. For gains $K_P > 2.25$, the roots are complex with a constant real part equal to -1.5. Figure 10.32 shows a plot of the closed-loop root locations in the complex plane. The two open-loop poles at $s = 0$ and $s = -3$ are indicated by an "×" and the thick lines and arrows indicate the location and "movement" of the closed-loop poles as the gain is increased from zero. The closed-loop poles begin at the open-loop poles with gain $K_P = 0$, as indicated on Fig. 10.32, and the two poles move toward each other

Table 10.5   Characteristic Roots of Eq. (10.37)

Control Gain, K_P	Characteristic Roots (Closed-Loop Poles)
0.001	$-2.9997, -0.0003$
1	$-2.6180, -0.3820$
2.25	$-1.5, -1.5$
4	$-1.5 \pm j1.3229$
10	$-1.5 \pm j2.7839$
100	$-1.5 \pm j9.8869$

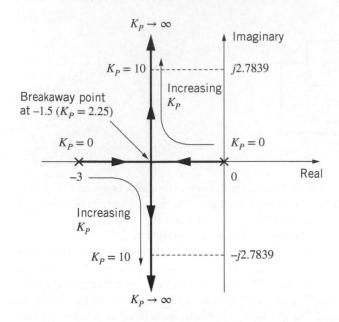

Figure 10.32 Characteristic root locations for Eq. (10.37).

as the gain is increased from 0 to 2.25. At gain $K_P = 2.25$, the two poles simultaneously occupy the real-axis location -1.5. As the gain is increased beyond 2.25, the two closed-loop poles move along a vertical line with a real part equal to -1.5. The magnitude of the imaginary part of the roots goes to infinity as the gain is increased to infinity. Recall that the distance from the origin to a complex root is equal to the undamped natural frequency ω_n and the cosine of the angle between the negative real axis and the radial line to the complex root is equal to the damping ratio ζ (see Fig. 7.17). Therefore, ω_n increases and ζ decreases as gain K_P is increased beyond 2.25. Finally, Fig. 10.32 shows that the root locations remain in the left-half of the complex plane (i.e., the roots always have negative real parts) for all positive gains and therefore the closed-loop system is always stable for $K_P > 0$.

A plot of the paths or loci followed by the n closed-loop poles (or roots) as a single parameter is varied is called the *root locus*. Figure 10.32 shows the root locus (the thick solid lines) for the two roots as the gain K_P is varied from zero to infinity. The root locus shown in Fig. 10.32 demonstrates that the two closed-loop roots start at the *open-loop poles* ($s = 0$ and $s = -3$) when $K_P = 0$, move toward each other along the negative real axis as K_P increases until they meet at $s = -1.5$, and then move along vertical asymptotes with a real part equal to -1.5 as gain K_P is increased to infinity.

The root-locus method is the graphical procedure for constructing the root locus (such as Fig. 10.32) and is based only on the knowledge of the open-loop poles and zeros. We can derive the fundamental condition that determines whether or not a particular complex value s_1 is on the root locus by considering the very general closed-loop system shown by Fig. 10.33. Recall that the root locus is a plot of the paths followed by the closed-loop roots (poles) as a single parameter is varied. In Fig. 10.33 we have included the very general case where the parameter to be varied, control gain K, is in the forward path. The reader should note that the forward transfer function $G(s)$ contains both the controller dynamics (such as PID) and the plant dynamics. Using Eq. (10.4), the closed-loop transfer function is

$$T(s) = \frac{Y(s)}{R(s)} = \frac{KG(s)}{1 + KG(s)H(s)} \tag{10.38}$$

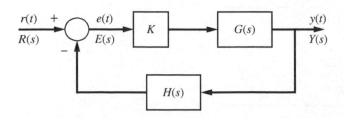

Figure 10.33 General closed-loop system with forward-path gain K.

The denominator of the closed-loop transfer function $T(s)$ is the characteristic equation that defines the location of the closed-loop poles (roots):

$$\text{Characteristic equation:} \quad 1 + KG(s)H(s) = 0$$

or, solving for the open-loop transfer function

$$G(s)H(s) = \frac{-1}{K} \tag{10.39}$$

In general, the open-loop transfer function $G(s)H(s)$ is a complex function of the complex variable s; that is, it consists of real and imaginary components. Therefore, we can rewrite Eq. (10.39) as two conditions: the *angle condition* and the *magnitude condition*. Because $G(s)H(s) = -1/K$ is a real negative number (for positive gain $K > 0$), the argument or phase angle of $G(s)H(s)$ must be 180°. Therefore, the angle condition is

$$\text{Angle condition:} \quad \angle \left(G(s)H(s) \right) = 180° + r360°, \quad r = 0, \pm 1, \pm 2, \ldots \tag{10.40}$$

Likewise, because $G(s)H(s) = -1/K$ is a real negative number, its magnitude is

$$\text{Magnitude condition:} \quad |G(s)H(s)| = \frac{1}{K} \tag{10.41}$$

Values of the complex variable s that satisfy both the angle condition (10.40) *and* the magnitude condition (10.41) lie on the root-locus plot (i.e., they are closed-loop roots). General rules for constructing a root locus can be derived from the angle and magnitude conditions. We will not derive these rules; instead we will simply summarize the rules for constructing a root-locus plot (see References 2–6 for additional details). In the rules that follow, let p_i ($i = 1, 2, \ldots, n$) denote the n open-loop poles of $G(s)H(s)$ and let z_j ($j = 1, 2, \ldots, m$) denote the open-loop zeros of $G(s)H(s)$; that is, z_j are the finite values of s that make $G(s)H(s)$ equal to zero.

Root-locus rules

1. The number of paths (loci) is n, the number of open-loop poles of $G(s)H(s)$.
2. The root locus is symmetric about the real axis.
3. The n loci begin at the open-loop poles p_i with gain $K = 0$.
4. The n loci terminate at the m (finite) open-loop zeros z_j or the $n - m$ "zeros at infinity" as the gain $K \to \infty$.
5. The $n - m$ loci that approach infinity do so along radial, straight-line asymptotes that intersect the real axis at

$$\sigma_a = \frac{\sum_{i=1}^{n} p_i - \sum_{j=1}^{m} z_j}{n - m} \quad \text{with angle} \quad \theta = \pm \frac{k180°}{n - m}, \quad k = 1, 3, 5, \ldots$$

6. A point on the real axis is on the root locus if there is an *odd* number of open-loop poles and zeros to the right of that point.
7. "Breakaway" and "break-in" points at which the root locus leaves/enters the real axis satisfy the equation

$$\frac{d}{ds}[G(s)H(s)] = 0$$

We can use the above mentioned root-locus rules to develop the recommended steps for sketching a root-locus plot:

1. Given the open-loop transfer function $G(s)H(s)$, determine the n open-loop poles p_i and the m open-loop zeros z_j.
2. Sketch the open-loop poles and zeros on the complex plane. Use "✗" for an open-loop pole, and "o" for an open-loop zero. The root loci start at the n open-loop poles with gain $K = 0$ (Rule 3). Open-loop poles act as "sources" and zeros act as "sinks."

3. Sketch the real-axis root locus using Rule 6.

4. Compute the asymptotes using Rule 5.

5. Complete the approximate root-locus sketch (experience helps here). With enough experience, one does not need to exactly compute the breakaway/break-in points using Rule 7.

Before the advent of computer software such as MATLAB, a great deal of emphasis was placed on sketching a root locus by using these rules and recommended construction steps (similarly, general rules were developed for sketching a Bode diagram). As we shall see shortly, MATLAB has a command that produces the root-locus plot from the open-loop transfer function $G(s)H(s)$. Therefore, we will not emphasize sketching the root locus by using a set of construction rules. Instead, we will focus on using the root locus with two specific goals in mind: (1) characterizing the effect that varying the parameter K has on the closed-loop transient response, and (2) understanding how adding a dynamic controller (i.e., transfer function $G_C(s)$ with poles and/or zeros) affects the closed-loop transient response.

Constructing the Root Locus Using MATLAB

As previously stated, several textbooks present rules for sketching the root locus. It is this author's opinion that it is more important for the control engineer to know how to use and interpret the root locus than it is to know how to sketch the root locus. This opinion is reinforced by the fact that the root locus can be easily constructed by a single MATLAB command. To illustrate, let us again consider the simple closed-loop control system present in Fig. 10.31, where the P-controller gain K_P is in the forward path. The open-loop transfer function is

$$G(s)H(s) = \frac{1}{s(s+3)} \tag{10.42}$$

In this example, the feedback transfer function $H(s)$ is unity. The required MATLAB commands to create the root locus are

```
>> sysGH = tf(1,[1 3 0])          % create the open-loop transfer function G(s)H(s)
>> rlocus(sysGH)                  % create and draw the root locus
```

The `rlocus` command draws the root locus to the screen. Executing these commands creates the root-locus plot shown in Fig. 10.34, which matches the diagram in Fig. 10.32. We can also use the basic `rlocus` command to determine the specific closed-loop poles (roots) for a desired gain; for example, $K_P = 2$:

```
>> sysGH = tf(1,[1 3 0])          % create G(s)H(s)
>> KP = 2                         % set the desired gain K_P
>> CL_roots = rlocus(sysGH,KP)    % compute closed-loop roots (no plot)
```

In this case where gain $K_P = 2$, the two closed-loop poles are $s_1 = -1$ and $s_2 = -2$.

Another extremely useful MATLAB command is `rlocfind`, which draws the root locus and then allows the user to place a "cross-hair target" cursor on a desired closed-loop root on the root locus. When the user clicks the mouse, the `rlocfind` command returns the gain K that produces the desired closed-loop root as well as the complete set of n closed-loop roots for gain K. Using the simple closed-loop system from Fig. 10.31, the following MATLAB commands illustrate the use of `rlocfind`

```
>> sysGH = tf(1,[1 3 0])          % create G(s)H(s)
>> rlocus(sysGH)                  % create and draw the root locus
>> [KP,CL_roots] = rlocfind(sysGH)   % allow cross-hair target on root locus
```

After the user clicks the cross-hair target on a desired loci branch, MATLAB returns the associated value of gain K and all n closed-loop roots. For example, if the user executes the MATLAB commands shown above and clicks the cross-hair target on -0.5, the command `rlocfind` returns $K_P = 1.25$ and the associated closed-loop roots $s_1 = -2.5$ and $s_2 = -0.5$ (the desired root from the cross-hair target).

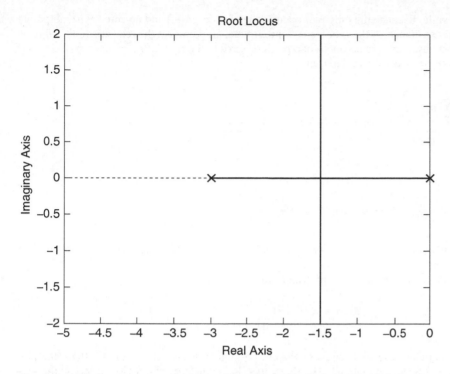

Figure 10.34 Root-locus plot created by MATLAB for $G(s)H(s) = 1/(s^2 + 3s)$.

The following examples illustrate the construction and use of the root-locus plot. Although this textbook does not emphasize sketching the root locus, the preliminary steps for sketching the root locus are presented and the corresponding root locus is created using MATLAB.

Example 10.10

Consider again the unity-feedback control system from Example 10.9. Figure 10.35 shows the closed-loop system with proportional control and plant dynamics represented by $G(s)$. Use the root-locus construction rules to develop the basic structure of the root locus and use MATLAB to create the root locus. Finally, use the root-locus plot to characterize the closed-loop transient response and closed-loop stability.

The root locus solely depends on the open-loop transfer function, which is

$$G(s)H(s) = \frac{1}{0.5s^3 + 4s^2 + 23s + 34} \tag{10.43}$$

The three open-loop poles are $p_1 = -2$ and $p_{2,3} = -3 \pm j5$. There are no finite open-loop zeros because the numerator of $G(s)H(s)$ is a constant. Hence, a sketch of the root locus would begin with three open-loop "×" markers in the complex plane at -2 (negative real axis) and $-3 \pm j5$ (complex). Rule 6 states that the real-axis root-locus branch exists if an odd number

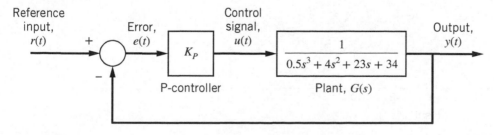

Figure 10.35 Closed-loop control system (Example 10.10).

of open-loop poles and zeros are to the right. Because the only *real* open-loop pole is $p_1 = -2$ and no zeros exist, all points on the real axis to the *left* of -2 constitute a branch of the root locus. Furthermore, all points on the real axis to the *right* of -2 can never be part of the root locus because there are no open-loop poles or zeros to the right of $p_1 = -2$ (zero is an even number). Rule 5 states that the asymptotes intersect the real axis at

$$\sigma_a = \frac{\sum\limits_{i=1}^{n} p_i - \sum\limits_{j=1}^{m} z_j}{n - m} = \frac{(-2) + (-3 + j5) + (-3 - j5)}{3} = \frac{-8}{3}$$

The three asymptote angles are

$$\theta_{1,2} = \pm\frac{180°}{3} = \pm 60° \quad \text{and} \quad \theta_3 = \pm\frac{3(180°)}{3} = \pm 180°$$

Hence, two asymptotes emanate outward from $\sigma_a = -8/3$ at angles $\theta_{1,2} = \pm 60°$ and the third asymptote ($\theta_3 = \pm 180°$) is along the negative real axis.

The following MATLAB commands create the accurate root-locus plot:

```
>> sysGH = tf(1,[0.5 4 23 34])          % create G(s)H(s)
>> rlocus(sysGH)                         % create and draw the root locus
```

Figure 10.36 shows MATLAB's construction of the root locus for the closed-loop control system in Fig. 10.35 (the plot has been enhanced by adding the asymptotes and text labels). The three open-loop poles are clearly identified by the "×" markers. One closed-loop root begins at the open-loop pole $p_1 = -2$ (for $K_P = 0$) and moves left along the negative real axis as the gain is increased. No closed-loop roots lie on the real axis to the right of $p_1 = -2$ as expected. The other two closed-loop roots begin at the open-loop complex conjugate pairs $p_{2,3} = -3 \pm j5$ (again with $K_P = 0$) and move to the right as the gain is increased. Eventually, these two complex closed-loop poles follow the $\pm 60°$ asymptotes.

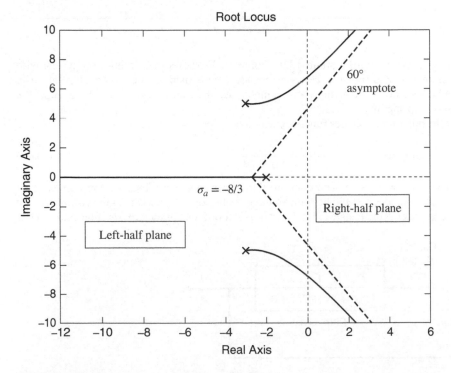

Figure 10.36 Root-locus plot for Example 10.10.

Once we have the root locus (either by sketch or from MATLAB), we can interpret the effect that varying the proportional gain K_P has on the closed-loop transient response and closed-loop stability. Figure 10.36 shows that as the gain is increased, the two complex roots move to the right, cross the imaginary axis, and eventually follow the $\pm 60°$ asymptotes. Because the real part of these two roots is approaching zero, the associated settling time is increasing, which means that the closed-loop response is becoming slower as the gain increases. Furthermore, the closed-loop damping ratio ζ associated with the two complex roots decreases (as gain increases) as the complex roots approach the imaginary axis. By contrast, the closed-loop pole emanating from $p_1 = -2$ moves to the left as gain increases and hence its contribution to the transient response exhibits a decreasing (shorter) settling time. As an example, we can use MATLAB to compute the closed-loop roots for the gain $K_P = 30$:

$$K_P = 30: \quad \text{closed-loop roots are} \quad s_1 = -4.2573, \quad s_{2,3} = -1.8714 \pm j5.1540$$

Hence, the damping ratio is $\zeta = 0.3413$ for the two complex poles and settling time for the slowest roots is 2.14 s. If we double the gain, the three closed-loop roots are

$$K_P = 60: \quad \text{closed-loop roots are} \quad s_1 = -5.7101, \quad s_{2,3} = -1.1450 \pm j5.6226$$

and the damping ratio is $\zeta = 0.1995$ and the slowest settling time is 3.49 s. When the gain is increased to $K_P = 150$, the three closed-loop roots are

$$K_P = 150: \quad \text{closed-loop roots are} \quad s_1 - -8, \quad s_{2,3} = \pm j6.7823$$

Therefore, the closed-loop system becomes *marginally stable* at gain $K_P = 150$ because the two complex roots lie on the imaginary axis. The root-locus plot in Fig. 10.36 tells us that the closed-loop system goes unstable at high gains as the two complex roots eventually cross the imaginary axis and move into the right-half plane. Our previous stability analysis, Example 10.9, shows that the closed-loop system goes unstable for gains $K_P > 150$ (see Table 10.4).

In summary, the root locus presented in Fig. 10.36 shows that the P-controller strategy used in the closed-loop system will provide very poor performance. There is no way to obtain a fast, well-damped closed-loop response by simply adjusting the P-gain K_P. Another type of controller must be used to improve the closed-loop response and system stability.

Example 10.11

Figure 10.37 shows a unity-feedback closed-loop system with proportional control and plant dynamics represented by $G(s)$. Use the root-locus construction rules to develop the basic structure of the root locus and use MATLAB to create the root locus. Finally, use the root-locus plot to characterize the closed-loop transient response and closed-loop stability.

As before, we start with the open-loop transfer function

$$G(s)H(s) = \frac{s+6}{2s^2 + 4s} = \frac{s+6}{2s(s+2)} \tag{10.44}$$

The two open-loop poles ($n = 2$) are $p_1 = 0$ and $p_2 = -2$, and the single open-loop zero ($m = 1$) is $z_1 = -6$. Consequently, a sketch of the root locus begins with two open-loop "×" markers at $s = 0$ and $s = -2$, and one open-loop "○" marker at $s = -6$. Because we have three open-loop poles and zeros on the real axis, Rule 6 tells us that a root-locus branch lies on the

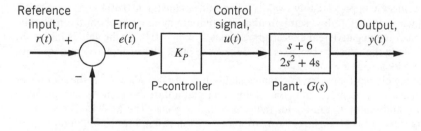

Figure 10.37 Closed-loop control system (Example 10.11).

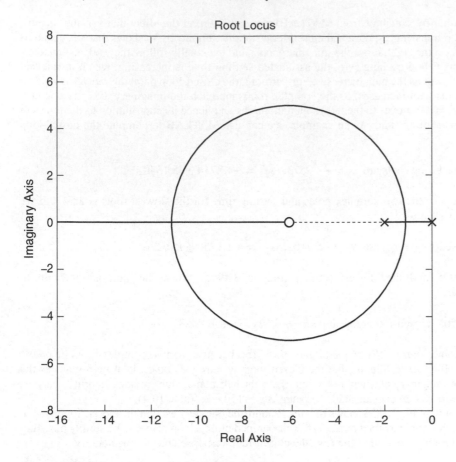

Figure 10.38 Root-locus plot for Example 10.11.

real axis to the *left* of −6 and in between −2 and the origin. Rule 5 states that we have $n − m$ or only one asymptote with the angle

$$\theta = \pm \frac{180°}{n − m} = \pm 180°$$

Hence, the single asymptote is the negative real axis and there is no need to compute the real-axis intersection point σ_a.

The following MATLAB commands create the root-locus plot:

```
>> sysGH = tf([1 6],[2 4 0])        % create G(s)H(s)
>> rlocus(sysGH)                     % create and draw the root locus
```

Figure 10.38 shows MATLAB's construction of the root locus for the closed-loop control system in Fig. 10.37. The two open-loop poles and single open-loop zero are clearly identified by the "×" and "○" markers. The two closed-loop poles begin at $s = 0$ and $s = −2$ for the gain $K_P = 0$ and move toward each other along the negative real axis as the gain increases. When the P-gain is $K_P = 0.404$, the two closed-loop roots meet at approximately $s = −1.1$. For gains in the range $0.404 < K_P < 39.596$, the two closed-loop roots are complex and their branches follow semicircular arcs that are symmetric about the real axis. When the gain is $K_P = 39.596$, the two roots enter the negative real axis at approximately $s = −10.9$. As the gain is increased beyond 39.6, one closed-loop root moves left along the negative real axis to −∞ and the other closed-loop root moves right and eventually terminates at the open-loop zero located at $s = −6$.

The root locus in Fig. 10.38 shows that the closed-loop system remains stable for all gains $K_P > 0$, as the loci never cross the imaginary axis into the right-half plane. Furthermore, the control engineer can adjust the gain K_P and obtain a wide range of undamped natural frequency ω_n and damping ratio ζ for the two complex poles. The MATLAB command sgrid can be used to overlay grid lines of constant ω_n and ζ on an existing root locus, as shown in Fig. 10.39. Figure 10.39 presents the same root locus as shown in Fig. 10.38 but with semicircular grid lines of constant ω_n and radial grid lines of constant ζ (recall that the distance from the origin to the root is ω_n and that ζ decreases as the radial line from the origin to the root approaches the imaginary axis). The radial ζ lines in Fig. 10.39 show that closed-loop damping ratio is always

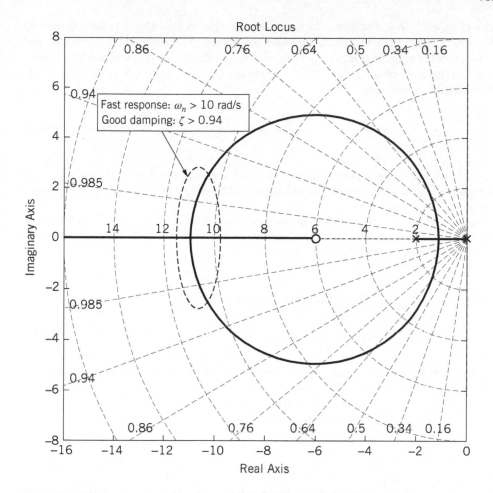

Figure 10.39 Root-locus plot with ω_n and ζ grid lines (Example 10.11).

greater than 0.5 for all complex roots. Furthermore, it is possible to select a gain K_P such that the closed-loop complex roots simultaneously have a fast response ($\omega_n > 10$ rad/s) with very good damping ($\zeta > 0.94$). This "wedge" location in the root locus for fast response and good damping is indicated in Fig. 10.39. In summary, the root locus in Fig. 10.39 shows that a simple P-controller can provide a desirable closed-loop transient response for the system shown in Fig. 10.37.

Controller Design Using Root Locus

The previous examples illustrate how the closed-loop root locations change as the control gain increases. In Example 10.10, a P-controller cannot provide a good closed-loop response for the given plant because the root locus (Fig. 10.36) shows that two branches move toward and eventually cross the imaginary axis as gain increases. By contrast, Example 10.11 and Fig. 10.38 demonstrate that a fast, well-damped transient response can be obtained (for a different plant) by a simple gain adjustment. The cause of this performance difference is the plant system dynamics. The plant in Example 10.10 did not have any open-loop zeros, while the second-order plant in Example 10.11 included a single open-loop zero at $s = -6$. Open-loop zeros act as "sinks" and tend to attract the root-locus branches (recall that a branch will terminate at the open-loop zero as the gain increases to infinity). Therefore, in many cases where the plant does not have sufficient damping, it is possible to introduce an open-loop zero by *adding* a PD controller to the forward path. Recall that the transfer function for a PD controller is

$$G_C(s) = K_P + K_D s \tag{10.45}$$

which has two adjustable gains K_P and K_D and one zero at $s = -K_P/K_D$. Another way to express the PD controller is

$$G_C(s) = K(s + z_D) \tag{10.46}$$

where $K = K_D$ is the single gain and $z_D = K_P/K_D$. Therefore, the open-loop zero is $s = -z_D$. Consequently, by adding the PD transfer function (10.46) to the forward path, the control engineer has altered the open-loop pole-zero map and changed the structure of the root locus. If the open-loop zero $-z_D$ is properly selected, then it may be possible to "bend" the root-locus branches to the left and therefore obtain a fast, well-damped closed-loop response. We demonstrate this aspect of root-locus design with the following example.

Example 10.12

Consider again the closed-loop position-control system presented in Fig. 10.40 (Examples 10.6 and 10.8). Use the root locus to design a controller $G_C(s)$ that provides a fast, well-damped closed-loop transient response to a reference step position command $x_{ref}(t) = 0.1$ m. The actuator gain is $K_A = 2$ N/V.

The reader should review Example 10.6, where the closed-loop responses were obtained with P- and PD-controllers. This example will illustrate how the root-locus plot allows the control engineer to visualize the effect of changing the controller dynamics without deriving the closed-loop characteristic equation.

Proportional controller

The open-loop transfer function with P-control or $G_C(s) = K_P$ and actuator gain $K_A = 2$ N/V is

$$G(s)H(s) = \frac{2}{s^2 + 0.3s} \tag{10.47}$$

Hence, the following MATLAB commands will create the root locus with natural-frequency and damping-ratio grid lines

```
>> sysGH = tf(2,[1 0.3 0])        % create G(s)H(s) for P-controller
>> rlocus(sysGH)                   % create and draw the root locus
>> sgrid                           % draw ωₙ and ζ grid lines on root locus
```

The root locus in Fig. 10.41 shows that as gain K_P increases, the two closed-loop roots move along the real axis starting from the open-loop roots at $s = 0$ and $s = -0.3$ until they meet at $s = -0.15$. The roots then move vertically along $\pm 90°$ asymptotes at $\sigma_a = -0.15$. Consequently, the complex roots will *always* have a real part equal to -0.15 and an increasing imaginary part as the gain K_P increases. Hence, the decay envelop term will always be $e^{-0.15t}$ with an associated settling time of about 26.67 s. Furthermore, the root locus shows that the damping ratio ζ will continually decrease with increasing gain as the roots move along the vertical asymptotes. These observations gleaned from the root-locus plot are illustrated by Fig. 10.19 (Example 10.6), which shows an underdamped closed-loop response for three K_P gains. All three closed-loop responses in Fig. 10.19 exhibit a settling time of 26.67 s and a decrease in damping as gain is increased. In summary, the root locus in Fig. 10.41 demonstrates that a P-control scheme will never provide a fast, well-damped transient response.

Proportional-derivative controller

Shifting the closed-loop root branches in Fig. 10.41 to the *left* will improve the transient response by speeding up the response and adding damping. Adding an open-loop zero to the forward path will act as a "sink" and "attract" the branches. Therefore,

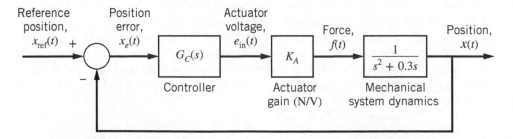

Figure 10.40 Closed-loop position control of a mechanical system (Example 10.12).

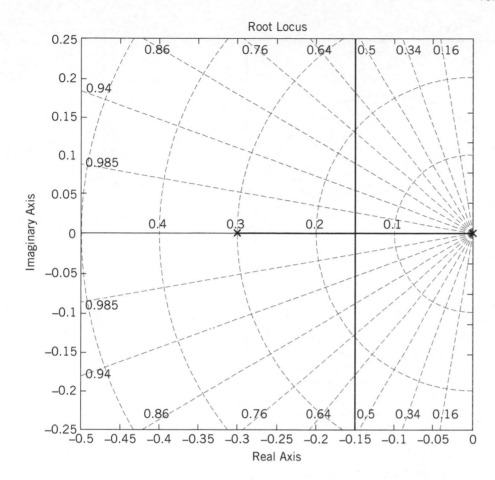

Figure 10.41 Root-locus plot for system with P-controller (Example 10.12).

inserting an open-loop zero to the *left* of the open-loop pole at $s = -0.3$ will shift the branches left. A PD controller will add an open-loop zero. To begin, let us test the following PD controller with an open-loop zero at $s = -3$

$$G_C(s) = K(s + 3) = K_P + K_D s \tag{10.48}$$

If we compare the PD controller in Eq. (10.48) to the "standard" format of Eq. (10.45), we see that the control gain K is equal to the derivative gain K_D and that $3K$ is equal to the proportional gain K_P (therefore, the open-loop zero is located at $s = -K_P/K_D = -3$). The open-loop transfer function consists of the PD controller and the mechanical system plant (with actuator gain $K_A = 2\,\text{N/V}$).

$$G(s)H(s) = \frac{2(s + 3)}{s^2 + 0.3s} \tag{10.49}$$

Hence, the following MATLAB commands will create the root-locus plot presented in Fig. 10.42:

```
>> sysGH = tf(2*[1 3],[1 0.3 0])      % create G(s)H(s) with PD controller
>> rlocus(sysGH)                       % create and draw the root locus
```

Figure 10.42 shows that as the single control gain K increases, the two closed-loop roots move toward each other along the negative real axis starting from the open-loop roots at $s = 0$ and $s = -0.3$ until they meet and break away from the real axis at approximately -0.154. As the gain K is increased, the closed-loop roots become complex and move along symmetric semicircles until they reenter the negative real axis at approximately -5.85. As the gain K is further increased to infinity, one closed-loop root follows an asymptote along the negative real axis to $-\infty$, while the other closed-loop root terminates at the

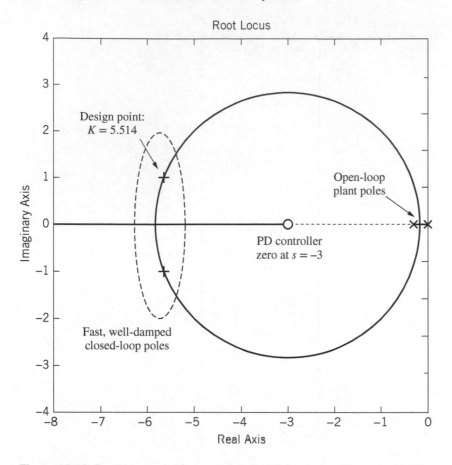

Figure 10.42 Root-locus plot for system with PD controller (Example 10.12).

open-loop zero at $s = -3$. Figure 10.42 shows a region near the real-axis break-in point where the closed-loop roots have large negative real parts (hence, the transient response decays quickly) and are close to the real axis (good damping). We may use the `rlocfind` MATLAB command to place the cross-hair target on the root-locus branch in this desired region in order to obtain the corresponding gain K. One such candidate design point is $K = 5.514$, which results in closed-loop poles at $s = -5.6640 \pm j1.0016$ (marked with cross "+" on Fig. 10.42). With these closed-loop poles, the undamped natural frequency is $\omega_n = \sqrt{5.664^2 + 1.0016^2} = 5.75$ rad/s and the damping ratio is $\zeta = 5.664/5.75 = 0.985$, which will result in a fast, well-damped transient response. Figure 10.43 shows the closed-loop response to a step position input $x_{\text{ref}}(t) = 0.1$ m (using design gain $K = 5.514$) and verifies that the transient response is indeed well damped and dies out quickly.

This example has essentially validated the results obtained in Example 10.6, which also used a PD controller (both examples use the same mechanical-system plant). In Example 10.6, the PD gains were set at $K_P = 16$ V/m and $K_D = 4$ V-s/m; in this example, the PD gains gleaned from the root-locus design point are $K = K_D = 5.514$ V-s/m and $K_P = 3K = 16.542$ V/m. Because these two PD gain settings are similar, the two corresponding closed-loop responses are similar (compare Figs. 10.20 and 10.43). However, this example has demonstrated how the root-locus method can be used to visualize the effect of adding a PD controller and select the gains for a good transient response.

As a final note, we briefly investigate the consequences of placing the open-loop zero of the PD controller even farther to the left. Suppose we double the open-loop zero and use a PD controller with a zero at $s = -6$

$$G_C(s) = K(s + 6) = K_P + K_D s$$

Hence, the open-loop transfer function is

$$G(s)H(s) = \frac{2(s + 6)}{s^2 + 0.3s}$$

Figure 10.44 shows the root locus for the new PD controller with an open-loop zero at $s = -6$. Note that the circular part of the root locus has a larger radius compared to the previous PD design (see Fig. 10.42) and hence the real-axis break-in point

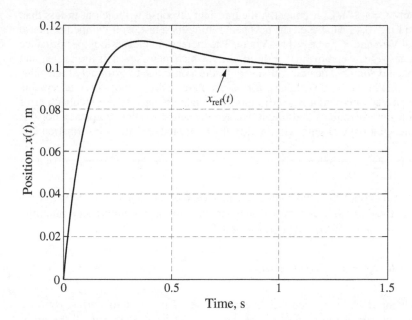

Figure 10.43 Closed-loop step response using PD controller with gain $K = 5.514$ (Example 10.12).

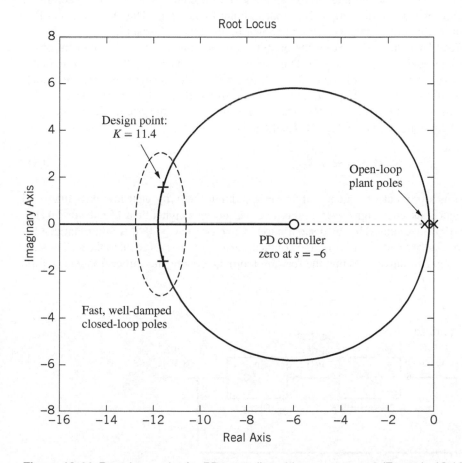

Figure 10.44 Root-locus plot for PD controller with zero at $s = -6$ (Example 10.12).

has been shifted farther to the left (approximately $s = -12$). Consequently, the transient response will die out in less than half the time compared to the response shown in Fig. 10.43. However, the new gain setting for a fast, well-damped response is about $K = 11.4$ and therefore $K_P = 6K = 68.4$ V/m and $K_D = K = 11.4$ V-s/m. The increase in closed-loop performance comes at a price: the higher PD gains result in larger voltage commands for the actuator, which may exceed its limitations and damage the device. By this reasoning, we see that a *theoretical* design with a very fast response due to extremely high gains must be checked to ensure that the *physical* signals in the loop (voltages, forces, etc.) are realistic and within acceptable boundaries. Another practical issue is that a high derivative gain K_D will amplify sudden changes in the feedback error (position error x_e, in this case) and produce large control signals that might damage the actuator. High-frequency noise is sometimes present in the feedback signal and hence a large D-gain will amplify the corrupted derivative information and produce an extremely large control signal.

All of the previous examples using PD or PID controllers have demonstrated good theoretical closed-loop performance. However, we should carefully reconsider the derivative control action when the reference input is a step function. Let us repeat Eq. (10.11), which presents the expression for the PID control signal

$$u(t) = K_P e(t) + K_I \int e(t)dt + K_D \dot{e}(t) \tag{10.50}$$

where $e(t)$ is the error between the reference input $r(t)$ and feedback measurement of the system output $y(t)$ (see Fig. 10.13). Note that if the reference input $r(t)$ is a step function, then the derivative $\dot{e}(t)$ will theoretically be infinity at time $t = 0$ when input $r(t)$ instantaneously steps up from zero to a constant value. This (theoretically) infinite derivative signal will be gained by K_D to produce an infinite control signal $u(t)$. Consequently, a PID or PD controller with a step command will produce and send an *infinite* (impulsive) control input to the plant. Therefore, PD and PID controllers are rarely implemented on real-world control systems as shown by the block diagram in Fig. 10.13. One possible solution is to measure and feed back the derivative of the output ($\dot{y}$) so that the derivative component of the control signal in Eq. (10.50) is $-K_D \dot{y}$ instead of $K_D \dot{e} = K_D (\dot{r} - \dot{y})$. This technique is called *rate feedback* and eliminates the discontinuous step-input derivative $\dot{r}(t)$ at time $t = 0$ (of course, $\dot{r}(t) = 0$ for $t > 0$ because the input is a step function).

Another method for circumventing the troublesome infinite error derivative and subsequent impulsive control spike is to insert a low-pass filter into the control loop. Figure 10.45 shows a feedback control system with a unity-gain low-pass filter inserted before the PID controller $G_C(s)$. In Fig. 10.45, the low-pass filter is

$$G_{\text{LP}}(s) = \frac{15}{s + 15} \tag{10.51}$$

which has a unity DC gain and a corner (or break) frequency of 15 rad/s. Adding the unity-gain low-pass filter will remove any high-frequency components in the error signal $e(t)$ (in this case, frequencies higher than 15 rad/s or 2.4 Hz), which includes removing the infinite derivative caused by the discontinuous step change at $t = 0$. It is important that the low-pass filter has a DC gain of one so that the steady-state amplitude of the filtered error $e_F(t)$ is equal to the steady-state error $e(t)$ (in other words, no additional gain is introduced into the forward loop). Of course, the corner frequency of the low-pass filter can be set to any desired value.

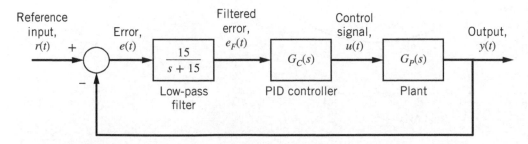

Figure 10.45 PID controller with low-pass filter 15/(s + 15).

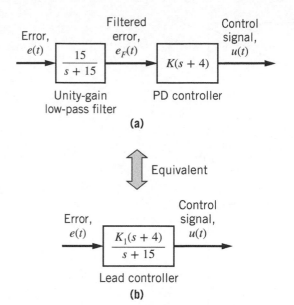

Figure 10.46 Equivalent controllers: (a) low-pass filter + PD controller and (b) lead controller.

Suppose we choose to implement a PD controller with an open-loop zero at $s = -4$ that is preceded by a low-pass filter with corner frequency $\omega_c = 15$ rad/s. Figure 10.46a shows this controller structure with the unity-gain low-pass filter $G_{LP}(s)$ and PD controller $G_C(s)$ in series. The product of the low-pass filter and PD controller is

$$G_{LP}(s)G_C(s) = \frac{15}{s+15}K(s+4) \tag{10.52}$$

We can factor the low-pass filter's numerator gain ($=15$) into the arbitrary control gain $K_1 = 15K$ to yield

$$G_{LF}(s) = \frac{K_1(s+4)}{s+15} \tag{10.53}$$

The controller presented by Eq. (10.53) and shown in Fig. 10.46b is called a *lead controller* or *lead filter*. The name arises from the fact that this transfer function adds phase lead at low frequencies. Figure 10.47 shows the Bode diagram for the following PD and lead controllers:

$$\text{PD controller:} \quad 0.25(s+4)$$

$$\text{Lead controller:} \quad \frac{3.75(s+4)}{s+15}$$

The control gains K and K_1 for the PD and lead controllers have been selected so that both controllers have unity DC gains and hence they both exhibit a low-frequency magnitude of 0 dB, as shown in Fig. 10.47. Because both controllers have a zero at $s = -4$, both phase plots exhibit an increase in phase ("phase lead") at frequencies near 4 rad/s. The lead controller's pole at $s = -15$ contributes phase lag as frequency approaches 15 rad/s and hence the lead controller's zero/pole pair results in a net zero phase change. The corner frequency of the numerator of the lead controller is always less than the corner frequency of the denominator so that the filter introduces phase lead at low frequencies. The PD controller, on the other hand, consists of a zero only and hence the phase continually increases from zero to $+90°$ at high frequencies. Note that the zero/pole pair of the lead controller causes the high-frequency magnitude to level off at $20\log_{10}(3.75) = 11.48$ dB. Because the PD controller's sinusoidal transfer function is $G_C(j\omega) = 0.25(j\omega + 4)$, its magnitude continues to increase unbounded with frequency, as shown in Fig. 10.47. The magnitude plot in Fig. 10.47 shows that a pure PD controller has the undesirable trait of greatly amplifying a feedback signal that contains high-frequency noise.

In summary, the lead controller demonstrates the desirable anticipatory characteristic of PD control as demonstrated by the positive phase lead shown in Fig. 10.47. Hence, a lead controller will add damping and reduce settling

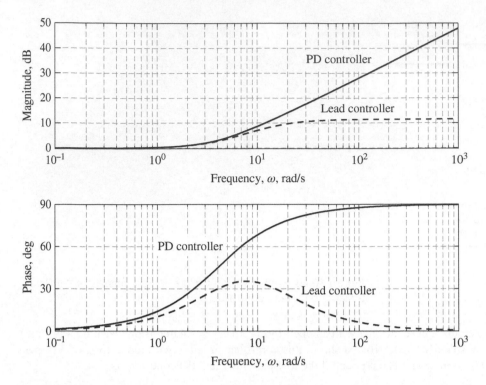

Figure 10.47 Bode diagrams of PD controller transfer function $0.25(s + 4)$ and lead-controller transfer function $3.75(s + 4)/(s + 15)$.

time just like a PD controller. However, unlike PD control, the lead controller does not amplify high-frequency signals. For these reasons, a lead controller (zero/pole pair) is often used instead of a pure PD controller.

Example 10.13

Consider again the closed-loop position-control system presented in Fig. 10.48 (Examples 10.6, 10.8, and 10.12). Use the root locus to design a lead controller $G_{LF}(s)$ that provides a fast, well-damped closed-loop transient response to a reference step position command $x_{ref}(t) = 0.1$ m. Compare the lead-controller design to the PD controller result from Example 10.12. The actuator gain is $K_A = 2$ N/V.

Let us use the following lead controller with a zero at $s = -3$ and pole at $s = -12$

$$G_{LF}(s) = \frac{K(s + 3)}{s + 12}$$

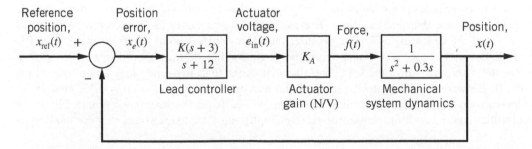

Figure 10.48 Closed-loop position control of a mechanical system (Example 10.13).

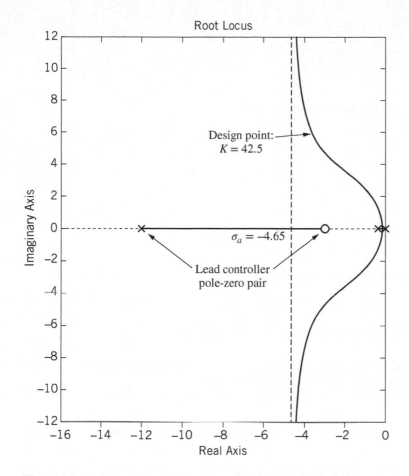

Figure 10.49 Root-locus plot for system with lead controller $G_{LF}(s) = (s + 3)/(s + 12)$ (Example 10.13).

The reader should remember that we have the freedom to select the zero and pole location of the lead controller. Combining the lead controller and mechanical system plant (with actuator gain $K_A = 2 \, \text{N/V}$), the open-loop transfer function is

$$G(s)H(s) = \frac{2(s + 3)}{(s + 12)(s^2 + 0.3s)}$$

Hence, the open-loop transfer function has three poles at $s = 0$, $s = -0.3$, and $s = -12$ and one zero at $s = -3$. The following MATLAB commands will create the root-locus plot presented in Fig. 10.49:

```
>> sysGc = tf([1 3],[1 12])          % create lead controller transfer function
>> sysGp = tf(2,[1 0.3 0])           % create plant transfer function
>> sysGH = sysGc*sysGp               % create open-loop transfer function
>> rlocus(sysGH)                     % create and draw the root locus
```

Because the lead controller has an open-loop pole at $s = -12$, two vertical asymptotes exist at $\sigma_a = -4.65$. If we compare the root-locus plots for the mechanical system with the lead controller (Fig. 10.49) to the system with the PD controller (Fig. 10.42), we see that the PD controller provides more damping because the semicircular root locus (Fig. 10.42) eventually intersects the negative real axis. A good lead-controller design point is indicated on Fig. 10.49 with gain $K = 42.5$. This gain setting results in the closed-loop poles $s_1 = -4.863$ and $s_{2,3} = -3.718 \pm j6.214$ for a closed-loop damping ratio of $\zeta = 0.514$.

Figure 10.50 shows the closed-loop step responses using both the PD and lead controllers. The closed-loop system using the PD controller (Example 10.12) exhibits better damping and a shorter transient response compared to the closed-loop

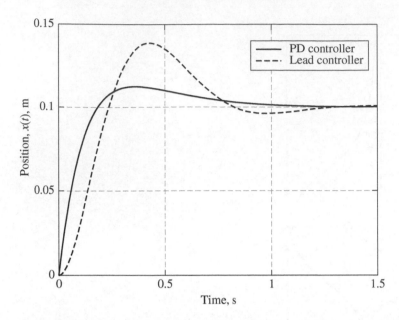

Figure 10.50 Closed-loop step response using lead and PD controllers (Example 10.13).

system using the lead controller. However, the lead controller has removed high-frequency feedback signals. Unlike the PD controller, the lead controller does not produce an impulsive force because of the step input.

On the basis of the previous discussion and Examples 10.12 and 10.13, we can make the following summary statements regarding PD and lead controllers:

1. Implementing a PD controller will result in an impulsive control signal if the reference input is a step function because the discontinuous step input produces an infinite derivative in the error signal at $t = 0$.

2. PD controllers must be used carefully because they greatly amplify the high-frequency component of a feedback signal (see the corresponding magnitude Bode diagram, Fig. 10.47). If noise is present in the feedback signal, it will be amplified by the PD controller.

3. Adding a low-pass filter before a PD controller creates a lead controller, which consists of a zero/pole pair and alleviates the high-frequency (noise) problem.

4. A lead controller is an approximation of the PD controller. Both controllers improve damping and reduce the settling time.

Example 10.13 shows that the lead controller does not add as much damping as the PD controller. The closed-loop effect of the lead controller's zero is magnified if the lead controller's pole is placed farther to the left. In fact, if the pole of the lead controller is pushed too far left, then the low-pass filtering benefit is lost and the lead controller essentially becomes a PD controller (this effect is demonstrated in an end-of-chapter problem). On the other hand, if the lead controller's zero and pole are too close together, they cancel each other and the lead controller has a diminished effect on the closed-loop response. A good rule-of-thumb is to make the lead controller's denominator corner frequency 3–5 times greater than the numerator corner frequency. For example, the lead controller in Example 10.13

$$G_{LF}(s) = \frac{K(s+3)}{s+12}$$

provides a good trade-off between low-pass filtering and an active zero for damping. However, the lead controller

$$G_{LF}(s) = \frac{K(s+3)}{s+4}$$

will *not* provide good damping because the zero at $s = -3$ is nearly cancelled by the pole at $s = -4$. Finally, the lead controller

$$G_{\mathrm{LF}}(s) = \frac{K(s+3)}{s+100}$$

will essentially act similar to a PD controller with a zero at $s = -3$ because the denominator corner frequency of $100 \, \mathrm{rad/s}$ is much greater than the numerator corner frequency of $3 \, \mathrm{rad/s}$. This lead filter does not provide low-pass filtering.

10.7   STABILITY MARGINS

One of the general performance requirements of a good control system is a measure of its closed-loop stability. Clearly, a good control system design must provide a stable closed-loop operation. However, it is also useful to quantify *relative stability*, that is, how "close" the current design point is from an unstable control system configuration. By current design point, we mean the controller selection (P, PD, lead filter, etc.) and the associated gains. The previous section has shown that the root locus tells us whether increasing a control gain will eventually lead to instability. For example, the root locus of Fig. 10.36 shows that the closed-loop system discussed in Example 10.10 will eventually go unstable as the P-gain is increased. However, the root-locus plot does not explicitly provide the gain information and therefore we cannot easily quantify the gain increase that drives a closed-loop system unstable.

It turns out that our other graphical analysis tool, the Bode diagram, can be used to quantify the so-called *stability margins* between a current design point and an unstable design. The *Nyquist stability criterion* is the fundamental method for analyzing relative stability in the frequency domain. However, we present a somewhat more intuitive approach for investigating stability margins using the frequency-response information provided by the Bode diagram. The interested reader can consult References 2–5 for details regarding the Nyquist criterion.

Figure 10.51 shows a general closed-loop control system where the forward transfer function $G(s)$ contains both the controller dynamics (such as PD, PID, lead filter, etc.) and the plant system dynamics. A single control gain K is factored out of $G(s)$. This gain might be any one of the PID gains or the gain of a lead controller. Consider the case where the reference input $r(t)$ is a sinusoidal function with an amplitude of two and frequency ω

$$\text{Reference input:} \quad r(t) = 2 \sin \omega t$$

This sinusoidal input is shown in Fig. 10.51. We know from the frequency-response analysis presented in Chapter 9 that the feedback signal $b(t)$ in Fig. 10.51 will also be a sine wave that depends on the magnitude and phase angle of the sinusoidal transfer function $KG(j\omega)H(j\omega)$. Suppose the magnitude of the open-loop sinusoidal transfer function $KG(j\omega)H(j\omega)$ is unity and the phase angle is $-180°$. For this special case, the feedback signal $b(t)$ will be

$$\text{Feedback signal:} \quad b(t) = 2 \sin(\omega t - \pi)$$

This sinusoidal feedback signal is shown in Fig. 10.51. Note that feedback $b(t)$ is the "mirror" opposite of the reference input $r(t)$; that is, $b(t)$ has the same magnitude as $r(t)$ but is $180°$ out of phase. Clearly, if this feedback scenario exists, then the error signal $e(t) = r(t) - b(t)$ will be a *doubling* of the reference signal $r(t)$. Subsequent feedback of the doubled signal will eventually produce a sinusoidal error signal with infinite amplitude. Therefore, the scenario depicted in Fig. 10.51 is unstable.

The critical condition that leads to the unstable frequency response shown in Fig. 10.51 can now be summarized. The feedback signal $b(t)$ is the "mirror opposite" of the reference signal $r(t)$ only if the magnitude $|KG(j\omega)H(j\omega)| = 1$ and the phase angle $\angle KG(j\omega)H(j\omega) = -180°$. This critical condition can be easily read from the Bode diagram of the *open-loop* transfer function. Recall that the magnitude plot of the Bode diagram is in decibels and hence the unity magnitude condition is $0 \, \mathrm{dB}$.

Let us illustrate the critical unstable condition with a simple example. Suppose we use a simple proportional controller $G_C(s) = K_P$ with the following third-order plant

$$\text{Plant:} \quad G(s) = \frac{1}{s(s+2)(s+3)} \tag{10.54}$$

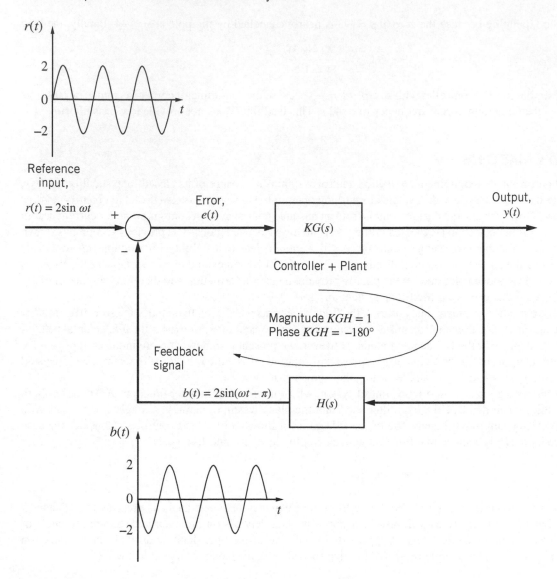

Figure 10.51 Unstable closed-loop feedback for a sinusoidal reference input.

For simplicity, we will assume unity feedback, $H(s) = 1$. Therefore, the closed-loop transfer function using P-control is

$$T(s) = \frac{K_P G(s)}{1 + K_P G(s)} = \frac{K_P}{s(s + 2)(s + 3) + K_P} \tag{10.55}$$

Using Eq. (10.55), we see that the characteristic equation is $1 + K_P G(s) = 0$, or $s^3 + 5s^2 + 6s + K_P = 0$. If the P-gain is $K_P = 30$, the characteristic equation is

$$K_P = 30: \quad s^3 + 5s^2 + 6s + 30 = 0$$

The corresponding closed-loop roots are $r_1 = -5$ and $r_{2,3} = \pm j2.45$. Because two roots are on the imaginary axis, the closed-loop system is marginally stable for $K_P = 30$. For gain $K_P > 30$, the closed-loop system is unstable. Figure 10.52 shows the Bode diagram of the open-loop transfer function $K_P G(s)H(s)$ with gain $K_P = 30$. The critical condition for marginal stability can be seen on the Bode diagram: when the control gain is $K_P = 30$, the critical condition of unity magnitude (0 dB) and $-180°$ phase occurs simultaneously at the frequency $\omega = 2.45$ rad/s. Hence, the Bode diagram in

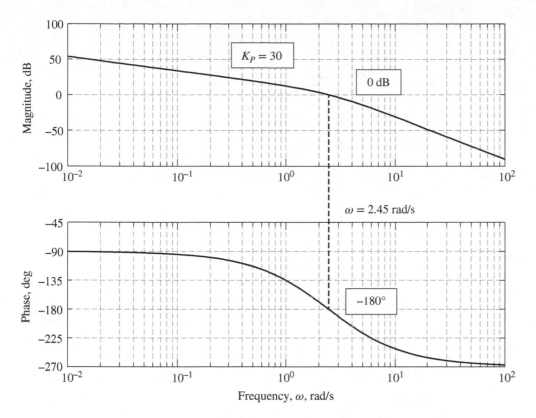

Figure 10.52 Bode diagram of open-loop transfer function $30/(s^3 + 5s^2 + 6s)$ showing a marginally stable closed-loop system with $K_P = 30$.

Fig. 10.52 shows a *marginally stable* system that oscillates at a frequency of 2.45 rad/s. This frequency matches the two marginally stable roots that lie on the imaginary axis, $r_{2,3} = \pm j2.45$, when the P-gain is $K_P = 30$.

Figure 10.53 shows the Bode diagram of $G(s)$ with P-gain $K_P = 2$. The Bode diagram shows that the closed-loop system is stable because the 0-dB and $-180°$ phase condition *do not* occur simultaneously at a common frequency. If we compare Bode diagrams in Figs. 10.52 and 10.53, we see that the magnitude plot in Fig. 10.53 has been shifted down because the gain K_P has decreased from 30 to 2 (the phase plot does not change with gain). We now define the first of the two important relative stability margins: the *gain margin* is the maximum factor by which the current gain setting can be multiplied by until the closed-loop system is driven unstable. For the system described by Eqs. (10.54) and (10.55) with a P-controller, we have shown that the closed-loop system becomes marginally stable for $K_P = 30$. Therefore, if the current gain is $K_P = 2$, then the gain margin is 15. We can easily read the gain margin from the Bode diagram using the following steps. First, we locate the "phase-crossing frequency" ω_{gm} where the phase angle is $-180°$ (the critical point of the phase plot). Next, we project up to the magnitude plot along frequency ω_{gm} and read the associated magnitude. This magnitude must be less than unity (i.e., negative decibel value) for a stable closed-loop system. The gain margin (GM_{dB}, in decibels) is shown in Fig. 10.53 and it is the difference between the magnitude at the phase-crossing frequency ω_{gm} and the 0-dB critical point. Because gain margin is defined as a multiplicative factor, we must convert the gain margin in decibels to an absolute value

$$\text{Gain margin} = 10^{GM_{dB}/20} \tag{10.56}$$

where GM_{dB} is the gain margin in decibels as determined from the Bode diagram (see Fig. 10.53). For our illustrative example, we estimate the gain margin (in decibels) from Fig. 10.53 to be $GM_{dB} = 23.5$ dB. Using Eq. (10.56), we find that the gain margin is 14.96 (we know that the exact gain margin is 15). In summary, ω_{gm} is the frequency where the gain margin is measured and it is easily identified from the $-180°$ phase crossing on the phase plot of the Bode diagram (see Fig. 10.53).

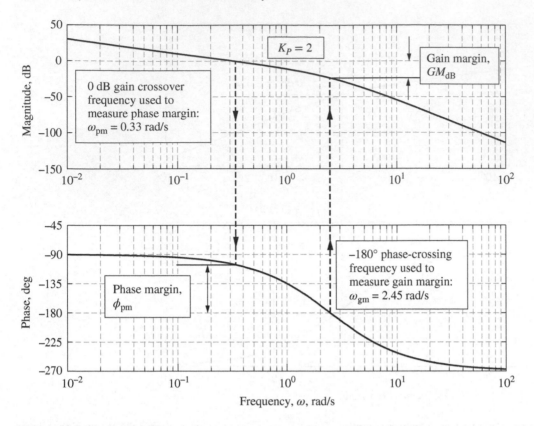

Figure 10.53 Bode diagram of open-loop transfer function $2/(s^3 + 5s^2 + 6s)$ showing gain and phase margins with $K_P = 2$.

The second important relative stability margin is the *phase margin* and it is defined as the maximum amount of phase lag that can be added to a system before driving it unstable. Similar to gain margin, we can easily read the phase margin from the Bode diagram using the following steps. First, we locate the "unity-gain crossover frequency" ω_{pm}, where the magnitude is 0 dB (the critical point of the magnitude plot). Next, we project down to the phase plot along frequency ω_{pm} and read the associated phase angle. This phase must be greater than $-180°$ for a stable closed-loop system. The phase margin ϕ_{pm} (in degrees) is shown in Fig. 10.53 and it is the difference between the phase at the unity-gain frequency and the critical phase angle of $-180°$. Using Fig. 10.53 we can estimate the 0-dB gain crossover frequency to be $\omega_{pm} = 0.33$ rad/s and the associated phase angle to be approximately $\phi = -105°$. Hence, the phase margin is $\phi_{pm} = 180° - 105° = 75°$, as shown in Fig. 10.53. Finally, it should be noted that if the P-gain was increased by a factor of 15 from $K_P = 2$ to $K_P = 30$, then the magnitude plot in Fig. 10.53 would be shifted up $20\log_{10}(15) = 23.5$ dB and consequently the unity-gain crossover frequency ω_{pm} would become equal to the phase-crossing frequency ω_{gm}. The resulting Bode diagram would match Fig. 10.52 and the closed-loop system would be marginally stable.

Phase margin is an indication of damping in the system. It has been shown that phase margin exhibits an approximate correlation with the damping ratio of the dominant second-order poles:

$$\zeta \cong \frac{\phi_{pm}}{100} \tag{10.57}$$

where phase margin ϕ_{pm} is expressed in degrees. Hence, a phase margin of 60° corresponds approximately to a closed-loop damping ratio $\zeta = 0.6$. Of course, if the closed-loop system exhibits dominant second-order roots, then we can relate the damping ratio (and undamped natural frequency ω_n) to time-response performance criteria such as maximum overshoot and peak time (see Table 7.4).

Gain margin indicates how much additional gain is acceptable before driving the system unstable. A good rule-of-thumb is that a gain margin should be at least a factor of 2 (or 6 dB) in order to accommodate uncertainties in the system model.

Gain and Phase Margins Using MATLAB

The MATLAB command `margin` will compute the gain and phase margins and the corresponding crossover frequencies of an LTI system. Let us demonstrate its use with the previous example where Eq. (10.54) presents the open-loop transfer function

```
>> sysG = tf(1,[1 5 6 0])          % create G(s) = 1/(s³ + 5s² + 6s)
>> Kp = 2;                          % P-gain setting
>> [Gm,Pm,Wgm,Wpm] = margin(Kp*sysG)  % compute gain and phase margins
```

The input to the command `margin` is the system's open-loop transfer function $KG(s)H(s)$ and the computed output values are gain margin `Gm` (as a multiplicative factor), phase margin `Pm` (in degrees), phase-crossing frequency for measuring the gain margin `Wgm` (in rad/s), and unity-gain crossover frequency for measuring the phase margin `Wpm` (in rad/s). Executing these commands yields

```
Gm = 15.0000
Pm = 74.4923  (degrees)
Wgm = 2.4495  (rad/s)
Wpm = 0.3270  (rad/s)
```

which is the same result obtained from the Bode diagram in Fig. 10.53. If we want to compute the gain margin in decibels, we must add the command

```
>> Gm_dB = 20*log10(Gm)
```

Another way to use the command `margin` is to execute it without the left-hand side arguments

```
>> margin(Kp*sysG)          % plot Bode diagram and display gain and phase margins
```

which creates the Bode diagram of LTI system `sysG` and displays the gain and phase margins on the magnitude and phase plots in a similar manner as in Fig. 10.53. Figure 10.54 shows the result of the `margin` command. Numerical values for gain and phase margins and their respective crossover frequencies are automatically displayed in the figure title.

Controller Design in the Frequency Domain

We can use the frequency domain (i.e., Bode diagrams) to design control schemes and References 2–5 each devote an entire chapter to control system design in the frequency domain. Because this textbook provides an introduction to feedback control systems, we will not pursue the details of control design in the frequency domain. However, we use our previous discussion of gain and phase margins to explain one concept behind control design using Bode diagrams. We previously noted that increasing the phase margin ϕ_{pm} will add damping to the closed-loop system. Figure 10.53 shows that phase margin is computed using the 0-dB crossover frequency. If the phase margin is insufficient (say, ϕ_{pm} = 15° or damping ratio $\zeta \cong 0.15$), then it may be possible to increase the phase margin by adding a positive phase angle near the unity-gain crossover frequency ω_{pm} shown in Fig. 10.53. Adding a *lead controller* (or lead filter) to the forward path will add phase angle. For example, consider again the lead controller discussed in the previous section

$$\text{Lead controller:} \quad \frac{K(s+4)}{s+15}$$

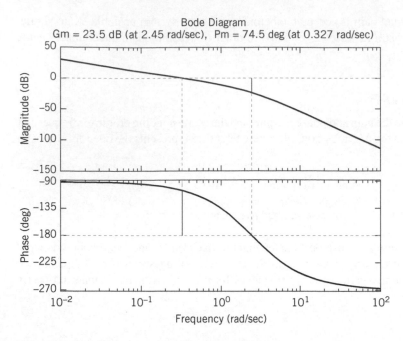

Figure 10.54 Bode diagram created by MATLAB's `margin` command for the open-loop transfer function $2/(s^3 + 5s^2 + 6s)$ showing gain and phase margins.

This controller has a zero at $s = -4$ and pole at $s = -15$. Figure 10.47 presents the Bode diagram of the lead controller, which exhibits a significant phase lead contribution ($+30°$) for the frequency range $4 < \omega < 15$ rad/s. This frequency range for additional phase lead roughly corresponds to the zero/pole locations for the lead controller. Hence, if we want to add damping (or, phase margin) to a system, we can introduce a lead controller with a zero placed near the unity-gain crossover frequency ω_{pm} of the uncompensated system. The following example illustrates controller design using the Bode diagram.

Example 10.14

Consider a unity-feedback control system with the plant transfer function

$$\text{Plant:} \quad G_P(s) = \frac{4}{s(s+1)(s+2)}$$

Check the stability margins of the "uncompensated" system (plant only with P-gain). If the damping is poor, improve it by adding a dynamic controller to the forward path.

Figure 10.55a shows the Bode diagram of the plant transfer function $G_P(s)$, which we consider the "uncompensated system" with P-gain $K_P = 1$. We see that both the gain and phase margins are poor: $GM_{dB} = 3.52$ dB and $\phi_{pm} = 11.4°$ (damping ratio $\zeta \cong 0.11$). The two crossover frequencies are close together, which indicates a closed-loop system with small stability margins and hence the P-gain cannot be increased. Decreasing the gain K_P will improve gain and phase margins but it will also slow down the response and degrade steady-state tracking errors. One remedy for the low phase margin (i.e., low damping) is to add a *lead controller* with a zero near the unity-gain crossover frequency of $\omega_{pm} = 1.14$ rad/s. We choose the following lead controller:

$$\text{Lead controller:} \quad G_C(s) = \frac{K(s+0.9)}{s+5}$$

We select the controller gain $K = 3.5$ so that its DC gain is close to unity. This gain selection ensures that the low-frequency gain of $G_C(s)G_P(s)$ is nearly equal to the low-frequency gain of the plant $G_P(s)$ and therefore the steady-state tracking

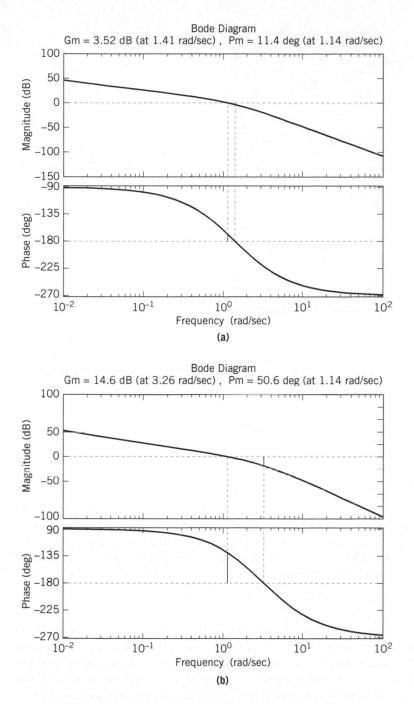

Figure 10.55 Bode diagram and stability margins for Example 10.14: (a) uncompensated system $G_P(s)$ and (b) compensated system with lead controller $G_C(s)$ and plant $G_P(s)$.

accuracy is not compromised. Figure 10.55b shows the Bode diagram of the compensated plant (with lead controller added). The gain margin has increased to 14.6 dB and the phase margin has increased to 50.6° (damping ratio $\zeta \cong 0.5$). If we compare the phase plots in Fig. 10.55, we see that the lead controller has added phase angle near $\omega = 1$ rad/s as designed, which has subsequently increased the phase margin. Figure 10.56 shows both Bode plots from Fig. 10.55 on the same diagram. The

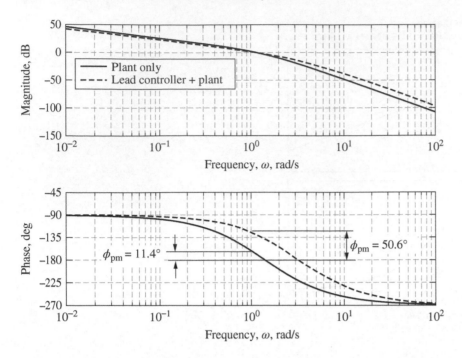

Figure 10.56 Bode diagram and phase margins for Example 10.14: uncompensated system $G_P(s)$ and compensated system with lead controller $G_C(s)$ and plant $G_P(s)$.

significant increase in phase due to the lead controller is apparent. Consequently, the phase margin (and damping) has been dramatically increased by introducing the lead controller.

10.8 IMPLEMENTING CONTROL SYSTEMS

We end this chapter with a brief discussion on some of the practical issues associated with implementing feedback controllers in physical systems. Much of the analysis in the previous sections has involved a barrage of controller transfer functions, which may lead the reader to wonder how "controller design" via root locus or Bode diagrams relates to the actualization of a control scheme in a physical system. This brief section attempts to explain how particular controllers and feedback structures are implemented as a combination of hardware and software.

Digital Control Systems

Before the advent of microprocessors, control systems were implemented using analog electrical circuits. For example, various filters and controllers (such as the lead controller) can be constructed from circuits using resistors and capacitors (see the end-of-chapter problems in Chapter 3). Electromechanical sensors are used to convert mechanical motion into electrical voltage signals, which are fed back and compared to a reference voltage signal that is proportional to the desired position or velocity. This voltage error is the input to an electrical network that represents the controller design. Hence, in the "old days" before digital microprocessors, the controller transfer functions (such as PI, PID, and lead controllers) designed by root-locus or Bode-diagram methods where implemented by analog RC circuits and op amps. The output voltage from the analog controller circuit would drive an actuator (such as a DC motor or solenoid), which in turn would ultimately drive the physical plant.

With the arrival of microprocessors, controller and filter designs are now implemented as digital algorithms (software). Figure 10.57 shows a simplified block diagram of a digital control system. The reader should note that the digital controller resides as programmable software inside the computer. For example, if we design a PID controller, it is ultimately implemented in the computer by a software algorithm that computes numerical values for the proportional, integral, and derivative terms. The reference input and measured plant output are the inputs to the digital computer, and therefore both signals must exist as digital signals. However, the measurement of the physical plant output is an analog

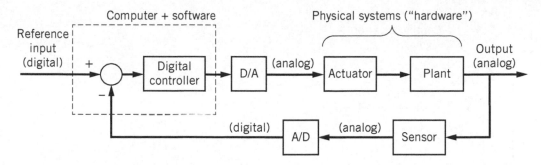

Figure 10.57 General structure of a digital control system.

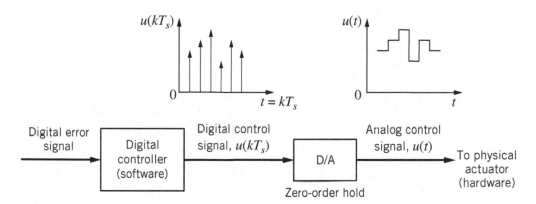

Figure 10.58 D/A converter using a zero-order hold.

or continuous-time signal that must be converted to a digital signal by the analog-to-digital (A/D) converter. The A/D process involves sampling the analog signal (i.e., making discrete measurements at regularly spaced intervals in time) and quantizing the discrete measurements to produce a digital signal with a finite level of numerical precision. After the digital signals are processed by the digital controller algorithm, the control signal must be converted from a digital signal to an analog signal to be used by the physical actuator. For example, an actuator such as a DC motor or a solenoid would require a continuous-time input voltage. A digital-to-analog (D/A) converter creates the continuous-time signal required by the actuator. One common D/A method is the *zero-order hold*, where a continuous signal is created by "holding" the digital value constant during the sample interval and then changing (and holding) the output amplitude at the next sample interval when a new digital signal is computed. Hence, if the digital controller produces a "train" of impulses with finite amplitudes, as shown in Fig. 10.58, then the output of the zero-order hold is a continuous-time "stair-step" signal. Note that the digital signal $u(kT_s)$ in Fig. 10.58 only exists at the sample time $t = kT_s$, where k is the sample index. It should be stressed that the selection of the appropriate sample period T_s of the digital system depends on the response speed (or time constants) associated with the physical systems.

Digital Controller Algorithms

In the previous subsection, we described the digital controller as an algorithm that resides in the computer or microprocessor. The reader might ask how a PID or lead controller transfer function is converted into an algorithm that exists as software. As a first example, consider the basic PI control logic

$$u(t) = K_P e(t) + K_I \int e(t)dt \tag{10.58}$$

Equation (10.58) involves two continuous-time (analog) signals: error $e(t)$ and control signal $u(t)$. Therefore, we must obtain a discrete-time representation of Eq. (10.58) for use in a digital PI algorithm. To begin, let us assume that the

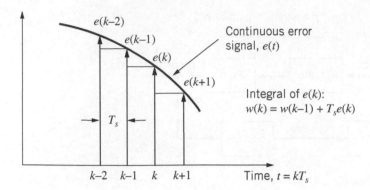

Figure 10.59 Numerical integration using the rectangle rule.

feedback error is a digital signal (note that the output of the summing junction in Fig. 10.57 is a digital signal). The digital error signal can be denoted as $e(kT_s)$, where k is the sample index and T_s is the sample period (or step size) in seconds. Hence, the digital error signal only exists at times $t = kT_s$, $k = 1, 2, \ldots$, etc. For convenience, let us omit the sample period T_s in the notation for the digital signal with the understanding that $e(k) = e(kT_s)$; therefore, $e(k-1)$ is the digital error signal one sample period before $e(k)$. Using this notation, a *digital* representation of the PI algorithm (10.58) is

$$u(k) = K_P e(k) + K_I w(k) \tag{10.59}$$

where the digital signal $w(k)$ is the numerical integral of the digital signal $e(k)$ as computed by rectangular-rule integration:

$$w(k) = w(k-1) + T_s e(k) \tag{10.60}$$

Clearly, the product $T_s e(k)$ is the rectangular area associated with the time step T_s, as shown by Fig. 10.59. Of course, when the algorithm is initiated and processes the first sample ($k = 1$), the initial value of the integral is $w(0) = 0$. Equations (10.59) and (10.60) are the discrete-time equations that represent the PI control logic (10.58). Implementing the PI controller in the digital computer requires two lines of computer code: Eq. (10.60) computes the numerical integral of the digital error signal, and Eq. (10.59) computes the digital control signal $u(k)$ using the PI gains K_P and K_I. Remember that the digital control signal $u(k)$ must be converted into a continuous (analog) signal $u(t)$ for use by the physical actuator and this D/A conversion is typically performed by a zero-order hold as shown in Fig. 10.58.

As a second example, consider the PID control logic

$$u(t) = K_P e(t) + K_I \int e(t) dt + K_D \dot{e}(t) \tag{10.61}$$

Now we must develop a discrete-time equation for the *derivative* of the error signal $\dot{e}(t)$. The simplest technique is to use the backward finite-difference equation

$$v(k) = \frac{e(k) - e(k-1)}{T_s} \tag{10.62}$$

where digital signal $v(k)$ is the numerical derivative of the error signal. Using Eqs. (10.62) and (10.60), the digital PID control algorithm is

$$u(k) = K_P e(k) + K_I w(k) + K_D v(k) \tag{10.63}$$

Implementing the PID controller requires three lines of computer code: Eq. (10.60) computes the numerical integral of the digital error $e(k)$, Eq. (10.62) computes the numerical derivative of $e(k)$, and Eq. (10.63) computes the digital control signal $u(k)$ using the PID gains. This particular PID algorithm is called a *first-order recursive method* because it requires the computer memory to only store the integral signal $w(k-1)$ and error signal $e(k-1)$ from the previous sample time.

Computing the derivative using the finite-difference equation (10.62) may produce very poor results if the signal $e(k)$ is corrupted by noise. This issue was briefly discussed in a previous section, and the simplest remedy is to replace the PD controller with a lead controller. For example, consider the continuous-time PD controller with gains $K_P = 12$ and $K_D = 4$:

$$\text{PD control:} \quad u(t) = 12e(t) + 4\dot{e}(t) \tag{10.64}$$

The PD controller transfer function is

$$\text{PD controller:} \quad G_{\text{PD}}(s) = 12 + 4s = 4(s+3) = \frac{U(s)}{E(s)} \tag{10.65}$$

This PD controller has a low-frequency (DC) gain of 12. We can replace the pure differentiation by inserting a low-pass filter in series with the original PD controller to yield a lead controller. For example, let us add a low-pass filter with a corner frequency of 20 rad/s

$$\text{Lead controller:} \quad G_{\text{LF}}(s) = \frac{80(s+3)}{s+20} = \frac{U(s)}{E(s)} \tag{10.66}$$

The lead controller transfer function (10.66) is a good approximation to the original PD controller (10.65) as it maintains the same zero location at $s = -3$ and the same DC gain of 12. As with the previous PID controller example, we need to determine a discrete-time equation for the lead controller in order to implement the control algorithm as software. The details of converting the continuous-time transfer function to a discrete-time transfer function are beyond the scope of this textbook (the interested reader can consult Reference 7 for details). However, the MATLAB command c2d will perform the conversion given the continuous-time system and sample period T_s. Using $T_s = 0.01$ s (or, sampling frequency of 100 Hz), the continuous-time lead controller (10.66) is converted to the following discrete-time algorithm:

$$u(k) = 0.8187u(k-1) + 80e(k) - 77.82e(k-1) \tag{10.67}$$

Consequently, the continuous-time lead controller (10.66) can be represented in a digital computer using one line of code: the discrete-time equation (10.67). It is important to remember that this digital lead controller only holds for a digital system with a sample period $T_s = 0.01$ s. If the sample period changes, then the discrete-time equation must be recomputed.

This section has presented a very brief introduction to some of the practical issues associated with implementing feedback control systems with a digital computer. One important point is that it is often possible to design the controller (PI, PID, lead filter, etc.) by assuming a purely continuous-time control system. That is, all system equations are ordinary differential equations and hence all transfer functions are in the Laplace (or "s") domain. After a satisfactory controller is designed in the continuous-time domain, it must be converted into a discrete-time (digital) control algorithm and this step requires knowledge of the sampling time T_s. Implementing the digital control algorithm as computer code emphasizes the principal advantage of digital control systems: changing the controller gains or controller structure is relatively easy as these tasks only require software changes.

SUMMARY

This chapter has presented an introduction to feedback control systems. First, we developed the closed-loop transfer function that allows us to calculate the closed-loop roots (or poles). Consequently, we can use the system-analysis methods developed in Chapter 7 to determine important *closed-loop* system response characteristics such as response time, damping, and steady-state tracking. Next, we investigated the use of the three-term PID control scheme. In summary, the proportional (P) control signal reduces the response time, the integral (I) signal reduces the steady-state tracking error, and the derivative (D) signal adds damping to the system. Pure differentiators, however, have two drawbacks: they are difficult to implement in practice and they amplify noise in the feedback signal. For these reasons, the PD control path is typically replaced by a lead controller. The root-locus method is a graphical technique for visualizing the effect that varying one control parameter (usually a gain) has on the locations of the closed-loop poles (or roots). We demonstrated that adding a zero to the control loop (such as a PD or lead controller) "bends" the root locus to the left and away from the imaginary axis, which improves the response time and system damping. We also presented the gain and phase margins, which are quantitative measures of how far the current closed-loop system design is from the point of instability. These stability margins can be read directly from the Bode diagram of the open-loop transfer function. Because phase margin is related to system damping, we can use the Bode diagram to design controllers that improve closed-loop damping characteristics. Finally, we briefly discussed how controllers are implemented as discrete-time algorithms in a digital computer.

REFERENCES

1. Ziegler, J.G., and Nichols, N.B., "Optimum Settings for Automatic Controllers," *Transactions of the ASME*, Vol. 64, 1942, pp. 759–768.
2. Phillips, C.L., and Parr, J.M., *Feedback Control Systems*, 5th ed., Prentice Hall, Upper Saddle River, NJ, 2011, pp. 209–218, 232–251, 328–347.
3. Franklin, G.F., Powell, J.D., and Emami-Naeini, A., *Feedback Control of Dynamic Systems*, 4th ed., Prentice Hall, Upper Saddle River, NJ, 2002, pp. 158–166, 277–310, 390–402.
4. Dorf, R.C., and Bishop, R.H., *Modern Control Systems*, 12th ed., Prentice Hall, Upper Saddle River, NJ, 2011, pp. 391–399, 443–467, 642–653.
5. Ogata, K., *System Dynamics*, 4th ed., Pearson Prentice Hall, Upper Saddle River, NJ, 2004, pp. 539–561, 630–636.
6. Palm, W.J., *System Dynamics*, McGraw-Hill, New York, NY, 2005, pp. 711–724.
7. Phillips, C.L., and Nagle, H.T., *Digital Control System Analysis and Design*, 3rd ed., Prentice Hall, Englewood Cliffs, NJ, 1995, pp. 430–462.
8. Wait, K.W., and Goldfarb, M., "Enhanced Performance and Stability in Pneumatic Servosystems with Supplemental Mechanical Damping," *ASME Journal of Dynamic Systems, Measurement, and Control*, Vol. 132, July 2010, pp. 041012-1–8.

Chapter 11

Case Studies in Dynamic Systems and Control

11.1 INTRODUCTION

This final chapter encapsulates the fundamental topics associated with the modeling, simulation, and control of dynamic systems by presenting case studies in engineering. Most of the case studies presented here are inspired by research that has been presented at professional conferences or published in the archival journals. All case studies involve physical engineering systems and many of the complexities associated with real-world systems. Each case study begins with the development of the mathematical modeling equations, followed by analysis of the system response using numerical simulations and/or analytical methods. In most cases, each study concludes with a design section, where the performance impact of varying system parameters is investigated.

11.2 VIBRATION ISOLATION SYSTEM FOR A COMMERCIAL VEHICLE

In this section, we analyze a vibration isolation system that improves the ride quality for a commercial vehicle such as a tractor-trailer used for long-distance transportation of freight. Figure 11.1 shows a schematic diagram of the seat-suspension system we investigated in Chapters 2 and 5. Travel over a rough road causes vibrations that are transmitted to the vehicle's cabin floor, and the floor vibrations, $z_0(t)$, are transmitted to the seat mass m_1 and ultimately to the driver (mass m_2). A properly designed seat-suspension system will suppress the road vibrations transmitted to the driver [1]. The seat suspension consists of a passive shock absorber, modeled by ideal (linear) damper and stiffness elements b_1 and k_1. Recall that the damping and stiffness of the seat cushion are modeled by ideal elements b_2 and k_2. The goal of this section is to analyze the dynamic response of the seat-suspension system utilizing the tools we have developed in the preceding chapters.

Mathematical Model

As we have seen throughout this textbook, developing an appropriate mathematical model is always the first step in analysis and design of dynamic systems. We developed the mathematical modeling equations of the seat-suspension system in Chapter 2, which are

$$m_1\ddot{z}_1 + b_1\dot{z}_1 + b_2(\dot{z}_1 - \dot{z}_2) + k_1z_1 + k_2(z_1 - z_2) = b_1\dot{z}_0(t) + k_1z_0(t) \tag{11.1a}$$

$$m_2\ddot{z}_2 + b_2(\dot{z}_2 - \dot{z}_1) + k_2(z_2 - z_1) = 0 \tag{11.1b}$$

Because we assumed ideal damping and stiffness elements, the system is linear. A state-space representation (SSR) approach is likely best suited for system analysis as the two second-order linear ordinary differential equations (ODEs) are coupled, which presents a challenge to developing transfer functions. In Chapter 5, we derived the SSR, which is composed of the state equation

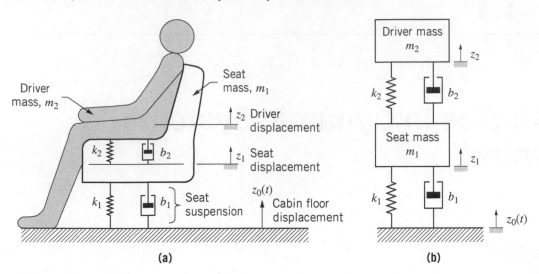

Figure 11.1 (a) Schematic diagram of the seat-suspension system and (b) mechanical model for the seat-suspension system.

$$\dot{\mathbf{x}} = \begin{bmatrix} 0 & 1 & 0 & 0 \\ -(k_1 + k_2)/m_1 & -(b_1 + b_2)/m_1 & k_2/m_1 & b_2/m_1 \\ 0 & 0 & 0 & 1 \\ k_2/m_2 & b_2/m_2 & -k_2/m_2 & -b_2/m_2 \end{bmatrix} \mathbf{x} + \begin{bmatrix} 0 & 0 \\ k_1/m_1 & b_1/m_1 \\ 0 & 0 \\ 0 & 0 \end{bmatrix} \mathbf{u} \qquad (11.2)$$

and the output equation

$$\mathbf{y} = \begin{bmatrix} 0 & 0 & 1 & 0 \\ k_2/m_2 & b_2/m_2 & -k_2/m_2 & -b_2/m_2 \end{bmatrix} \mathbf{x} + \begin{bmatrix} 0 & 0 \\ 0 & 0 \end{bmatrix} \mathbf{u} \qquad (11.3)$$

In our SSR, the state vector is $\mathbf{x} = \begin{bmatrix} z_1 & \dot{z}_1 & z_2 & \dot{z}_2 \end{bmatrix}^T$, the input variables are $u_1 = z_0(t)$ and $u_2 = \dot{z}_0(t)$, and the two system outputs (or measurements) are position and acceleration of the driver mass, or $y_1 = z_2$ and $y_2 = \ddot{z}_2$. Recall that this SSR is relatively easy to derive because we defined *two* independent input variables when, in reality, the system has *one* independent input, $z_0(t)$, because $u_2 = \dot{u}_1$. Therefore, we must be careful that our second input u_2 is indeed the time derivative of the first input u_1. If the mathematical model and SSR are unclear, the reader may want to review Examples 2.3 and 5.8.

Table 11.1 presents nominal numerical values for the inertia, stiffness, and damping elements. For a real experimental test rig of the seat-suspension system, the stiffness elements (suspension spring k_1 and seat cushion spring k_2) are estimated by loading the seat suspension and cushion and measuring their static deflections [1]. In this case, the force–deflection relationships are nearly linear, and hence k_1 and k_2 are determined by a least-squares fit through the force–deflection data [1]. Damping elements b_1 and b_2 are estimated by a dynamic test where a sinusoidal input (vibration) is applied to the test rig.

Table 11.1 Parameters for the Seat-Suspension System

System Parameter	Value
Seat mass, m_1	20 kg
Driver mass, m_2	50 kg
Suspension stiffness, k_1	7410 N/m
Suspension friction coefficient, b_1	1430 N-s/m
Seat cushion stiffness, k_2	8230 N/m
Seat cushion friction coefficient, b_2	153 N-s/m

Impulse Response

In Chapter 7, we investigated a linear system's homogeneous or free response, which depends on the "natural dynamics" of the system, which in turn can be determined by evaluating the roots of the characteristic equation. For a system composed of a single transfer function, the free response depends on the poles of transfer function. Because the seat-suspension system has multiple states, the state-space method is likely the best format for the system dynamics. Hence, the natural dynamics of the seat-suspension system can be determined by evaluating the eigenvalues of the state matrix $\mathbf{A}$. The reader should recall that the poles of a system's transfer function and the eigenvalues of the system's state matrix $\mathbf{A}$ are both equivalent to the roots of the characteristic equation.

The state matrix of the seat-suspension system can be computed by using the numerical parameters in Table 11.1, and the result is

$$\mathbf{A} = \begin{bmatrix} 0 & 1 & 0 & 0 \\ -782 & -79.15 & 411.5 & 7.65 \\ 0 & 0 & 0 & 1 \\ 164.6 & 3.06 & -164.6 & -3.06 \end{bmatrix}$$

The system eigenvalues are computed using the MATLAB command

```
>> eig(A)
```

The four eigenvalues (or characteristic roots) are

$$r_1 = -67.5958 \quad r_2 = -7.2675 \quad r_3 = -3.6733 + j10.5189 \quad r_4 = -3.6733 - j10.5189$$

Two roots are real and negative, and two roots are complex with negative real parts. Consequently, the free response will eventually decay to zero as the system reaches its steady-state value. The general form of the free response (for either output) is

$$v_H(t) = c_1 e^{-67.5958t} + c_2 e^{-7.2675t} + c_3 e^{-3.6733t} \cos(10.5189t + \phi) \tag{11.4}$$

Equation (11.4) shows that the free response consists of two damped exponential functions and a damped sinusoidal function. Root $r_1 = -67.5958$ is the "fastest" root, as its exponential function decays to zero in about 0.06 s, and therefore its contribution to the total response is extremely short-lived. Root $r_2 = -7.2675$ corresponds to an exponential function that decays to zero in about 0.55 s. The complex roots are the "slowest" roots because their exponential function decays to zero in about 1.09 s. Therefore, real root r_2 and complex roots r_3 and r_4 are the dominant roots.

We can construct the fourth-order characteristic equation from the four eigenvalues or roots:

$$(r + 67.5958)(r + 7.2675)(r + 3.6733 - j10.5189)(r + 3.6733 + j10.5189) = 0 \tag{11.5}$$

Therefore, the *underdamped* part of the characteristic equation can be determined by multiplying the last two (or complex) terms of Eq. (11.5)

$$r^2 + 7.3467r + 124.1403 = 0 \tag{11.6}$$

The undamped natural frequency is $\omega_n = \sqrt{124.1403} = 11.14\,\text{rad/s}$ and the damping ratio is $\zeta = 7.3467/(2\omega_n) = 0.3297$. Because the complex roots are dominant roots, we expect the free response of the seat-suspension system to exhibit underdamped response characteristics.

Next, we use Simulink to determine the seat-suspension system's response to inputs, and we will first consider a "spike" input for the cabin floor displacement $z_0(t)$. We assume that the vehicle passes over a bump, which causes a sudden displacement of the cabin floor. The floor displacement $z_0(t)$ is modeled by a "triangular pulse" with a peak displacement of 0.03 m (3 cm) and a constant vertical "bump rate" of $\dot{z}_0 = 5.4$ m/s (rising). The floor displacement is a symmetrical triangular pulse, and therefore the constant "bump rate" is $\dot{z}_0 = -5.4$ m/s (descending) after the pulse reaches its peak value of 0.03 m. Hence, the duration of half of the triangular pulse is $\Delta t = z_{0\,\text{max}}/\dot{z}_0 = 0.0056$ s or 5.6 ms. Because the total duration of the triangular pulse is about 11 ms, we can consider the "spike" input to essentially be an impulse input.

Figure 11.2 shows the Simulink model of the seat-suspension system with a triangular pulse input $z_0(t)$. The triangular pulse input is created by integrating a sequence of constant-velocity pulses. The initial positive "bump rate" $\dot{z}_0 = 5.4$ m/s is created by the first step function (step time = 0.5 s). The second step function has magnitude -10.8 m/s (step time = $0.5 + \Delta t$ s) and is added to the first step input to create a negative velocity input of -5.4 m/s. The third step function has magnitude $+5.4$ m/s (step time = $0.5 + 2\Delta t$ s), which, when added to the other two step functions, results in $\dot{z}_0 = 0$. Figure 11.3 shows zoomed-in views of the pulse input $\dot{z}_0(t)$ created by the three step functions, and the triangular pulse input $z_0(t)$ created by integrating the velocity pulses. The SSR matrices $\mathbf{A}$, $\mathbf{B}$, $\mathbf{C}$, and $\mathbf{D}$ that are required in the Simulink model are determined by Eqs. (11.2) and (11.3), and the system parameters found in Table 11.1. Finally, the initial state vector is set to a zero 4×1 column vector.

Figure 11.4a,b shows the driver displacement (mm) and acceleration responses to a 3-cm triangular bump or impulse input. The impulse input is applied at time $t = 0.5$ s. Figure 11.4a shows that the driver displacement $z_2(t)$ exhibits the characteristics of an underdamped second-order response. The transient response dies out to zero in about 1.1 s after the impulse input, which corresponds to the slowest of the four roots or eigenvalues. We can use the first two peak values of $z_2(t)$ and the log decrement method to estimate the damping ratio of the seat-suspension system. The first peak is 1.52 mm (at $t = 0.11$ s after the impulse), and the second peak value is 0.19 mm (at $t = 0.70$ s after the impulse). Therefore, the log decrement is

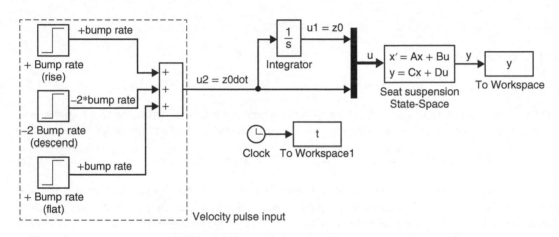

Figure 11.2 Simulink diagram for the seat-suspension system: triangular pulse input.

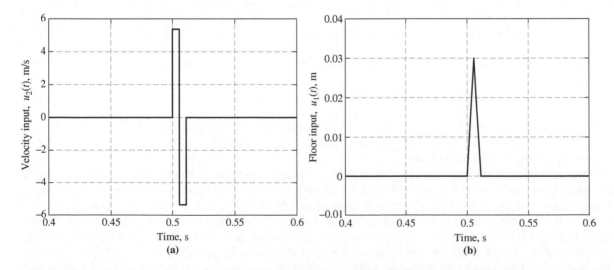

Figure 11.3 Seat-suspension system inputs: (a) velocity-pulse input $u_2(t)$ and (b) triangular-pulse input $u_1(t)$.

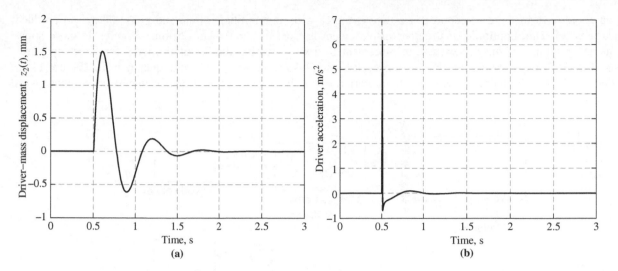

Figure 11.4 Impulse response of the seat-suspension system: (a) driver displacement and (b) driver acceleration.

$$\delta = \ln \frac{1.52}{0.19} = 2.0694$$

The approximate damping ratio is

$$\zeta = \frac{\delta}{\sqrt{4\pi^2 + \delta^2}} = 0.3128$$

The log decrement method is based on the impulse response of a second-order system, and hence $\zeta = 0.3128$ is an approximation of the system's damping ratio. As previously shown, the damping ratio of the two complex roots of the fourth-order seat-suspension system is 0.3297. The log decrement method thus provides a fairly accurate estimation of the system's damping ratio. The frequency of the impulse response is 2π rad divided by the time between the first two peaks $(0.70 - 0.11\,\text{s})$, or 10.65 rad/s. Because damped frequency for a second-order system is $\omega_d = \omega_n \sqrt{1 - \zeta^2}$, the approximate undamped natural frequency is 11.21 rad/s, where the actual undamped frequency of the complex roots is 11.14 rad/s. The accuracy of the approximate damping ratio and undamped frequency implies that we can develop an approximate, reduced-order (second-order) input–output (I/O) equation relating the driver displacement z_2 and input $z_0(t)$.

Frequency Response

The objective of the vibration isolation system is to suppress the motion of the vehicle cabin floor $z_0(t)$ that is transmitted to the driver. The previous case showed that the seat-suspension system damps out an impulsive input in about 1.1 s, and the impulse response for driver displacement $z_2(t)$ exhibits two peaks during the transient response with a period of about 0.6 s. The impulse response for acceleration of the driver exhibited an initial spike when the impulse is applied, followed by a quickly damped response to zero.

Travel over a rough road will result in a repeating (periodic) input. For example, a vehicle moving with constant speed over equally spaced road bumps will experience a periodic road displacement with fixed frequency, which may be modeled as a sinusoidal input function. Hence, we can use frequency-response methods to analyze the performance of the seat-suspension system. The periodic road vibrations will produce vibrations in the cabin floor, which can be modeled as $z_0(t) = a \sin \omega t$, where a is the amplitude of the floor vibrations (m), and ω is the input frequency (rad/s). Because the driving speed and the spacing between bumps both influence the input frequency ω, we analyze the vibration isolation system's response to a range of frequencies.

Figure 11.5 shows the Simulink diagram for the seat-suspension system with a sinusoidal input for floor displacement $z_0(t) = a \sin \omega t$ (m). Note that the second input (floor velocity $\dot{z}_0$) is created by differentiating the displacement input (we could have inserted a second `Sine Wave` source with an amplitude of ωa and phase of $\pi/2$ to produce $\dot{z}_0(t) = \omega a \cos \omega t$). The amplitude of the floor displacement is fixed at $a = 0.02$ m (20 mm). Figures 11.6a–c show the frequency response of the driver-mass displacement z_2 from executing the Simulink model for three input frequencies of 0.25, 1, and 4 Hz. Recall that hertz is cycles/second, so ω (in rad/s) $= 2\pi \times$ frequency in Hz. Figure 11.6a shows that for the low-frequency case (0.25 Hz, or period $= 4$ s), the driver and floor displacements are in phase with

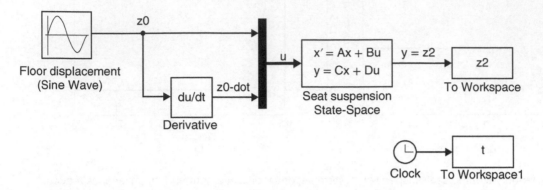

Figure 11.5 Simulink diagram for the seat-suspension system: sinusoidal input.

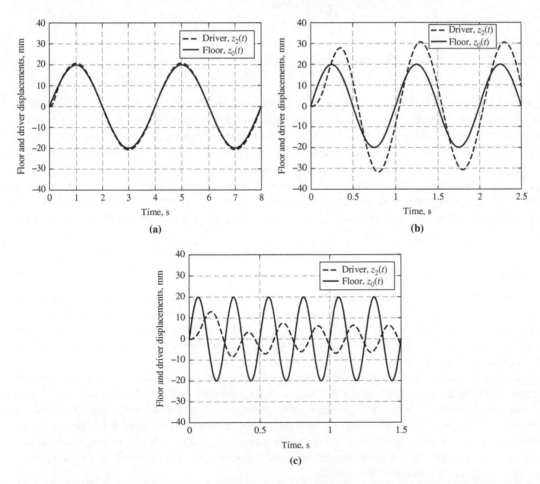

Figure 11.6 Frequency response of driver-mass displacement z_2: (a) input frequency $= 0.25$ Hz, (b) input frequency $= 1$ Hz, and (c) input frequency $= 4$ Hz.

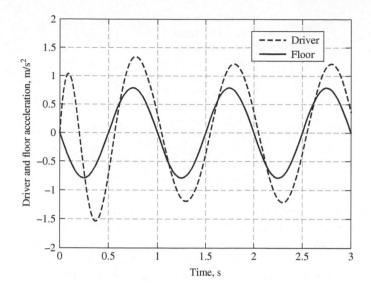

Figure 11.7 Frequency response of driver acceleration for input frequency = 1 Hz.

essentially the same amplitude (20 mm). Therefore, if we consider the transfer function $G(s)$ that relates output z_2 (driver) to input $z_0(t)$ (floor), the magnitude of the corresponding sinusoidal transfer function $G(j\omega)$ is essentially unity when the input frequency is 0.25 Hz (or, $\omega = 1.57$ rad/s). Furthermore, the phase angle of $G(j1.57)$ is nearly zero (for 0.25 Hz), as there is no phase difference between the two sine waves in Fig. 11.6a. Figure 11.6b shows that when the input frequency is 1 Hz (or, $\omega = 6.28$ rad/s), the amplitude ratio of the output/input sinusoids is about $31/20 = 1.55$, and the steady-state driver displacement is 55% greater than the floor displacement. Figure 11.6b also shows that the driver-mass sinusoid lags behind the input (floor) sinusoid. Figure 11.6c shows that when the input frequency is 4 Hz (or, $\omega = 25.13$ rad/s), the amplitude ratio of the output/input sinusoids is about $6.6/20 = 0.33$, and the steady-state driver displacement is 33% less than the floor displacement. The peaks of z_2 are nearly aligned with the valleys of z_0. Hence, the phase lag between the two sinusoids is nearly 180°.

Figure 11.7 shows the frequency response for driver acceleration along with the input floor acceleration $\ddot{z}_0(t)$ for an input frequency of 1 Hz. Because floor displacement is $z_0(t) = a \sin \omega t$, floor velocity and acceleration are $\dot{z}_0(t) = \omega a \cos \omega t$ and $\ddot{z}_0(t) = -\omega^2 a \sin \omega t$, respectively. The magnitude of the floor acceleration is $\omega^2 a$, or 0.79 m/s^2 when $a = 0.02$ m and $\omega = 6.28$ rad/s. The amplitude ratio of output (driver) and floor (input) acceleration is $1.21/0.79 = 1.53$, which is nearly the same as the amplitude ratio of the driver/floor displacements shown in Fig. 11.6b for the same 1-Hz input frequency. The steady-state phase lag between the driver and floor accelerations shown in Fig. 11.7 is nearly the same as the phase lag in Fig. 11.6b.

The frequency-response results presented by Figs. 11.6a–c can be succinctly summarized by a Bode diagram of the seat-suspension system, where the output is defined as the driver displacement z_2. Because the Bode diagram displays the frequency-response information for a single input and single output, we must re-derive the SSR in terms of the sole independent input, which is floor displacement $z_0(t)$. In this case, the definition of the second state variable x_2 must be altered, and subsequently the 4×1 input matrix **B** differs from the 4×2 input matrix presented by Eq. (11.2) for the case of two (dependent) inputs (for details, see Problem 5.32, Chapter 5). The following MATLAB commands define the desired SSR (with $u = z_0(t)$ as the single input and z_2 as the single output) and then plot the Bode diagram:

```
>> m1 = 20;                              % seat mass, kg
>> m2 = 50;                              % driver mass, kg
>> k1 = 7410;                            % suspension stiffness, N/m
>> k2 = 8230;                            % cushion stiffness, N/m
>> b1 = 1430;                            % suspension friction, N-s/m
>> b2 = 153;                             % cushion friction, N-s/m
>> Arow1 = [0 1 0 0];                    % row 1 of matrix A
>> Arow2 = [(-k1-k2)/m1 (-b1-b2)/m1 k2/m1 b2/m1];   % row 2 of matrix A
```

```
>> Arow3 = [0 0 0 1];                           % row 3 of matrix A
>> Arow4 = [k2/m2 b2/m2 -k2/m2 -b2/m2];         % row 4 of matrix A
>> A = [Arow1 ; Arow2 ; Arow3 ; Arow4];         % system matrix A
>> Brow1 = b1/m1;                               % row 1 of matrix B
>> Brow2 = (-b1^2 - b1*b2 + k1*m1)/m1^2;        % row 2 of matrix B
>> Brow3 = 0;                                   % row 3 of matrix B
>> Brow4 = b1*b2/(m1*m2);                       % row 4 of matrix B
>> B = [Brow1 ; Brow2 ; Brow3 ; Brow4];         % input matrix B
>> C = [0 0 1 0];                               % output y = x_3 = z_2
>> D = 0;                                       % direct-link (null)
>> sys = ss(A,B,C,D);                           % build SSR sys
>> bode(sys)                                    % plot Bode diagram
```

Figure 11.8 shows the Bode diagram of the vibration isolation system. For low-frequency inputs (1–2 rad/s or 0.16–0.32 Hz), the magnitude is 0 dB and the phase is nearly zero. Recall that a magnitude of 0 dB is equal to an amplitude ratio of unity, so for low-frequency inputs the steady-state driver position $z_2(t)$ essentially matches the input sinusoid $z_0(t)$. Figure 11.6a demonstrates how the output $z_2(t)$ matches the input $z_0(t)$ when input frequency is 0.25 Hz. The Bode diagram in Fig. 11.8 also shows that the maximum amplitude ratio (resonance) is about 6 dB, which occurs at an input frequency of about 10 rad/s. An exact value of the frequency response for a desired input frequency can be obtained using the MATLAB commands:

```
>> w = 10;                          % input frequency, ω, rad/s
>> [mag,phase] = bode(sys,w)        % compute magnitude and phase
```

Magnitude is returned as the output/input amplitude ratio, and phase angle is in degrees. Using these MATLAB commands, we find that the peak amplitude ratio is about 2.0643 at the corresponding resonant frequency of 10.07 rad/s (or, 1.60 Hz). The peak magnitude in decibels is $20 \log_{10}(2.0643) = 6.3$ dB. The Bode diagram also shows that the amplitude ratio drops off at a rate of about 40 dB/decade for frequencies higher than the resonant frequency of 10 rad/s.

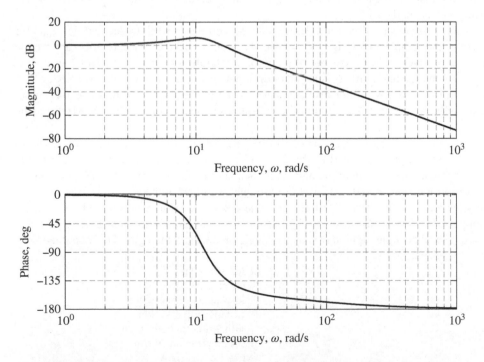

Figure 11.8 Bode diagram of seat-suspension system for output z_2 and input z_0.

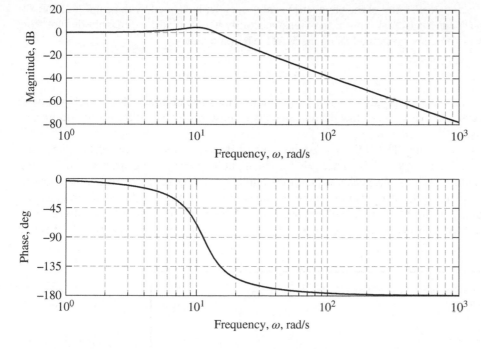

Figure 11.9 Bode diagram of the approximate second-order model for the seat-suspension system with output z_2 and input z_0.

The constant low-frequency magnitude, followed by a single peak at a resonant frequency, and a -40 dB/decade magnitude drop for high-frequency inputs all indicate that we can develop an approximate, *second-order* transfer function relating the driver displacement z_2 and input z_0. Because the low-frequency magnitude is unity (or, 0 dB), an approximate transfer function must have a unity DC gain. Therefore, an *approximate* transfer function $\overline{G}(s)$ would have the standard second-order form

$$\overline{G}(s) = \frac{\omega_n^2}{s^2 + 2\zeta\omega_n s + \omega_n^2} = \frac{Z_2(s)}{Z_0(s)} \tag{11.7}$$

Our previous analysis of the impulse response showed that the undamped natural frequency ω_n is approximately 11.21 rad/s, and the damping ratio ζ is approximately 0.3128, and therefore the approximate transfer function is

$$\overline{G}(s) = \frac{125.66}{s^2 + 7.01s + 125.66} \tag{11.8}$$

Figure 11.9 shows the Bode diagram of the approximate transfer function relating output z_2 to input z_0. The resonant peak is about 4.5 dB (amplitude ratio of 1.68), which is less than the 6.3 dB peak for the actual fourth-order system. However, the approximate frequency response shown in Fig. 11.9 closely resembles the frequency response of the true fourth-order system shown in Fig. 11.8.

Parametric Sensitivity Analysis

The objective of the vibration isolation system is to suppress the motion of the vehicle cabin floor $z_0(t)$ that is transmitted to the driver. The previous sections presented the impulse and frequency responses for the nominal seat-suspension parameters in Table 11.1. It is useful for the design engineer to understand the effect each parameter has on the performance of the system. A *parametric sensitivity analysis* is a standard engineering method where each free parameter is varied and the sensitivity (or lack thereof) of the system's performance metric is determined. Such information will aid the design engineer in optimizing system performance in the context of operating constraints.

Transmissibility is the performance measure of the parametric analysis, and it is defined by the amplitude ratio of the frequency-response output $z_2(t)$ and sinusoidal input $z_0(t)$. Transmissibility is essentially the magnitude plot from the Bode diagram for output $y = z_2$. It can be computed using the MATLAB command

```
>> [mag,phase] = bode(sys,w)                    % compute magnitude and phase
```

where w is the input frequency in rad/s. MATLAB M-file script 11.1 computes the transmissibility for input frequencies ranging from 0.1 to 5 Hz. Previous studies have shown that the human body is most sensitive to relatively low-frequency vibrations in the 0.5–5 Hz range [2].

Figure 11.10 shows the transmissibility $|z_2(t)|/|z_0(t)|$ for three parametric variations in seat-cushion stiffness: nominal $k_2 = 8230$ N/m, 50% reduction in k_2, and 50% increase in k_2. The input frequencies range from 0.1 to 5 Hz (or, 0.63–31.41 rad/s). Stiffer seat cushions suppress the peak transmissibility and increase the resonant frequency where the peak transmissibility occurs. Increasing stiffness k_2 also increases the transmitted vibrations at the higher frequencies. Vehicle vibrations are predominantly in the 2.5 Hz (15.7 rad/s) range [2], and therefore the seat-cushion stiffness must be selected so that a good trade-off exists between low peak transmissibility and low transmissibility around 2.5 Hz. In

MATLAB M-file 11.1

```
%  M-file for computing transmissibility for
%  the seat-suspension system

%  mechanical system parameters
m1 = 20;          %  seat mass, kg
m2 = 50;          %  driver mass, kg
k1 = 7410;        %  suspension stiffness, N/m
k2 = 8230;        %  seat cushion stiffness, N/m
b1 = 1430;        %  suspension friction, N-s/m
b2 = 153;         %  seat cushion friction, N-s/m

%  State-space representation:
%    x = [ z1 ; z1dot-b1*z0/m1 ; z2 ; z2dot ]'
%    u = z0(t) (floor displacement)
Arow1 = [ 0 1 0 0 ];
Arow2 = [(-k1-k2)/m1 (-b1-b2)/m1 k2/m1 b2/m1 ];
Arow3 = [ 0 0 0 1 ];
Arow4 = [ k2/m2 b2/m2 -k2/m2 -b2/m2 ];
A = [ Arow1 ; Arow2 ; Arow3 ; Arow4 ];
B = [ b1/m1 ; (-b1*b1 - b1*b2 + k1*m1)/m1^2 ; 0 ; b1*b2/(m1*m2) ];

%  output y = z2 = x3 (driver displacement)
C = [ 0 0 1 0 ];
D = 0;

%  build SSR system
sys = ss(A,B,C,D);

%  Loop for computing TR for range of input frequency
Npts = 500;
w_Hz = linspace(0.1,5,Npts);       %  range of frequency: 0.1 --> 5 Hz

for i=1:Npts
    w_in = w_Hz(i)*2*pi;           %  input frequency in rad/s
    [mag,phase] = bode(sys,w_in);
    TR(i) = mag;                   %  transmissibility = |z2|/|z0|
end

%  Plot TR vs input frequency
plot(w_Hz,TR)
grid
xlabel('Input frequency, Hz')
ylabel('Transmissibility')
```

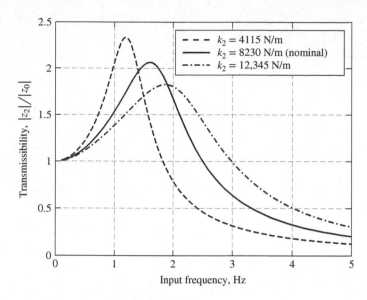

Figure 11.10 Transmissibility $|z_2|/|z_0|$ for variations in seat-cushion stiffness k_2.

light of these constraints, the nominal k_2 appears to provide good performance, as transmissibility at 2.5 Hz is about unity when $k_2 = 8230$ N/m. If reducing the transmitted vibrations at 2.5 Hz is more important than attenuating the peak transmissibility, then decreasing the seat-cushion stiffness is likely a good option because Fig. 11.10 shows that transmissibility at 2.5 Hz is highly sensitive to k_2.

Figure 11.11 shows the transmissibility for three parametric variations in suspension friction coefficient: nominal $b_1 = 1430$ N-s/m, a 50% reduction, and a 50% increase. Higher damping suspensions suppress the peak transmissibility and increase the resonant frequency. Increasing friction b_1 also increases the transmitted vibrations at the higher frequencies, but decreasing b_1 below the nominal value has very little effect on transmitted vibrations at higher frequencies. Figure 11.11 also shows that the peak transmissibility is more sensitive to decreasing suspension damping b_1 when compared to the peak transmissibility in Fig. 11.10 for variations in seat-cushion stiffness. Therefore, options for decreasing suspension friction b_1 must be considered carefully, as there is relatively little

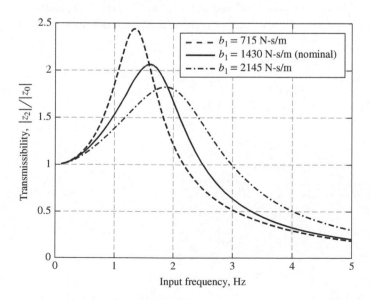

Figure 11.11 Transmissibility $|z_2|/|z_0|$ for variations in suspension friction b_1.

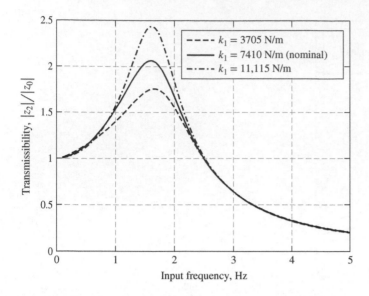

Figure 11.12 Transmissibility $|z_2|/|z_0|$ for variations in suspension stiffness k_1.

improvement in transmissibility at 2.5 Hz, and the corresponding increase in peak transmissibility at low frequencies is magnified.

Finally, the sensitivity to variations in suspension stiffness k_1 was investigated. Figure 11.12 shows the transmissibility for three values of suspension stiffness: nominal $k_1 = 7410$ N/m, a 50% reduction, and a 50% increase. Clearly, stiffer suspension springs increase the peak transmissibility, and varying k_1 has very little effect on the resonant frequency, or the transmissibility at higher frequencies. Varying suspension stiffness k_1 has almost no effect on transmissibility at 2.5 Hz, and therefore the lightest possible suspension spring k_1 will provide the best ride comfort for the driver.

11.3 SOLENOID ACTUATOR–VALVE SYSTEM

The second case study involves the analysis and design of a solenoid actuator that is often used to position valves for metering flow [3, 4]. The basic principles of a solenoid have been described and analyzed in Chapters 2, 3, 5, and 6. Figure 11.13 shows the electromechanical system, which consists of a solenoid coil, an armature mass (plunger), a spool

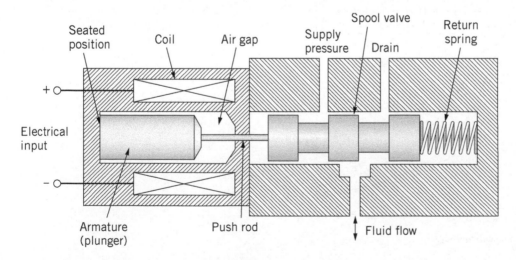

Figure 11.13 Solenoid actuator–valve system.

valve, and a return spring. When current flows through the coil, the resulting electromagnetic force pulls the plunger (to the right) toward the center of the coil, which pushes on the spool valve in order to properly meter hydraulic fluid flow. When the voltage source is switched off and current is dissipated, the electromagnetic force goes to zero and the compressed spring returns the armature to the seated position.

Our objective is to determine the solenoid system parameters that minimize the settling time for the armature–valve mass to reach a displacement of 2 mm. We will modify our integrated Simulink model for the solenoid actuator that was developed in Chapter 6 and include physical models for the electrical inductance and electromagnetic force of the solenoid coil.

Mathematical Model

We developed the mathematical modeling equations of the solenoid actuator–valve system in Chapters 2, 3, and 5. The complete modeling equations are

$$L(x)\dot{I} + RI = e_{in}(t) - L_x I \dot{x} \tag{11.9}$$

$$m\ddot{x} + b\dot{x} + F_{dry}\text{sgn}(\dot{x}) + kx = F_{em} - F_{PL} + F_C \tag{11.10}$$

Recall that the coil current is I, resistance is R, $e_{in}(t)$ is the input voltage, and coil inductance is L. The mechanical system is composed of armature–valve mass m, x is the position of m, b is the viscous friction coefficient, k is the return spring constant, F_{em} is the electromagnetic force from the solenoid coil, F_{PL} is the preload force in the spring, and F_C is the wall-contact force when the mass is seated. If the mathematical model of the solenoid is unclear, the reader may want to review Example 2.5, Section 3.5, and Example 5.3.

Note that we have assumed that the coil inductance $L(x)$ is a function of armature position x. Inductance L increases as the armature moves toward the center of the coil, which is evident in the accepted modeling equation for solenoid inductance [3, 4]

$$L(x) = \frac{c}{d - x} = \frac{L_0}{1 - x/d} \tag{11.11}$$

Because armature displacement x is measured positive to the right from the seated position (see Fig. 11.13), inductance is minimum when $x = 0$. The constants c and d depend on geometry and material properties of the solenoid coil, such as the number of coil turns N, area of the air gap A, coil length l, and magnetic permeabilities of air and the iron core μ. The inductance when $x = 0$ is

$$L_0 = \frac{\mu A N^2}{l} \tag{11.12}$$

Recall that in Chapter 3 we derived the "back-emf" voltage e_b due to motion of the armature mass relative to the coil by taking the time derivative of the magnetic flux, $\lambda = L(x)I$, which resulted in

$$e_b = L_x I \dot{x} = \frac{dL}{dx} I \dot{x} \tag{11.13}$$

This back-emf voltage term appears on the right-hand side of the solenoid coil equation (11.9) with a negative sign. Consequently, positive velocity of the armature decreases the net voltage in the coil. Using Eq. (11.11), the derivative dL/dx is

$$\frac{dL}{dx} = L_x = \frac{L_0}{d(1 - x/d)^2} \tag{11.14}$$

The electromagnetic force F_{em} of the solenoid was also derived in Chapter 3, and it is a nonlinear function of current

$$F_{em} = \frac{1}{2}\frac{dL}{dx}I^2 \tag{11.15}$$

Equations (11.13) and (11.15) show that both the back emf and electromagnetic force depend on the derivative L_x. In order to simplify our simulation model, we assume that the change in inductance L is constant with position x, which

Table 11.2 Parameters for the Solenoid Actuator

System Parameter	Value
Coil resistance, R	$3\,\Omega$
Iron core/air permeability, μ	$800\pi(10^{-7})$ N/A^2
Length of coil, l	0.04 m
Area of the air gap, A	$\pi(10^{-4})$ m^2
Characteristic displacement, d	0.0078 m
Armature–valve mass, m	0.05 kg
Viscous friction coefficient, b	10 N-s/m
Dry-friction force, F_{dry}	0 N
Preload spring force, F_{PL}	2 N

is a reasonable assumption for a 2-mm displacement. Hence, we define the constant $K = dL/dx$ and compute K using Eq. (11.14) with a nominal displacement $x_{nom} = 0.001$ m (1 mm) and the initial inductance L_0 using Eq. (11.12).

Equation (11.12) shows that the number of turns N (coil-wire wraps) determines the initial inductance L_0, which is needed to determine the electromagnetic force and back emf. In addition, Eq. (11.10) shows that the spring force must balance the electromagnetic force at equilibrium (i.e., $\ddot{x} = \dot{x} = 0$) when the valve has reached its full stroke. Therefore, N and k are the free parameters that will determine the best solenoid design. Table 11.2 summarizes the numerical values of the physical parameters for the solenoid actuator. Note that we have neglected dry friction for this case, but we have assumed a 2-N preload force from the spring when the armature is seated ($x = 0$).

Simulink Model

Figure 11.14 shows the Simulink model of the integrated solenoid actuator, which is a modified version of the solenoid model developed for Example 6.9. Figure 11.14 is slightly different from Fig. 6.28 as the armature position from the mechanical subsystem is fed back as an input to the electrical subsystem because inductance $L(x)$ increases with x. Figure 11.15 shows the inner details of the electrical subsystem. The reader should be able to identify the back emf and electromagnetic force computations, Eqs. (11.13) and (11.15), in Fig. 11.15, as well as the summation of all voltage terms that appear in Eq. (11.9). A user-defined function Fcn divides the net voltage (from the summing junction) by the inductance $L(x)$ in order to determine the time rate of current, $\dot{I}$. Note that inductance is simply $L(x) = L_0 + Kx$ as we assumed that $K = dL/dx$ is constant.

Figure 11.16 shows the inner details of the mechanical subsystem. The reader should be able to identify the six force components from the mathematical model (11.10) that are summed to produce the net force acting on the armature mass (note that dry-friction force is included in the Simulink model, even though the constant Fdry is set to zero as indicated in Table 11.2). Recall that the wall-contact force F_C balances the difference between the spring preload and electromagnetic forces if the armature is seated:

$$F_C = \begin{cases} F_{PL} - F_{em} & \text{if} \quad F_{em} < F_{PL} \\ 0 & \text{if} \quad F_{em} \geq F_{PL} \end{cases} \tag{11.16}$$

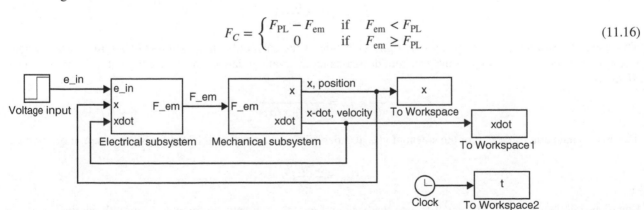

Figure 11.14 Simulink diagram for the solenoid actuator.

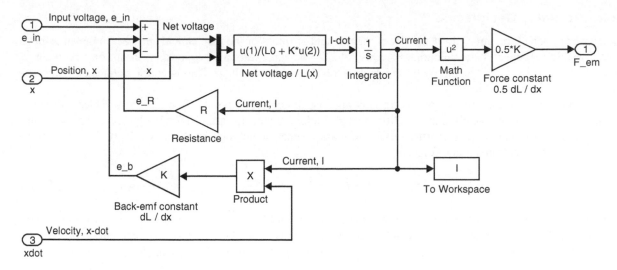

Figure 11.15 Simulink diagram for the solenoid: electrical subsystem.

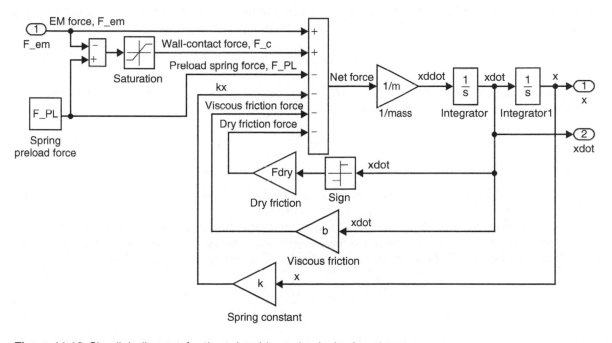

Figure 11.16 Simulink diagram for the solenoid: mechanical subsystem.

The contact force is determined in the Simulink model according to Eq. (11.16) by subtracting the electromagnetic force from the preload spring force and sending the result to the Saturation block from the Discontinuities library. A Saturation block will set its output equal to the input if the input is between upper and lower limits set by the user. Otherwise, the output is set either to the upper limit if the input exceeds the upper limit or to the lower limit if the input is less than the lower limit. Equation (11.16) is satisfied if the upper limit of the Saturation block is set to a large positive number and the lower limit is set to zero. In other words, the wall-contact force is the positive force difference $F_{PL} - F_{em}$ required to balance the preloaded spring (when $F_{em} < F_{PL}$ and the mass is still seated), but the contact force cannot be negative.

Solenoid–Actuator Design

Our objective is to select the strength of the electromagnet and stiffness of the mechanical spring in order to achieve the fastest possible response of the solenoid. For this solenoid design, we require a constant 2-mm displacement of the armature–valve mass when we supply the coil with a 12-V step input voltage. This scenario defines our expected nominal operation of the solenoid.

As previously stated, the number of coil turns N and spring constant k are the free design variables. We begin our analysis by observing the steady-state conditions of the solenoid actuator after the valve has reached $x = 2$ mm. Equation (11.10) shows that when the armature–valve mass reaches static equilibrium ($\ddot{x} = \dot{x} = 0$), the total spring force must balance the electromagnetic force, or

$$k\bar{x} + F_{\text{PL}} = \overline{F}_{\text{em}} \tag{11.17}$$

where $\bar{x}$ is the required steady-state stroke of the armature ($\bar{x} = 2$ mm) and $\overline{F}_{\text{em}}$ is the steady-state electromagnetic force. The reader should note that the wall-contact force F_C is zero because the armature has been pulled off its seat. Equation (11.15) shows that the steady-state electromagnetic force depends on the steady-state coil current $\bar{I}$:

$$\overline{F}_{\text{em}} = \frac{1}{2}K\bar{I}^2 \tag{11.18}$$

Steady-state current can be determined from the mathematical model of the coil. Equation (11.9) shows that when the actuator reaches static equilibrium ($\dot{I} = \dot{x} = 0$), the mathematical model of the solenoid coil is reduced to Ohm's law:

$$R\bar{I} = e_{\text{in}}(t) \tag{11.19}$$

Because the step input voltage is $e_{\text{in}}(t) = 12$ V and $R = 3\,\Omega$, the steady-state current is $\bar{I} = 4$ A. Finally, the steady-state electromagnetic force equation (11.18) can be expanded by substituting Eq. (11.14) for $K = dL/dx$ and Eq. (11.12) for L_0. The result is

$$\overline{F}_{\text{em}} = \frac{\mu A N^2 \bar{I}^2}{2dl(1 - x_{\text{nom}}/d)^2} \tag{11.20}$$

where $x_{\text{nom}} = 1$ mm is the nominal displacement required for the computation of the constant $K = dL/dx$. Every parameter in Eq. (11.20) is fixed (see Table 11.2) except the number of turns N.

The selection of the design parameters N and k can now be summarized: choosing N determines the steady-state electromagnetic force via Eq. (11.20), and this applied force must be balanced by the steady-state spring force shown in Eq. (11.17). Therefore, Eq. (11.17) shows that the spring constant k must be

$$k = \frac{\overline{F}_{\text{em}} - F_{\text{PL}}}{\bar{x}} \tag{11.21}$$

which is solely a function of $\overline{F}_{\text{em}}$. In other words, the solenoid parameters N and k are *not* independent because of the 2-mm displacement constraint for a 12-V step input. The solenoid design problem outlined here has only one free variable: number of turns N. Figures 11.17a,b show the variation of steady-state electromagnetic force $\overline{F}_{\text{em}}$ and spring constant k for a range of N, as computed by using Eqs. (11.20) and (11.21) and the fixed parameters in Table 11.2. Both the applied force and spring stiffness increase dramatically with N. The minimum number of turns is $N = 28$, as $\overline{F}_{\text{em}} < 2$ N for coils with less than 28 turns. Thus, the electromagnetic force cannot overcome the preload spring force F_{PL} and unseat the armature. Figure 11.17b shows that k is negative (infeasible) for $N < 28$.

We can now simulate a few trial solenoid designs using the integrated Simulink model shown in Fig. 11.14 with $N = 40, 50,$ and 60. Equation (11.20) and Fig. 11.17a show that the steady-state electromagnetic force is between 4.3 (for $N = 40$) and 9.7 N (for $N = 60$). The spring constant k required to balance $\overline{F}_{\text{em}}$ for a 2-mm stroke is computed using Eq. (11.21), and the inductance L_0 and dL/dx values are computed using Eqs. (11.12) and (11.14), respectively. Figure 11.18 shows the valve position $x(t)$ for a 12-V step input for these three values of N. Clearly, all three solenoid

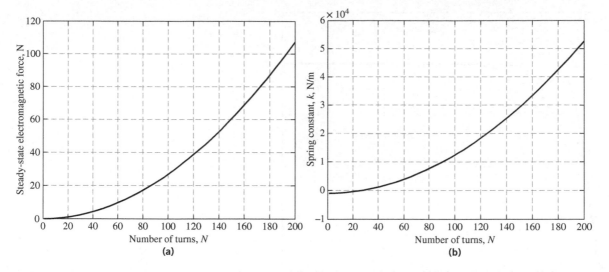

Figure 11.17 Solenoid parameters vs. number of turns N: (a) steady-state electromagnetic force and (b) spring constant k.

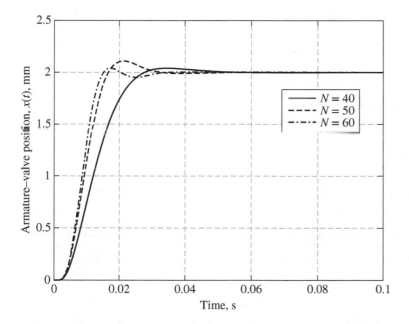

Figure 11.18 Armature–valve position $x(t)$ for $N = 40$, 50, and 60 turns.

designs show a 2-mm stroke at steady state because the return spring is perfectly matched to balance the electromagnetic force. The solenoid actuator with $N = 60$ has the smallest response time, while the actuator with $N = 40$ took the longest time to reach steady state. Next, we investigated solenoid designs with more powerful electromagnets with $N = 70$, 80, and 90 turns. Figure 11.19 shows the valve responses for these three cases. While all three solenoid designs show good initial movement toward the steady-state position, all three show "dips" in $x(t)$ before reaching $x = 2$ mm and subsequently an extended settling time for the transient response. Figure 11.20 shows that the solenoid responses become even slower for even stronger electromagnets with $N = 100$, 110, and 120.

The poor performance of the solenoid designs with stronger electromagnets (i.e., large N) can be explained by the corresponding increase in coil inductance L_0. Equation (11.12) illustrates that the initial coil inductance L_0 is proportional to N^2. While increasing inductance L_0 increases the strength of the magnetic field and the electromagnetic force, it also

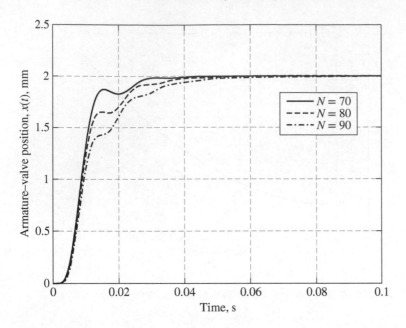

Figure 11.19 Armature–valve position $x(t)$ for $N = 70$, 80, and 90 turns.

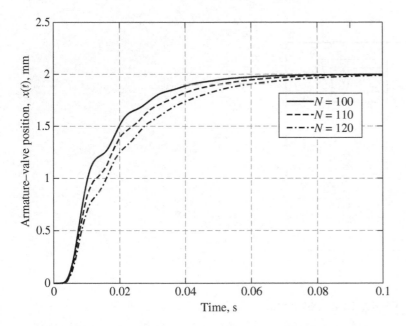

Figure 11.20 Armature–valve position $x(t)$ for $N = 100$, 110, and 120 turns.

slows down the response of the solenoid circuit. To show this effect, consider the mathematical model of the solenoid coil with nominal inductance L_0 and *without* the back-emf term

$$L_0 \dot{I} + RI = e_{\text{in}}(t) \tag{11.22}$$

The time constant for this first-order I/O equation is $\tau = L_0/R$. Because resistance is fixed at $3\,\Omega$, the time constant τ will increase at a quadratic rate with N. Hence, a strong electromagnet (large N) will result in a large time constant and consequently a slow current response. The slow current response slows down the buildup of electromagnetic force,

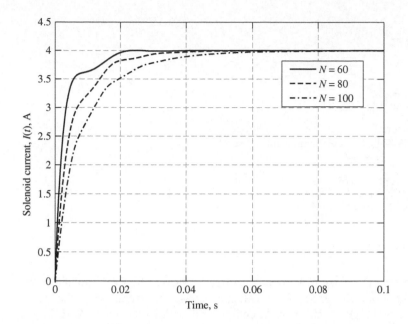

Figure 11.21 Solenoid current $I(t)$ for $N = 60$, 80, and 100 turns.

Table 11.3 Optimal Solenoid Design Characteristics

Solenoid Characteristic	Value
Number of turns, N	58
Initial inductance, L_0	0.0066 H
Force/back-emf constant, K	1.128 N/A^2
Return spring constant, k	3510.4 N/m
Settling time, t_S	0.02 s
Percent overshoot	3.2%

which in turn slows down the response of the armature–valve mass. The back-emf coefficient $K = dL/dx$ also slows down the current response of the solenoid. Equation (11.14) shows that dL/dx is proportional to inductance L_0, and therefore stronger electromagnets enhance the back-emf effect. Figure 11.21 displays the solenoid coil current response for designs with $N = 60$, 80, and 100. All three cases show a "dip" in the current response before reaching the steady-state value, which is due to the back-emf voltage induced by the high velocity of the armature during the out-stroke phase. However, the solenoid with $N = 60$ shows the fastest current response to the step voltage input, and therefore this solenoid design reaches its steady-state electromagnetic force sooner than any other design. A large electromagnetic force is important during the initial out-stroke phase because it is important to accelerate the mass. However, the magnitude of the steady-state electromagnetic force is not important as it ultimately is balanced by the return spring when $x = 2$ mm.

Figures 11.18–11.20 indicate that the best solenoid design has about 60 coil turns. Several more Simulink simulations were run, where N varied from 52 to 64 turns in increments of 2, and it was found that $N = 58$ provided the fastest valve response. Table 11.3 summarizes the characteristics and performance metrics of the best solenoid design.

11.4 PNEUMATIC AIR-BRAKE SYSTEM

Our third case study analyzes a pneumatic actuator for an air-brake system for large commercial vehicles such as tractor-trailers and buses. Most commercial vehicles in the United States use an "S-cam" drum brake mechanism that is actuated by compressed air [5]. Figure 11.22 shows a schematic diagram of the air-brake system, which is composed of pneumatic and mechanical subsystems. The pneumatic subsystem includes the supply pressure P_S (charged by a compressor),

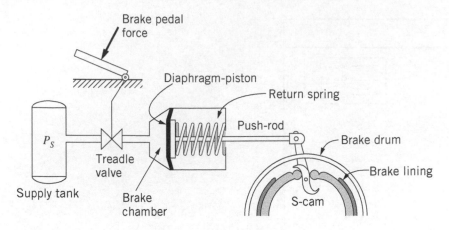

Figure 11.22 Schematic of the pneumatic air-brake system.

which is connected to the brake chamber. Pressing the brake pedal opens the treadle valve. The valve modulates the high-pressure air flow from the supply tank to the brake chamber. The mechanical subsystem includes the diaphragm (piston) and push-rod, return spring, and S-cam drum brake mechanism. As compressed air flows into the brake chamber, the increase in air pressure provides a force on the diaphragm-piston and moves the push-rod to the right. The transmitted force from the push-rod rotates the S-cam, which presses the brake lining against the inside of the drum, thus providing the braking friction for the wheel.

Our objective is to accurately model the air-brake system and simulate its response to a step input (opening) of the treadle valve. An accurate integrated model would allow brake designers to predict pressure transients in the brake chamber over a range of partial brake applications. This capability may be used as a diagnostic tool for inspecting the health of the air-brake system [5].

Mathematical Model

The complete mathematical model consists of the pneumatic and mechanical subsystems. Figure 11.23 shows a free-body diagram of the mechanical subsystem, consisting of the piston/push-rod mass m. Piston displacement x is positive to the right, and the seat of the diaphragm constrains x to positive displacements only. The forces acting on mass m include the air-pressure forces, the seat contact force, the spring force (including the preload), the viscous friction force, and the reactive load force due to a spring in the S-cam mechanism. Applying Newton's second law with a positive sign convention to the right yields

$$+ \rightarrow \sum F = PA_b + F_C - P_{\text{atm}}A_b - kx - F_{\text{PL}} - b\dot{x} - F_{\text{load}} = m\ddot{x} \tag{11.23}$$

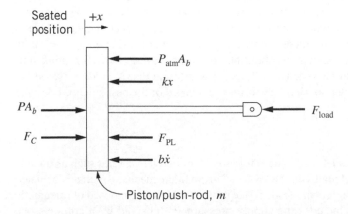

Figure 11.23 Free-body diagram of the mechanical subsystem.

where P is the air pressure in the brake chamber, A_b is the area of the diaphragm-piston, P_{atm} is atmospheric pressure, F_C is the contact force with the seat, k is the return spring constant, b is the viscous friction coefficient, F_{PL} is the preload spring force, and F_{load} is the load force from actuating the S-cam in the brake drum. Rearranging Eq. (11.23) so that all terms involving displacement x are on the left-hand side yields

$$m\ddot{x} + b\dot{x} + kx = (P - P_{atm})A_b + F_C - F_{PL} - F_{load} \qquad (11.24)$$

As with the solenoid actuator, the contact force only exists when the preload spring force exceeds the differential pressure force and the piston is seated with $x = 0$

$$F_C = \begin{cases} F_{PL} - \left(P - P_{atm}\right)A_b & \text{if} \quad (P - P_{atm})A_b < F_{PL} \text{ and } x = 0 \\ 0 & \text{if} \quad (P - P_{atm})A_b \geq F_{PL} \text{ or } x > 0 \end{cases} \qquad (11.25)$$

In all other scenarios (differential pressure force exceeds preload, or $x > 0$), the contact force is zero. The load force required to actuate the S-cam is modeled as a nonlinear function of push-rod displacement

$$F_{load} = k_1 x - k_2 x^2 \qquad (11.26)$$

where constants k_1 and k_2 are the linear and quadratic stiffness terms. We choose $k_2 > 0$ so that the force required to engage the S-cam mechanism decreases with large displacement x. Equations (11.24)–(11.26) constitute the mathematical model of the mechanical subsystem.

Figure 11.24 shows the pneumatic subsystem, which consists of a single chamber (pressure P), the supply pressure P_S, and the valve. Brake chamber pressure will vary because of the input mass flow w and change in volume. Using the basic modeling equation for pneumatic systems derived in Chapter 4, the chamber pressure equation is

$$\dot{P} = \frac{nRT}{V}\left(w - \frac{P}{RT}\dot{V}\right) \qquad (11.27)$$

where R is the gas constant, T is the air temperature, n is the exponent of the polytropic expansion process, and V is the volume of the brake chamber. We assume an isothermal process, and therefore $n = 1$. Brake chamber volume is a function of piston position x

$$V = V_0 + A_b x \qquad (11.28)$$

where V_0 is the volume when the diaphragm is seated (i.e., $x = 0$). The time derivative of the chamber volume is simply a function of piston velocity, or $\dot{V} = A_b \dot{x}$. Clearly, the pressure dynamics of the brake chamber are nonlinear.

Input mass-flow rate w through the valve is modeled by the orifice-flow equations for pneumatic systems, which we presented in Chapter 4 and are repeated below

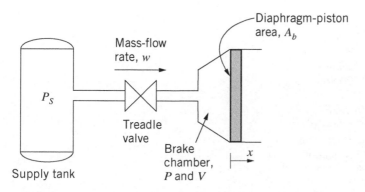

Figure 11.24 Air-brake pneumatic subsystem.

$$w = C_d A_v P_S \sqrt{\frac{2\gamma}{(\gamma - 1)RT} \left[\left(\frac{P}{P_S}\right)^{\frac{2}{\gamma}} - \left(\frac{P}{P_S}\right)^{\frac{\gamma+1}{\gamma}} \right]} \quad \text{if} \quad \frac{P}{P_S} > C_r \tag{11.29}$$

$$w = C_d A_v P_S \sqrt{\frac{\gamma}{RT} C_r^{\frac{\gamma+1}{\gamma}}} \quad \text{if} \quad \frac{P}{P_S} \le C_r \tag{11.30}$$

where γ is the ratio of specific heats ($=1.4$ for air), C_d is the discharge coefficient for losses associated with flow through the orifice, and A_v is the orifice area due to the opening in the treadle valve. Orifice area is assumed to be equal to the product of orifice height h and valve displacement y, or $A_v = hy$. Equation (11.29) pertains to subsonic ("unchoked") flow, where the ratio of downstream and upstream pressures P/P_S is greater than the critical pressure ratio C_r. "Choked" flow involves sonic (Mach 1) flow at the valve orifice, and Eq. (11.30) shows that choked flow occurs when the upstream (supply) pressure is sufficiently larger than the downstream (chamber) pressure. The critical pressure ratio that divides the flow regimes is a function of γ

$$C_r = \left(\frac{2}{\gamma + 1}\right)^{\frac{\gamma}{\gamma-1}} \tag{11.31}$$

and is equal to 0.528 for air. The reader should note that when a constant supply pressure P_S is sufficiently high such that $P/P_S \le 0.528$, Eq. (11.30) shows that the choked mass-flow rate is constant if valve area A_v and temperature T remain constant. Equations (11.27)–(11.31) constitute the mathematical model of the pneumatic subsystem. Table 11.4 presents the numerical values for all parameters for the air-brake system. Note that the maximum stroke (distance) of the push-rod is 0.04 m; that is, the brake lining is fully engaged with the drum when $x = 0.04$ m.

Simulink Model

The complete air-brake system model is fairly complex as indicated by Eqs. (11.24)–(11.30) and the multitude of system parameters summarized in Table 11.4. Before constructing the Simulink model, it is useful to identify and understand the state and input variables of the complete system. Equation (11.24) shows that the mechanical subsystem is second-order, and consequently can be defined by two state variables, piston position (x) and velocity ($\dot{x}$). Chamber pressure P is the input to the mechanical subsystem, as it provides the actuation force. Equation (11.27) shows that the pneumatic system is first-order, and therefore involves a single state variable, chamber pressure P. Piston position and velocity are required to compute the chamber volume and its time derivative, which are both needed in the pressure-rate

Table 11.4   Parameters for the Pneumatic Air-Brake System

System Parameter	Value
Diaphragm/push-rod mass, m	10 kg
Return spring, k	1250 N/m
Viscous friction coefficient, b	12 N-s/m
Preload spring force, F_{PL}	450 N
S-cam load force constant, k_1	2650 N/m
S-cam load force constant, k_2	23,700 N/m^2
Maximum push-rod stroke, x_{max}	0.04 m
Diaphragm area, A_b	0.0129 m^2
Minimum chamber volume (seated), V_0	$1.64(10^{-4})$ m^3
Atmospheric pressure, P_{atm}	$1.0133(10^5)$ N/m^2
Gas constant, R	287 N-m/kg-K
Air temperature, T	298 K
Discharge coefficient, C_d	0.8
Valve orifice height, h	0.002 m

equation (11.27). Input mass-flow rate w is also needed in Eq. (11.27). It is determined by the appropriate orifice-flow equation, Eq. (11.29) or (11.30). Both orifice-flow equations depend on the supply pressure P_S and the valve area, $A_v = hy$. Hence, the supply pressure P_S and valve displacement y are the two free input variables for the air-brake system. In reality, the valve displacement y is proportional to the driver's brake-pedal force (see Fig. 11.22).

Figure 11.25 shows the Simulink model of the integrated air-brake system. Note that the two system inputs are the driver's brake-pedal force (step input) and the constant supply pressure. Valve displacement y is proportional to the brake-pedal force. The orifice-flow subsystem, Eqs. (11.29) and (11.30), has y, P_S, and P as its input variables and mass-flow rate w as the single output. The chamber pressure subsystem, Eq. (11.27), has mass flow w, piston position x, and piston velocity $\dot{x}$ as its input variables and chamber pressure as the single output. Finally, the mechanical subsystem, Eq. (11.24), has pressure P as the single input and piston displacement and velocity as its two output variables.

Figure 11.26 shows the inner details of the orifice subsystem block. Because the orifice-flow equations (11.29) and (11.30) are complicated, they are contained in an M-file instead of explicitly shown as a simulation diagram. The `Interpreted MATLAB Fcn` block, found in the `User-Defined Functions` Simulink library, allows the user to write a customized script (M-file) that is a function of several input variables. Note that the three inputs (P_S, y, and P) in Fig. 11.26 are sent to the multiplexer `Mux`, which combines the three inputs into a 3×1 column vector (`Mux` is found in the `Signal Routing` library). MATLAB M-file 11.2 shows the customized M-file `Valve.m`, which contains the necessary equations, (11.29)–(11.31), to compute either choked or unchoked flow.

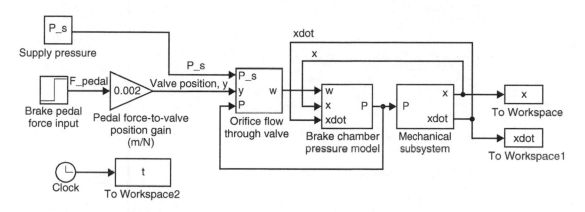

Figure 11.25 Simulink diagram for the pneumatic air-brake system.

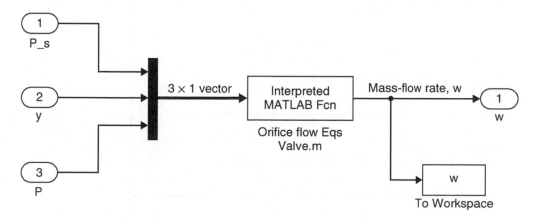

Figure 11.26 Orifice-flow subsystem for the pneumatic air brake.

MATLAB M-file 11.2

```
%   Valve.m
%
%   This M-file models the mass-flow rate of air in/out
%   of the brake chamber.  Assumes flow through the
%   valve is air flow through a sharp-edged orifice.
%
%   Inputs:  u (3x1 vector) = [ P_s y P ]'
%                   P_s = supply pressure, Pa
%                   y = valve displacement, m
%                   P = brake chamber pressure, Pa
%
%   Output:  w = in/out mass-flow rate, kg/s
%

function w = Valve(u)

%  system parameter
h = 2e-3;        %  height of valve opening, m

%  pneumatic constants (air)
gamma = 1.4;          % = cp/cv = ratio of specific heats
Cd = 0.8;             %  discharge coefficient
R = 287;              %  gas constant (air), N-m/kg-K
T = 298;              %  temperature, K
P_atm = 1.0133e5;     %  ambient (atmospheric) pressure, Pa

%  System inputs
P_s = u(1);           %  supply pressure, Pa
y   = u(2);           %  valve displacment, m
P   = u(3);           %  pressure in brake chamber, Pa

%  Compute valve orifice area
Av = abs(y)*h;        %  valve orifice area, m^2

%  Determine if flow is from supply tank (y > 0), or if flow
%  is out to atmospheric (ambient) pressure (y < 0)
if y >= 0
    Pv = P_s;         %  use supply pressure
else
    Pv = P_atm;       %  use atmospheric pressure
end

%  find up/down stream pressure
P_hi = max(P,Pv);     %  highest pressure (upstream)
P_lo = min(P,Pv);     %  lowest pressure (downstream)

%  critical pressure ratio (for choked flow)
Cr = (2/(gamma+1))^(gamma/(gamma-1));   %  = 0.528 for air

%  Determine whether or not flow is choked (sonic at throat)
PR = P_lo/P_hi;       %  pressure ratio downstream/upstream of orifice

%  mass-flow rate equations for compressible flow, kg/s
if PR > Cr
    %  flow is not choked
    w = sign(Pv-P)*Cd*Av*P_hi*sqrt(((2*gamma/(gamma-1))/(R*T))*(PR^(2/gamma)...
        - PR^((gamma+1)/gamma)));
else
    %  flow is choked (Mach=1 at throat)
    w = sign(Pv-P)*Cd*Av*P_hi*sqrt((gamma/(R*T))*Cr^((gamma+1)/gamma));
end
```

Figure 11.27 shows the inner details of the chamber pressure subsystem block, which also utilizes a user-defined M-file. MATLAB M-file 11.3 shows the customized M-file Pdot.m, which contains the pressure-rate equation (11.27) with the appropriate calculations for volume V and the time rate of volume, $\dot{V}$. Note that the single integrator in Fig. 11.27

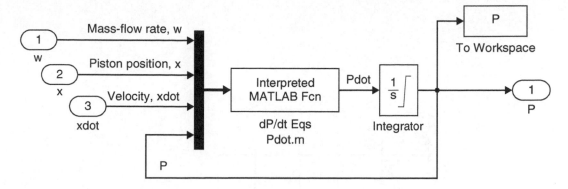

Figure 11.27 Brake chamber pressure subsystem for the pneumatic air brake.

MATLAB M-file 11.3

```
%
%  M-file for computing the pressure-rate (P-dot)
%  for the air-brake chamber
%
%  Inputs:   u (4x1 vector) = [ w x xdot P ]'
%                w = mass-flow rate in/out chamber (kg/s)
%                x = diaphragm-piston position, m
%                xdot = diaphragm-piston velocity, m/s
%                P = brake chamber pressure, Pa
%
%  Output: dPdt = dP/dt, Pa/s
%

function dPdt = Pdot(u)

%  brake chamber parameters
Ab = 0.0129;          %  area of diaphragm, m^2
V0 = 1.64e-4;         %  volume of brake chamber when x=0, m^3

R = 287;              %  gas constant (air), N-m/kg-K
n = 1;                %  polytropic expansion index
T = 298;              %  air temperature, K

%  system inputs
w    = u(1);          %  in/out mass-flow rate of air(+ or -), kg/s
x    = u(2);          %  diaphragm-piston position, m
xdot = u(3);          %  diaphragm-piston velocity, m/s
P    = u(4);          %  pressure of brake chamber, Pa

%  compute chamber volume and dV/dt
V = V0 + Ab*x;        %  volume of brake chamber, m^3
Vdot = Ab*xdot;       %  time-rate of volume, m^3/s

%  pressure-rate for brake chamber, Pa/s
dPdt = ((n*R*T)/V)*(w - P*Vdot/(R*T));
```

contains the "limit" symbol, which indicates that limits have been imposed on the output of the integrator, which in this case is chamber pressure. A lower limit of zero has been set to the integrator output so that pressure can never be negative. This limit is simply a safeguard, because normal simulations never result in negative chamber pressure.

Finally, Fig. 11.28 shows the inner details of the mechanical subsystem block. This block diagram has many features similar to the mechanical subsystem for the solenoid actuator shown in Fig. 11.16. Both mechanical subsystems involve a contact force F_C due to the preload spring force pushing the mass against its seat. A dashed box in Fig. 11.28

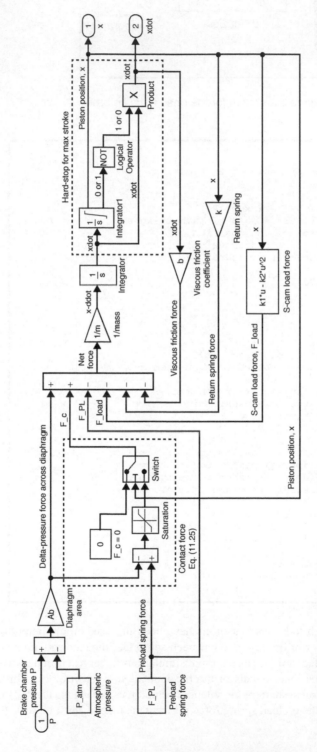

Figure 11.28 Mechanical subsystem for the pneumatic air brake.

indicates the part of the simulation diagram that computes the contact force. As shown in Eq. (11.25), F_C is determined by subtracting the differential pressure force from the preload spring force and sending the result to a `Saturation` block with a lower limit of zero and upper limit of $+\infty$. Therefore, the contact force can never be negative (it cannot "pull" on the mass) and can only be a positive value that balances the difference between the preload force and the pressure force. However, Eq. (11.25) also shows that the contact force only exists when the piston is seated, or $x = 0$, and therefore a `Switch` block (from the `Signal Routing` library) is used to switch the contact force to zero when $x > 0$. Piston position x is fed back to input port 2 in the `Switch` block and is used as the "decision signal" for switching between the positive contact force and a value of zero. The `Switch` block has the following logic: if $x > 0$, then pass the "top" signal (`F_c = 0`); otherwise, pass the "bottom" signal, which is the positive difference between the preload and differential pressure forces.

The second dashed box in Fig. 11.28 shows the logic for enforcing the hard-stop limit for the piston/push-rod stroke, or $x \leq x_{max}$. Recall that the push-rod displacement x cannot exceed $x_{max} = 0.04$ m (4 cm) because of the full engagement of the S-cam and brake drum. This hard-stop limit is enforced by setting the upper limit in the velocity integrator block to 0.04. While setting the upper limit on the output of the velocity integrator will limit the stroke x, it will not provide accurate velocity information (which should become zero when the stroke x has reached its constant, maximum value). Therefore, the "saturation port" of the velocity integrator is enabled in order to provide a binary output signal that indicates whether the integrator output has reached its limit: if the binary signal is zero, the integral has not reached its limit; if the signal is unity, the upper limit is being applied to the integral. When the integral has reached its limit (i.e., $x = x_{max}$ and binary signal $= 1$), we wish to set the velocity signal-path variable $\dot{x}$ to zero. Hence, we send the saturation port binary signal (0 or 1) to the `Logical Operator` block from the `Logic and Bit Operations` library. We select the Boolean operation `NOT` so that its output is 0 (`FALSE`) when the input is 1 (`TRUE`), and vice versa. In other words, the Boolean `NOT` operation takes a binary input signal (0 or 1) and produces an opposite binary output signal (1 or 0). This binary output signal is multiplied by the velocity signal $\dot{x}$ to produce the velocity information to be used in the simulation. To summarize, the hard-stop logic shown in Fig. 11.28 passes the acceleration-integrator output (velocity) without alteration if $x < x_{max}$, but sets the velocity signal to zero if the stroke x has reached its hard-stop upper limit.

Air-Brake System Analysis

Next, we present a simulation of the air-brake system under nominal operating conditions with a constant supply pressure $P_S = 5.84(10^5)$ Pa and step brake-pedal force of 1 N (applied at time $t = 0.5$ s). Note that the supply pressure is 5.76 times greater than atmospheric pressure. The Simulink diagram in Fig. 11.25 shows that a 1-N pedal force produces a 0.002 m (2 mm) step displacement of the treadle valve y. Because the height of the valve opening is $h = 0.002$ m (2 mm), the step pedal-force input produces a step valve area of 4 mm^2. Figure 11.29 shows the piston/push-rod response $x(t)$ (in centimeters) to the step valve opening. The push rod reaches its hard-stop limit (4 cm) in less than 0.22 s. Brake chamber pressure P, shown in Fig. 11.30, exhibits an oscillatory behavior during the out-stroke phase as the push rod is displaced to the right. These pressure transients are the result of oscillations in the chamber volume, or piston velocity, which can be observed in the push-rod response $x(t)$ shown in Fig. 11.29. When the push rod reaches its hard-stop limit of 4 cm at approximately $t = 0.71$ s, the subsequent brake chamber pressure response no longer exhibits oscillations because the chamber volume remains constant. Figure 11.31 shows the mass-flow rate of the compressed air through the treadle valve, which begins with choked flow at $w = 0.00438$ kg/s when the valve is stepped open. The flow remains choked until the brake chamber pressure P exceeds $0.528P_S = 3.084(10^5)$ Pa at time $t = 1.02$ s (see Figs. 11.30 and 11.31), at which point the flow becomes unchoked and is governed by Eq. (11.29). The reader should note that the chamber pressure exhibits a *constant* rate of pressure increase between $t = 0.71$ s (when the push rod reaches its limit) and $t = 1.02$ s (when the flow rate becomes unchoked) as demonstrated by Eq. (11.27) and Fig. 11.30. Equation (11.27) shows that pressure rate $\dot{P}$ is constant in this time span because w is constant (choked flow) and $\dot{V} = 0$ (the piston/push rod has reached its hard-stop limit).

Figures 11.30 and 11.31 show that the brake chamber pressure reaches its steady-state pressure ($= P_S$) at approximately $t = 1.8$ s, and consequently mass-flow rate through the treadle valve becomes zero at this time. In summary, the piston/push-rod mass takes about 0.21 s to reach its hard-stop limit and fully rotate the S-cam after the treadle valve is opened. The brake chamber continues to pressurize after the drum brake is engaged, and the chamber pressure takes a total of 1.3 s to reach its steady-state pressure. At steady state, the total force transmitted to the brake drum is

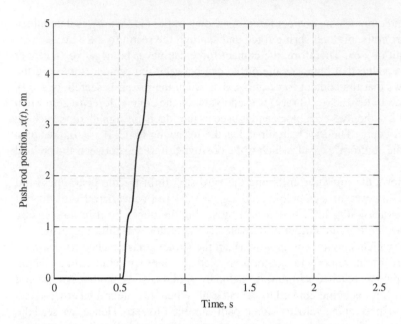

Figure 11.29 Piston/push-rod position for $P_S = 5.84(10^5)$ Pa.

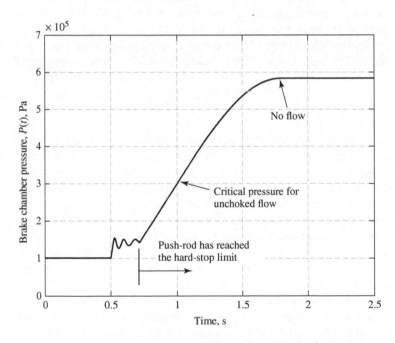

Figure 11.30 Brake chamber pressure for $P_S = 5.84(10^5)$ Pa.

$$F_{\text{drum}} = (P_S - P_{\text{atm}})A_b - F_{\text{PL}} - kx_{\text{max}} - F_{\text{load}} \tag{11.32}$$

where the load force required to rotate the S-cam mechanism is

$$F_{\text{load}} = k_1 x_{\text{max}} - k_2 x_{\text{max}}^2$$

Hence, for the case with supply pressure $P_S = 5.84(10^5)$ Pa, the steady-state force transmitted to the drum brake is 5658 N.

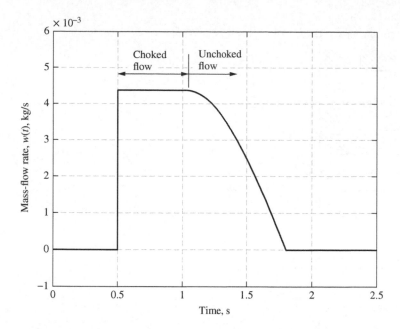

Figure 11.31 Mass-flow rate through the treadle valve for $P_S = 5.84(10^5)$ Pa.

11.5 HYDRAULIC SERVOMECHANISM CONTROL

Our fourth case study involves a feedback control system design for a hydraulic actuator. Hydraulic actuators have many applications ranging from robotics, earth-moving machinery, construction equipment, and aerospace [6, 7]. Figure 11.32 shows a schematic diagram of an electrohydraulic actuator (EHA) that consists of an electromechanical actuator (solenoid), spool valve, and hydraulic cylinder with piston. An input voltage signal is applied to the solenoid actuator (not shown in Fig. 11.32), which in turn moves the spool valve left or right in order to meter flow in and out of the hydraulic cylinder. If the spool-valve displacement y is positive (to the right) as shown in Fig. 11.32, fluid flows from the supply pressure P_S through the valve orifice and into the right-hand side of the cylinder. Consequently, if right-side cylinder pressure P_1 is higher than pressure P_2, the piston moves to the left resulting in a positive displacement x for the piston. When $y > 0$, fluid flows from the left-side cylinder (pressure P_2) to the reservoir (drain) pressure P_r. Volumetric flow into and out of the right-side cylinder is Q_1, while Q_2 is the volumetric flow out of and into the left-side cylinder.

Our objective is to develop a feedback control system for the EHA that will automatically adjust the input voltage so that the piston stroke x reaches a desired target position. We want a fast piston response with good damping characteristics and very little overshoot. As an example, this case study might represent a hydraulic actuator for positioning an aerodynamic surface such as an airplane's elevator, aileron, or rudder. This type of actuator is called a servomechanism.

Mathematical Model

The complete mathematical model consists of the electromechanical (solenoid), hydraulic, and mechanical subsystems. Figure 11.33 shows a free-body diagram of the mechanical subsystem, which consists of the piston and load mass m. Piston displacement x is positive to the left as measured from the right end of the cylinder (see Fig. 11.32). Applying Newton's second law with a positive sign convention to the left yields

$$+ \leftarrow \sum F = P_1 A - P_2 A - b\dot{x} = m\ddot{x} \tag{11.33}$$

where P_1 and P_2 are the chamber pressures of the right and left sides of the cylinder, A is the area of the piston, and b is the viscous friction coefficient. Rearranging Eq. (11.33) so that all terms involving displacement x are on the left-hand side yields

$$m\ddot{x} + b\dot{x} = (P_1 - P_2)A \tag{11.34}$$

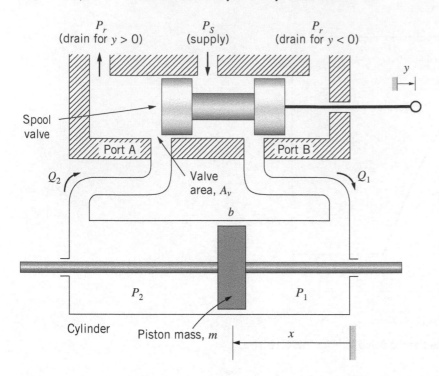

Figure 11.32 Schematic diagram of an electrohydraulic actuator.

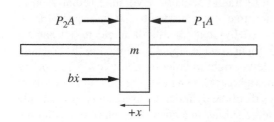

Figure 11.33 Free-body diagram of the mechanical subsystem.

Two pressure-rate equations are required for the two cylinder chambers

$$\dot{P}_1 = \frac{\beta}{V_1}(Q_1 - \dot{V}_1) \tag{11.35}$$

$$\dot{P}_2 = \frac{\beta}{V_2}(Q_2 - \dot{V}_2) \tag{11.36}$$

where β is the fluid bulk modulus, Q_1 and Q_2 are the volumetric-flow rates for chambers 1 and 2, and V_1 and V_2 are the volumes of cylinder chambers 1 and 2. Equations (11.35) and (11.36) are the basic pressure-rate equations derived in Chapter 4 for hydraulic systems with compressible fluids. The instantaneous volumes for cylinder chambers 1 and 2 depend on piston position x

$$V_1 = V_0 + Ax \tag{11.37}$$

$$V_2 = V_0 + A(L - x) \tag{11.38}$$

where V_0 is the volume when $x = 0$ (piston is at the right end of the cylinder). The total stroke of the piston is L. The time rate of change of the two cylinder volumes depends on the piston velocity: $\dot{V}_1 = A\dot{x}$ and $\dot{V}_2 = -A\dot{x}$.

Volumetric flow through the spool valve between the supply pressure P_S and the cylinder (for either chamber 1 or 2) is modeled by the orifice-flow equation for hydraulic systems

$$Q_{1,2} = C_d A_v \mathrm{sgn}(P_S - P_{1,2})\sqrt{\frac{2}{\rho}|P_S - P_{1,2}|} \tag{11.39}$$

where A_v is the valve area, C_d is the discharge coefficient, and ρ is the fluid density. When spool-valve position $y > 0$, the supply pressure P_S is connected to cylinder chamber P_1 and Eq. (11.39) is used to compute Q_1. When $y < 0$, the supply pressure P_S is connected to P_2 and Eq. (11.39) is used to compute Q_2. In general, the supply pressure P_S is always greater than the cylinder pressure P_1 or P_2, but the possibility for fluid flow *from* the cylinder back to the supply pressure is included by using the signum function. Volumetric flow through the spool valve between the cylinder (chamber 1 or 2) and the reservoir (drain) pressure P_r is also modeled by the orifice-flow equation

$$Q_{1,2} = -C_d A_v \mathrm{sgn}(P_{1,2} - P_r)\sqrt{\frac{2}{\rho}|P_{1,2} - P_r|} \tag{11.40}$$

When $y < 0$, Eq. (11.40) models flow Q_1 from chamber 1 to the drain and when $y > 0$ Eq. (11.40) models flow Q_2 from chamber 2 to the drain. Equation (11.40) shows that Q_1 (or Q_2) is negative when P_1 (or P_2) is greater than the reservoir pressure and the fluid flows from the cylinder to the reservoir tank.

To complete the mathematical model, we must show the relationship for the electromagnetic solenoid used to position the spool valve. One common modeling method [6] is to use an underdamped, second-order transfer function to relate spool-valve position y to voltage input $e_{\mathrm{in}}(t)$

$$G(s) = \frac{Y(s)}{E_{\mathrm{in}}(s)} = \frac{K_v \omega_n^2}{s^2 + 2\zeta\omega_n s + \omega_n^2} \tag{11.41}$$

where ω_n is the undamped natural frequency, ζ is the damping ratio, and K_v is the DC gain of the solenoid. In many cases, we can determine the second-order modeling parameters ω_n, ζ, and K_v by open-loop experimental trials where we provide a step voltage signal $e_{\mathrm{in}}(t)$ and measure the resulting valve position $y(t)$.

Finally, the spool-valve orifice area A_v is assumed to be a linear function of spool-valve position y

$$A_v = h|y| \tag{11.42}$$

where h is the height of the valve opening. The absolute value of valve position y must be used because y can be positive (right) or negative (left). Valve orifice area is required in the nonlinear in/out-flow equations (11.39) and (11.40).

Equations (11.34)–(11.42) constitute the complete mathematical model of the EHA. Table 11.5 presents the numerical values for all parameters for the EHA system.

Table 11.5 Parameters for the EHA

System Parameter	Value
Piston and load mass, m	12 kg
Viscous friction coefficient, b	250 N-s/m
Piston area, A	$6.33(10^{-4})$ m^2
Minimum chamber volume, V_0	$1.64(10^{-4})$ m^3
Piston stroke, L	0.6 m
Supply pressure, P_S	$17.2(10^6)$ Pa
Reservoir (drain) pressure, P_r	$1.0133(10^5)$ Pa
Fluid bulk modulus, β	$689(10^6)$ Pa
Valve discharge coefficient, C_d	0.62
Fluid density, ρ	875 kg/m^3
Valve orifice height, h	0.008 m
Solenoid–valve undamped natural frequency, ω_n	350 rad/s
Solenoid–valve damping ratio, ζ	0.9
Solenoid–valve DC gain, K_v	$2(10^{-5})$ m/V

Simulink Model

The complete EHA model is nonlinear and fairly complex, as indicated by Eqs. (11.34)–(11.42) and the multitude of system parameters summarized in Table 11.5. Just as we did with the pneumatic air-brake system, it is useful to identify and understand the state and input variables of the complete system before constructing the Simulink model. The mechanical modeling equation (11.34) is second-order and requires two state variables, piston position (x) and velocity ($\dot{x}$). Cylinder pressures P_1 and P_2 are the two inputs to the mechanical subsystem, as the differential pressure across the piston provides the actuation force. Equations (11.35) and (11.36) show that the complete hydraulic system is composed of two first-order nonlinear ODEs with two additional state variables P_1 and P_2. Piston position and velocity are required to compute the chamber volumes and time derivatives, which are both needed in the pressure-rate equations (11.35) and (11.36). Volumetric-flow rates Q_1 and Q_2 are also needed in the pressure-rate equations, and they are determined by the appropriate valve-orifice-flow equations, Eq. (11.39) or (11.40). The orifice-flow equations depend on the supply pressure P_S, reservoir pressure P_r, cylinder pressures, and the valve area $A_v = h|y|$. Finally, the solenoid–valve dynamics are modeled by the linear second-order transfer function (11.41) with source voltage $e_{in}(t)$ as the input, and spool-valve displacement y as the output. Hence, the supply pressure P_S, reservoir pressure P_r, and voltage input $e_{in}(t)$ are the three input variables for the EHA system.

Figure 11.34 shows the Simulink model of the integrated EHA system. Note that the three system inputs are the voltage input $e_{in}(t)$ and the supply and reservoir pressures. Spool-valve displacement y, supply and reservoir pressures, and cylinder pressures are the four inputs to the two orifice-flow subsystems (recall that a valve displacement y will produce two orifice flows, one from P_S to the cylinder and one flow from the cylinder to the reservoir). The two outputs of the orifice-flow subsystems are volumetric-flow rates Q_1 and Q_2, which are inputs to the two pressure-rate subsystems. Piston position and velocity, x and $\dot{x}$, are required for the volume and volume-rate terms in the two pressure-rate subsystems and consequently are also input variables. Cylinder pressures P_1 and P_2 are the two output variables from the pressure-rate subsystem that become the two inputs for the mechanical subsystem.

Because the valve-orifice-flow equations (11.39) and (11.40) are complicated nonlinear functions, they are contained in a user-defined M-file instead of as a simulation diagram. MATLAB M-file 11.4 shows the customized M-file `flow_Q1.m`, which computes the volumetric flow Q_1 either from P_S to P_1 (for $y > 0$), or from cylinder P_1 to the reservoir P_r (for $y < 0$). The customized M-file `flow_Q2.m` is identical to `flow_Q1`, except that it computes the flow in/out of cylinder chamber 2 and hence uses pressure P_2. The inner components of either orifice-flow subsystem shown in Fig. 11.34 consist of an `Interpreted MATLAB Fcn` block (M-file `flow_Q1` or `flow_Q2`) with four inputs (P_S, P_r, y, and P_1 or P_2) and a single output signal (Q_1 or Q_2). Therefore, both hydraulic-flow subsystems are similar to the pneumatic valve-flow subsystem shown in Fig. 11.26. Figure 11.35 shows the inner details of the pressure-rate subsystem for the right-side chamber P_1. The reader should be able to trace the signals and computations for volume V_1, volume-rate $\dot{V}_1$, and ultimately pressure-rate $\dot{P}_1$. The second pressure-rate subsystem (for $\dot{P}_2$) is nearly identical to Fig. 11.35, except that the volume computation is governed by Eq. (11.38). Finally, Fig. 11.36 shows the inner details of the mechanical subsystem. The reader should be able to identify the mechanical modeling equation (11.34) in the subsystem block diagram.

Pulse Response of the EHA

Next, we demonstrate the open-loop response of the EHA for a 10-V pulse input. We assume that the initial cylinder pressures are $P_1 = P_2 = 0.1P_S$ (the value of the initial pressure has little effect on the response as long as the system is initially in equilibrium, or $P_1 = P_2$). The piston is initially at rest at the center of the cylinder, or $x(0) = 30$ cm. A constant (step) input voltage ($e_{in} = 10\,\text{V}$) is applied at time $t = 0.5\,\text{s}$ and then stepped to zero at time $t = 1\,\text{s}$ in order to create a half-second pulse input.

Because the solenoid–valve subsystem is linear, it is relatively easy to compute its pulse response. Transfer function (11.41) shows that the DC gain of the solenoid is K_v and consequently the steady-state valve position is $e_{in} \times K_v$, or $10\,\text{V} \times 2(10^{-5})\,\text{m/V} = 2(10^{-4})\,\text{m}$ (0.2 mm). The settling time of the valve is $4/(\zeta\omega_n) = 0.013\,\text{s}$ (13 ms). Hence, the spool valve reaches its 0.2 mm steady-state position very quickly and with little overshoot because its damping ratio is $\zeta = 0.9$. The valve closes in 13 ms after time $t = 1\,\text{s}$ when the pulse voltage is stepped to zero. Consequently, the valve response $y(t)$ is very nearly a pulse itself with a magnitude of 0.2 mm for $0.5 < t < 1\,\text{s}$.

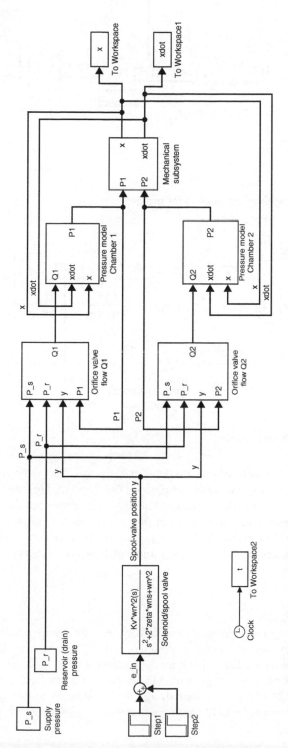

Figure 11.34 Simulink diagram for the EHA system.

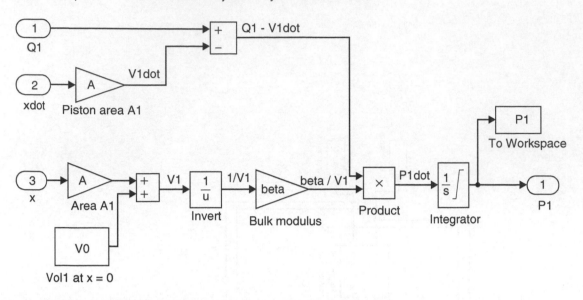

Figure 11.35 Cylinder pressure subsystem (chamber 1) for the EHA.

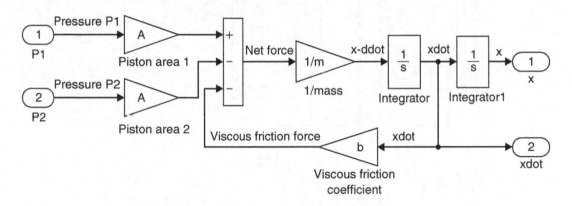

Figure 11.36 Mechanical subsystem for the EHA.

Figure 11.37 shows the piston response $x(t)$ for the 10-V pulse input. The piston stroke exhibits a nearly linear increase from its initial position (30 cm) to its steady-state position (40.7 cm) during the 0.5 s pulse input. Figure 11.38 shows the volumetric-flow rates Q_1 (into cylinder chamber 1) and Q_2 (out of cylinder chamber 2). Note that during the "steady-state" phase of the valve opening, the magnitudes of the in- and out-flow rates are equal, which indicates that the differential pressure across the piston has reached a constant value. This phenomenon is confirmed by Fig. 11.39, which shows that the differential pressure exhibits a damped sinusoidal decay to a steady-state value of about 85,000 Pa during the pulse opening of the valve.

Linear EHA Model

Recall that our overall goal is to design an automatic feedback system for precise position control of the EHA. The full EHA model presented in the previous subsections is complex and highly nonlinear. Designing a feedback control system becomes systematic when we have a *linear* plant model because we can make use of linear analysis tools such as the root-locus and frequency-response methods presented in Chapter 10. Therefore, it is to our advantage to develop a linear EHA model purely for the sake of control system design. It should be noted that we intend to test any potential control scheme designs with the full *nonlinear* EHA system dynamics.

MATLAB M-file 11.4

```
%    flow_Q1.m
%
%    This M-file models the volumetric-flow rate of fluid
%    in/out of chamber 1.  Assumes valve flow is modeled
%    by flow through a sharp-edged orifice
%
%    Inputs:   u (4x1 vector) = [ P_s P_r y P1 ]'
%                   P_s = supply pressure, Pa
%                   P_r = reservoir pressure (drain), Pa
%                   y = valve displacement, m
%                   P1 = pressure of chamber 1, Pa
%
%    Output:   Q1 = in/out volumetric-flow rate, m^3/s
%

function Q1 = flow_Q1(u)

%  system parameter
h = 0.008;              %   height of valve opening, m

%  hydraulic constants
Cd = 0.62;             %   discharge coefficient
rho = 875;             %   fluid density, kg/m^3

%  System inputs
P_s = u(1);       %   supply pressure, Pa
P_r = u(2);       %   reservoir pressure, Pa
y   = u(3);       %   valve displacement, m
P1  = u(4);       %   pressure in chamber 1, Pa

%  Compute valve orifice area
Av = abs(y)*h;         %   valve orifice area, m^2

%  Determine if flow is from supply (y > 0), or if flow
%  is out to reservoir pressure (y < 0)
if y >= 0
    %  Chamber 1 is connected to the supply, P_s
    %    (flow is positive if P_s > P1)
    Q1 = Cd*Av*sign(P_s - P1)*sqrt( 2*abs(P_s - P1)/rho );
else
    %  Chamber 1 is connected to the reservoir, P_r
    %    (flow is negative if P1 > P_r)
    Q1 = -Cd*Av*sign(P1 - P_r)*sqrt( 2*abs(P1 - P_r)/rho );
end
```

The following linearization steps are similar to the process presented by Ogata [8] for linearizing a hydraulic actuator. To begin the analysis, rewrite Eqs. (11.39) and (11.40) for the *magnitudes* of the volumetric flows with valve displacement $y > 0$

$$\text{In-flow to chamber 1:} \quad Q_1 = C_d A_v \sqrt{\frac{2}{\rho}(P_S - P_1)} \tag{11.43}$$

$$\text{Out-flow from chamber 2:} \quad Q_2 = C_d A_v \sqrt{\frac{2}{\rho}(P_2 - P_r)} \tag{11.44}$$

If we assume steady, incompressible flow where $Q_1 = Q_2$, then we can equate the two pressure differences contained in the radicands in Eqs. (11.43) and (11.44)

$$P_S - P_1 = P_2 - P_r \tag{11.45}$$

Let us define $\Delta P = P_1 - P_2$ as the differential pressure across the piston ($\Delta P > 0$ because fluid is flowing from chamber 2 to the reservoir). Substituting $P_2 = P_1 - \Delta P$ into Eq. (11.45) and solving for the supply pressure yields

$$P_S = P_1 + P_1 - \Delta P - P_r \tag{11.46}$$

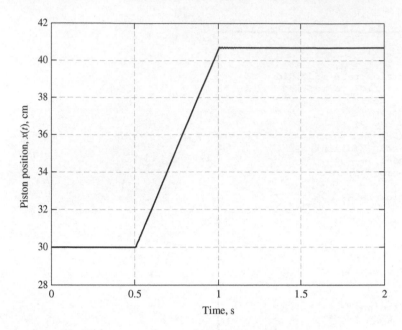

Figure 11.37 Piston stroke for 10-V pulse input.

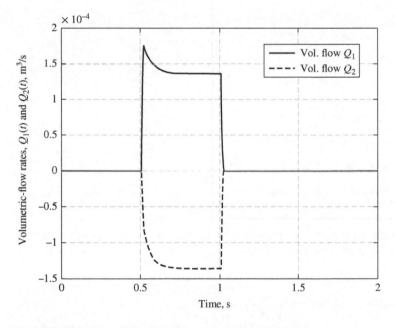

Figure 11.38 Volumetric-flow rates for 10-V pulse input.

Usually, the reservoir pressure P_r is much less than the other pressures and can be neglected. With this assumption, we obtain the following expression for cylinder pressure P_1 from Eq. (11.46)

$$P_1 = \frac{P_S + \Delta P}{2} \tag{11.47}$$

Substituting Eq. (11.47) for pressure P_1 in Eq. (11.43) yields the following expression for flow rate Q_1

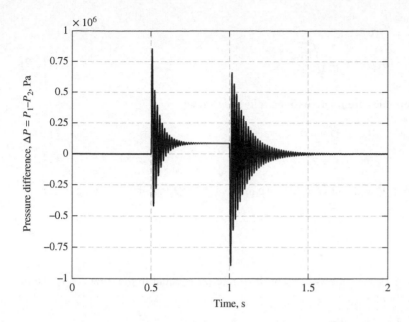

Figure 11.39 Differential pressure across the piston for 10-V pulse input.

$$Q_1 = C_d A_v \sqrt{\frac{2}{\rho}\left(P_S - \frac{P_S + \Delta P}{2}\right)}$$

Applying simple algebra and substituting $A_v = hy$ yields

$$Q_1 = C_d hy \sqrt{\frac{P_S - \Delta P}{\rho}} = f(y, \Delta P) \tag{11.48}$$

Equation (11.48) is a nonlinear function of valve position y and differential pressure ΔP. We can linearize flow rate about the reference states y^* and ΔP^*

$$\delta Q_1 = \left.\frac{\partial f}{\partial y}\right|_* \delta y + \left.\frac{\partial f}{\partial (\Delta P)}\right|_* \delta(\Delta P) \tag{11.49}$$

where the partial derivatives of Eq. (11.48) are

$$\left.\frac{\partial f}{\partial y}\right|_* = C_d h \sqrt{\frac{P_S - \Delta P}{\rho}}\Bigg|_* \tag{11.50}$$

$$\left.\frac{\partial f}{\partial (\Delta P)}\right|_* = \frac{-C_d hy}{2\rho}\left(\frac{P_S - \Delta P}{\rho}\right)^{-\frac{1}{2}}\Bigg|_* \tag{11.51}$$

We select the reference (or nominal) states as $y^* = 0$ and $\Delta P^* = 0$ (no flow), and consequently the partial derivatives evaluated at the operating point are

$$\left.\frac{\partial f}{\partial y}\right|_* = C_d h \sqrt{\frac{P_S}{\rho}}$$

$$\left.\frac{\partial f}{\partial (\Delta P)}\right|_* = 0$$

Hence, the linear flow equation (11.49) becomes

$$\delta Q_1 = C_d h \sqrt{\frac{P_S}{\rho}} \delta y \tag{11.52}$$

The reader should note that the perturbation variables are relative to the reference values

$$\delta Q_1 = Q_1 - Q_1^* \tag{11.53}$$

$$\delta y = y - y^* \tag{11.54}$$

and because the reference condition is zero flow ($Q_1^* = 0$ and $y^* = 0$), we can use $\delta Q_1 = Q_1$ and $\delta y = y$ in the linearized flow equation (11.52)

$$Q_1 = C_d h \sqrt{\frac{P_S}{\rho}} y \tag{11.55}$$

If we assume steady, incompressible flow where $\dot{P}_1 = 0$, the volumetric-flow rate Q_1 is equal to the time derivative of chamber volume ($\dot{V}_1 = A\dot{x}$) and Eq. (11.55) becomes

$$C_d h \sqrt{\frac{P_S}{\rho}} y = A\dot{x} \tag{11.56}$$

Solving Eq. (11.56) for the piston velocity, we obtain

$$\dot{x} = K_{\text{HA}} y \tag{11.57}$$

where the "hydraulic actuator" gain is

$$K_{\text{HA}} = \frac{C_d h}{A} \sqrt{\frac{P_S}{\rho}} \tag{11.58}$$

The linearized solution for the piston stroke $x(t)$ is simply the integral of Eq. (11.57)

$$x(t) = x_0 + K_{\text{HA}} \int y \, dt \tag{11.59}$$

where x_0 is the initial stroke at time $t = 0$. In other words, the complex, nonlinear EHA model shown in Fig. 11.34 can be replaced by a single integrator block. Figure 11.40 shows a Simulink model of the linearized EHA system, which consists of the voltage input $e_{\text{in}}(t)$, the second-order linear solenoid–valve model, and the single integrator that represents the

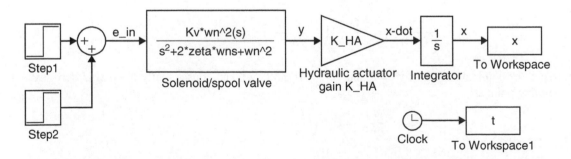

Figure 11.40 Simulink model of the linearized EHA system.

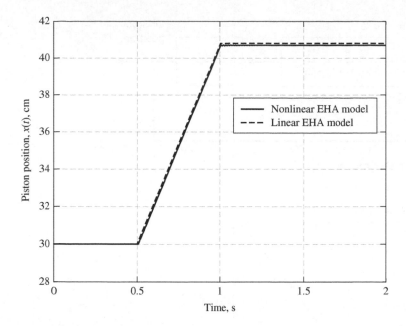

Figure 11.41 Piston position responses to a pulse input for nonlinear and linear EHA models.

linear I/O equation relating spool-valve position y and piston stroke x. Of course, the linearized EHA model does not provide information about cylinder pressures.

We can compare the piston responses $x(t)$ from the full nonlinear EHA simulation (Fig. 11.34) and the extremely simple linear model (Fig. 11.40) for a 10-V pulse input. Using the nominal EHA parameters in Table 11.5, the hydraulic actuator gain is determined to be $K_{HA} = 1080.9\,\text{s}^{-1}$. Hence, the linear model (11.57) predicts that the piston velocity is the product of K_{HA} and valve displacement y, which yields $\dot{x} = 0.2162\,\text{m/s}$ for $y = 2(10^{-4})\,\text{m}$. Figure 11.41 shows the piston stroke $x(t)$ for a 10-V pulse input from the full nonlinear EHA model and the linear EHA model. The linear model response shows an excellent match with the nonlinear model response.

Feedback Control System Design

The goal of the feedback system is precise position control for a reference position (stroke) command. We begin with a proportional control scheme, where the solenoid-voltage signal $e_{in}(t)$ is proportional to the position error. Figure 11.42 shows a proportional feedback control system where x_{ref} is the reference position command for the piston rod. Note that we are using the simple linear hydraulic actuator model, which is an integrator block with numerator $K_{HA} = 1080.9\,\text{s}^{-1}$. The proportional-control gain is K_P, and it has units of V/m because it converts a position error (m) to a voltage signal. We can check the steady-state accuracy of the proposed control scheme by calculating the closed-loop transfer function

$$T(s) = \frac{K_P G(s)}{1 + K_P G(s) H(s)} \tag{11.60}$$

where $G(s)$ is the forward transfer function

$$G(s) = \frac{2648.15}{s(s^2 + 630s + 122{,}500)} \tag{11.61}$$

and $H(s)$ is the feedback transfer function (unity in this case). Consequently, the closed-loop transfer function in Eq. (11.60) becomes

$$T(s) = \frac{2648.15 K_P}{s^3 + 630s^2 + 122{,}500s + 2648.15 K_P} \tag{11.62}$$

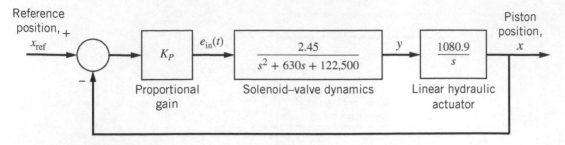

Figure 11.42 Proportional control for the linear EHA model.

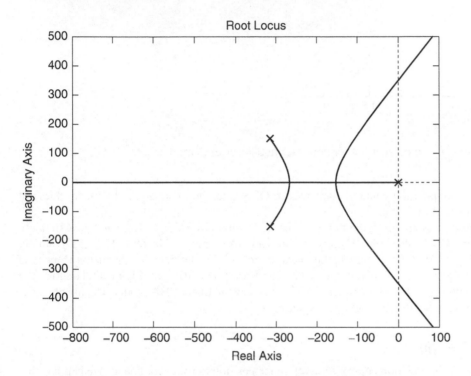

Figure 11.43 Root locus for the linear EHA model with proportional control.

Because the DC gain of the closed-loop transfer function, $T(s = 0)$, is unity for *any* positive value of the proportional gain K_P, the proportional-control system will exhibit zero steady-state error for a constant reference position command. Recall that our steady-state error analysis in Chapter 10 showed that a type 1 system (i.e., one integrator in the forward transfer function) exhibits zero steady-state error for a constant input, and a finite steady-state error for a ramp input. Clearly, Fig. 11.42 shows that our linearized EHA model is a type 1 system because the hydraulic actuator model is an integrator.

The variation of the closed-loop response with gain K_P can be determined by using the root-locus method. We can easily produce the root locus by using the following MATLAB commands:

```
>> sysG = tf(2648.15,[ 1 630 122500 0 ]);     % forward transfer function
>> rlocus(sysG)                                % plot root locus
```

The first command builds the forward transfer function $G(s)$, and the second command creates the root locus, which is shown in Fig. 11.43. The closed-loop poles begin at the open-loop roots (poles) when the proportional gain K_P is zero. In this case, the open-loop poles are $s = 0$ (the integrator) and $s = -315 \pm j152.56$ (the solenoid–valve dynamics). Figure 11.43 shows that two closed-loop poles move from the two complex open-loop poles to the negative real axis, and a single closed-loop pole moves to the left from the origin to a break-away point near -153 on the negative real axis. If the proportional gain K_P is too high, two closed-loop poles follow $\pm 60°$ asymptotes and eventually cross the imaginary axis, rendering the closed-loop system unstable.

The root-locus diagram shown in Fig. 11.43 indicates that with the proper gain selection the three closed-loop poles can be large negative values, and consequently the closed-loop response will be extremely fast and overdamped (no oscillations in the transient response). As an example, let the proportional gain be $K_P = 2850$ V/m and compute the closed-loop poles using MATLAB's `rlocus` command

```
>> Kp = 2850;                    % set proportional gain
>> CLpoles = rlocus(sysG,Kp)     % compute closed-loop poles for gain Kp
```

The three closed-loop poles for this gain setting are $s = -140.3$, $s = -166.4$, and $s = -323.3$. Using the "slowest" closed-loop pole, $s = -140.3$, the "slowest" component of the closed-loop transient response will be $e^{-140.3t}$, which decays to a steady-state value in about 0.029 s. However, this idealized closed-loop response has its limitations owing to the large control gain. Figure 11.42 shows that the solenoid-voltage input $e_{in}(t)$ is the position error $x_{ref} - x$ multiplied by control gain K_P. Consequently, if the position error is 10 cm (0.1 m) and $K_P = 2850$ V/m, the voltage input will be $e_{in} = 285$ V, which likely exceeds the capability of the solenoid. Hence, the proportional gain K_P is limited by the voltage capacity of the solenoid–valve subsystem.

Assuming that 60 V is the maximum acceptable voltage input to the solenoid, we can select a feasible control gain for the closed-loop system for a "nominal" position error. If the nominal position error is 0.15 m, the control gain is $K_P = (60\,\text{V})/(0.15\,\text{m}) = 400$ V/m. Using this gain, the closed-loop poles are $s = -9.06$ and $s = -310.5 \pm j143.1$. Therefore, we expect the closed-loop response to reach its steady-state value in about 0.44 s. Figure 11.44 shows the closed-loop responses of the piston position for a reference position command $x_{ref} = 45$ cm and control gain $K_P = 400$ V/m. The initial piston position is 30 cm. The closed-loop responses of the nonlinear and linearized EHA models were simulated with Simulink using the proportional control scheme (the reader should note that the nonlinear hydraulic actuator shown in Fig. 11.34 is inserted as a subsystem in place of the simple integrator $1080.9/s$ in Fig. 11.42). Despite the complexity of the nonlinear EHA model, the nonlinear and linear closed-loop responses $x(t)$ shown in Fig. 11.44 are indistinguishable from each other. As the closed-loop system is well damped, there is little need for derivative feedback

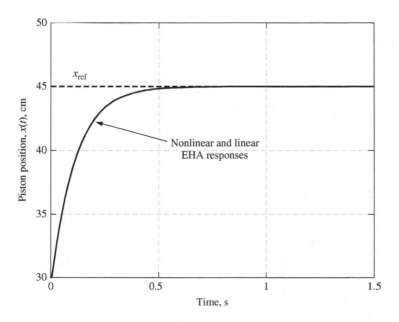

Figure 11.44 Closed-loop responses of the nonlinear and linear EHA models with $x_{ref} = 45$ cm and proportional gain $K_P = 400$ V/m.

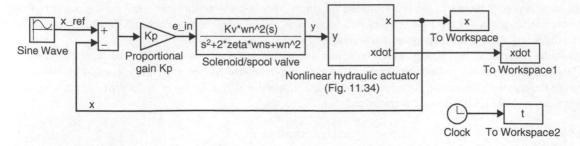

Figure 11.45 Closed-loop Simulink model with proportional control scheme and sinusoidal reference input.

(i.e., a proportional-derivative or PD control scheme), and therefore the proportional control scheme provides adequate performance for a step reference input.

A second performance test for the control system is the ability to track a dynamic, periodic reference input. Figure 11.45 shows a Simulink diagram of the closed-loop control system with a sinusoidal reference input for $x_{\text{ref}}(t)$ (note that the *nonlinear* hydraulic actuator subsystem from Fig. 11.34 is included instead of the simple integrator actuator). The reference position $x_{\text{ref}}(t)$ is a sine function with a ±15-cm amplitude about the piston's starting position of 30 cm. Note that we could construct a linear closed-loop Simulink model using the linearized EHA plant; that is, we would replace the nonlinear EHA plant in Fig. 11.45 with the integrator model $X(s)/Y(s) = K_{\text{HA}}/s$.

Figure 11.46 shows the closed-loop responses of the nonlinear and linear hydraulic actuator models for the sinusoidal reference position $x_{\text{ref}}(t)$ with a frequency of 2 Hz. The proportional gain is $K_P = 500$ V/m. Figure 11.46 clearly shows that the closed-loop frequency responses from the nonlinear and linear EHA models are indistinguishable from each other. Furthermore, the closed-loop response $x(t)$ using proportional control exhibits significant lag with respect to the reference input. The voltage signal $e_{\text{in}}(t)$ oscillates between ±60 V when the P-gain is $K_P = 500$ V/m and therefore increasing the gain will cause the voltage input to exceed its limits.

We can use the following MATLAB commands to compute the magnitude and phase of the *closed-loop* frequency response of the linear EHA model:

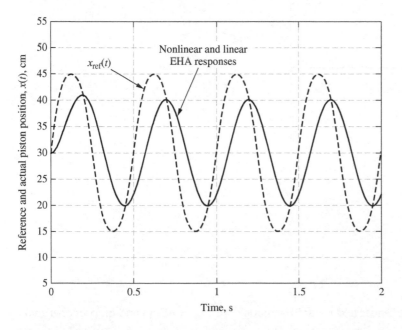

Figure 11.46 Closed-loop responses of the nonlinear and linear EHA models with sinusoidal $x_{\text{ref}}(t)$ and proportional gain $K_P = 500$ V/m.

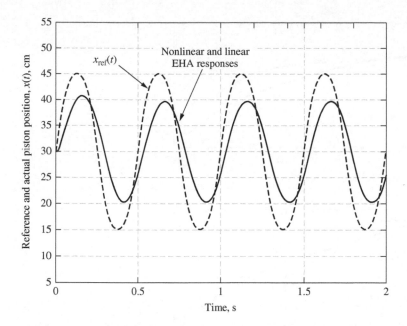

Figure 11.47 Closed-loop responses of the nonlinear and linear EHA models with sinusoidal $x_{\text{ref}}(t)$ and lead controller with gain $K_{\text{LF}} = 2000$.

```
>> Kp = 500;                                          % P-gain
>> sysT = tf(2648.15*Kp,[1 630 122500 2648.15*Kp])   % closed-loop transfer function T(s)
>> w = 2*2*pi;                                        % 2 Hz frequency, rad/s
>> [mag,phase] = bode(sysT,w)                         % closed-loop freq response
```

The magnitude and phase are `mag = 0.6737` and `phase = −51.46°` (=−0.8982 rad). The magnitude value matches the simulation results shown in Fig. 11.46 because the output/input amplitude ratio is roughly $(10\,\text{cm})/(15\,\text{cm}) = 0.6667$ (note that the amplitude is measured relative to the 30-cm mid-point of the hydraulic cylinder). We can use the phase lag to compute the time lag as $0.8982\,\text{rad}/(4\pi\,\text{rad/s}) = 0.072\,\text{s}$, which is the time delay between the peaks of the input and output sinusoids in Fig. 11.46.

One way to improve the tracking performance would be to replace the proportional controller with a *lead controller*. Recall that a lead controller approximates PD control and hence "anticipates" the reference signal due to the derivative term. We can replace the gain K_P in Fig. 11.45 with the following lead controller

$$G_C(s) = \frac{K_{\text{LF}}(s + 10)}{s + 40} \tag{11.63}$$

where K_{LF} is the "lead filter" gain. Figure 11.47 shows the closed-loop responses of the nonlinear and linear hydraulic actuator models using the lead controller with gain $K_{\text{LF}} = 2000$ (the lead controller gain was selected so that its DC gain matches the P-gain $K_P = 500\,\text{V/m}$). Note that the output/input amplitude ratio is roughly the same as the closed-loop response using P-control because the DC gain of the lead controller matches the P-gain K_P. However, adding the lead controller has improved the closed-loop response because the time lag between input and output peaks has been reduced. We can use the MATLAB commands to compute the magnitude and phase angle:

```
>> K_LF = 2000;                              % lead filter gain
>> sysGc = tf([1 10],[1 40]);               % lead filter Gc(s)
>> sysG = tf(2648.15,[ 1 630 122500 0 ]);   % plant transfer function G(s)
>> sysT = feedback(K_LF*sysGc*sysG,1)       % closed-loop transfer function T(s)
>> w = 2*2*pi;                              % 2 Hz frequency, rad/s
>> [mag,phase] = bode(sysT,w)              % closed-loop frequency response
```

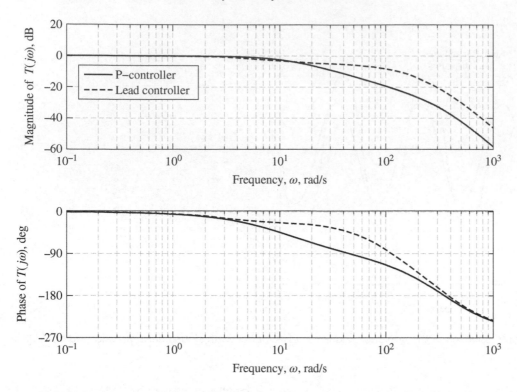

Figure 11.48 Bode diagram of the closed-loop EHA systems using proportional and lead controllers.

The magnitude is 0.6532 (roughly 10/15, as expected) and the phase angle is $-25.34°$ ($-0.4423\,\text{rad}$), which is less than half of the phase lag for the closed-loop system using proportional control. Hence, the time lag using the lead controller is $0.4423\,\text{rad}/(4\pi\,\text{rad/s}) = 0.035\,\text{s}$, which is half the time lag for the P-control system.

Another way to illustrate the benefit of adding the lead controller is to observe the Bode diagrams of the respective *closed-loop* transfer functions. Figure 11.48 shows the Bode diagram of the EHA closed-loop transfer function $T(s)$ for the P-controller [i.e., Eq. (11.62)] with gain $K_P = 500\,\text{V/m}$ and the lead controller (with gain $K_{\text{LF}} = 2000$). Both controllers show the ability to closely track low-frequency input signals as the closed-loop magnitude is $0\,\text{dB}$ (unity output/input ratio) and the phase angle is small. However, for input frequencies greater than $4\,\text{rad/s}$ (about 0.6 Hz), the phase angle of the lead-controller system is greater than the phase of the P-control system. Therefore, the closed-loop EHA system with a lead controller can track a sinusoidal input with smaller phase (or time) lag when compared to the closed-loop P-control system. Note that we can read the magnitude and phase for $\omega = 12.57\,\text{rad/s}$ (2 Hz) from the Bode diagram and estimate the closed-loop frequency responses shown in Figs. 11.46 and 11.47.

11.6   FEEDBACK CONTROL OF A MAGNETIC LEVITATION SYSTEM

Our fifth and last case study involves a feedback control system design for a magnetic levitation ("maglev") system. Maglev systems are used to suspend and propel trains without contact with a rail. Magnetic levitation is also used to support rotating bearings for high-performance machines in order to eliminate friction and the need for lubrication [9]. A laboratory experiment demonstrating magnetic levitation is shown in Fig. 11.49 (see also Problems 3.33 and 5.34). Passing a current through the coil produces an electromagnet that provides an attraction force on the metal ball. The electromagnetic force balances the gravity force at steady state. The constant d in Fig. 11.49 is the nominal air gap between the electromagnet tip and the ball for a nominal electromagnetic force. Ball displacement z is measured positive upward from a fixed static equilibrium position that corresponds to the nominal electromagnetic force, and consequently $d - z$ is the air gap between the ball and the electromagnet. Our goal is to design a control system for the coil circuit so that the ball can be held in static equilibrium at a desired reference position; or, moved from one position to another position.

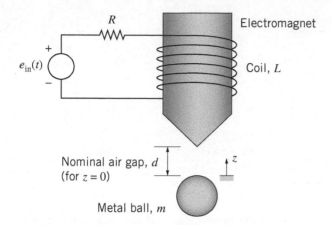

Figure 11.49 Magnetic levitation system.

Mathematical Model

The complete mathematical model of the electromagnetic system consists of the electrical (coil) and mechanical (ball) subsystems. We will assume that the electrical system shown in Fig. 11.49 is a linear RL circuit comprised of coil resistance R, constant inductance L, and source voltage $e_{in}(t)$. Applying Kirchhoff's voltage law around the loop yields

$$-e_R - e_L + e_{in}(t) = 0 \tag{11.64}$$

where the voltage drops across the resistor and inductor are $e_R = RI$ and $e_L = L\dot{I}$, respectively. Substituting the voltage drops across the passive elements in Eq. (11.64) we obtain

$$L\dot{I} + RI = e_{in}(t) \tag{11.65}$$

Equation (11.65) is the mathematical model of the electrical coil and it matches the solenoid coil modeling equation (11.9) for the case with constant inductance L.

Figure 11.50 shows a free-body diagram of the mechanical subsystem, which consists of the ball mass m. The only forces acting on the ball are the electromagnetic force F_{em} and gravity. Applying Newton's second law with a positive sign convention upward yields

$$+\uparrow \sum F = F_{em} - mg = m\ddot{z} \tag{11.66}$$

The electromagnetic force is

$$F_{em} = \frac{K_F I^2}{(d-z)^2} \tag{11.67}$$

where K_F is a force constant that depends on the number of coil wraps, material properties of the electromagnetic core, and electromagnet geometry. It is again noted that $d - z$ is the air gap (distance between the ball and the electromagnet) and hence the electromagnetic force exhibits an inverse-square relationship with the air gap. Clearly, the force is a nonlinear function of current I and position z. Substituting Eq. (11.67) into Eq. (11.66) and rearranging yields the mechanical subsystem model

$$m\ddot{z} = \frac{K_F I^2}{(d-z)^2} - mg \tag{11.68}$$

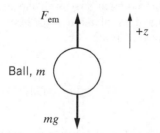

Figure 11.50 Free-body diagram of the mechanical ball.

Table 11.6   Parameters for the Magnetic Levitation System

System Parameter	Value
Ball mass, m	0.003 kg
Coil resistance, R	5 Ω
Coil inductance, L	0.018 H
Electromagnetic force constant, K_F	$2.6487(10^{-5})$ N-m^2/A^2
Gravitational acceleration, g	9.81 m/s^2
Nominal source voltage, $e_{in}^*(t)$	4 V
Nominal coil current, I^*	0.8 A
Nominal air-gap distance, d	0.024 m

Equations (11.65) and (11.68) constitute the mathematical model of the maglev system. The electrical coil model is linear, whereas the mechanical subsystem model is nonlinear. Table 11.6 presents the numerical values for all parameters for the maglev system.

Linear Maglev Model

We wish to design a feedback control scheme for position control of the ball using the electromagnetic force. In Chapter 10, we investigated various methods for analyzing and designing control systems. All of these techniques are based on linear, time-invariant (LTI) plant models. Therefore, we must first develop a linear model of the maglev system; once we have an LTI system, we can apply control-design techniques such as the root-locus method. If we can arrive at a satisfactory control system design using the approximate LTI plant model, we must eventually test its performance with the full *nonlinear* maglev system.

Our nonlinear maglev system is third order, so let us begin by defining three nonlinear state-variable equations for the state vector $\mathbf{x} = [I \quad z \quad \dot{z}]^T$ with the input vector $\mathbf{u} = \left[e_{in}(t) \quad g\right]^T$. Using Eqs. (11.65) and (11.68) we can write three first-order ODEs

$$\dot{x}_1 = \frac{-R}{L}x_1 + \frac{1}{L}u_1 = f_1(\mathbf{x}, \mathbf{u}) \tag{11.69}$$

$$\dot{x}_2 = x_3 = f_2(\mathbf{x}, \mathbf{u}) \tag{11.70}$$

$$\dot{x}_3 = \frac{K_F x_1^2}{m(d - x_2)^2} - u_2 = f_3(\mathbf{x}, \mathbf{u}) \tag{11.71}$$

Note that we have treated source voltage as the first input, $u_1 = e_{in}(t)$, and gravitational acceleration as the second input, $u_2 = g$. Of course, we have the freedom to modulate the first input u_1 but we cannot adjust the second input $u_2 = g$. The reader should see that Eqs. (11.69)–(11.71) match the mathematical modeling equations (11.65) and (11.68). Next, we define the perturbation vector from the nominal (reference) state $\delta\mathbf{x} = \mathbf{x} - \mathbf{x}^*$, where $\mathbf{x}^* = [I^* \quad z^* \quad \dot{z}^*]^T$ is the reference state vector. We will define our reference state as static equilibrium (i.e., a levitating ball) for the nominal source voltage input $u_1^* = e_{in}^*(t) = 4$ V. Equations (11.69) and (11.65) show that the steady-state coil current is $I_{ss} = e_{in}/R$ and therefore the nominal current is $I^* = e_{in}^*/R = (4\,\text{V})/(5\,\Omega) = 0.8\,\text{A}$. We will define our nominal (or reference) static-equilibrium position for the nominal coil current to be $z^* = 0$ and $\dot{z}^* = 0$. Therefore, the nominal state vector is $\mathbf{x}^* = [0.8 \quad 0 \quad 0]^T$ and the nominal input vector is $\mathbf{u}^* = [4 \quad 9.81]^T$. Finally, we can solve Eq. (11.71) for the nominal air-gap distance d when the ball is in equilibrium (i.e., $\dot{x}_3 = \ddot{z} = 0$) with $x_1^* = I^* = 0.8\,\text{A}$ and $x_2^* = z^* = 0$ and we obtain $d = 0.024\,\text{m}$ (24 mm). The reference or nominal operating point values for $e_{in}^*(t)$, I^*, and d are summarized in Table 11.6.

Our standard linearization process described in Chapter 5 will yield

$$\delta\dot{\mathbf{x}} = \left.\frac{\partial\mathbf{f}}{\partial\mathbf{x}}\right|_* \delta\mathbf{x} + \left.\frac{\partial\mathbf{f}}{\partial\mathbf{u}}\right|_* \delta\mathbf{u} \tag{11.72}$$

where $\mathbf{f}(\mathbf{x}, \mathbf{u}) = \begin{bmatrix} f_1(\mathbf{x}, \mathbf{u}) & f_2(\mathbf{x}, \mathbf{u}) & f_3(\mathbf{x}, \mathbf{u}) \end{bmatrix}^T$ is the 3×1 vector of the right-hand sides of Eqs. (11.69)–(11.71). Carrying out the partial derivatives of $\mathbf{f}(\mathbf{x}, \mathbf{u})$ yields

$$\delta\dot{\mathbf{x}} = \begin{bmatrix} -R/L & 0 & 0 \\ 0 & 0 & 1 \\ 2K_F x_1/m(d-x_2)^2 & 2K_F x_1^2/m(d-x_2)^3 & 0 \end{bmatrix}_* \delta\mathbf{x} + \begin{bmatrix} 1/L & 0 \\ 0 & 0 \\ 0 & -1 \end{bmatrix}_* \delta\mathbf{u} \tag{11.73}$$

Next, we evaluate the two matrices using $x_1^* = 0.8\,\mathrm{A}$ (nominal current), $x_2^* = 0$ (nominal position), and the numerical parameters for R, L, K_F, m, and d. The linearized state equation (11.73) becomes

$$\delta\ddot{\mathbf{x}} = \begin{bmatrix} -277.7778 & 0 & 0 \\ 0 & 0 & 1 \\ 24.525 & 817.5 & 0 \end{bmatrix} \delta\mathbf{x} + \begin{bmatrix} 55.5556 & 0 \\ 0 & 0 \\ 0 & -1 \end{bmatrix} \delta\mathbf{u} \tag{11.74}$$

The 3×3 square matrix in Eq. (11.74) is the system matrix $\mathbf{A}$ and the 3×2 matrix is the input matrix $\mathbf{B}$. We can check the eigenvalues of the open-loop system by computing the determinant

$$\det[\lambda\mathbf{I} - \mathbf{A}] = \det \begin{bmatrix} \lambda + 277.7778 & 0 & 0 \\ 0 & \lambda & -1 \\ -24.525 & -817.5 & \lambda \end{bmatrix} = 0$$

Using MATLAB we determine that the three eigenvalues are $\lambda_1 = -277.7778$, and $\lambda_{2,3} = \pm 28.5920$. Hence, the linearized maglev system is *unstable* owing to the positive eigenvalue (root) at $\lambda_2 = +28.5920$. This result is not surprising because the maglev system has no natural feedback mechanism (such as stiffness or damping) that will return the ball back to its original equilibrium position if it is disturbed.

Although we have derived the linearized maglev system using state-space methods, it is advantageous to construct the linear system in terms of transfer functions so that we can use root-locus methods. To do this, let us rewrite the first and third linearized state equations in Eq. (11.74) with the substitutions $\delta x_1 = \delta I$, $\delta x_2 = \delta z$, $\delta x_3 = \delta\dot{z}$, and $\delta u_1 = \delta e_{\mathrm{in}}$

$$\text{First state equation:} \quad \delta\dot{I} = -277.778\delta I + 55.5556\delta e_{\mathrm{in}} \tag{11.75}$$

$$\text{Third state equation:} \quad \delta\ddot{z} = 24.525\delta I + 817.5\delta z - \delta g \tag{11.76}$$

Note that the perturbation in the second input, $\delta u_2 = \delta g$, is zero because $\delta g = g - g^* = 0$. In other words, there can never be a change in gravitational acceleration from its nominal value as g is a constant. Hence, Eqs. (11.75) and (11.76) are both I/O equations

$$\delta\dot{I} + 277.778\delta I = 55.5556\delta e_{\mathrm{in}} \quad \text{and} \quad \delta\ddot{z} - 817.5\delta z = 24.525\delta I$$

which can be written as two transfer functions:

$$\text{RL circuit:} \quad G_1(s) = \frac{55.5556}{s + 277.7778} = \frac{\delta I(s)}{\delta E_{\mathrm{in}}(s)} \tag{11.77}$$

$$\text{Mechanical ball:} \quad G_2(s) = \frac{24.525}{s^2 - 817.5} = \frac{\delta Z(s)}{\delta I(s)} \tag{11.78}$$

Figure 11.51 shows an open-loop block diagram of the linearized maglev system described by transfer functions (11.77) and (11.78). Note that the "linearized" RL circuit transfer function (11.77) is identical to the original RL circuit model (11.65) because it was linear to begin with. The mechanical ball transfer function (11.78) is a linearized version of

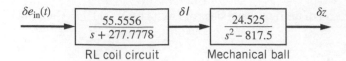

$\delta e_{in}(t)$ $\rightarrow$ $\boxed{\dfrac{55.5556}{s + 277.7778}}$ $\xrightarrow{\delta I}$ $\boxed{\dfrac{24.525}{s^2 - 817.5}}$ $\xrightarrow{\delta z}$

RL coil circuit Mechanical ball

Figure 11.51 Open-loop block diagram of linearized maglev system.

Eq. (11.68) and has been linearized about a nominal air gap of 24 mm and coil current of 0.8 A. Finally, the reader should note that the poles of the two transfer functions are

$$\text{RL circuit:} \quad s + 277.7778 = 0 \quad \rightarrow s = -277.7778$$

$$\text{Mechanical ball:} \quad s^2 - 817.5 = 0 \quad \rightarrow s = \pm 28.5920$$

Therefore, the three poles are identical to the three eigenvalues of the system matrix $\mathbf{A}$.

At this point, it would be useful to compare open-loop simulations of the nonlinear and linearized maglev systems in order to determine the accuracy of the linearization results. However, this system is *unstable*, so any "standard" input (step, impulse, etc.) would result in an unstable response. We will test and compare the *closed-loop* system using the nonlinear and linearized plant in the next section as a means to check the accuracy of the linearization process. To do this, we must develop a closed-loop control system that produces a stable maglev system.

Maglev Control System Design

The goal is to design a feedback control system for position control of the metal ball. As previously mentioned, we will use the linearized maglev system so that we can use our standard control system analysis tools (root locus and Bode plots) which only apply to LTI systems. Figure 11.52 shows a control system where δz_{ref} is the reference position command and $G_C(s)$ is the controller transfer function. The reader should note that every variable in the closed-loop control system is a *perturbation* from the nominal value. Recall that a transfer function may only be used when a system has zero initial conditions. In the case of the maglev system, we assume that the system begins in static equilibrium: source voltage is $e_{in}^*(t)$, coil current is I^*, and ball position is $z^* = 0$ (levitating). Therefore, all perturbations variables ($\delta \mathbf{x} = \mathbf{x} - \mathbf{x}^*$) are initially zero.

Let us begin the control system design with a simple proportional controller $G_C(s) = K_P$. Figure 11.53 shows the root-locus plot for a P-controller. Note that the two open-loop poles originating at $s = \pm 28.5920$ (mechanical ball transfer function) move toward each other, meet at the origin, break away from the real axis, and follow $\pm 60°$ asymptotes as the P-gain is increased from zero to infinity. Because one root-locus branch remains in the right-half plane for all gains, the closed-loop system with proportional control is *always* unstable. A simple P-controller will not stabilize the maglev system.

Adding a zero to the controller will "bend" the unstable root-locus branches to the left. The simplest option is to add a PD controller (a single zero); however, a pure differentiator is physically difficult (if not impossible) to implement in a real system (see the discussion of PD vs. lead controllers in Chapter 10). In order to keep the control design feasible, we will select a pole-zero pair lead controller:

$$\text{Lead controller:} \quad G_C(s) = \frac{K(s + 40)}{s + 160} \tag{11.79}$$

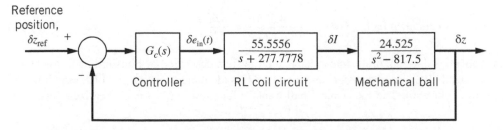

Figure 11.52 Closed-loop control of the linearized maglev system.

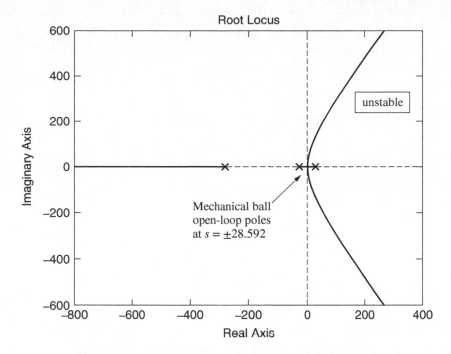

Figure 11.53 Root locus for linearized maglev system with P-controller.

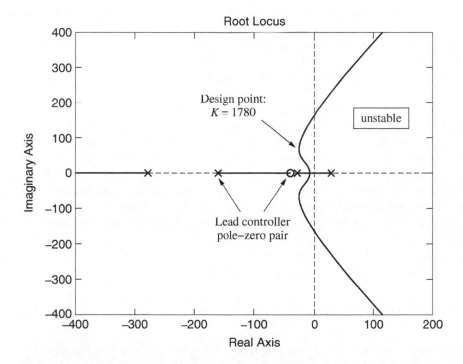

Figure 11.54 Root locus for linearized maglev system with lead controller.

We choose the lead controller's zero location ($s = -40$) to be left of the open-loop pole at $s = -28.5920$. If the added zero is to the right of $s = -28.5920$, then the overall loop gain is decreased, which degrades steady-state tracking. The pole of the lead controller is selected to be four times the zero location. Figure 11.54 shows the root-locus plot for the maglev system with the lead controller (11.79). Note that the two loci branches emanating from $s = -28.5920$ and $s = +28.5920$ meet together in the left-half (stable) plane and are bent to the left by the addition of the lead controller

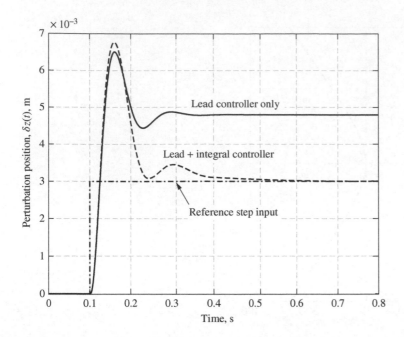

Figure 11.55 Closed-loop step responses of the linearized maglev system: lead controller and lead-plus-integral controller.

zero at $s = -40$. As the gain K is further increased, these two branches eventually cross the imaginary axis and the closed-loop system becomes unstable. However, the addition of the lead controller stabilizes the maglev system for a range of gain K. The "best" design point for gain K is selected using MATLAB's `rlocus` and `rlocfind` commands. We choose the gain so that the complex closed-loop poles are at the "inflection point," as indicated in Fig. 11.54. This design point ($K = 1780$) provides good damping and good response speed because it is simultaneously close to the real axis and far from the origin.

Figure 11.55 shows the closed-loop step response of the linearized maglev system (Fig. 11.52) with a lead controller and gain setting $K = 1780$. The reference position command δz_{ref} is a 0.003 m (3 mm) step input that is applied at $t = 0.1$ s. The step response with the lead controller shows a peak value of about 0.0065 m, which is a 36% overshoot of its steady-state value of 0.0048 m (4.8 mm). The lead controller provides good damping as the step response reaches the 4.8-mm steady-state value in about 0.2 s after the step is applied. However, the steady-state ball position is 1.8 mm above the 3-mm reference position (note that positive z is upward; see Fig. 11.49). It is not surprising that the lead controller cannot provide zero steady-state error if we examine the closed-loop control system. Figure 11.52 shows that if the position tracking error goes to zero, then the lead controller will command zero change in source voltage (i.e., $\delta e_{in} = 0$) and therefore the change in the coil current is also zero (i.e., $\delta I = 0$). Consequently, the maglev system is back to its original nominal operation, which causes the ball to levitate at $z^* = 0$ (or, an air-gap distance of 0.024 m).

The clear solution to the steady-state tracking problem is to add an integrator to the control loop:

$$\text{Lead + integral controller:} \quad G_C(s) = K\left(\frac{s+40}{s+160} + \frac{K_I}{s}\right) \tag{11.80}$$

Equation (11.80) presents a lead-plus-integral controller with the common gain $K = 1780$ and integral gain K_I. Figure 11.56 shows a Simulink model of the linearized maglev system with a lead controller in parallel with an integral control loop. Note that the control gain $K = 1780$ is applied to both control loops. We could use the root-locus plot to select the integral gain K_I with the common gain K fixed at 1780 but the resulting root locus becomes cluttered near the origin. Instead, we can run a few simulation trials varying the integral gain K_I until we determine a satisfactory closed-loop response. Figure 11.55 shows the closed-loop step response of the lead-plus-integral controller with $K = 1780$ and integral gain $K_I = 1.2$. Note that the initial transient response is similar to the lead-only step response; however, after the peak overshoot, the integral controller adjusts the coil voltage so that the ball moves to the desired

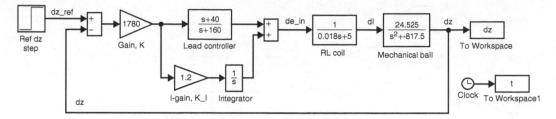

Figure 11.56 Simulink model of the linearized maglev system with lead-plus-integral controller.

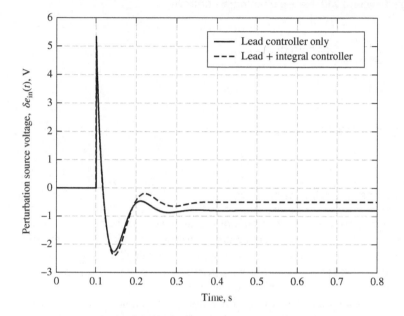

Figure 11.57 Voltage commands for the linearized maglev system: lead controller and lead-plus-integral controller.

3-mm reference position. The closed-loop response with the lead-plus-integral controller reaches the 3-mm reference position in about 0.4 s after the step is applied. Hence, the addition of the integral control has slightly slowed down the response, but it has provided perfect steady-state tracking.

Figure 11.57 shows the voltage commands $\delta e_{in}(t)$ that are produced by the controller $G_C(s)$. Note that both the lead-only and lead-plus-integral controllers begin with a large voltage spike $\delta e_{in}(t) \approx 5.3$ V in order to increase the coil current, which in turn increases the electromagnetic force and causes the ball to accelerate upward toward the $\delta z_{ref} = 0.003$ m reference position. However, when the ball approaches the electromagnet tip (reducing the air gap), the attractive electromagnetic force *increases* even if current remains constant, as shown by Eq. (11.67). When the ball overshoots the new reference position, both feedback controllers reduce the source voltage (and thus current) to values below the nominal voltage (i.e., $\delta e_{in}(t) < 0$). Figures 11.55 and 11.57 show that the ball settles at about $\delta z = 0.0048$ m (4.8 mm) and $\delta e_{in} = -0.8$ V for the lead-only controller. This steady-state position can be determined by equating the electromagnetic force (11.67) with the ball's weight mg.

$$\frac{K_F I_{ss}^2}{(d - z_{ss})^2} = mg \tag{11.81}$$

For steady-state position $z_{ss} = 0.0048$ m, the steady-state current computed using Eq. (11.81) is $I_{ss} = 0.64$ A, and the corresponding steady-state source voltage is $e_{in}(\infty) = R I_{ss} = 3.2$ V. Hence, the steady-state *perturbation* voltage is $\delta e_{in} = e_{in}(\infty) - e_{in}^* = -0.8$ V, as shown in Fig. 11.57. For the desired reference position $z_{ss} = 0.003$ m, Eq. (11.81) shows that the steady-state current is $I_{ss} = 0.7$ A, and therefore the steady-state voltage is 3.5 V. Hence, the required steady-state perturbation voltage is $\delta e_{in} = -0.5$ V, which is the steady-state voltage command of the lead-plus-integral controller system, as shown in Fig. 11.57.

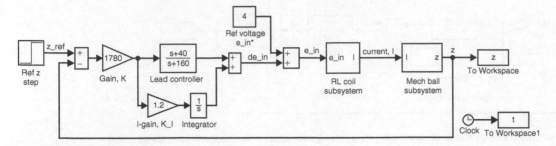

Figure 11.58 Simulink model of nonlinear maglev system with lead-plus-integral controller.

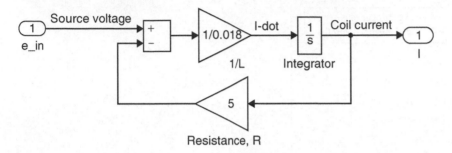

Figure 11.59 Nonlinear maglev system: electromagnet coil subsystem.

It is important at this point to summarize our approach and the results of our analysis. We have linearized the maglev system so that we can utilize the root-locus method to design the controller transfer function. Using the root locus we have successfully designed a lead controller for a good transient response; later, we added an integral control loop to eliminate the steady-state tracking error. All analyses and simulations were performed in terms of the perturbation variables (i.e., δz, δI, δe_{in}), which is a necessary by-product of the linearization approach.

The final test is to use the lead-plus-integral controller in a closed-loop simulation of the maglev system using the complete *nonlinear* mathematical model. The previous linearized results are of little use if the controller design cannot stabilize or adequately control the nonlinear maglev system. The first step is to develop a numerical simulation of the nonlinear maglev system dynamics represented by Eqs. (11.65) and (11.68). Figure 11.58 shows a Simulink model of the closed-loop nonlinear maglev system using subsystem blocks for the electromagnet coil and mechanical ball. It is important to note that the "true" dynamic variables (i.e., z, I, and e_{in}) are used in the nonlinear simulation instead of the perturbation variables δz, δI, and δe_{in}. However, Fig. 11.58 shows that the input to the lead-plus-integral controller is position error $z_{\text{ref}} - z$ (in meters), which is the same error signal in the linearized closed-loop block diagram shown in Fig. 11.56. Furthermore, both controllers in Figs. 11.58 and 11.56 develop the command $\delta e_{\text{in}}(t)$, which is an incremental change in source voltage. Because the nonlinear Simulink model (Fig. 11.58) involves the "full" dynamic variables, we must add the incremental voltage command $\delta e_{\text{in}}(t)$ to the nominal voltage ($e_{\text{in}}^* = 4\,\text{V}$) to produce the full voltage command $e_{\text{in}}(t)$ that is applied to the electromagnet coil (recall that by definition $\delta e_{\text{in}} = e_{\text{in}} - e_{\text{in}}^*$). Figures 11.59 and 11.60 show the inner details of the electromagnet coil and mechanical ball subsystems, respectively. Note that the electromagnet coil dynamics (Fig. 11.59) are actually linear. However, we cannot use a transfer function (unless we revert to perturbation variables δI and δe_{in}) because the electromagnet coil has a nonzero initial current. Therefore, the integrator in Fig. 11.59 must be initialized with current $I^* = 0.8\,\text{A}$ (recall that an 0.8-A current produces an electromagnetic force that levitates the ball at air gap distance $d = 24\,\text{mm}$). Figure 11.60 shows the nonlinear nature of the force F_{em}, which involves terms I^2 and $(d - z)^2$. Initial conditions for both integrators are set to zero because the ball begins in static equilibrium ($z(0) = \dot{z}(0) = 0$). The reader should be able to correlate the governing mathematical model (11.65) and (11.68) with the subsystem block diagrams of Figs. 11.59 and 11.60.

Figure 11.61 shows the closed-loop step responses of the nonlinear and linearized maglev models. As before, the reference position $z_{\text{ref}}(t)$ is a 3-mm step function applied at time $t = 0.1\,\text{s}$. The nonlinear model response shows slightly more peak overshoot than the linearized model response but otherwise the two responses are very similar. The closed-loop response of the nonlinear maglev model (the true test for the controller design) is stable and tracks the reference

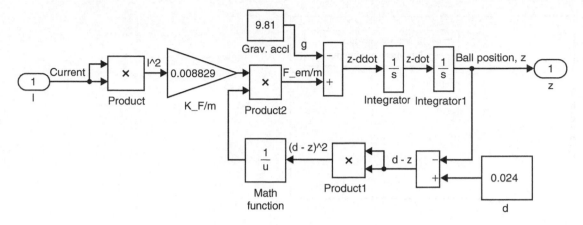

Figure 11.60 Nonlinear maglev system: mechanical ball subsystem.

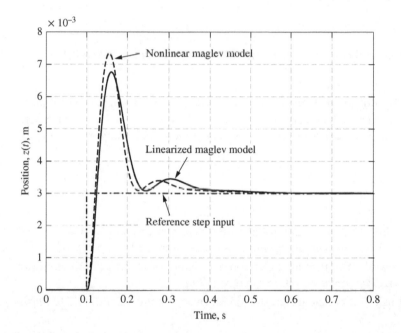

Figure 11.61 Closed-loop step responses of the linearized and nonlinear maglev systems using the lead-plus-integral controller.

command at steady state. Figure 11.62 shows that the controller voltage commands for the nonlinear and linearized models are very similar.

This example has demonstrated the extraordinary power of developing approximate linear models of nonlinear systems and using these LTI models for control system analysis and design. Despite the extreme difference between the LTI and nonlinear models (e.g., compare the LTI transfer function (11.78) with the nonlinear block diagram in Fig. 11.60), the linearization method can produce good results as demonstrated by the closed-loop simulations presented here.

SUMMARY

This chapter has served as a "capstone" for the textbook, where we have demonstrated the concepts of modeling, analysis, simulation, and control of dynamic systems by presenting case studies. These case studies involve integrated engineering systems with "mixed-discipline" components, such as mechanical, electrical, and fluid subsystems. The case studies

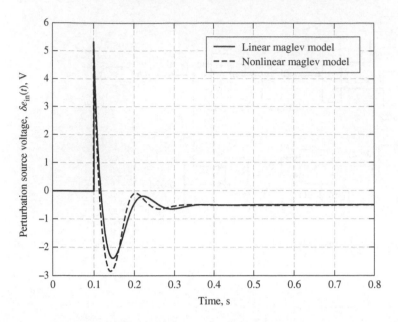

Figure 11.62 Voltage commands for the linearized and nonlinear maglev systems using the lead-plus-integral controller.

illustrate the steps that are common to nearly all problems in dynamic systems: (1) developing the mathematical models, (2) predicting the system's behavior using analytical and numerical methods, and (3) selecting the important system parameters in order to improve performance. Typically, these steps involve several iterations, becoming the design process in engineering.

REFERENCES

1. Choi, S.B., Choi, J.H., Lee, Y.S., and Han, M.S., "Vibration Control of an ER Seat Suspension for a Commercial Vehicle," *ASME Journal of Dynamic Systems, Measurement, and Control*, Vol. 125, March 2003, pp. 60–68.

2. Gouw, G.J., Rakheja, S., Sankar, S., and Afework, Y., "Increased Comfort and Safety of Drivers of Off-Highway Vehicles Using Optimal Seat Suspension," *SAE Paper No. 901646*, September 1990.

3. Yuan, Q., and Li, P.Y., "Self-Calibration of Push-Pull Solenoid Actuators in Electrohydraulic Valves," *ASME Paper No. 2004-62109*, November 2004.

4. Chladny, R.R., Koch, C.R., and Lynch, A.F., "Modeling Automotive Gas-Exchange Solenoid Valve Actuators," *IEEE Transactions on Magnetics*, Vol. 41, No. 3, March 2005, pp. 1155–1162.

5. Subramanian, S.C., Darbha, S., and Rajagopal, K.R., "Modeling the Pneumatic Subsystem of an S-cam Air Brake System," *ASME Journal of Dynamic Systems, Measurement, and Control*, Vol. 126, March 2004, pp. 36–46.

6. Karpenko, M., and Sepehri, N., "Quantitative Fault Tolerant Control Design for a Leaking Hydraulic Actuator," *ASME Journal of Dynamic Systems, Measurement, and Control*, Vol. 132, September 2010, pp. 36–43.

7. Fales, R., Spencer, E., Chipperfield, K., Wagner, F., and Kelkar, A., "Modeling and Control of a Wheel Loader with a Human-in-the-Loop Assessment Using Virtual Reality," *ASME Journal of Dynamic Systems, Measurement, and Control*, Vol. 127, September 2005, pp. 415–423.

8. Ogata, K., *System Dynamics*, 4th ed., Pearson Prentice Hall, 2004, pp. 344–345.

9. Yeh, T.-J., Chung, Y.-J., and Wu, W.-C., "Sliding Control of Magnetic Bearing Systems," *ASME Journal of Dynamic Systems, Measurement, and Control*, Vol. 123, September 2001, pp. 353–362.

CHAPTER 2: PROBLEMS

Conceptual Problems

2.1 Figure P2.1 shows a single-mass translational mechanical system. Both springs are undeflected when $z = 0$ and $f_a(t) = 0$. Derive the mathematical model of the mechanical system.

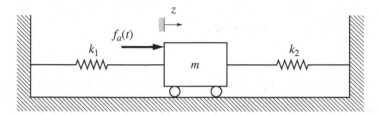

Figure P2.1

SS **2.2** Figure P2.2 shows a vertical mechanical system. All displacements are measured relative to their static equilibrium positions when the applied force $f_a(t) = 0$. Derive the mathematical model of this system.

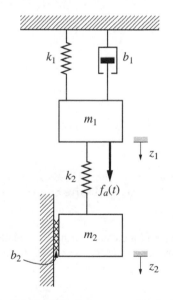

Figure P2.2

2.3 Figure P2.3 shows a mechanical system driven by the displacement of the left end, $x_{in}(t)$, which could be supplied by a rotating cam and follower. When displacements $x_{in}(t) = 0$ and $x = 0$, the spring k is neither compressed nor stretched. Derive the mathematical model of the mechanical system.

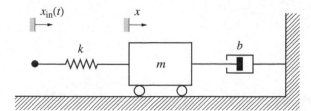

Figure P2.3

SS **2.4** Figure P2.4 shows the mechanical system from Problem 2.3 (Fig. P2.3) where the damper b and spring k are switched. The system is driven by the displacement of the left end, $x_{in}(t)$, which could be supplied by a rotating cam and follower. When displacement $x = 0$, the spring k is neither compressed nor stretched. Derive the mathematical model of the mechanical system.

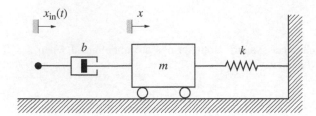

Figure P2.4

2.5 Figure P2.5 shows a two-mass model of a mechanical damper. Mass m_1 is the piston and rod, and mass m_2 is the cylinder case. The cylinder is attached to the wall by the spring element k. Viscous friction interaction between the piston and cylinder is modeled by coefficient b. External force $f_a(t)$ is applied directly to the piston/rod mass m_1. Displacements z_1 and z_2 are absolute positions of masses m_1 and m_2, respectively, measured relative to the static equilibrium positions with $f_a(t) = 0$. Derive the mathematical model of the mechanical system.

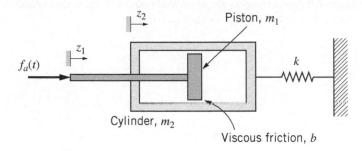

Figure P2.5

2.6 Consider the two-mass mechanical damper presented in Fig. P2.5. Suppose a spring with stiffness k_1 is inserted between the piston and the cylinder (redefine the spring between the cylinder and wall as k_2). Derive the mathematical model of the mechanical system.

2.7 A simple 1-DOF mechanical system is presented in Fig. P2.7. Displacement z is measured from the undeflected spring position. External force $f_a(t)$ is applied directly to mass m. The spring force obeys a nonlinear relationship with displacement

$$F_k = k_1 z + k_3 z^3$$

Derive the mathematical model of the mechanical system.

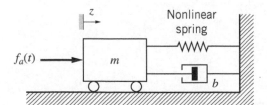

Figure P2.7

SS 2.8 A 2-DOF mechanical system with a damper and spring in series is shown in Fig. P2.8. The independent displacement of the mass m is z_1 and z_2 is the independent displacement of the node point between the damper b and spring k. The spring is undeflected when $z_1 = z_2 = 0$. An external force $f_a(t)$ acts on mass m. Derive the mathematical model of the mechanical system. [Hint: draw FBDs for the node and mass m and neglect the inertia of the "massless node"].

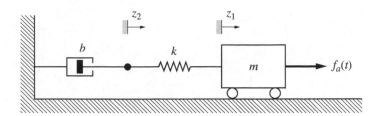

Figure P2.8

2.9 Figure P2.9 shows the same 2-DOF mechanical system as in Fig. P2.8 but with the series connection of the damper and spring reversed. The independent displacement of the mass m is z_1 and z_2 is the independent displacement of the node point between spring k and damper b. The spring is undeflected when $z_1 = z_2 = 0$. An external force $f_a(t)$ acts on mass m. Derive the mathematical model of the mechanical system. [Hint: draw FBDs for the node and mass m and neglect the inertia of the "massless node"].

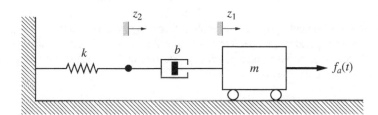

Figure P2.9

2.10 Figure P2.10 shows a two-mass mechanical system. The nonlinear spring force obeys the equation

$$F_k = k_1 d + k_3 d^3$$

where d is the relative displacement between the two masses. The spring is undeflected when $z_1 = z_2 = 0$. An external force $f_a(t)$ acts on mass m_2. Derive the mathematical model of the mechanical system.

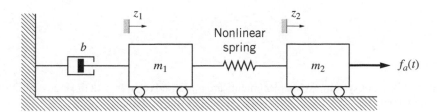

Figure P2.10

2.11 Figure P2.11 shows a mass m sliding on an oil film with viscous friction coefficient b. The mass is moving toward the stiffness element k and at time $t = 0$ it has position $x(0) = 0$, velocity $\dot{x}(0) = 0.4$ m/s, and it is 0.5 m from the stiffness element. Derive the mathematical model for this mechanical system (note that the mathematical model will require *two* equations—one for the case of "no contact" with spring k, and one for the case of contact with the spring. Furthermore, the spring cannot "pull" in tension on mass m).

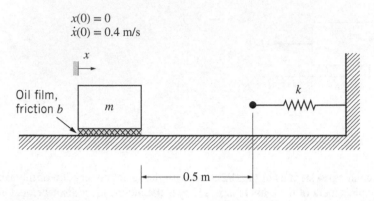

Figure P2.11

2.12 Figure P2.12 shows a two-mass translational mechanical system. The applied force $f_a(t)$ acts on mass m_1. Displacements z_1 and z_2 are absolute positions of masses m_1 and m_2, respectively, measured relative to fixed coordinates (the static equilibrium positions with $f_a(t) = 0$). An oil film with viscous friction coefficient b separates masses m_1 and m_2. Derive the mathematical model of the mechanical system.

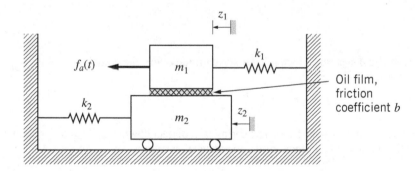

Figure P2.12

2.13 Figure P2.13 shows a two-mass mechanical system. All displacements are measured relative to their static equilibrium positions when the applied force $f_a(t) = 0$. Derive the mathematical model of this system.

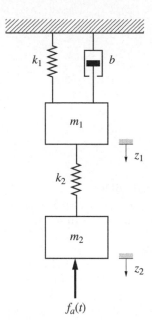

Figure P2.13

2.14 Figure P2.14 shows a mechanical system. The connecting link has moment of inertia J about its pivot point, and rotation angle θ is positive clockwise. Position of mass m is positive to the right. Both the angular and translational displacements are measured from the equilibrium position where all springs are undeflected. Derive the mathematical model of this system assuming small rotation angle θ.

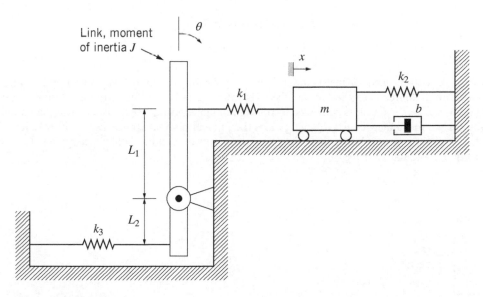

Figure P2.14

2.15 Figure P2.15 shows a frictionless 2-DOF mechanical system that consists solely of inertia and stiffness elements. Displacements x_1 and x_2 are measured from the equilibrium position (undeflected springs).

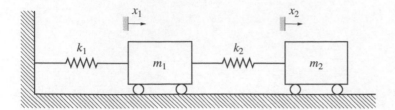

Figure P2.15

a. Derive the mathematical model of the complete system using Newton's laws and FBDs.
b. Derive the mathematical model of the system using the fact that total mechanical energy of the system remains constant [Hint: write an expression for total energy of the system and set its time derivative to zero].

2.16 A two-mass translational mechanical system has the following mathematical model:

$$m_1\ddot{x}_1 + b_1\dot{x}_1 + k_1(x_1 - x_2) = f_a(t)$$
$$m_2\ddot{x}_2 + b_2\dot{x}_2 + k_1(x_2 - x_1) + k_2x_2 = 0$$

The displacements x_1 and x_2 are measured from their respective equilibrium positions. An external force $f_a(t)$ is applied to the system. Sketch a possible configuration of the two-mass mechanical system. Label all elements and show the positive convention for displacements on your sketch. Sketch the FBDs for each mass and verify the system model by applying Newton's laws.

2.17 Repeat Problem 2.16 if the mechanical model of the two-mass translational system is

$$m_1\ddot{x}_1 + k_1x_1 + k_2(x_1 - x_2) = 0$$
$$m_2\ddot{x}_2 + k_2(x_2 - x_1) = 0$$

2.18 Figure P2.18 shows a disk with moment of inertia J that is initially rotating on an axis supported by bearings. No input torque from an external source is applied to the disk. The rotating disk is immersed in a stationary cylinder that is filled with hydraulic fluid. The disk is subjected to friction, modeled using linear viscous friction coefficient b. Derive an equation for the instantaneous *rate* of energy loss (i.e., power dissipated) because of the rotation of the disk in the viscous friction using two methods: (1) use the time derivative of total mechanical energy and (2) use the definition power = torque × angular velocity.

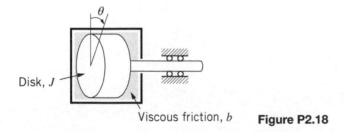

Disk, J

Viscous friction, b **Figure P2.18**

2.19 Figure P2.19 shows a mechanical system that consists of a pulley on a rigid shaft. The pulley has moment of inertia J and radius r. An external torque T_{in} is applied directly to the pulley. Derive the mathematical model.

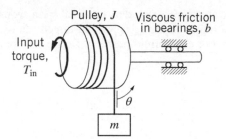

Figure P2.19

2.20 Figure P2.20 shows a rotational mechanical system driven by the displacement of the left end of the flexible shaft, $\theta_{in}(t)$. When displacements $\theta_{in}(t) = 0$ and $\theta = 0$, the torsional spring k is not twisted. Derive the mathematical model of the mechanical system.

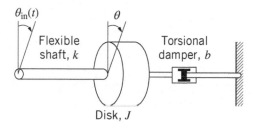

Figure P2.20

SS **2.21** Figure P2.21 shows a two-disk rotational mechanical system. The left end of the flexible shaft is driven by the input displacement, $\theta_{in}(t)$. When displacements $\theta_{in}(t) = \theta_1 = \theta_2 = 0$, the torsional spring k is not twisted. Derive the mathematical model of the system.

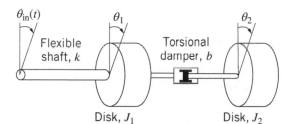

Figure P2.21

2.22 A two-disk rotational mechanical system is shown in Fig. P2.22. An external motor (not shown) delivers the input torque T_{in} to the input gear 1 (radius r_1). The moment of inertia of disk 1 and gear 2 is J_1 and J_2 is the moment of inertia of the second disk. The gear train is ideal. The inertia of gear 1 can be neglected, and gear 2 is larger than gear 1 ($r_2 > r_1$). The two disks are connected by a flexible shaft with torsional spring constant k. Disk J_2 is immersed in a viscous fluid modeled by friction coefficient b. Derive the mathematical model of the complete mechanical system.

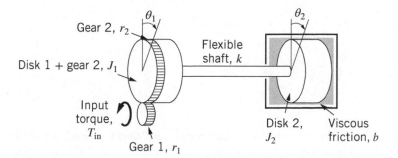

Figure P2.22

2.23 Figure P2.23 shows a dual-disk mechanical system with two flexible shafts represented by torsional spring constants k_1 and k_2, respectively. Disk J_1 experiences viscous friction b and input torque T_{in} is applied directly to disk J_1. Both angular positions (θ_1 and θ_2) are measured from the untwisted or equilibrium position. Derive the complete mathematical model of this system.

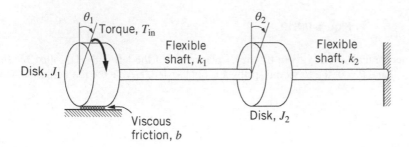

Figure P2.23

2.24 Figure P2.24 shows a simple lever system driven by an external force $f_a(t)$. The lever has moment of inertia J about the pivot axis. When the lever angle $\theta = 0$, the spring is undeflected. Derive the mathematical model of the lever system assuming that the angular rotation angle θ remains small.

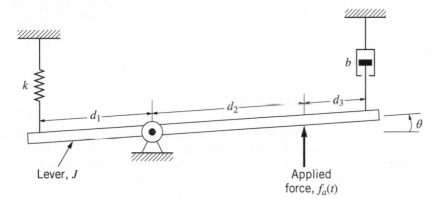

Figure P2.24

MATLAB Problems

2.25 An engineer wants to develop mathematical models for two mechanical springs. She loads the springs (in tension and compression), measures their static deflections, and compiles the results in Table P2.25. Use MATLAB's `polyfit` command to determine spring-force modeling equations for each spring. Plot both spring-model equations for deflections ranging from −8 to 8 mm and include the data points from Table P2.25 on your plot. Comment on whether the springs exhibit linear or nonlinear relationships.

Table P2.25

Load force (N)	Spring #1 deflection (mm)	Spring #2 deflection (mm)
0	0	0
10	1.2482	1.1682
15	1.8822	1.7523
25	3.1938	2.9206
35	4.6099	4.0888
45	6.2354	5.2570
50	7.2055	5.8411
−10	−1.2482	−1.1682
−15	−1.8822	−1.7523
−25	−3.1938	−2.9206
−35	−4.6099	−4.0888
−45	−6.2354	−5.2570
−50	−7.2055	−5.8411

2.26 Accurately modeling friction in mechanical systems is often challenging because of the "stick-slip" characteristic at very low relative velocities. A nonlinear friction model that accommodates the stick-slip phenomena is

$$F_f = [F_C + (F_{st} - F_C)\ \exp(-|\dot{x}|/c)]\text{sgn}(\dot{x}) + b\dot{x}$$

where F_f is the total friction force, F_{st} is the static friction (or "stiction") force near zero velocity, F_C is the Coulomb ("dry") friction force, c is a velocity coefficient, b is the viscous friction coefficient, and $\dot{x}$ is the relative velocity between the sliding mass and the surface. Let the friction model parameters be $F_{st} = 10$ N, $F_C = 7$ N, $b = 70$ N-s/m. Use MATLAB to plot the total friction force versus relative velocity for $-0.05 \leq \dot{x} \leq 0.05$ m/s for three values of velocity coefficient: $c = 0.001, 0.002,$ and 0.005 m/s. Describe the friction force at very small relative velocities and at the (relatively) high velocity of ± 0.05 m/s. In addition, explain how varying the velocity coefficient c alters the so-called "Stribeck friction effect" when relative velocity is very small.

2.27 The damping force for shock absorbers for heavy trucks often exhibits a nonlinear relationship with the relative velocity between the piston and cylinder of the damper. A typical damper force relationship is

$$F_d = \frac{4500\dot{x}}{\sqrt{\dot{x}^2 + v^2}}$$

where F_d is the damper force (in N), $\dot{x}$ is the relative velocity between the piston and the cylinder (in m/s), and $v = 0.2$ m/s. Use MATLAB to plot the damper force versus relative velocity for $-1.5 \leq \dot{x} \leq 1.5$ m/s. Describe the nature of the damper force for "small" and "large" values of relative velocities.

2.28 The nonlinear damping force for automotive shock absorbers often exhibits a larger force during the extension stroke $(\dot{x} > 0)$ compared to the compression stroke $(\dot{x} < 0)$ for the same magnitude in relative velocity $\dot{x}$. Therefore, the nonlinear damper model from Problem 2.27 is modified

$$F_d = \frac{3389(\dot{x} - v_1)}{\sqrt{(\dot{x} - v_1)^2 + v_2^2}} + 1020.84 \text{ N}$$

where F_d is the damper force (in N), $\dot{x}$ is the relative velocity between the piston and the cylinder (in m/s), $v_1 = 0.06$ m/s, and $v_2 = 0.19$ m/s. Write a MATLAB M-file to plot the damper force versus relative velocity for $-1.5 \leq \dot{x} \leq 1.5$ m/s. Does this damper model accurately represent the friction force when $\dot{x} \approx 0$? Describe the nature of the damper force for "small" and "large" ranges for relative velocities.

Engineering Applications

2.29   Figure P2.29 shows a round-wire spring made of stainless steel. Search the engineering literature to find the appropriate equation for the spring constant and compute the spring constant k (in N/m).

1.5 cm

$N = 5$ coils

Figure P2.29

2.30   Repeat Problem 2.29 for a five-coil, stainless steel spring with a square cross-sectional area of 0.8 mm^2. The outer diameter of the square-wire coil is identical to the round-wire spring shown in Fig. P2.29 (i.e., 1.5 cm).

SS   **2.31**   Figure P2.31 shows a simplified mechanical model of a nanopositioner [4]. A piezoelectric actuator produces input position $z_{in}(t)$. Mass m_1 and spring constant k_1 represent the mass and stiffness of the piezoelectric device. Mass m_2 is the sample mass to be positioned by the actuator, and it is connected to the massless lever by stiffness element k_2. Displacements θ and z_2 are measured from their static equilibrium positions. Derive a linear model of the mechanical system by assuming that lever rotation angle θ remains small at all times.

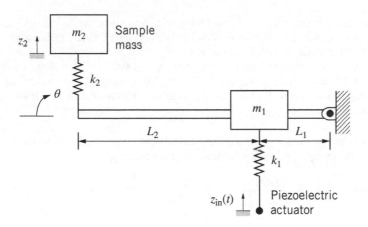

Figure P2.31

2.32   Figure P2.32 shows a gear train driving a robot arm through a flexible shaft. An external motor (not shown) delivers the input torque T_{in} to the input gear 1 (radius r_1). The moment of inertia of disk 1 and gear 2 is J_1. The gear train is ideal. Gear 2 is larger than gear 1 ($r_1 < r_2$) and the inertia of gear 1 can be neglected. The gear train and robot arm are connected by a flexible shaft with torsional spring constant k. The robot arm has inertia J_{cm} about its center-of-mass (c.m.), and its c.m. is distance d from the rotation axis. The weight of the robot arm (concentrated at the c.m.) will provide an opposing torque on the arm about the rotation axis for angular displacement $0 < \theta_2 < 180°$. The system has negligible friction. Derive the mathematical model of the complete mechanical system [Hint: the moment of inertia of the arm about the rotation axis is $J_2 = J_{cm} + md^2$ via the parallel-axis theorem].

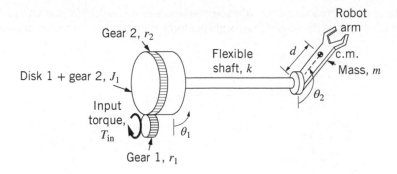

Figure P2.32

SS **2.33** Figure P2.33 shows a rotational mechanical system that represents the torsional powertrain of a hybrid electric vehicle during engine restart [5]. Three moments of inertia (J_1, J_2, and J_3) represent the internal combustion engine, electric motor, and clutch plate, respectively. The moments of inertia are connected by flexible shafts that are modeled by torsional stiffness (k_1 and k_2) and torsional viscous friction elements (b_1 and b_2). All stiffness and damping torques depend on relative displacements and relative velocities. The angular displacements of the three disks are measured from the untwisted shaft positions. Electric motor torque T_m acts directly on the motor inertia J_2. Derive the mathematical model of powertrain system.

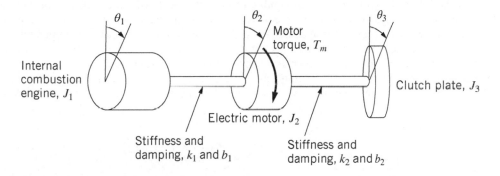

Figure P2.33

2.34 Figure P2.34 shows a schematic representation of a semi-submersible platform for supporting a floating offshore wind turbine [6]. The platform is rigidly connected to a submerged pontoon, which provides the buoyancy force for floatation. Two columns, masses m_1 and m_2, are connected to the ends of the platform via identical suspensions modeled by ideal stiffness and damping elements k and b, respectively. The columns damp out platform vibrations induced by ocean waves. The two columns may move independently with positions z_1 and z_2 measured upward from their respective static equilibrium positions. For this simple model, assume that the platform and pontoon motion is restricted to rotation (θ) about its center-of-mass (c.m.). The platform and pontoon have moment of inertia J. The ocean waves produce an external torque on the platform (T_a) and vertical forces on the two columns (F_{a1} and F_{a2}), respectively. Two restoring buoyancy forces, F_{res1} and F_{res2}, act on the columns

$$F_{res1} = -\rho g A z_1 \qquad F_{res2} = -\rho g A z_2$$

where ρ is the density of water, g is gravitational acceleration, and A is the cross-sectional area of the column (note that if Column 1 is deflected "down," $z_1 < 0$, the restoring buoyancy force F_{res1} pushes "up"). A restoring buoyancy torque T_{res} acts on the platform

$$T_{res} = \rho g V D \theta$$

where V is the volume of water displaced by the pontoon and D is the so-called meta-centric height (distance from c.m. to metacenter). Derive the mathematical model of the semi-submersible platform assuming small rotation angle θ.

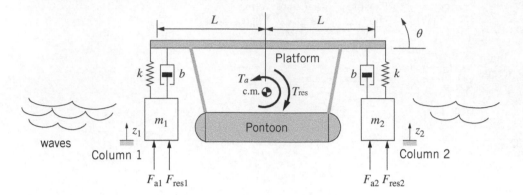

Figure P2.34

SS **2.35** Figure P2.35a shows a schematic diagram of an optical disk drive system. The spindle motor (contained in the chassis) rotates the disk and the cart motor translates the pick-up head (PUH) along the radial direction of the spinning disk so that the focused laser "reads" the desired data track on the optical disk. Note that although the cart servo motor can position the cart in the radial direction, the cart is rigidly attached to the chassis (see Ref. [7] for additional details). Figure P2.35b shows the optical disk drive as a simplified two-mass mechanical system. The PUH mass m_1 is connected to the cart with stiffness k_1 and friction coefficient b_1, while the chassis and cart are lumped into mass m_2. A series of rubber mounts connects the chassis mass m_2 to the frame and these mounts have lumped stiffness k_2 and friction coefficient b_2. The mounts are used to suppress transmitted vibrations from the motion of the frame. The absolute displacements (as measured from the static equilibrium position) of the PUH and chassis/cart are x_1 and x_2, respectively. The absolute displacement of the frame is $x_{in}(t)$. Derive the mathematical model of the optical disk drive system.

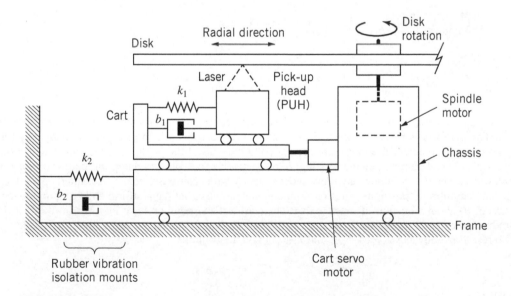

Figure P2.35a

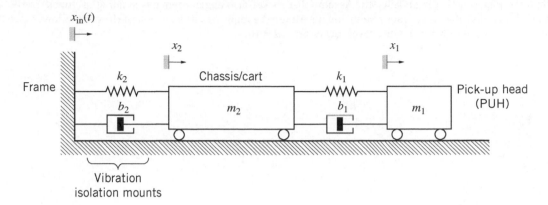

Figure P2.35b

2.36 High-speed electric trains use a mechanical arm called a pantograph to transfer electric current from an overhead wire to the train (see Fig. P2.36a). The pantograph typically consists of a two-arm frame linkage that provides an upward force in order to maintain contact between a small pan-head and the catenary wire [8]. Figure P2.36b shows a two-mass lumped mechanical model of the pantograph where m_1 is the head mass, m_2 is the frame mass, and k_1 is the stiffness of the "shoe" contact between the head and catenary wire. The head suspension is modeled by lumped stiffness k_2 and lumped friction coefficient b_1 while the frame suspension only involves a lumped friction coefficient b_2. A pneumatic piston provides the force $f_a(t)$ that pushes up on the frame so that the shoe remains in contact with the wire. Displacements z_1 and z_2 are measured from the static equilibrium positions, and $z_w(t)$ is the displacement of the overhead wire. Derive the mathematical model of the pantograph system (note that the stiffness element k_1 can only be in compression; that is, the wire cannot "pull" in tension on the head mass).

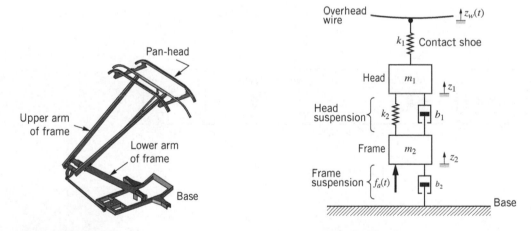

Figure P2.36a   **Figure P2.36b**

2.37 Figure P2.37 shows a schematic of an automotive valve train. The rotating cam moves the pushrod (follower), and the rocker-arm rotates to move the valve in the vertical direction. A simplified mechanical model is shown adjacent to the valve train schematic diagram. Moment of inertia J represents the rocker-arm inertia about the pivot point. Deflection of the pushrod is modeled by spring k_1, and the return spring on the valve is k_2. Friction in the rocker-arm pivot is modeled by viscous coefficient b. The vertical motion of the cam follower is $x_C(t)$, the input to the system. Angular position of the rocker-arm is θ, which is positive in the clockwise direction. When the rocker-arm angle is level ($\theta = 0$), the return spring has a compressive preload force of F_L. When cam follower position $x_C = 0$ and $\theta = 0$, then the pushrod (k_1) is undeflected. Assume that rocker-arm angle θ remains small at all times. Derive the mathematical model of the valve train system and determine the required displacement of the cam follower x_C when the system is at rest (equilibrium) with a level rocker arm ($\theta = 0$).

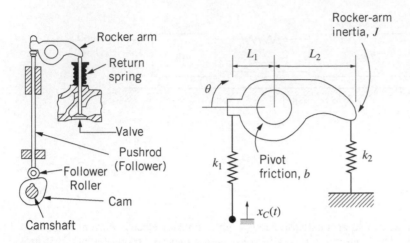

Figure P2.37

2.38 Figure P2.38 shows a model of a locomotive pulling two railroad cars [9]. The two couplers are modeled by stiffness coefficient k and friction coefficient b. Displacements z_i are the absolute positions of each mass (car) measured from a static equilibrium condition where no forces exist in the couplers. In order to derive a linear model, we assume that the rolling friction of each mass is equal to the product of friction coefficient b_r and its absolute velocity $\dot{z}_i$. The locomotive's propulsion system provides external force $F_a(t)$ to mass m_1. Derive the complete mathematical model of the railroad-car system.

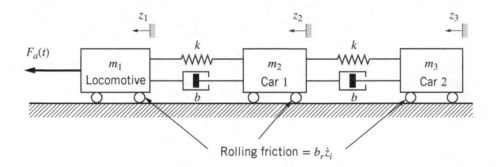

Figure P2.38

SS **2.39** Figure P2.39 shows a DC motor (moment of inertia J_1) driving a rotational load (moment of inertia J_2) through an ideal gear-train transmission. Motor torque T_m acts directly on the motor inertia J_1. Moments of inertia J_1 and J_2 are rigidly connected to their respective gears, and the gear ratio is $N = r_2/r_1 > 1$. Both the motor and load inertias experience viscous friction modeled by coefficients b_1 and b_2, respectively. In addition, an external load torque T_L acts directly on inertia J_2 as shown in Fig. P2.39. Derive the mathematical model of the motor-load rotational system in terms of angular displacement of the load inertia, θ_2.

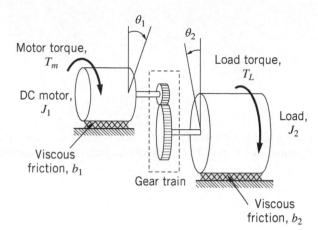

Figure P2.39

2.40 Consider again the piezoelectric MEMS actuator discussed in Example 2.6. Derive the complete mathematical model for the additional scenario where the friction between masses m_1 and m_2 causes them to "stick together" and, therefore, the system can be modeled as a single (joined) mass. In this case, we have "static friction" or "stiction" and there is zero relative motion between the two masses ($\dot{x}_1 = \dot{x}_2$). Derive an expression for the magnitude of the static friction (or "stiction") force so that $\dot{x}_1 = \dot{x}_2$ and $\ddot{x}_1 = \ddot{x}_2$. Note that the complete mathematical model will consist of three sets of ODEs for three cases: (1) sliding friction (contact, $F_{PZT} > 0$), (2) stiction (joined, $F_{PZT} > 0$), and (3) no contact (release with $F_{PZT} = 0$).

2.41 Figure P2.41 shows a lumped-parameter representation of a concept for a MEMS "tuning fork gyroscope" for measuring angular velocity. Displacements x_1, x_2, and x_3 are absolute positions of the respective masses and are measured from their static equilibrium positions where there is no deflection in the springs. The system friction is represented by three ideal lumped dashpot elements b_1, b_2, and b_3. Two external forces f_1 and f_2 are applied to masses m_1 and m_2, respectively. Derive the mathematical model of the MEMS gyroscope.

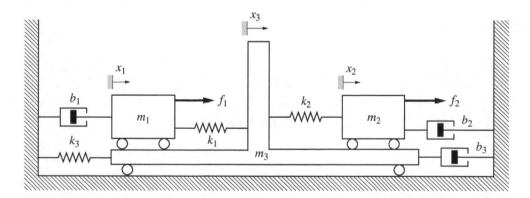

Figure P2.41

CHAPTER 3: PROBLEMS

Conceptual Problems

3.1 Figure P3.1 shows a single-loop electrical circuit. The dashed box denotes a single energy-storage element. Derive the mathematical model in terms of the appropriate dynamic variables if the energy-storage element is an inductor, L.

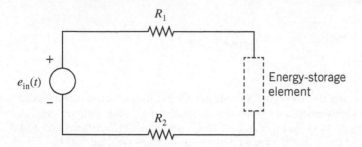

Figure P3.1

3.2 Repeat Problem 3.1 if the energy-storage element in Fig. P3.1 is a capacitor, C.

3.3 Derive the mathematical model of the electrical system shown in Fig. P3.3. The model should be in terms of the appropriate dynamic variables.

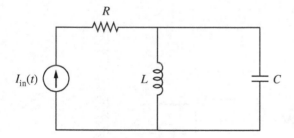

Figure P3.3

SS   **3.4** An electrical circuit is shown in Fig. P3.4. The current source provides the input current $I_{in}(t)$. Derive the mathematical model in terms of the appropriate dynamic variables.

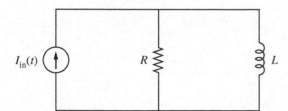

Figure P3.4

3.5   Figure P3.5 shows a parallel LC circuit driven by current source $I_{in}(t)$. Derive the mathematical model in terms of the appropriate dynamic variables.

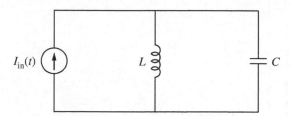

Figure P3.5

3.6   Figure P3.6 shows a parallel RC circuit driven by current source $I_{in}(t)$. Derive the mathematical model in terms of the appropriate dynamic variables.

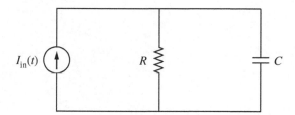

Figure P3.6

3.7   Figure P3.7 shows the same electrical system as Fig. P3.6 except that a resistor R_2 is added in series with the capacitor C. Derive the mathematical model in terms of the appropriate dynamic variables.

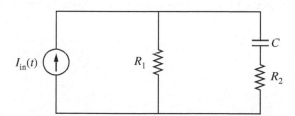

Figure P3.7

3.8   Derive the mathematical model of the electrical system shown in Fig. P3.8 in terms of the appropriate dynamic variables.

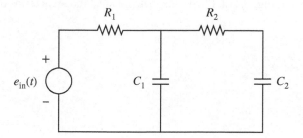

Figure P3.8

3.9 An electrical system is shown in Fig. P3.9. Derive the mathematical model in terms of the appropriate dynamic variables. The source provides the input voltage $e_{in}(t)$.

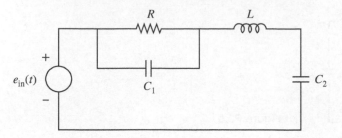

Figure P3.9

3.10 An electrical system is shown in Fig. P3.10. Derive the mathematical model in terms of the appropriate dynamic variables. The source provides the input voltage $e_{in}(t)$.

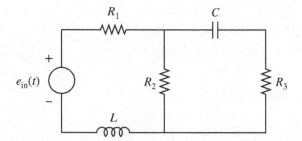

Figure P3.10

3.11 Figure P3.11 shows an electrical circuit with a current source $I_{in}(t)$. Derive the mathematical model in terms of the appropriate dynamic variables.

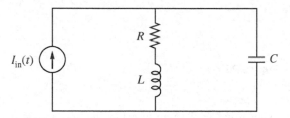

Figure P3.11

3.12 An RLC circuit with a parallel bypass resistor is shown in Fig. P3.12. Derive the mathematical model in terms of the appropriate dynamic variables.

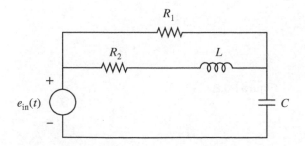

Figure P3.12

SS **3.13** Figure P3.13 shows an electrical circuit with three energy-storage elements. Derive the mathematical model in terms of the appropriate dynamic variables.

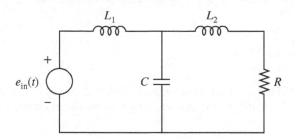

Figure P3.13

3.14 Figure P3.14 shows a simple series RL circuit with a nonlinear inductor L. An engineer performs a series of tests and measures magnetic flux linkage λ and current I_L across the inductor. After curve-fitting the experimental results, he develops the following nonlinear function for inductor current:

$$I_L(\lambda) = 97.3\lambda^3 + 4.2\lambda \quad \text{(amps, A)}$$

where λ is the flux linkage in weber (Wb). Derive the mathematical model of the RL circuit with flux linkage λ as the dynamic variable.

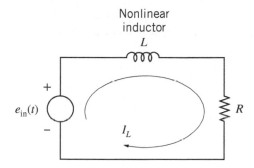

Figure P3.14

3.15 Suppose we have an electrical circuit that consists of one resistor R, one capacitor C, and one voltage source $e_{in}(t)$ connected in series. The mathematical model of the system is

$$RC\dot{e}_O + e_O = RC\dot{e}_{in}(t)$$

Is output voltage e_O the voltage across the resistor or the capacitor? Explain your answer.

3.16 Figure P3.16 shows an electrical system. The switch is open for time $t < 0$. For time $0 \leq t \leq 1$ s, the switch is in position "1" and the voltage source $e_{in}(t)$ is connected to the RL loop. For time $t > 1$ s, the switch is in position "2." Derive the complete mathematical model of the electrical system that can account for both switch positions (you will need two ODE modeling equations).

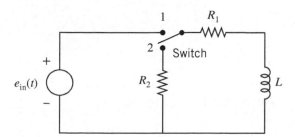

Figure P3.16

3.17 Figure P3.17 shows an RC circuit with a switch. For time $t < 0$, the switch is in position 1 and the capacitor is connected to the voltage source $e_{in}(t) = 3$ V. At time $t = 0$, the switch is moved to position 2 thus disconnecting the voltage source from the RC circuit and the capacitor begins to discharge. As we see in Chapter 7, the discharge phase of the capacitor ($t \geq 0$) is governed by the equation

$$e_C(t) = 3e^{-t/RC} \text{ V}$$

where the initial capacitor voltage is $e_C(0) = e_{in}(t) = 3$ V. Derive an equation for the time-rate of energy (i.e., power) dissipated by the resistor and show that the total energy dissipated by the resistor is equal to the initial energy stored by the capacitor at time $t = 0$.

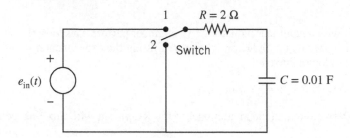

Figure P3.17

3.18 Figure P3.18 presents an op-amp circuit. Determine the relationship between input and output voltages.

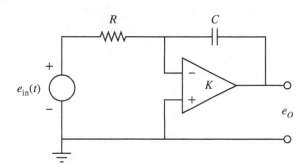

Figure P3.18

3.19 Figure P3.19 presents an op-amp circuit. Determine the relationship between input and output voltages.

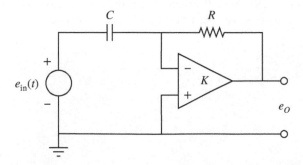

Figure P3.19

3.20   Figure P3.20 presents an op-amp circuit. Determine the relationship between input and output voltages.

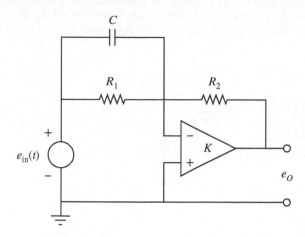

Figure P3.20

SS **3.21**   Figure P3.21 shows an op-amp circuit. Determine the relationship between input and output voltages.

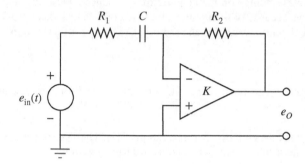

Figure P3.21

3.22   Figure P3.22 presents an op-amp circuit. Determine the relationship between input and output voltages.

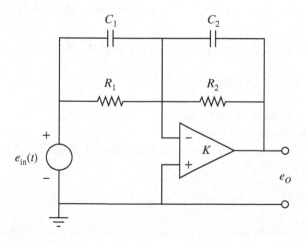

Figure P3.22

SS **3.23**   Figure P3.23 shows a mass–spring system that is attracted by the electromagnetic force F_{em} produced by the current-carrying coil. The electromagnet force F_{em} depends on current and position of the mass

$$F_{em} = \frac{K_F I^2}{(d - x)^2}$$

where K_F is a "force constant" (units of N-m^2/A^2) that depends on the number of coil wraps, material properties of the electromagnetic core, and geometry of the electromagnet. Position x of the mass is measured from a fixed reference (where the spring is undeflected), and d is a constant equal to the distance between the electromagnet tip and the at-rest mass when $F_{em} = 0$. Derive the mathematical model of this system in terms of the appropriate dynamic variables.

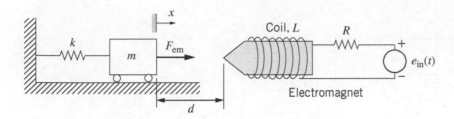

Figure P3.23

3.24   A microphone consists of a diaphragm with mass m connected to a circular coil of wire that moves back and forth relative to a fixed magnetic held. Sound (pressure) waves impinge on the diaphragm to produce a net force, and the diaphragm experiences both stiffness and friction forces because of its motion. The wire coil has resistance R and its output current is amplified by a constant gain. List the dynamic variables and the input variable for this system.

MATLAB Problems

3.25   Consider again the RC circuit in Problem 3.17. Plot the energy stored by the capacitor versus time during the discharge phase $t \geq 0$.

3.26   A series RL circuit is driven by a constant 2-V voltage source (see Fig. 3.7 in Example 3.1). The system parameters are $L = 0.08$ H and $R = 4$ Ω. As we see in Chapter 7, the resulting current response for this system is

$$I_L(t) = 0.5 \left(1 - e^{-50t}\right) \text{ A}$$

Use MATLAB to compute the energy stored by the inductor for time $0 \leq t \leq 0.2$ s. Plot stored energy versus time.

SS **3.27**   In Problem 3.14, the following empirical relationship relating inductor current I_L and magnetic flux linkage λ was used:

$$I_L(\lambda) = 97.3\lambda^3 + 4.2\lambda \text{ (amps, A)}$$

a. Use MATLAB to plot current I_L as a function of flux linkage λ for the range $-0.4 \leq \lambda \leq 0.4$ Wb.
b. Use the data from part (a) to plot flux linkage λ as a function of current I_L.
c. Write an M-file that uses the plot data from part (b) to estimate the inductance L and plot inductance as a function of flux linkage λ [Hint: recall that inductance is $L = d\lambda/dI_L$].

Engineering Applications

3.28   A simple capacitor consists of two parallel plates separated by a silicon insulator. The area of one plate is 300 mm^2 and the separation distance between the plates is 0.5 mm. Search the engineering literature to find the appropriate equation for the capacitance and compute C for this particular capacitor.

3.29   A simple "air core" inductor is constructed by wrapping a wire into a coil. The coil has 12 turns, the total length of the coil is 2.4 cm, and the (inner) diameter of the coil is 0.8 cm. Search the engineering literature to find the appropriate equation for the inductance and compute L for this particular inductor.

SS **3.30** Figure P3.30 shows a circuit diagram of a photovoltaic solar cell. The solar cell is modeled as an ideal current source $I_{in}(t)$ in parallel with a *diode* and shunt resistor R_{SH}. An ideal diode has high internal resistance so that it conducts current in only one direction ("down" as shown in Fig. P3.30). The shunt resistance R_{SH} is a low-resistance path that models the unwanted short circuit between the front and back surfaces of a solar cell. Derive an equation for the solar-cell output current I_O in terms of $I_{in}(t)$, diode current I_D, and solar-cell output voltage e_O.

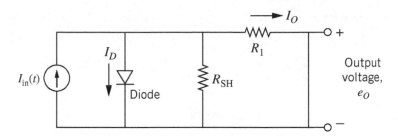

Figure P3.30

SS **3.31** Figure P3.31 shows a high-pass filter circuit with a "tweeter" speaker. The speaker is modeled by lumped resistance R (typically 8 Ω). Derive the mathematical model in terms of the appropriate dynamic variables.

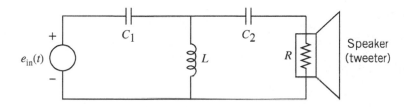

Figure P3.31

3.32 Figure P3.32 shows a simplified circuit for the discharge phase that produces a camera flash. During the charging phase the capacitor C is charged by a 1.5-V battery connected to an oscillator circuit that includes a transformer for boosting voltage (not shown in Fig. P3.32). The voltage across a fully charged capacitor is about 200 V. Pressing the shutter button activates the circuit shown in Fig. P3.32 and the capacitor discharges energy to the RC circuit. A second transformer (shown in Fig. P3.32) boosts the voltage by a factor of 10 in order to ionize xenon gas and produce a flash. Derive the mathematical model of the discharge circuit (neglect the effect of the transformer).

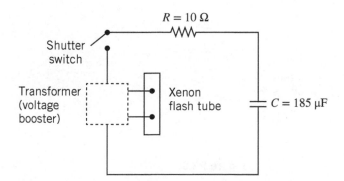

Figure P3.32

3.33 Figure P3.33 shows an electrical system known as a *buck converter*, which is a circuit used to step the source voltage $e_{in}(t)$ "down" to a lower desired output voltage $e_O = e_C$ (voltage across the capacitor). The step-down voltage converter uses a switch to connect and disconnect the voltage supply $e_{in}(t)$ from the remainder of the circuit until e_C is equal to the desired voltage. Derive the mathematical model of this system in terms of the appropriate dynamic variables and include the switching effect.

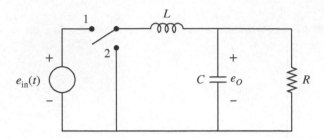

Figure P3.33

3.34 A series RC circuit known as a *washout filter* is shown in Fig. P3.34. This circuit (or filter) will "pass" the initial part of the dynamic output voltage response e_O (voltage across the resistor) but will eventually "wash out" (i.e., reduce to zero) a constant input voltage. These filters are often used in control systems such as an aircraft's automatic flight control system. Derive the mathematical model with e_O (voltage across the resistor) as the dynamic variable and source voltage $e_{in}(t)$ as the input.

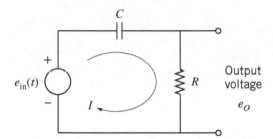

Figure P3.34

3.35 Figure P3.35 shows an electrical circuit known as a *band-stop filter* or *notch filter*, which attenuates (reduces) the amplitude of input signals that have a certain frequency band and "passes" all other signals without alteration. These electrical circuits are used to remove signals with undesirable frequencies; for example, notch filters are used on aerospace vehicles to remove mechanical vibrations that are transmitted to sensors such as gyroscopes and accelerometers.

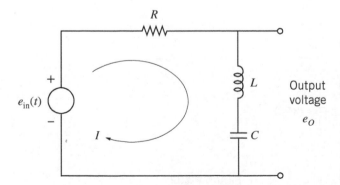

Figure P3.35

a. Derive the mathematical model of the band-stop filter where current I is the sole dynamic-response variable and voltage source $e_{in}(t)$ is the input variable.

b. Derive the mathematical model of the band-stop filter where output voltage e_O is the sole dynamic-response variable and $e_{in}(t)$ is the system input.

3.36 The RC electrical system shown in Fig.P3.36 is known as a *lead filter* and is used as a "compensator" in feedback control systems in order to improve the damping characteristics. Voltage source $e_{in}(t)$ is the input variable.

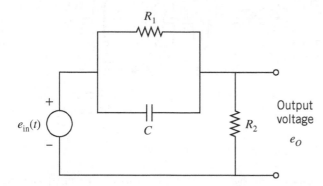

Figure P3.36

a. Derive the mathematical model of the lead filter in terms of the appropriate dynamic variables associated with the energy-storage elements.

b. Derive the mathematical model of the lead filter where output voltage e_O is the sole dynamic-response variable and $e_{in}(t)$ is the system input.

3.37 Figure P3.37 shows an RC electrical system known as a *lag filter* that is used as a "compensator" in feedback control systems in order to reduce errors between the reference and feedback variables. Voltage source $e_{in}(t)$ is the input variable.

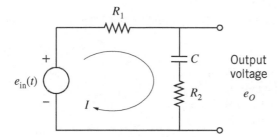

Figure P3.37

a. Derive the mathematical model of the lag filter in terms of the appropriate dynamic variables associated with the energy-storage elements.

b. Derive the mathematical model of the lag filter where output voltage e_O is the sole dynamic-response variable and $e_{in}(t)$ is the system input.

3.38 Consider the DC motor modeled by Eqs. (3.90a) and (3.90b). The motor parameters are rotor moment of inertia $J = 0.005$ kg-m^2, rotor viscous friction torque coefficient $b = 8(10^{-4})$ N-m-s/rad, motor-torque constant $K_m = 0.26$ N-m/A, back-emf constant $K_b = 0.26$ V-s/rad, coil inductance $L = 9$ mH, coil resistance $R = 6$ Ω, and load torque $T_L = 0.4$ N-m. Compute the constant voltage input, e_{in}, so that the motor eventually achieves the *constant* angular velocity $\omega = 52.36$ rad/s (about 500 rpm). In addition, determine the corresponding constant armature current I and motor torque T_m [Hint: the modeling equations (3.90a) and (3.90b) become algebraic equations when the dynamic variables achieve constant values].

3.39 Figure P3.39 shows the DC motor with gear-train transmission from Problem 2.39 in Chapter 2.

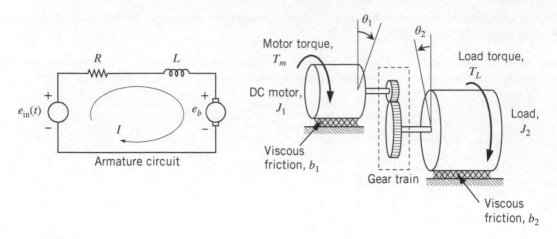

Figure P3.39

a. Derive the complete mathematical model of the electromechanical system in terms of current (I) and angular displacement of the load inertia (θ_2).

b. The desired angular acceleration of the load inertia and the known load torque for time $0 \leq t \leq 0.2$ s are

$$\ddot{\theta}_2(t) = 20 \sin \omega t \text{ rad/s}^2$$

$$T_L(t) = 2.5 \sin 0.5 \omega t \text{ N-m}$$

where the frequency is $\omega = 10\pi$ rad/s (i.e., the period is 0.2 s). Determine the desired angular velocity and angular position of the load inertia (assume that the load inertia is initially at rest and $\theta_2 = 0$ at $t = 0$). Use MATLAB to plot angular velocity and angular position of the load for $0 \leq t \leq 0.2$ s.

c. Determine explicit equations for the required motor torque $T_m(t)$, armature current $I(t)$, and input voltage $e_{in}(t)$ so that the load follows the desired acceleration prescribed in part (b) (this is an example of *inverse dynamics*; i.e., computing the input required to produce the desired dynamic response).

d. Use MATLAB to create plots of motor torque, current, input voltage, and input power versus time. Use the following numerical values for the DC motor and load.

Motor inertia:	$J_1 = 3.6(10^{-5})$ kg-m^2
Load inertia:	$J_2 = 0.04$ kg-m^2
Motor viscous friction:	$b_1 = 1.5(10^{-4})$ N-m-s/rad
Load viscous friction:	$b_2 = 7.2(10^{-3})$ N-m-s/rad
Gear ratio:	$N = 50$
Inductance:	$L = 4$ mH
Resistance:	$R = 3 \, \Omega$
Motor-torque constant:	$K_m = 0.1$ N-m/A
Back-emf constant:	$K_b = 0.1$ V-s/rad

3.40 Magnetic levitation ("maglev") is used to suspend and propel trains without contact with a rail. Magnetic levitation is also used to support rotating bearings for high-performance machines in order to eliminate friction and the need for lubrication [7]. A laboratory experiment demonstrating magnetic levitation is shown in Fig. P3.40. The current-carrying coil produces an electromagnet that provides an attraction force F_{em} on the metal ball:

$$F_{em} = \frac{K_F I^2}{(d - x)^2}$$

where K_F is a "force constant" (units of N-m^2/A^2) that depends on the number of coil wraps, material properties of the electromagnetic core, and geometry of the electromagnet [7]. Position x of the ball is measured upward from a fixed at-rest position and distance d is a constant equal to the nominal air-gap distance from the electromagnet tip and the ball for a nominal coil current.

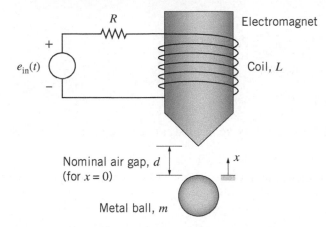

Figure P3.40

Derive the complete mathematical model of the magnetic levitation system where voltage source $e_{in}(t)$ is the input variable.

3.41 Consider a comb-drive MEMS actuator with the following physical dimensions:

   Comb width $w = 5$ μm
   Gap between fingers $d = 2$ μm
   86 fingers
   Source voltage $e_{in} = 30$ V (constant)
   Actuator stiffness $k = 0.04$ N/m

Compute the displacement of the comb drive in microns at "steady state" where $e_C = e_{in}$ and the comb mass is not accelerating and stationary (i.e., $\ddot{x} = \dot{x} = 0$).

3.42 Figure P3.42 shows an electrostatic actuator that uses a simple parallel-plate capacitor instead of the more complicated interlocking comb-drive fingers. The top plate is movable and the bottom plate is fixed. Displacement x is measured downward from the static equilibrium position when the plates are uncharged. The capacitance of the parallel plates is

$$C(x) = \frac{\varepsilon_0 A}{d - x}$$

where ε_0 is the dielectric constant in air, A is the plate area, and d is the nominal gap between the plates when they are uncharged. The voltage input is $e_{in}(t)$, the circuit resistance is R, the upper plate has mass m, and the stiffness and friction forces of the actuator are represented by spring constant k and friction coefficient b. Derive the complete mathematical model of the electromechanical system.

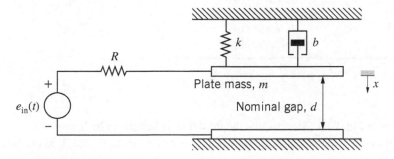

Figure P3.42

CHAPTER 4: PROBLEMS

Conceptual Problems

4.1 A hydraulic tank has the following pressure–mass relationship:

$$P = 1.013(10^5) + 3.6m \, \text{N/m}^2$$

where fluid mass m is in kg. The tank holds water (density $\rho = 1000 \, \text{kg/m}^3$) and has a constant circular cross-sectional area A. Compute the fluid capacitance C of the tank.

4.2 An engineer measures the water-level height h and volumetric-flow rate Q through a valve that is draining a hydraulic tank. Table P4.2 summarizes the measurements.

Table P4.2

Liquid height (m)	Volumetric-flow rate (m³/s)
4.10	8.9689 (10^{-4})
3.24	7.9730 (10^{-4})
2.68	7.2513 (10^{-4})
1.79	5.9262 (10^{-4})
1.05	4.5388 (10^{-4})
0.33	2.5445 (10^{-4})

Is the flow through the valve laminar or turbulent? Explain your answer.

SS **4.3** Table P4.3 presents experimental flow measurements for a high-pressure valve. Develop a model for volumetric flow as a function of the pressure difference across the valve.

Table P4.3

Pressure difference across valve (Pa)	Volumetric-flow rate (m³/s)
0	0
255,783	3.519 (10^{-4})
504,082	4.936 (10^{-4})
979,231	6.886 (10^{-4})
1,429,048	8.314 (10^{-4})
2,134,862	10.160 (10^{-4})

SS **4.4** A two-tank hydraulic system is shown in Fig. P4.4. Assume that the fluid flow is laminar in both valves.
   a. Derive the complete mathematical model of the system with the base pressure of the tanks P_1 and P_2 as the dynamic variables.
   b. Derive the complete mathematical model of the system with tank heights h_1 and h_2 as the dynamic variables.

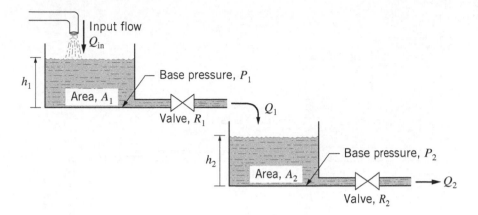

Figure P4.4

4.5 Figure P4.5 shows a two-tank hydraulic system. The pump takes water from the sump (reservoir) at atmospheric pressure P_{atm} and increases the pressure by the constant amount $\Delta P = \rho g H$ where H is the "pressure head." Assume that the constant pressure head H is greater than the vertical-wall height of tank 1. Flow through all valves is laminar. Derive the complete mathematical model with tank pressures P_1 and P_2 as the dynamic variables and head H as the input variable.

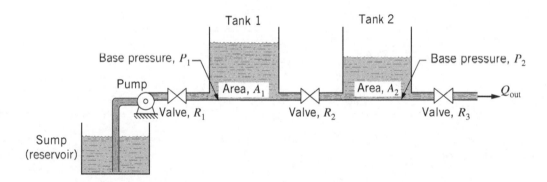

Figure P4.5

4.6 Figure P4.6 shows a centrifugal pump that supplies input volumetric-flow rate Q_{in} to the hydraulic tank. The pump takes water from the sump (reservoir) at atmospheric pressure and increases the pressure according to the equation

$$\Delta P = a\omega^2 \text{ Pa}$$

where ω is the angular velocity of the centrifugal pump in revolutions per minute (rpm) and a is a constant derived from empirical data. Flow through output valve 2 is assumed to be laminar.

a. Assuming that the flow through valve 1 is laminar (i.e., $R_1 = R_L$) derive the mathematical model with pressure P as the dynamic variable and pump speed ω as the input variable.

b. Repeat part (a) assuming that the flow through valve 1 is turbulent where $R_1 = R_T$.

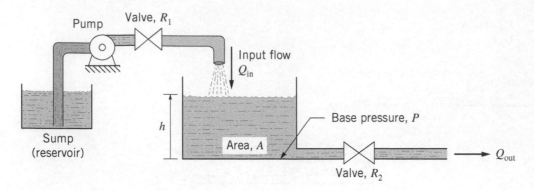

Figure P4.6

4.7 Figure P4.7 shows a pneumatic system that consists of a rigid vessel, and an inlet pipe with valve that can be connected to two separate supply tanks (air or hydrogen). Suppose the fluid resistance across the valve is laminar and independent of the type of gas. Both supply tanks have the same constant pressure, and the process is assumed to be isothermal and at 40 °C in both cases. If the rigid vessel is independently pressurized with either supply tank, will the corresponding pressure response profiles of the vessel be identical or different? Explain your answer.

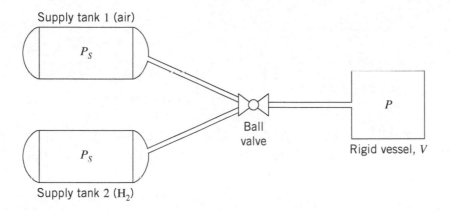

Figure P4.7

4.8 Figure P4.8 shows a rigid vessel with input volumetric-flow rate, Q_{in}. The hydraulic fluid has fluid bulk modulus $\beta = 1.2(10^9)$ Pa. At the instant shown, the pressure in the chamber is $P = 1.8(10^6)$ Pa, temperature is $T = 345$ K, and the volumetric-flow rate is $Q_{in} = 2.4(10^{-6})$ m³/s.

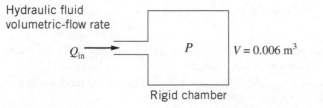

Figure P4.8

    a. Compute the fluid capacitance C of the hydraulic system.

    b. Compute the time-rate of pressure dP/dt at this instant.

4.9 Repeat Problem 4.8 if the same rigid vessel is used for a pneumatic system with air as the working fluid. Assume that the input mass-flow rate is $w_{in} = 0.0035$ kg/s, temperature is $T = 298$ K, chamber pressure is $P = 2.5(10^5)$ Pa, and the process is isothermal.

4.10 A spring-loaded hydraulic accumulator is shown in Fig. P4.10. Hydraulic fluid flows into the accumulator with volumetric-flow rate Q_{in}. The "spring side" of the accumulator has constant atmospheric pressure P_{atm}. Derive an expression for the total capacitance C of the hydraulic accumulator [Hint: the goal is to find an expression in the form $C\dot{P} = Q_{in}$. Begin with the basic pressure-rate equation for the CV, Eq. (4.40), and replace $\dot{V}$ by an equation in terms of $\dot{P}$ by noting that a differential change in pressure is balanced by a differential displacement of the piston or plate with area A. This analysis neglects the friction and inertia of the accumulator piston].

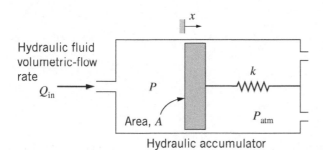

Hydraulic accumulator   **Figure P4.10**

4.11 A spring-loaded pneumatic accumulator is shown in Fig P4.11. Air flows into the accumulator with mass-flow rate w_{in}. The "spring side" of the accumulator has constant atmospheric pressure P_{atm}. Derive an expression for the total capacitance C of the pneumatic accumulator [Hint: the goal is to find an expression in the form $C\dot{P} = w_{in}$. Begin with the basic pressure-rate equation for the pneumatic CV, Eq. (4.74), and replace $\dot{V}$ by an equation in terms of $\dot{P}$ by noting that a differential change in pressure is balanced by a differential displacement of the piston or plate with area A. This analysis neglects the friction and inertia of the accumulator piston].

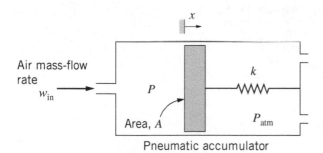

Pneumatic accumulator   **Figure P4.11**

4.12 A container with a volumetric capacity of 0.0023 m³ is filled with water. At time $t = 0$, the water temperature is 288 K and the ambient temperature is 308 K. The specific heat capacity of water at constant pressure is $c_p = 4186$ J/kg-K. At $t = 0$, the rate of heat transfer to the water is $q = 3.1$ J/s (watts).

    a. Compute the thermal capacitance of the container of water.

    b. Determine the thermal resistance of the container.

    c. Determine the time-rate of the water temperature at $t = 0$.

Engineering Applications

4.13 Figure P4.13 shows laminar flow through a pipe. The Darcy equation relates pressure drop for a liquid flowing in a pipe (laminar or turbulent flow):

$$\text{Darcy equation}: \Delta P = \frac{1}{2}\rho v^2 \frac{L}{d} f$$

where f is the nondimensional Darcy friction factor. For laminar flow, the friction factor is a simple function of Reynolds number, Re: $f = 64/\text{Re}$. Search the engineering literature to determine an expression for Reynolds number and show that for laminar flow the Darcy equation is equivalent to the Hagen–Poiseuille law for laminar flow resistance, Eq. (4.3).

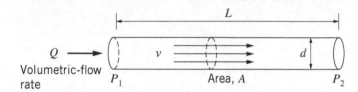

Figure P4.13

4.14 A steel pipe is carrying water (density $\rho = 1000$ kg/m^3). The pipe length is 2 m and its diameter is 4 cm. The flow is turbulent, and the Moody diagram is used to compute the Darcy friction factor, and it is found to be $f = 0.038$. Use the Darcy equation in Problem 4.13 to compute the turbulent fluid resistance R_T coefficient for this pipe (include the units).

4.15 An engineer wants to model the low-speed flow of water ($\rho = 1000$ kg/m^3) through a length of pipe. After conducting experiments and measuring volumetric-flow rate and pressure drop across the pipe, the data are tabulated, as shown in Table P4.15:

Table P4.15

Trial	Pressure drop (Pa)	Volumetric-flow rate (m^3/s)
1	7.2	$6\,(10^{-5})$
2	16.8	$1.4\,(10^{-4})$
3	25.2	$2.1\,(10^{-4})$

Compute the laminar-flow fluid resistance coefficient R_L for the pipe.

4.16 A hydraulic valve manufacturer provides the following equation relating pressure drop (in N/m^2 or Pa) and volumetric-flow rate (in m^3/s) when the valve is fully open:

$$\Delta P = 2.1646(10^{11})Q^2 \ \text{Pa}$$

a. If the density of the hydraulic oil is $\rho = 864$ kg/m^3 and the discharge coefficient is $C_d = 0.62$, compute the valve area A_0 when the valve is fully open.

b. Using MATLAB, plot flow rate Q versus pressure drop ΔP for three cases: (1) a fully open valve, (2) valve is half-open, and (3) valve is one-quarter open. Put all three plots on the same figure, and let the pressure drop range from 0 to 1.4 MPa.

SS **4.17**   Figure P4.17 shows a simple pneumatic shock absorber that consists of a piston (mass m), spring with stiffness k, and an air-filled chamber with pressure P. The chamber is sealed so that air cannot leak to the environment or past the piston. The rubber seals on the piston cause a friction force F_{dry} that is modeled by Coulomb (dry) friction. The base of the shock absorber is rigidly fixed. The position of the piston x is measured downward relative to its static equilibrium position. The volume of the air chamber is $V = V_0 - Ax$, where $V_0 =$ volume with $x = 0$, and A is the area of the piston. Derive the complete mathematical model of this system.

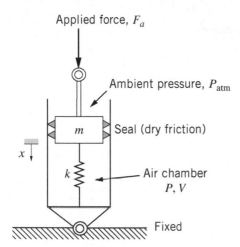

Applied force, F_a

Ambient pressure, P_{atm}

m Seal (dry friction)

x

k Air chamber
P, V

Fixed

Figure P4.17

SS **4.18**   Figure P4.18 shows a hydraulic system where a hose is connected to a chamber and a low-pressure reservoir. The hose and chamber both have constant volumes, V_h and V_c, respectively. A pump (not shown) supplies input volumetric flow rate Q_{in}. Two valves (with orifice areas A_{v1} and A_{v2}, respectively) control flow out of the hose. Derive the complete mathematical model of the hydraulic system.

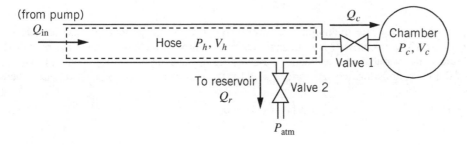

(from pump)
Q_{in}

Hose P_h, V_h

Q_c

Chamber
P_c, V_c

Valve 1

To reservoir
Q_r Valve 2

P_{atm}

Figure P4.18

4.19   Figure P4.19 shows a motorcycle shock absorber [3]. The main shock absorber cylinder has a "bound/rebound" valve mounted on its head that allows flow to and from the compensation cylinder. For example, when the piston and rod move upward ($x > 0$), hydraulic fluid (with pressure P_3) flows into the fluid side of the compensation cylinder with volumetric-flow rate Q_1. The compensation cylinder consists of a hydraulic chamber (with pressure P_2) and an air chamber (with pressure P_1) separated by the floating piston (m_1). Both piston masses experience viscous friction modeled by coefficients b_1 and b_2, respectively. Displacements y and x are measured from the equilibrium positions. The compensator cylinder chambers have volumes V_{01} (air) and V_{02} (fluid) when $y = 0$; and the shock absorber cylinder chambers have volumes V_{03} and V_{04} when $x = 0$. The orifice area for each valve is A_0. Derive the complete mathematical model of the shock absorber.

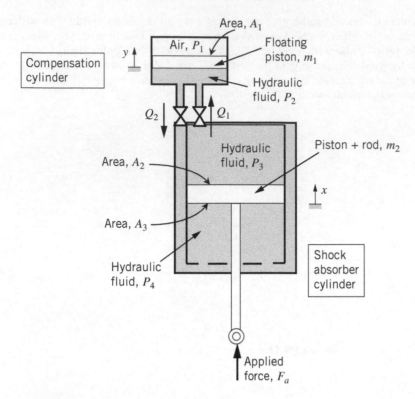

Figure P4.19

4.20 Consider again the motorcycle shock absorber presented in Problem 4.19 and Figure P4.19. Assuming that the hydraulic fluid is *incompressible* (i.e., the fluid flow rate in/out of a chamber is always equal to the time-rate of chamber volume), determine the relationship that must hold between the velocities of the floating piston and shock-absorber piston.

SS **4.21** Figure P4.21 shows a poppet valve used for hydraulic flow control. When $x = 0$, the poppet valve is "seated" and the metered flow is zero, $Q_2 = 0$. However, flow Q_1 to the upper chamber (through a channel inside the poppet valve) is feasible when $x = 0$ due to the variable-area orifice. The area of the upper-chamber orifice is equal to $w(x_0 + x)$, where w is the width of the rectangular orifice. The orifice area for the metered flow Q_2 is equal to $w_o x$, where the "area gradient" term is

$$w_o = \pi \left(d - \frac{x}{2}\right) \sin \alpha$$

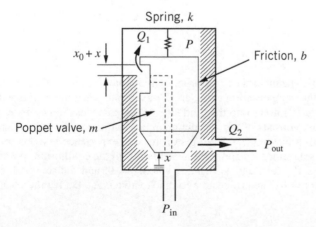

Figure P4.21

In this expression, d is the outer diameter of the poppet valve and α is its cone angle. The volume of the upper chamber is $V_0 - Ax$, where A is the area of the "top" of the poppet valve. The input and output pressures, P_{in} and P_{out}, are assumed to be constants. Derive the complete mathematical model of the poppet valve system.

4.22 Figure P4.22 shows a double-acting piston-cylinder hydraulic servomechanism. A positive spool-valve displacement $z > 0$ connects the high-pressure supply P_S with the left cylinder chamber 1 and chamber 2 is connected with the low-pressure drain P_r. Consequently for $z > 0$, volumetric flow Q_1 is into chamber 1 and flow Q_2, is out of chamber 2 to the drain. When $z < 0$ the flow is reversed. The orifice area of the spool valve is $A_0 = |z|h$, where h is the height of the valve opening. Piston position x is measured from the left side and hence the volume of chamber 1 is $V_1 = A_1 x$ and the volume of chamber 2 is $V_2 = A_2(L - x)$, where L is the total stroke length of the piston. Derive the complete mathematical model of the hydromechanical system.

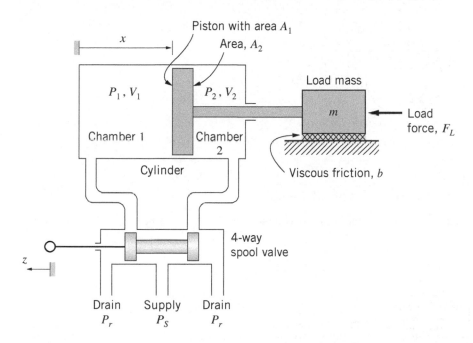

Figure P4.22

4.23 Figure P4.23 presents a hydraulic pump device. A scotch-yoke mechanism provides sinusoidal input to the piston, $y(t) = r \sin \omega t$, where r is the radius and ω is the constant rotation rate in rad/s. The pump chamber has pressure P and volume $V = V_0 - Ay$, where V_0 is the pump volume when $y = 0$ and A is the piston area. The pump chamber is connected to constant supply pressure P_S and constant load pressure P_L through two "ball" check valves, which only allow flow in one direction as shown in Fig. P4.23. Each check valve has orifice area A_0 when opened. Derive the complete mathematical model of the hydraulic pump.

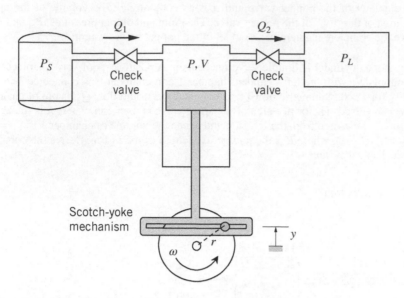

Figure P4.23

SS **4.24** Figure P4.24 shows a coke oven with pressure P, fixed volume $V = 25$ m^3, and temperature T. Ambient pressure is $P_{atm} = 1.013(10^5)$ Pa.

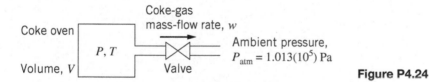

Figure P4.24

a. Derive the mathematical model of the pneumatic system with compressible flow through the sharp-edged orifice (valve) to atmospheric conditions.

b. At the instant the valve is opened, the system has the following characteristics: chamber temperature $T = 373$ K, chamber pressure $P = 9(10^5)$ Pa, valve discharge coefficient $C_d = 0.8$, and valve area $A_0 = 5(10^{-4})$ m^2. Assume an isothermal expansion process where the ratio of specific heats for coke gas is 1.41 and its gas constant is $R = 300$ N-m/kg-K. Compute the pneumatic capacitance C and the initial mass-flow rate through the valve w. Is the initial mass-flow rate choked or unchoked?

c. Eventually, the interior pressure of the coke oven P will match atmospheric pressure and the mass-flow rate will go to zero. Use MATLAB to plot mass-flow rate w as a function of oven pressure P.

4.25 An "air spring" is shown in Fig. P4.25. The column of air at nominal volume V_0 and nominal pressure P_0 supports the piston mass at static equilibrium. Displacement of the piston (x) is measured positive downward from the static equilibrium position. Derive an equation for the equivalent "air spring constant" k_{air} (N/m) assuming that the pneumatic reaction force from the compressed air is $F = k_{air}x$ [Hint: apply a positive differential displacement dx and assume that the gas undergoes a polytropic expansion when it is compressed].

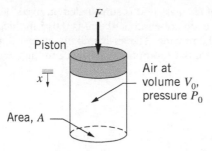

Figure P4.25

4.26 Figure P4.26 shows an electropneumatic clutch actuation system for heavy-duty trucks [4]. An electronic control unit (not shown) sends a signal to fully open either the supply or exhaust valve. When the supply valve is open, high-pressure air from the supply tank flows through the valve to the cylinder chamber. When the exhaust valve is open, air flows from the cylinder chamber to the surroundings. The two valves cannot be open at the same time. The constant supply pressure is P_S and the ambient pressure is P_{atm}. Fully open orifice area is A_0 for both valves. For normal operation the supply pressure is much greater than the chamber pressure P and the chamber pressure is significantly greater than ambient pressure P_{atm}. Therefore, we can assume that the supply and exhaust valve flows are always choked. Force F_L is the reaction force caused by displacing the clutch compression spring when engaging the clutch plates. The chamber volume is $V = V_0 + A_1 x$ where V_0 is the volume when the piston displacement is zero. Derive the complete mathematical model of the electropneumatic system.

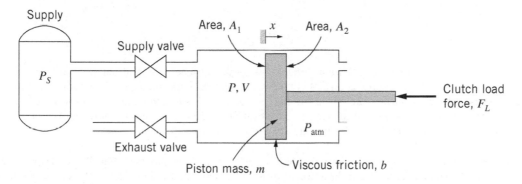

Figure P4.26

SS **4.27** Figure P4.27 shows an industrial furnace. The inner object (to be heat treated) has temperature T_1, thermal capacitance C_1, and thermal resistance R_1. The air inside the furnace has temperature T_2 and thermal capacitance C_2. The furnace is surrounded by the insulation with thermal resistance R_2. A heating unit supplies input heat flow $q_{in}(t)$ to the furnace. Derive the complete mathematical model, and identify all dynamic variables and input variables.

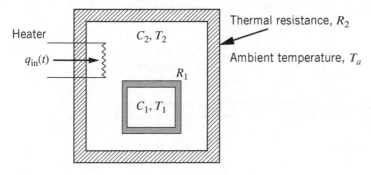

Figure P4.27

4.28 Figure P4.28 shows a cross-sectional view of a rectangular volume of air mass that is surrounded on five sides by a perfect insulator (zero heat transfer). The sixth (right) side is a square area of cardboard that is 0.015 m thick. The dimensions of the air-mass volume are 0.4 m long, 0.2 m high, and 0.2 m wide. The air is initially at 35 °C and the ambient temperature is $T_a = 25$ °C. Search the engineering literature for the appropriate physical constants required for this problem and then answer the following questions:

a. Derive the mathematical model of the thermal system.

b. Compute the initial heat-transfer rate $q(0)$ to the surroundings.

c. Compute the initial time-rate of the internal air temperature $\dot{T}(0)$.

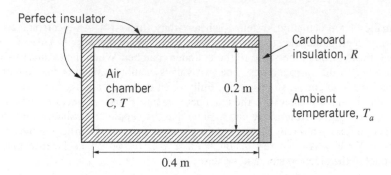

Figure P4.28

4.29 A kiln is constructed as a cube: four vertical walls and two horizontal surfaces (top and bottom). All surfaces have equal areas. The kiln is modeled as a simple thermal system with a thermal capacitance C (the interior chamber) and the four vertical walls and the top surface each have thermal resistance R. Assume that the bottom surface (the "floor" of the kiln) is a perfect insulator. A heating device provides heat flow rate q_{in} to the interior of the kiln. Derive the mathematical model for the kiln and define all dynamic variables and input variables.

4.30 A simple thermal system consisting of two thermal capacitances C_1 and C_2 is shown in Fig. P4.30. Thermal resistances R_1 separate the two capacitances from the ambient surroundings and thermal resistance R_2 is between the two capacitances. A heating device provides input heat flow q_{in} to thermal chamber 1. The top and bottom boundaries of the system have perfect insulation (infinite thermal resistance). Derive the complete mathematical model and identify all dynamic variables and input variables.

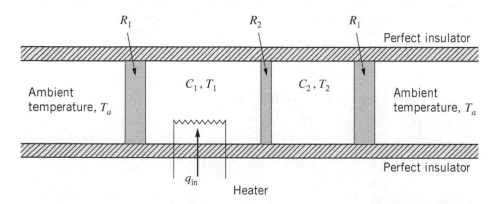

Figure P4.30

4.31   Figure P4.31 shows a thermal system where input fluid flows to a chamber, is mixed and heated, and then flows out of the chamber. The input and output mass-flow rates are equal, that is, $w_{in} = w_{out} = w$. The temperature of the input flow is T_{in} and the temperature of the output flow is the same as the chamber temperature (T). The heater supplies heat input q_{in}. The chamber is covered with insulation material with thermal resistance R. Derive the complete mathematical model of the system and identify all dynamic variables and input variables.

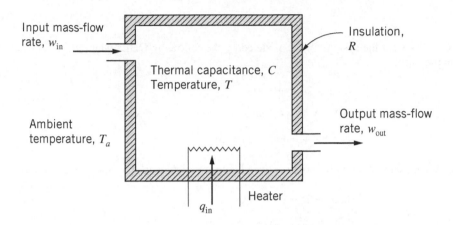

Figure P4.31

4.32   A simple model of a heat exchanger is shown in Fig. P4.32 [5]. Steam enters the chamber with temperature $T_{in,1}$ and mass-flow rate $w_{in,1}$ and leaves the chamber with temperature $T_{out,1} = T_1$, and mass-flow rate $w_{out,1} = w_{in,1} = w_1$. The temperature of the steam in the chamber is T_1, which is equal to the temperature of the outgoing steam. Cold water flows through copper tubes that enter the chamber with temperature $T_{in,2}$ and mass-flow rate $w_{in,2}$. The temperature of the water inside the chamber is T_2, which is equal to the temperature of the outgoing hot water. Heat is transferred from the steam to the water through the copper tubes, which have thermal resistance R_2, and the hot water leaves the chamber with the same mass-flow rate as the incoming cold water. Thermal insulation R_1 surrounds the heat exchanger and provides a thermal barrier with the ambient temperature T_a. Derive the complete mathematical model of the thermal system.

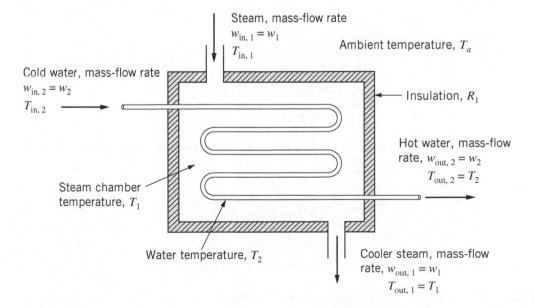

Figure P4.32

CHAPTER 5: PROBLEMS

Conceptual Problems

5.1 Derive the state-variable equations for the system that is modeled by the following ODEs where α, w, and z are the dynamic variables and v is the input.

$$0.4\dot{\alpha} - 3w + \alpha = 0$$
$$0.25\dot{z} + 4z - 0.5zw = 0$$
$$\ddot{w} + 6\dot{w} + 0.3w^3 - 2\alpha = 8v$$

5.2 Obtain a complete SSR for the given system, with input $u = v$ and output variables $y_1 = \dot{w}$ and $y_2 = w - z$.

$$0.25\dot{z} + 2z - 0.6w = 0$$
$$0.5\ddot{w} + 3\dot{w} + 20w - 4z = 2v$$

SS **5.3** Obtain a complete SSR for the following system, where the inputs are $u_1 = \alpha$ and $u_2 = \beta$, and the single output is $y = \theta$.

$$\ddot{\delta} + 8\delta - 0.5\dot{\theta} = 4\beta - 0.6\alpha$$
$$0.2\ddot{\theta} + 4\dot{\theta} + 12\theta - 3\dot{\delta} = 0.1\beta$$

5.4 Given the nonlinear first-order system

$$\dot{x} + 0.4\sqrt{x} = 2u$$

Derive the linear model by performing the linearization about the static equilibrium state x^* that results when the nominal input is $u^* = 0.25$.

SS **5.5** Given the following state-variable equations

$$\dot{x}_1 = -3x_1 + 6x_3$$
$$\dot{x}_2 = -0.5x_1^3 + 2x_1x_2 - x_2 + 8u$$
$$\dot{x}_3 = -4x_3 + 3x_2x_3^2$$

Derive the linear state equation for $\delta\mathbf{x} = \mathbf{x} - \mathbf{x}^*$ and $\delta u = u - u^*$, and express the state equation in matrix-vector form.

SS **5.6** Given the state-variable equations

$$\dot{x}_1 = -3x_1 + 2x_2 + 0.5u$$
$$\dot{x}_2 = x_1 - 12x_2$$

Derive the input–output (I/O) equation where u is the input and x_2 is the output.

5.7 Derive the transfer function $G(s) = Y(s)/U(s)$ of the given I/O equation

$$2\ddot{y} + 10\dot{y} + 48y = 0.8u$$

5.8 The mathematical model of a system is

$$\dot{w} + 6w = 4u$$

$$3\ddot{y} + 8\dot{y} + 20y = 16w$$

Derive the overall system transfer function where y is the output and u is the input.

5.9 The mathematical model of a system is

$$\dot{z} + 2z = 4u + y$$

$$\ddot{y} + 7\dot{y} + 16y = 3z$$

Derive the transfer function $G(s) = Y(s)/U(s)$ for this system.

5.10 Given the SSR

$$\dot{\mathbf{x}} = \begin{bmatrix} 0 & 1 \\ -20 & -4 \end{bmatrix} \mathbf{x} + \begin{bmatrix} 0 \\ 0.2 \end{bmatrix} u \quad y = \begin{bmatrix} 1 & 0 \end{bmatrix} \mathbf{x}$$

a. Obtain the I/O equation for this system where y is the output and u is the input.
b. Obtain the transfer function for this system.

5.11 Given the SSR

$$\dot{\mathbf{x}} = \begin{bmatrix} 0 & 1 \\ -20 & -4 \end{bmatrix} \mathbf{x} + \begin{bmatrix} 0.3 \\ 1 \end{bmatrix} u \quad y = \begin{bmatrix} 1 & 0 \end{bmatrix} \mathbf{x}$$

a. Obtain the I/O equation for this system where y is the output and u is the input.
b. Obtain the transfer function.

SS **5.12** A simple 1-DOF mechanical system is shown in Fig. P5.12 (see Problem 2.3). The system is driven by the displacement of the left end, $x_{in}(t)$, which could be supplied by a rotating cam and follower. When displacements $x_{in}(t) = 0$ and $x = 0$, the spring k is neither compressed nor stretched. Derive the system transfer function with position x as the output variable and displacement $x_{in}(t)$ as the input variable.

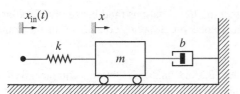

Figure P5.12

5.13 An RLC circuit with a parallel bypass resistor (Problem 3.12) is shown in Fig. P5.13. Obtain a complete SSR with the source voltage $e_{in}(t)$ as the input and voltage across the capacitor e_C as the output.

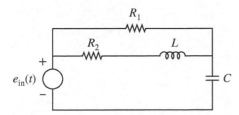

Figure P5.13

5.14   Consider again the solenoid actuator that is analyzed in Example 5.3. Can we express the state-variable equations obtained in Example 5.3 in a matrix-vector SSR? Explain your answer.

5.15   Figure P5.15 shows the 2-DOF single-mass mechanical system from Problem 2.8 in Chapter 2. The independent displacement of the mass m is z_1 and z_2 is the independent displacement of the node point between the damper b and spring k.

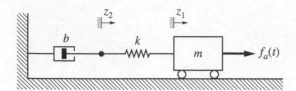

Figure P5.15

a. Obtain an SSR of this mechanical system where position of the mass is the system output and the applied force is the input.

b. Derive the I/O equation using the D-operator method.

c. Use the result from part (b) to derive the transfer function for this system.

5.16   Figure P5.16 shows the 2-DOF single-mass mechanical system from Problem 2.9 in Chapter 2. Note that the series connection of the damper and spring is reversed from Problem 5.15.

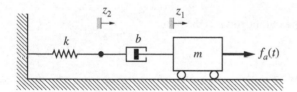

Figure P5.16

a. Obtain an SSR of this mechanical system where position of the mass is the system output and the applied force is the input.

b. Derive the I/O equation using the D-operator method. Show that the I/O equations (and hence transfer functions) for the systems shown in Figs. P5.15 and P5.16 are identical. Hence, either series connection of the damper or spring results in the same system dynamics.

5.17   Figure P5.17 shows the two-mass mechanical damper from Problem 2.5 in Chapter 2. Obtain a complete SSR where the external force $f_a(t)$ is the input and piston position z_1 is the output.

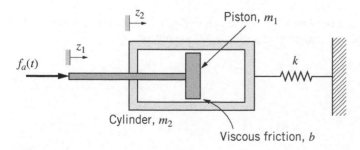

Figure P5.17

5.18   Derive the input–output (I/O) equation using the D-operator method for the two-mass mechanical damper shown in Fig. P5.17. Piston position z_1 is the output and applied force $f_a(t)$ is the input. In addition, determine the system transfer function.

SS **5.19**   Figure P5.19 shows a two-mass mechanical system. Vertical displacements z_1 and z_2 are measured relative to their equilibrium positions. Obtain a complete SSR where the applied force $f_a(t)$ is the input and the force transmitted to the base is the output.

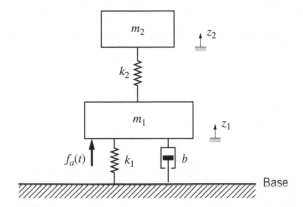

Figure P5.19

5.20   An electrical system is presented in Fig. P5.20 (see Problem 3.9). Obtain a complete SSR where source voltage $e_{in}(t)$ is the input and the two output variables are the voltage drops across the resistor R and capacitor C_2, respectively.

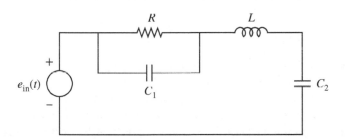

Figure P5.20

5.21   Sketch the block diagram for the second-order system given below using the integrator-block method. Label all blocks and signal-path variables.

$$2\ddot{y} + 6\dot{y} + 34y = u$$

5.22   Sketch the block diagram for the third-order system in Problem 5.7 using the integrator-block method. Label all blocks and signal-path variables.

SS **5.23**   Sketch the block diagram for the system defined by the state-variable equations in Problem 5.6. Use the integrator-block method, and label all blocks and signal-path variables (i.e., states x_1 and x_2, input u, etc.).

5.24   The modeling equations of a linear system are given below. The overall system input and output variables are u and y, respectively.

$$2\dot{z} + z = u$$
$$\dot{y} + 6y = 4z$$

a. Sketch the simplest possible block diagram for the case where all variables have zero initial conditions. Label all blocks and signal-path variables.

b. Sketch the block diagram using integrator blocks for the case where nonzero initial conditions are present; that is, $z(0) = 2$ and $y(0) = -3$. Label all blocks and signal-path variables.

5.25   Sketch the simplest possible block diagram for the following system:

$$w = K(u - y)$$

$$\ddot{y} + 6\dot{y} + 20y = 3w$$

All dynamic variables have zero initial conditions. The parameter K is a constant or "gain." Label all signal-path variables.

MATLAB Problems

5.26   A centrifugal pump has the nonlinear pressure-flow relation

$$P = 3.645 \left(10^5\right) \sqrt{1 - \frac{Q}{0.019}} \quad \text{Pa}$$

where Q is the volumetric-flow rate (in m^3/s) and P is the pressure output of the pump (in Pa). The pump model is valid for $0 < Q \leq 0.0175$ m^3/s. The nominal (operating) volumetric-flow rate is 0.008 m^3/s. Derive a linear model for the pump pressure about the operating (nominal) point. Plot the true (nonlinear) pump pressure and approximate (linearized) pump pressure versus volumetric-flow rate for $0 < Q < 0.0175$ m^3/s. Comment on the range of accuracy for the linear pump model.

5.27   The inductance of a solenoid actuator varies with armature position (or stroke) x and can be modeled by the nonlinear expression

$$L(x) = \frac{L_0}{1 - x/d}$$

For a particular solenoid coil, the constant $d = 7.8$ mm and the inductance at zero stroke is $L_0 = 0.006$ H. Note that inductance $L(x)$ increases with stroke x as the armature moves toward the center of the coil.

a. Develop a linearized approximation for inductance $L(x)$ about a nominal stroke $x^* = 1$ mm.

b. Plot the (true) nonlinear inductance $L(x)$ and the approximate (linearized) inductance for a stroke $0 < x < 3$ mm.

c. Plot the percent error between the nonlinear and linear inductances versus stroke and comment on the accuracy of the linear approximation.

Engineering Applications

SS   5.28   Figure P5.28 shows the hydromechanical actuator from Example 4.2 in Chapter 4. Obtain a complete set of state-variable equations for this system (note that the piston position is redefined as z so that x may be used as the state variable). Identify the state and input variables.

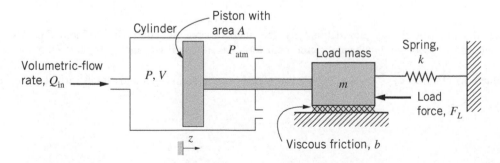

Figure P5.28

SS **5.29**   Figure P5.29 shows a high-pass filter circuit with a "tweeter" speaker (see Problem 3.31 in Chapter 3). Obtain a complete SSR of the speaker circuit where the states are the dynamic variables associated with the energy-storage elements, speaker voltage e_R is the system output, and $e_{in}(t)$ is the input.

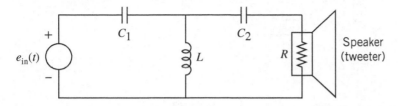

Figure P5.29

5.30   Figure P5.30 shows an electrical system known as a *buck converter*, which is a circuit used to step the source voltage $e_{in}(t)$ "down" to a lower desired output voltage (see Problem 3.33 in Chapter 3). The step-down voltage converter uses a switch to connect and disconnect the voltage supply $e_{in}(t)$ from the remainder of the circuit until output voltage $e_O = e_C$ is equal to the desired voltage. Obtain a complete SSR of the buck converter circuit where the states are the dynamic variables associated with the energy-storage elements, e_O is the system output, and $e_{in}(t)$ is the input.

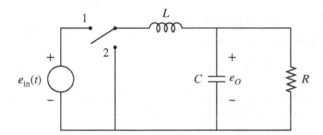

Figure P5.30

5.31   Figure P5.31 shows the *washout filter* from Problem 3.34. These circuits are used to "wash out" (i.e., reduce to zero) a constant input voltage. Derive the transfer function $G(s)$ of the washout filter where voltage e_O is the desired output and source voltage $e_{in}(t)$ is the input.

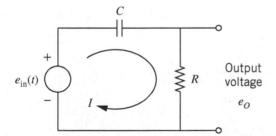

Figure P5.31

SS **5.32**   Figure P5.32 shows the electrical system (*band-stop* or *notch filter*) from Problem 3.35. These circuits are used to "filter out" or attenuate input signals within a particular frequency band. For example, notch filters are used on aerospace vehicles to remove the mechanical vibrations that are transmitted to the onboard sensors such as gyroscopes and accelerometers. Derive the transfer function $G(s)$ of the band-stop or notch filter where voltage e_O is the desired output and source voltage $e_{in}(t)$ is the input.

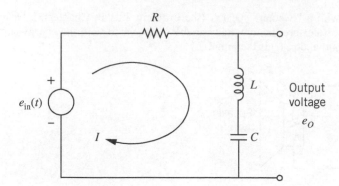

Figure P5.32

5.33 Obtain a complete SSR of the band-stop filter described in Problem 5.32 where the states are the dynamic variables associated with the energy-storage elements, e_O is the system output, and $e_{in}(t)$ is the input.

5.34 A simplified, linear representation of an electrohydraulic actuator (EHA) consists of a power amplifier, a solenoid model, and a mechanical valve model. The linear I/O equations for each subsystem are

$$\text{Power amplifier: } \tau_a \dot{e}_0 + e_0 = K_a e_{in}(t)$$

$$\text{Solenoid actuator: } \tau_s \dot{f} + f = K_s e_0$$

$$\text{Spool valve: } m\ddot{z} + b\dot{z} + kz = f$$

where $e_{in}(t)$ is the low-power voltage input to the amplifier, e_0 is the amplifier voltage output, f is the output force of the solenoid actuator (in N), and z is the position of the spool-valve mass m (in m).

a. Obtain a complete SSR with valve position z as the single output variable.

b. Derive the transfer functions for each subsystem of the EHA. Sketch a block diagram of the complete EHA system. Assume that all dynamic variables have zero initial conditions. Label all blocks and signal-path variables (with units).

5.35 Using the power amplifier and solenoid modeling equations from the EHA system in Problem 5.34, derive the I/O equation with low-power voltage $e_{in}(t)$ as the input and solenoid force f as the output.

5.36 The turning dynamics of an automobile are [1]

$$\dot{v} = -\frac{C_F + C_R}{mu}v + \left(\frac{C_R L_R - C_F L_F}{mu} - u\right)r + \frac{C_F}{m}\delta_F$$

$$\dot{r} = \frac{C_R L_R - C_F L_F}{Ju}v - \frac{C_R L_R^2 + C_F L_F^2}{Ju}r + \frac{C_F L_F}{J}\delta_F$$

where u and v are the forward and lateral components of the vehicle's velocity (in m/s), r is the vehicle's yaw (turning) rate (in rad/s), and δ_F is the steering angle of the front tire (in rad). For this problem, forward speed u is constant. The remaining system parameters are vehicle mass m, vehicle moment of inertia (yaw axis) J, front- and rear-tire cornering stiffness coefficients C_F and C_R, and moment arms from the center of mass to front and rear axles L_F and L_R.

Obtain a complete SSR. The three outputs are lateral velocity v, yaw rate r, and lateral acceleration $a_{lat} = \dot{v} + ru$. The single input is the front-wheel steering angle δ_F.

5.37 An electric motor drives a fixed-displacement hydraulic pump, which affects the time-rate of pressure P in a cylinder. The linearized dynamics for this hydromechanical system are

$$\text{DC motor circuit:} \qquad L\dot{I} + RI = e_{\text{in}}(t) - K_b\omega$$

$$\text{Motor/pump dynamics:} \qquad J\dot{\omega} + b_{\text{mp}}\omega = K_m I - d_p(P - P_{\text{atm}})$$

$$\text{Hydraulic cylinder:} \qquad \dot{P} = \frac{\beta}{V_0}(d_p\omega - A\dot{z})$$

$$\text{Mechanical load:} \qquad m\ddot{z} + b\dot{z} = PA - P_{\text{atm}}A$$

where $e_{\text{in}}(t)$ is voltage input to the motor; I, L, R, and K_b are the current, inductance, resistance, and back-emf constant of the motor; ω is the angular velocity of the motor/pump; J and b_{mp} are the motor + pump moment of inertia and viscous friction; K_m is the motor-torque constant; β is the fluid bulk modulus; V_0 is the nominal cylinder volume; d_p is the pump volumetric displacement; A is the hydraulic cylinder piston area; m is the mass of the mechanical load; b is the translational friction coefficient, z is the position of the mass, and P_{atm} is atmospheric pressure.

   Obtain a complete SSR. The output variables are mechanical position z, cylinder pressure P, and motor/pump velocity ω. The input variables are voltage $e_{\text{in}}(t)$ and atmospheric pressure P_{atm}.

5.38 Figure P5.38 shows the dual-disk mechanical system from Example 2.9 in Chapter 2. Recall that this system has been proposed as an efficient generator for hybrid vehicles. Obtain a complete SSR with relative angular displacement $\theta_2 - \theta_1$ as the output variable and input torque $T_{\text{in}}(t)$ as the input variable.

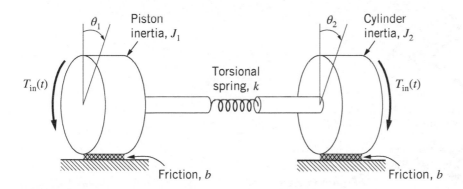

Figure P5.38

5.39 A simplified model of a solenoid actuator is

$$0.02\ddot{z} + 8\dot{z} + 3000z = 4.5I$$

$$0.01\dot{I} + 5I = e_{\text{in}}(t)$$

where z is the position of the mechanical mass–spring–damper subsystem and I is the current in the solenoid armature circuit. The single input to the system is source voltage $e_{\text{in}}(t)$. Obtain a complete SSR with position of the mass (z) as the only output variable.

5.40 Figure P5.40 shows the railroad-car system described in Problem 2.38 (Chapter 2). Obtain a complete SSR where the locomotive force $F_a(t)$ is the input and the two measured outputs are the velocity of the locomotive (mass m_1) and the relative displacement between the locomotive and Car 1 (i.e., $z_1 - z_2$).

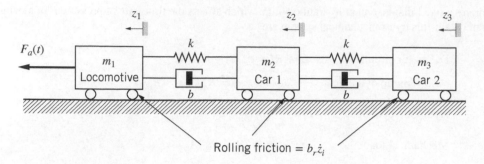

Figure P5.40

5.41 Consider again the seat-suspension system presented in Example 5.8 and Fig. 5.2. Obtain a complete SSR using the single input $u = z_0(t)$. The two output variables are driver position and acceleration, z_2 and $\ddot{z}_2$, respectively. [Hint: define the second state variable as $x_2 = \dot{z}_1 - b_1 z_0(t)/m_1$.

5.42 Figures P5.42a and P5.42b show the pantograph system described in Problem 2.36 (Chapter 2). Obtain a complete SSR where the two inputs are overhead wire displacement $z_w(t)$ and piston force $f_a(t)$ and the two measured outputs are the (compressive) contact force between the wire and head mass and the relative displacement between the head and frame masses (i.e., $z_1 - z_2$). You may assume that the spring k_1 is always in compression.

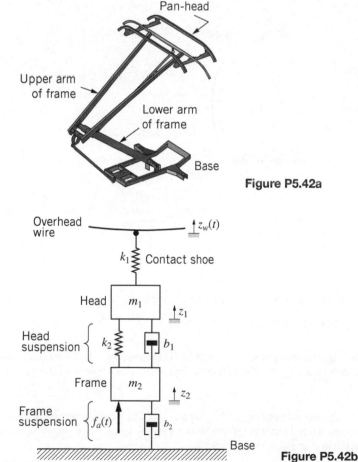

Figure P5.42a

Figure P5.42b

5.43 Figure P5.43 shows the magnetic levitation system described in Problem 3.40 (Chapter 3). Recall that the electro-magnetic force is

$$F_{em} = \frac{K_F I^2}{(d-z)^2}$$

where K_F is a "force constant" that depends on the number of coil wraps, material properties of the electromagnetic core, and electromagnet geometry; I is the coil current; and z is the position of the metal ball measured upward from a fixed at-rest position. Distance d is a constant equal to the nominal air gap from the electromagnet tip and the ball for a nominal coil current of 0.8 A. The voltage source $e_{in}(t)$ is the input to the electromagnet circuit. The system has the following parameters:

Coil inductance $L = 0.018$ H

Resistance $R = 5\ \Omega$

Force constant $K_F = 2.6487\left(10^{-5}\right)$ N-m^2/A^2

Metal ball mass $m = 0.003$ kg

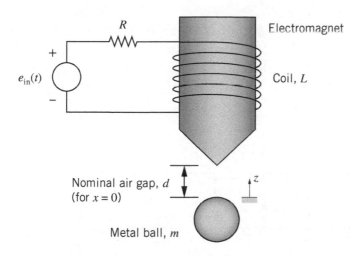

Figure P5.43

a. Derive the complete state-variable equations [Hint: treat gravitational acceleration g as the second input variable].

b. If the nominal input voltage is $e_{in}^*(t) = 4$ V and the nominal coil current is $I^* = 0.8$ A, compute the nominal (constant) levitation distance d (the air gap) at static equilibrium and $z^* = 0$.

c. Linearize the state-variable equations from part (a) and obtain an SSR with perturbation of the ball from the nominal position as the system output. What happens to the gravitational force in the linearized system?

5.44 Figure P5.44a shows a schematic diagram of the optical disk drive system described in Problem 2.35 (Chapter 2), and Fig. P5.44b shows a simplified lumped-parameter mechanical representation of this system.

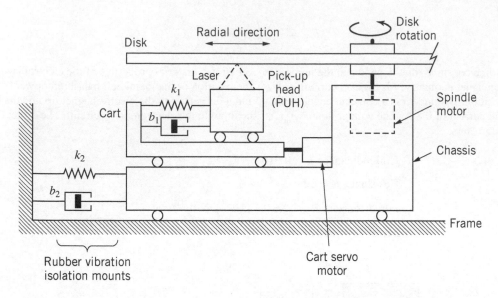

Figure P5.44a

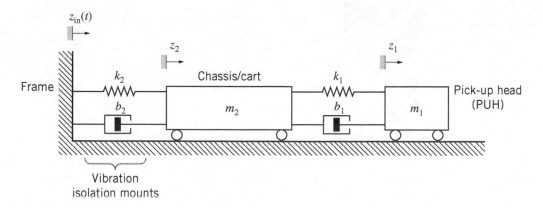

Figure P5.44b

a. Derive an I/O equation of the optical disk drive system with pick-up head mass displacement z_1 as the system output and frame displacement $z_{in}(t)$ as the input.

b. Derive the transfer function of the optical disk drive system where pick-up head mass displacement z_1 is the system output and frame displacement $z_{in}(t)$ is the input.

CHAPTER 6: PROBLEMS

Conceptual Problems

SS **6.1** Figure P6.1 shows a flywheel with moment of inertia $J = 0.5$ kg-m^2 that is initially rotating at an angular velocity $\dot{\theta}_0 = 40$ rad/s. The flywheel is subjected to friction, which is modeled by linear viscous friction torque $b\dot{\theta}$, with friction coefficient $b = 0.06$ N-m-s/rad. Use Simulink to obtain the dynamic response and plot the angular position $\theta(t)$ (in rad) and angular velocity $\dot{\theta}(t)$ (in rad/s). In addition, use the simulation to integrate the rate of energy dissipation $\dot{\xi}$ and plot dissipated energy (in J) versus time. Show that the total dissipated energy as computed by the simulation is equal to the system's initial energy.

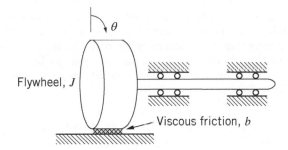

Figure P6.1

6.2 Repeat Problem 6.1 with the addition of dry (Coulomb) friction torque, $T_{dry}\,\text{sgn}\,(\dot{\theta})$, where $T_{dry} = 0.1$ N-m. Linear viscous friction torque ($b = 0.06$ N-m-s/rad) also acts on the flywheel.

SS **6.3** Figure P6.3 shows a mass–spring–damper mechanical system. The system is initially at rest. At time $t = 0$, a pulse force is applied to the mass. The pulse force is $F_a(t) = 8$ N for $0 \leq t \leq 0.05$ s and $F_a(t) = 0$ for $t > 0.05$ s. The system parameters are $m = 2$ kg, $b = 5$ N-s/m, and $k = 200$ N/m. Simulate the dynamic response using MATLAB commands and plot the response $z(t)$.

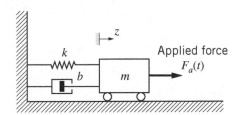

Figure P6.3

6.4 Given a system's SSR

$$\dot{\mathbf{x}} = \begin{bmatrix} 0 & 1 \\ -34 & -2 \end{bmatrix}\mathbf{x} + \begin{bmatrix} 0 \\ 2.5 \end{bmatrix}u \quad y = [1 \ \ 0]\,\mathbf{x}$$

Use MATLAB commands to obtain the dynamic response for a step input $u(t) = 0.6U(t)$. The initial states are $x_1(0) = 2$ and $x_2 = -1.5$. Plot the output versus time, $y(t)$.

6.5   Figure P6.5 shows a mechanical system driven by the displacement of the left end, $x_{in}(t)$, which could be supplied by a rotating cam and follower (see Problem 2.3). The mechanical parameters are mass $m = 0.5$ kg, $b = 6$ N-s/m, and $k = 500$ N/m, and the system is initially at rest. The displacement of the left end is $x_{in}(t) = 0.05 \sin \omega t$ (in m), where ω is the input frequency in rad/s. Using Simulink, obtain and plot the response $x(t)$ for input frequency $\omega = 20$ rad/s (or, $10/\pi$ cycles per second = 3.183 Hz). Show the input $x_{in}(t)$ on the same plot.

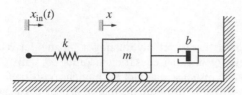

Figure P6.5

6.6   Write a MATLAB script (M-file) that executes the Simulink simulation for the mechanical system in Problem 6.5 for input frequencies ranging from $\omega = 1$ rad/s (0.16 Hz) to $\omega = 100$ rad/s (15.92 Hz) in increments of 1 rad/s. Store the ratios of the output/input amplitudes but use only the last one-third of the simulation response data (the so-called *steady-state* response) for each input frequency [Hint: to do this, compute the ratio x_{end}/x_{in} by using the `command` `max (x_end)/max(x_in)` where x_end is the last one-third of the system response $x(t)$]. Plot the output/input amplitude ratio versus input frequency, and based on your plot, summarize how the mechanical system's displacement response $x(t)$ varies with input frequency ω. This problem is an example of a system's *frequency response*, where the system input is a periodic function (such as a sinusoidal input function).

SS   **6.7**   Figure P6.7 shows the two-mass mechanical damper from Problem 2.5 (in addition, see Problem 5.17). Displacements z_1 and z_2 (in m) are measured from their equilibrium position when external force $f_a(t)$ is zero. The external force is applied directly to the rod/piston mass m_1. The system parameters are $m_1 = 0.2$ kg, $m_2 = 0.1$ kg, $k = 800$ N/m, and $b = 25$ N-s/rad.

   a.   Use MATLAB or Simulink to obtain the response of the piston (position z_1) when the external force is a step input with a magnitude of 10 N. Does this response realistically represent the physical system? Explain your answer.

   b.   Now, add a spring with stiffness $k_2 = 100$ N/m between the piston and the cylinder. Use MATLAB or Simulink to obtain the step response of the piston (with added spring k_2) if the external force is 10 N. In simple terms, explain why the two step responses (with and without k_2) are different.

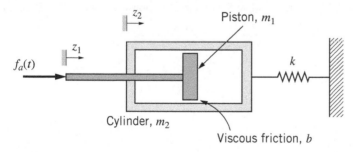

Figure P6.7

SS   **6.8**   Figure P6.8 shows the 1-DOF mechanical system from Problem 2.7 in Chapter 2. Displacement z is measured from the static equilibrium position (undeflected spring), and the spring force is governed by a nonlinear relationship with displacement:

$$F_k = k_1 z + k_3 z^3$$

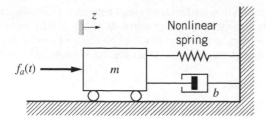

Figure P6.8

Use Simulink to obtain the response to a sinusoidal input force, $f_a(t) = 5 \sin \omega t$ N, where the input frequency is 0.75 Hz (i.e., $\omega = 1.5\pi$ rad/s). The system parameters are mass $m = 1.5$ kg, viscous friction coefficient $b = 3$ N-s/m, linear stiffness coefficient $k_1 = 20$ N/m, and nonlinear stiffness coefficient $k_3 = 50$ N/m^3. Plot the displacement $z(t)$.

6.9 An RLC circuit with a parallel bypass resistor (Problems 3.12 and 5.13) is shown in Fig. P6.9. At time $t = 0$ the circuit has zero current in both loops and the capacitor C has a stored charge of 0.01 C. The system parameters are $R_1 = 0.4$ Ω, $R_2 = 0.2$ Ω, $C = 0.04$ F, and $L = 0.01$ H. Use Simulink to obtain the system response where the source voltage is a sinusoidal function, $e_{in}(t) = 0.5 \sin 10t$ V. A voltmeter is used to measure the voltage across the capacitor C. Plot Simulink's prediction of the voltmeter output for a simulation time of 1.5 s.

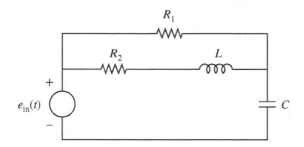

Figure P6.9

6.10 Use the MATLAB command `lsim` to obtain the voltage response for the system in Problem 6.9.

6.11 A series RL circuit with a nonlinear inductor is shown in Fig. P6.11. Recall that the following nonlinear function for inductor current was used in Problem 3.14 in Chapter 3:

$$I_L(\lambda) = 97.3\lambda^3 + 4.2\lambda \quad \text{(amps, A)}$$

where λ is the flux linkage. Use Simulink to obtain the dynamic response for current $I_L(t)$ if the source voltage is a 4-V step function, that is, $e_{in}(t) = 4U(t)$ V. The RL circuit has zero energy stored at time $t = 0$ and the resistance is $R = 1.2$ Ω. Plot current I_L versus time.

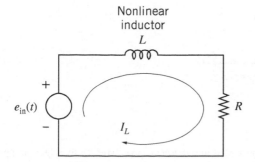

Figure P6.11

6.12 Figure P6.12 shows a mass m sliding on an oil film with viscous friction coefficient b (see Problem 2.11 in Chapter 2). The mass is moving toward the stiffness element k and at time $t = 0$ it has position $x(0) = 0$, velocity $\dot{x}(0) = 0.4$ m/s, and it is 0.5 m from the stiffness element. The system parameters are $m = 1.8$ kg, $b = 0.75$ N-s/m, and $k = 2$ N/m. Use Simulink to determine the system response and plot $x(t)$, $\dot{x}(t)$, and spring force. Does the mass return to its starting position ($x = 0$)?

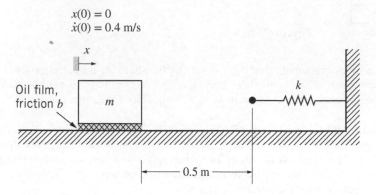

Figure P6.12

6.13 Consider the nonlinear system:

$$\dot{x}_1 - x_2 = 0$$

$$\dot{x}_2 + 2x_1^{1/4} + 3x_2 = u$$

The initial conditions are $x_1(0) = 0.08$, $x_2(0) = 0.02$, and the input is $u = 1.01$ (constant).

a. Simulate the nonlinear system using Simulink to obtain the state responses $\mathbf{x}(t) = [x_1(t) \ x_2(t)]^T$. Plot $x_1(t)$ and $x_2(t)$ on the same figure.

b. Linearize the system about the static equilibrium state vector $\mathbf{x}^*$ that arises when the nominal input is $u^* = 1$. Use Simulink to simulate the linear model and obtain the approximate state response $\mathbf{x}(t) = \mathbf{x}^* + \delta\mathbf{x}(t)$. Plot the nonlinear state solutions [from part (a)] and linearized state solutions on the same figure. Comment on the accuracy of the linear solution.

SS **6.14** Figure P6.14 shows a mass-spring system that is attracted by the electromagnetic force F_{em} produced by the current-carrying coil (see Problem 3.23). The electromagnet force F_{em} depends on current and position of the mass

$$F_{em} = \frac{K_F I^2}{(d - x)^2}$$

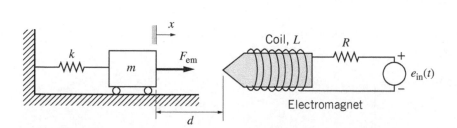

Figure P6.14

where K_F is a "force constant" that depends on the number of coil wraps and material properties of the electromagnetic core. Mass position x is measured from a fixed reference (where the spring is undeflected), and d is a constant equal to the distance between the electromagnet tip and the at-rest mass when $F_{em} = 0$.

The system parameters are

Mass $m = 0.15$ kg
Spring constant $k = 200$ N/m
Coil inductance $L = 0.03$ H
Coil resistance $R = 8\ \Omega$
Zero-force gap distance $d = 0.04$ m
Electromagnetic force constant $K_F = 0.006$ N-m^2/A^2

Use Simulink to determine the position response of the mass $x(t)$ for a sinusoidal voltage input $e_{in}(t) = 4 \sin \omega t$ V where ω is the input frequency. The mass starts at rest with $x(0) = 0$. Let the simulation time be 10 s, and plot three responses for input frequencies $\omega = 1, 5$, and 10 rad/s. Which of these input frequencies produces the largest total displacement of the mass?

Engineering Applications

6.15 Recall that Problem 2.26 in Chapter 2 presented a nonlinear model of "stick-slip" friction for mechanical systems. This problem will demonstrate how the dynamic response of a simple mechanical system is affected by the choice of the friction force model. The mathematical model of a simple 1-DOF mechanical system is

$$m\ddot{x} + F_f + kx = F_a(t)$$

where m is the mass, F_f is the friction force, k is the stiffness (spring) coefficient, x is the displacement of mass m from static equilibrium (in m), and $F_a(t)$ is the applied force. Obtain the dynamic responses using Simulink for the two friction models:

1) Linear viscous friction: $F_f = b\dot{x}$

2) Nonlinear stick-slip friction: $F_f = \left[F_C + (F_{st} - F_C) \exp(-|\dot{x}|/c)\right] \operatorname{sgn}(\dot{x}) + b\dot{x}$

The system parameters are $m = 2$ kg, $k = 800$ N/m, $b = 25$ N-s/m, $F_{st} = 1.2$ N (stiction force), $F_C = 1$ N (Coulomb friction force), $c = 0.002$ m/s (velocity coefficient). The external force $F_a(t)$ is a 15-N step function applied at time $t = 0.2$ s. The mass is initially at rest in static equilibrium. Plot the dynamic responses $x(t)$ obtained using both friction models on the same plot. In addition, plot the friction force $F_f(t)$ from both simulations on the same plot. Let the total simulation time be 1.8 s and use the fixed-step, fourth-order Runge–Kutta solver (ode4) with a step size of 0.001 s. On the basis of your simulation results describe the differences between the responses with the two friction models.

SS **6.16** Figure P6.16 shows a high-pass filter circuit with a "tweeter" speaker (see Problem 3.31 in Chapter 3). Use Simulink to obtain the speaker's voltage output $e_R(t)$ if the input voltage is $e_{in}(t) = 2 \sin \omega t$ V. Plot the input and output voltages on the same figure for input frequency $\omega = 4000$ rad/s (about 637 Hz). The circuit parameters are $R = 8\ \Omega$, $C_1 = 14$ µF, $C_2 = 42$ µF, and $L = 1$ mH.

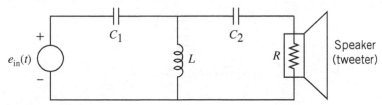

Figure P6.16

6.17 Figure P6.17 shows the *washout filter* circuit described in Problem 3.34 in Chapter 3 with capacitor $C = 0.01$ F and resistor $R = 2\ \Omega$. Use Simulink to obtain the dynamic response of the output voltage $e_O(t)$ if the input voltage is $e_{in}(t) = 1500t^2$ V for $0 \le t \le 0.04$ s until it reaches a maximum input voltage of 2.4 V. Therefore, $e_{in}(t) = 2.4$ V for $t > 0.04$ s. The circuit has zero stored energy at time $t = 0$. Plot output and input voltages $e_O(t)$ and $e_{in}(t)$ on the same plot. On the basis of your simulation results explain why this circuit is called a *washout filter*. [Hint: use the `Saturation` block from the `Discontinuities` library to limit the input voltage so it never exceeds 2.4 V].

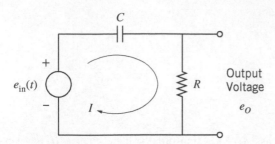

Figure P6.17

6.18 Use the MATLAB command `lsim` to obtain the voltage response for the washout filter in Problem 6.17. Plot output and input voltages $e_O(t)$ and $e_{in}(t)$ on the same plot.

6.19 Figure P6.19 shows the wind turbine generator system from Example 2.8 in Chapter 2. The system inputs are aerodynamic torque T_{aero} (from the wind) and generator (or electrical) torque T_{gen}. The system parameters for a 3000 kW wind turbine generator are

Turbine moment of inertia $J_1 = 1.26(10^7)$ kg-m^2
Turbine radius (blade tip to hub) $R = 45.6$ m
Turbine friction coefficient $b_1 = 1100$ N-m-s/rad
Generator moment of inertia $J_2 = 240$ kg-m^2
Generator friction coefficient $b_2 = 0.1$ N-m-s/rad
Gear ratio $N = r_2/r_1 = 1/93$

The aerodynamic torque is a function of air density ρ (in kg/m^3), turbine radius R (in m), wind speed V_w (in m/s), and torque coefficient C_q:

$$T_{aero} = \frac{1}{2}\rho\pi R^3 V_w^2 C_q$$

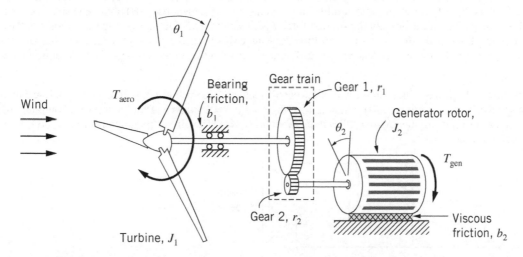

Figure P6.19

Assume that air density is $\rho = 1.225$ kg/m^3 and the torque coefficient is $C_q = 0.075$. At time $t = 0$, the wind speed is $V_w = 13$ m/s. The generator torque is a function of generator shaft speed: $T_{gen} = 158.7\dot\theta_2$ (N-m).

a. Compute the constant angular velocities of the turbine and generator shafts at time $t = 0$.

b. Obtain the dynamic response using Simulink. At time $t = 2$ s, the wind speed ramps up from 13 to 15 m/s at a constant acceleration of 0.25 m/s^2. Use the initial conditions computed in part (a) and set the simulation time to 90 s. Plot the angular velocities of the turbine and generator shafts versus time.

6.20 The dynamics of the turning motion of an automobile are

$$\dot{v} = -\frac{C_F + C_R}{mu}v + \left(\frac{C_R L_R - C_F L_F}{mu} - u\right)r + \frac{C_F}{m}\delta_F$$

$$\dot{r} = \frac{C_R L_R - C_F L_F}{Ju}v - \frac{C_R L_R^2 + C_F L_F^2}{Ju}r + \frac{C_F L_F}{J}\delta_F$$

where u and v are the forward and lateral components of the vehicle's velocity (in m/s), r is the vehicle's yaw (turning) rate (in rad/s), and δ_F is the steering angle of the front tire (in rad). The system parameters of the automobile are [1]

Vehicle mass $m = 1000$ kg
Vehicle moment of inertia (yaw axis) $J = 15,000$ kg-m^2
Constant forward speed $u = 10$ m/s (22 mph)
Front-tire cornering stiffness $C_F = 55,000$ N/rad
Rear-tire cornering stiffness $C_R = 45,000$ N/rad
Moment arm from c.m. to front axle $L_F = 1$ m
Moment arm from c.m. to rear axle $L_R = 1.5$ m

Use MATLAB or Simulink to numerically simulate the turning response to a ramped step input for the steering angle: $\delta_F(t) = t$ for $0 < t < 0.1$ s, $\delta_F(t) = 0.1$ rad for $t \geq 0.1$ s. Let the simulation time be 5 s. Plot the lateral velocity $v(t)$, yaw rate $r(t)$, and lateral acceleration $a_{\text{lat}}(t) = \dot{v} + ru$.

6.21 Figures P6.21a and P6.21b show the pantograph system described in Problems 2.36 (Chapter 2) and 5.42 (Chapter 5). The pantograph system parameters [2] are

Head mass $m_1 = 9$ kg
Frame mass $m_2 = 17$ kg
Contact shoe stiffness $k_1 = 8.2(10^4)$ N/m
Head suspension stiffness $k_2 = 7000$ N/m
Head suspension friction coefficient $b_1 = 130$ N-s/m
Frame suspension friction coefficient $b_2 = 30$ N-s/m

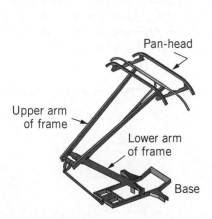

Figure P6.21a

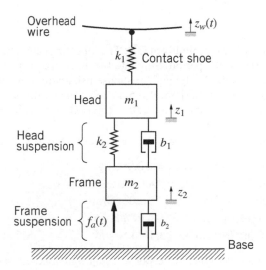

Figure P6.21b

The displacement of the overhead wire depends on the speed of the train and the spacing between towers that support the catenary wire. The wire displacement is modeled by the following sinusoidal function:

$$z_w(t) = z_{w0} + 0.02 \sin 3t \text{ m}$$

where $z_{w0} = -0.0010976$ m is the initial displacement of the wire.

a. Verify that the pantograph is in static equilibrium at time $t = 0$ if the piston force is $f_a(0) = 90$ N, $z_1(0) = 0$, $z_2(0) = 0.0128571$ m, and $z_w(0) = -0.0010976$ m. Compute the contact force at time $t = 0$.

b. Use Simulink to numerically simulate the pantograph response using the static equilibrium initial condition in part (a), a constant piston force $f_a(t) = 90$ N, and the sinusoidal wire displacement.

Let the simulation time be 4 s. Plot the two measured output variables: $y_1 =$ shoe contact force and $y_2 = z_1 - z_2$. Does the pantograph head remain in contact with the overhead wire? Explain your answer.

6.22 Figure P6.22 shows the dual-disk mechanical system from Example 2.9 in Chapter 2 and Problem 5.38. The system parameters are [3]

Piston moment of inertia $J_1 = 0.3$ kg-m^2
Cylinder moment of inertia $J_2 = 0.3$ kg-m^2
Torsional spring constant $k = 6000$ N-m/rad
Piston/cylinder friction coefficient $b = 2$ N-m-s/rad

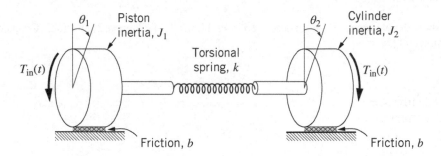

Figure P6.22

The system is initially at rest. The input torque is a periodic pulse with a magnitude of 5500 N-m and pulse width that is 10% of the period. Simulate two pulse frequencies: (a) period = 0.02 s (50 Hz) and (b) period = 0.0314 s (31.85 Hz). Obtain the dynamic response using Simulink. Plot the relative angular displacement $\Delta\theta = \theta_2 - \theta_1$ versus time for a simulation time of 0.5 s and comment on the effect of the pulse period [Hint: use the `Pulse Generator` (`Sources` library) to create the periodic torque input].

6.23 The railroad-car system discussed in Problems 2.38 and 5.40 is shown in Fig P6.23. The system parameters are [4]

Locomotive mass $m_1 = 12{,}000$ kg
Car mass $m_2 = m_3 = 9000$ kg
Coupler stiffness $k = 5.2(10^6)$ N/m
Coupler friction $b = 2.6(10^4)$ N-s/m
Rolling friction $b_r = 3(10^4)$ N-s/m

Obtain the dynamic response using Simulink. The system is initially at rest (equilibrium). The locomotive force is modeled by a 200-kN step function applied at time $t = 0$ s plus a quadratic term:

$$F_a(t) = \begin{cases} 2(10^5) + 5000t^2 & \text{for } 0 < t < 20 \\ 2.2\ (10^6) & \text{for } t \geq 20 \end{cases} \quad \text{(in N)}$$

The locomotive force increases at a quadratic rate until $t = 20$ s and then remains constant at $2.2(10^6)$ N. Plot the velocity of the locomotive versus time and the relative displacement of the first coupler $(z_1 - z_2)$ versus time for a simulation time of 25 s. What is the final speed of the locomotive in mph? [Hint: use the `Saturation` block from the `Discontinuities` library to limit the locomotive force so it never exceeds $2.2(10^6)$ N].

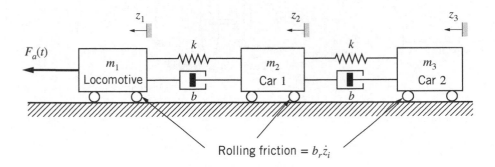

Figure P6.23

6.24 Figure P6.24 shows the DC motor electromechanical system from Problem 3.39 in Chapter 3. The motor and load parameters are

Motor inertia:	$J_1 = 3.6(10^{-5})$ kg-m^2
Load inertia:	$J_2 = 0.04$ kg-m^2
Motor viscous friction:	$b_1 = 1.5(10^{-4})$ N-m-s/rad
Load viscous friction:	$b_2 = 7.2(10^{-3})$ N-m-s/rad
Gear ratio:	$N = 50$
Inductance:	$L = 4$ mH
Resistance:	$R = 3\ \Omega$
Motor-torque constant:	$K_m = 0.1$ N-m/A
Back-emf constant:	$K_b = 0.1$ V-s/rad

For Problem 3.39, the input voltage was determined to be the harmonic function:

$$e_{\text{in}}(t) = C_1 \cos \omega t + C_2 \sin \omega t + C_3 \cos 0.5\omega t + C_4 \sin 0.5\omega t + C_5 \text{ V}$$

where the coefficients are $C_1 = -3.2637$, $C_2 = 1.5661$, $C_3 = 0.0314$, $C_4 = 1.5$, and $C_5 = 3.3291$. The input frequency is $\omega = 10\pi$ rad/s. The load torque is

$$T_L(t) = 2.5 \sin 0.5\omega t \text{ N-m}$$

Obtain the dynamic response using Simulink for simulation time $0 \le t \le 0.2$ s. The system is initially at rest (zero initial conditions). The desired output variables are angular position and angular acceleration of the load. Plot the two output variables and verify that the load's angular acceleration is

$$\ddot{\theta}_2(t) = 20 \sin \omega t \text{ rad/s}^2$$

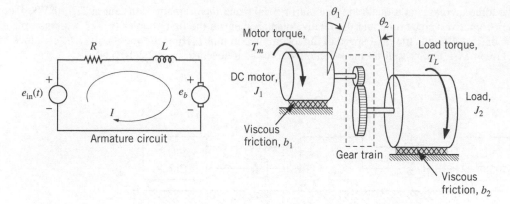

Figure P6.24

SS **6.25** Figure P6.25 shows a pneumatic servomechanism [5]. The total length of the cylinder is 10 cm and the piston position x is measured relative to the mid-point of the cylinder (hence, when $x = 0$ the piston is at the middle of the cylinder). This highly nonlinear system has been linearized about a nominal pressure, volume, and piston position ($x = 0$) and the resulting system transfer function is

$$G(s) = \frac{7.003\left(10^6\right)}{s\left(s^2 + 28.3s + 15,890\right)} = \frac{X(s)}{U(s)}$$

where $x(t)$ is the piston/load mass position (in m) and $u(t)$ is the spool-valve position (in m). See Reference [5] for details in developing the linear model.

a. Use Simulink to obtain the dynamic response of the piston/load mass $x(t)$ for a pulse input for valve position: $u(t) = 0.001$ m (or 1 mm) for $0 < t \le 0.05$ s, and $u(t) = 0$ for $0.05 < t \le 0.5$ s. At time $t = 0$ the piston is initially at rest at the mid-point position. Plot the time response of the piston/load mass position (in cm) for the 1-mm pulse input.

b. Use your simulation to determine the maximum pulse duration of a 1-mm valve displacement (i.e., a feasible piston position response). Plot the piston response $x(t)$ for this maximum pulse input.

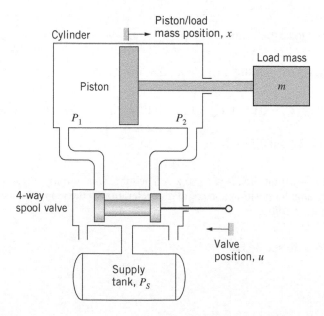

Figure P6.25

6.26 Figure P6.26 shows the thermal system (interior office room with baseboard heater) from Example 4.7 in Chapter 4. The system parameters are [6]

Total thermal capacitance $C = 4.83(10^5)$ J/°C
North wall thermal resistance $R_1 = 0.041$ °C-s/J
South wall thermal resistance $R_2 = 0.151$ °C-s/J
East wall thermal resistance $R_3 = 0.108$ °C-s/J
West wall thermal resistance $R_4 = 0.209$ °C-s/J
Ceiling thermal resistance $R_5 = 0.240$ °C-s/J
Floor thermal resistance $R_6 = 0.159$ °C-s/J

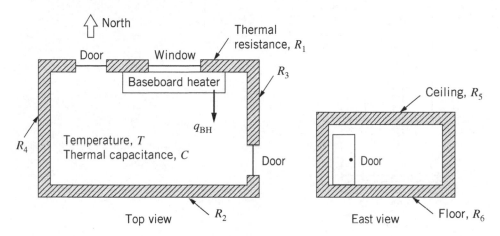

Figure P6.26

The interior room temperature is initially equal to the ambient temperature, which is $T_a = 10$ °C (or, 50 °F). At time $t = 0$, the baseboard heater supplies a constant heat input $q_{BH} = 1000$ W (1 kW). Obtain the temperature response using Simulink and plot room temperature T versus time (in hrs) for a simulation time of 18 hrs. What is the room temperature (in °C and °F) at 18 hrs?

6.27 Figure P6.27 shows a hydraulic pump device (see Problem 4.23). A scotch-yoke mechanism provides sinusoidal input to the piston, $y(t) = r \sin \omega t$, where r is the crank radius and ω is the constant rotation rate in rad/s. The pump chamber has pressure P and volume $V = V_0 - Ay$, where V_0 is the pump volume when $y = 0$ and A is the piston area. The pump chamber is connected to constant supply pressure P_S and constant load pressure P_L through two "ball" check valves, which only allow flow in one direction as shown in Fig. P6.27. Each check valve has orifice area A_0 when opened. The system parameters are

Area $A = 3(10^{-4})$ m^2
Crank radius $r = 0.05$ m
Valve orifice area $A_0 = 4(10^{-6})$ m^2
Discharge coefficient $C_d = 0.62$
Hydraulic fluid density $\rho = 860$ kg/m^3
Fluid bulk modulus $\beta = 690(10^6)$ Pa
Volume $V_0 = 1.25(10^{-4})$ m^3
Supply pressure $P_S = 4(10^5)$ Pa
Load pressure $P_L = 1(10^6)$ Pa

At time $t = 0$ the initial pump-chamber pressure is $1.1(10^5)$ Pa. The rotation rate of the crank is $\omega = 6$ rad/s. Develop an integrated Simulink model and simulate the pump's response to the sinusoidal piston displacement. Plot pressure $P(t)$. In addition, plot volumetric-flow rates $Q_1(t)$, $Q_2(t)$, and volume-rate $\dot{V}(t)$ on the same figure. Simulate 5 cycles. Use a fixed-step integration method with a very small step size. Two critical physical constraints must be enforced during the simulation: (a) the pump-chamber pressure P cannot drop below zero (vacuum), and (b) fluid can only flow left to right through both check valves as shown in Fig. P6.27. The pressure constraint can be enforced by setting the lower saturation limit on the $\dot{P}$ integrator to zero. Enforcing the check valve behavior can be achieved by using the `Saturation` block (see the `Discontinuities` library); that is, when simulating the valve-flow equations use two `Saturation` blocks to limit the pressure differences $P_S - P$ and $P - P_L$ to be between zero (lower limit) and infinity (upper limit). Hence, Q_1 will be zero (no flow) for $P > P_S$ and fluid will flow from P_S to the pump chamber when $P < P_S$. Similarly, Q_2 will be zero for $P < P_L$ and fluid will flow from the pump chamber to P_L only when $P > P_L$.

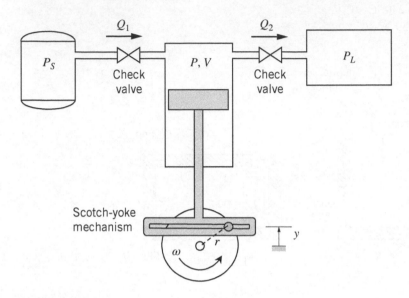

Figure P6.27

6.28 A hydromechanical system consists of an electric motor driving a fixed-displacement hydraulic pump, which changes the hydraulic cylinder pressure. The linearized dynamics for this system are (see Problem 5.37 in Chapter 5)

$$\text{DC motor circuit:} \qquad L\dot{I} + RI = e_{\text{in}}(t) - K_b\omega$$

$$\text{Motor/pump dynamics:} \qquad J\dot{\omega} + b_{\text{mp}}\omega = K_m I - d_p\left(P - P_{\text{atm}}\right)$$

$$\text{Hydraulic cylinder:} \qquad \dot{P} = \frac{\beta}{V_0}\left(d_p\omega - A\dot{z}\right)$$

$$\text{Mechanical load:} \qquad m\ddot{z} + b\dot{z} = PA - P_{\text{atm}}A$$

The system parameters are

Piston + mechanical load mass $m = 20$ kg
Viscous friction $b = 2000$ N-s/m
Piston area $A = 5(10^{-4})$ m^2
Fluid bulk modulus $\beta = 6.8(10^8)$ Pa
Nominal cylinder volume $V_0 = 3(10^{-4})$ m^3
Pump volumetric displacement $d_p = 1.7(10^{-7})$ m^3/rad
Motor-torque constant $K_m = 0.6$ N-m/A
Back-emf constant $K_b = 0.6$ V-s/rad
Motor inductance $L = 0.004$ H
Motor resistance $R = 0.5$ Ω
Motor + pump inertia $J = 1.6(10^{-3})$ kg-m^2

Motor + pump friction $b_{mp} = 2(10^{-4})$ N-m-s/rad

Atmospheric pressure $P_{atm} = 1.0133(10^5)$ Pa

At time $t = 0$, all states (except cylinder pressure P) are zero (i.e., zero stored energy). The initial hydraulic pressure is equal to atmospheric pressure P_{atm}. The input voltage is a 1-second pulse: $e_{in}(t) = 25$ V for $0 \leq t \leq 1$ s, $e_{in}(t) = 0$ for $t > 1$ s. Use MATLAB's lsim command to simulate the hydromechanical system's response to a 1-second pulse input. Plot motor/pump angular velocity $\omega(t)$ (in rpm), cylinder pressure $P(t)$, and mechanical load position $z(t)$ (in cm) for a total simulation time of 2 s. Determine the constant angular velocity of the motor/pump (in rpm) during the pulse voltage input, and determine the final position of the mechanical load (in cm) at $t = 2$ s.

6.29 Figure P6.29 shows the electropneumatic clutch actuator described in Problem 4.26 in Chapter 4. The system parameters are [7]

Rod and piston mass $m = 10$ kg

Viscous friction $b = 2000$ N-s/m

Clutch load force $F_L = 4000 (1 - e^{-500x}) - 20,000x$ N $(x$ in m$)$

Area $A_1 = 0.0123$ m^2

Area $A_2 = 0.0115$ m^2

Valve orifice area (open) $A_0 = 7(10^{-6})$ m^2

Discharge coefficient $C_d = 0.8$

Chamber volume $V = V_0 + A_1 x$

Half-cylinder volume $V_0 = 1.48(10^{-4})$ m^3

Gas constant (air) $R = 287$ N-m/kg-K

Air temperature $T = 298$ K

Isothermal process with choked flow through both valves

Supply pressure $P_S = 9.5(10^5)$ Pa

Ambient pressure $P_{atm} = 1.013(10^5)$ Pa

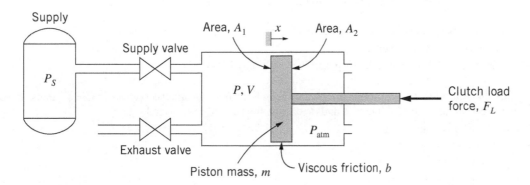

Figure P6.29

The initial piston position is $x = 0$ and the initial chamber pressure is $P = P_{atm}A_2/A_1$. Develop an integrated Simulink model and simulate the pneumatic clutch actuator response to a supply valve pulse opening that lasts 0.1 s (the exhaust valve remains closed). Plot piston position $x(t)$ and chamber pressure $P(t)$ responses. Use the pressure response to validate the assumption that the valve flow is always choked.

CHAPTER 7: PROBLEMS

Conceptual Problems

7.1 Given the input–output (I/O) equation

$$2\dot{y} + 10y = 3u(t)$$

Sketch the response $y(t)$ for a step input $u(t) = 6U(t)$ and the initial condition $y(0) = -2$.

SS **7.2** Given the I/O equation

$$0.4\dot{y} + 0.2y = 0.8u$$

Sketch the response $y(t)$ for a step input $u = -1.5U(t)$ and the initial condition $y(0) = 1$.

SS **7.3** Given the following homogeneous ODE

$$2\ddot{y} + 12\dot{y} + 68y = 0 \quad \text{with initial conditions} \quad y(0) = 3, \dot{y}(0) = 0$$

   a. Does the homogeneous response exhibit oscillations?
   b. Estimate the time to reach steady state.
   c. Describe the nature of the homogeneous response (a sketch may help).

7.4 Given the following homogeneous ODE

$$4\ddot{y} + 22\dot{y} + 18y = 0 \quad \text{with initial conditions} \ y(0) = 3, \dot{y}(0) = 0$$

   a. Does the homogeneous response exhibit oscillations?
   b. Estimate the time to reach steady state.
   c. Describe the nature of the homogeneous response (a sketch may help).

7.5 The DC gain and characteristic root locations are given for three different stable second-order LTI systems. In each case derive the corresponding transfer function and determine the time to reach steady state.
   a. $r_1 = -2.5$, $r_2 = -0.2$, DC gain $= 0.5$
   b. $r_1 = -3$, $r_2 = -10$, DC gain $= 6$
   c. $r_{1,2} = -2 \pm j4$, DC gain $= 125$
   d. $r_{1,2} = -0.4 \pm j1.6$, DC gain $= 0.02$

7.6 A system is represented by the following transfer function:

$$G(s) = \frac{Y(s)}{U(s)} = \frac{84}{\left(3s^2 + 21s + 36\right)\left(s^2 + 2s + 9\right)}$$

The input is a step function $u(t) = 4U(t)$.
   a. Determine the steady-state output $y(\infty)$.
   b. Determine the settling time to reach steady state.

7.7 Figure P7.7 shows a mechanical rotor with input motor torque $T_{in}(t)$ and viscous friction b. The motor parameters are moment of inertia $J = 0.06$ kg-m^2 and viscous friction coefficient $b = 0.008$ N-m-s/rad. At time $t = 0$, the rotor is initially spinning with angular velocity $\omega(0) = 20$ rad/s when a constant motor torque $T_{in}(t) = 0.4$ N-m is applied.

a. Determine the steady-state angular velocity of the rotor.

b. At what time does the rotor's speed increase by 20% relative to its initial value?

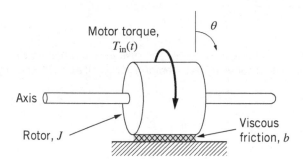

Figure P7.7

7.8 Figure P7.8 shows an electrical system. Input voltage is $e_{in}(t)$ and the output voltage is e_O. The electrical components are resistance $R = 2\ \Omega$ and capacitance $C = 0.005$ F. At time $t = 0$, there is zero energy stored by the capacitor.

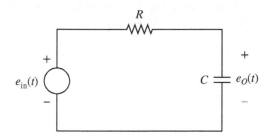

Figure P7.8

Sketch the step response of the output voltage $e_O(t)$ if the input is a constant 0.4 V. Label the important quantities on your sketch.

SS **7.9** Consider again the RC circuit presented in Problem 7.8 (see Fig. P7.8). At time $t = 0$, there is no energy stored by the capacitor. At $t = 0$, the voltage source supplies a 12-V pulse that lasts for $2(10^{-4})$ s. Sketch the output voltage $e_O(t)$ and label the important quantities on your sketch.

7.10 Figure P7.10 shows an electrical system (see Problem 3.16). The numerical values of the electrical parameters are $R_1 = 1.6\ \Omega$, $R_2 = 4\ \Omega$, and $L = 0.2$ H. The switch is open for time $t < 0$. For time $0 \le t \le 1$ s, the switch is in position "1" and the voltage source $e_{in}(t)$ is connected to the RL loop. For time $t > 1$ s, the switch is in position "2." Sketch the complete current response $I(t)$ if the voltage source is $e_{in}(t) = 4$ V.

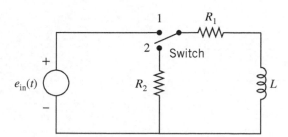

Figure P7.10

7.11   A second-order mechanical system has the following transfer function:

$$G(s) = \frac{1}{2s^2 + 3s + 24}$$

where the output $y(t)$ is the displacement and the input $u(t)$ is the applied force. The system is at rest at time $t = 0$. If a 3-N step input is applied to the system, does the transient response exhibit vibrations or oscillations? Explain your answer.

SS **7.12**   A system has the following transfer function:

$$G(s) = \frac{4}{s^2 + 2s + 36} = \frac{Y(s)}{U(s)}$$

A step input $u(t) = 18U(t)$ is applied. Compute the maximum value of the response $y(t)$.

7.13   Figure P7.13 shows a unit-step response $y(t)$. Which of the three transfer functions best represents the input–output relationship presented by Fig. P7.13? Justify your answer with the appropriate calculations.

a. $G(s) = \dfrac{0.9}{s^2 + 2.5s + 36}$

b. $G(s) = \dfrac{1}{s^2 + 4.8s + 40}$

c. $G(s) = \dfrac{1.5}{s^2 + 1.2s + 60}$

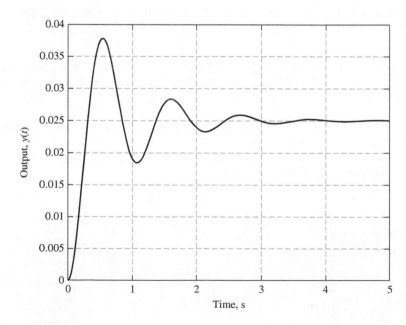

Figure P7.13

SS **7.14** Figure P7.14 shows a unit-step response $y(t)$. Which of the three transfer functions best represents the input–output relationship presented by Fig. P7.14? Justify your answer with the appropriate calculations.

a. $G(s) = \dfrac{10}{s^2 + 0.6s + 40}$

b. $G(s) = \dfrac{8}{0.2s^3 + 2.8s^2 + 15.4s + 32}$

c. $G(s) = \dfrac{8}{s^3 + 1.4s^2 + 40.48s + 32}$

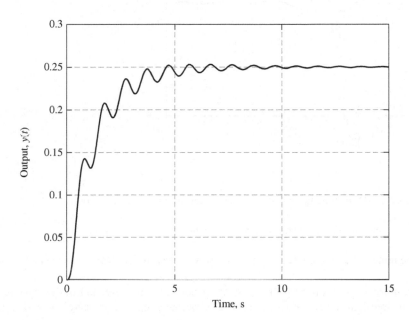

Figure P7.14

7.15 A vibration isolation system has an adjustable damper that can be "tuned." Three different damper settings produce the three following characteristic equation roots (or eigenvalues) for the system:

a. $s = -3 \pm j2$

b. $s = -2 \pm j3$

c. $s = -3 \pm j3$

Which damper setting provides the greatest damping ratio? Explain your answer.

7.16 A mechanical system is modeled as a 1-DOF, linear mass–spring–damper system with elements mass m, spring constant k, and friction coefficient b. An 18-N step-input force is applied to the system, which is initially at rest. The peak displacement is $y_{max} = 0.050$ m, and the steady-state displacement is $y_{ss} = 0.036$ m. If the mass is 0.1 kg, determine the spring constant k and friction coefficient b.

7.17 A mechanical system is given an impulsive force input, and the subsequent impulse response for displacement is shown in Fig. P7.17. Estimate the damping ratio ζ and the undamped natural frequency ω_n.

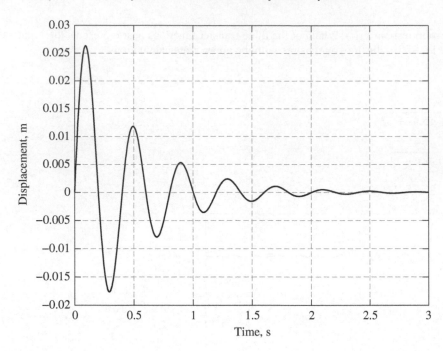

Figure P7.17

7.18   After developing an I/O equation for a 2-DOF mechanical system, we determine its characteristic roots to be

$$r_{1,2} = -1.2 \pm j6.3, \quad r_{3,4} = -0.4 \pm j4.2$$

   a. Write a general equation for the homogeneous (or free) response $y_H(t)$ (you may use constants c_1, c_2, etc. to denote undetermined coefficients).

   b. If the mechanical system has stored energy at time $t = 0$ (because of the positions and velocities of the two masses) estimate the time required for the transient response to die out (i.e., the mechanical system reaches the zero-energy state).

7.19   Consider a mechanical system that has so little damping that we can consider it to be frictionless and therefore we model it as a mass–spring system. After applying an initial displacement of 6 cm, we observe a sinusoidal displacement response:

$$x(t) = 6 \cos 50t \ \text{cm}$$

   If the mass is $m = 0.2$ kg, what is the stiffness (spring constant) of the system?

7.20   Given the system I/O equation

$$2\ddot{y} + 8\dot{y} + 6y = 3u$$

   a. Compute the characteristic roots.

   b. Compute the poles of the transfer function.

   c. Compute the eigenvalues of the system matrix from an SSR.

   d. Qualitatively describe the complete system response when initial conditions are $y(0) = 5$, $\dot{y}(0) = 0$, and $u(t) = 0$ (no input).

SS **7.21**   An engineer strikes a mechanical system with an impulse hammer and records the response. Table P7.21 shows the peak displacements and their associated time values. In a second test, the engineer applies a constant 15-N force to the mechanical system and measures a 0.012 m steady-state displacement. Determine the transfer function $G(s)$ for this mechanical system where the output is displacement and the input is force.

Table P7.21

Peak number	Time, s	Peak displacement, m
1	0.008	0.1091
2	0.044	0.0337
3	0.080	0.0104
4	0.116	0.0032

7.22 Figure P7.22 shows a linear third-order system and a linear first-order system. Analytically and numerically show that the first-order transfer function $G_2(s)$ is an excellent low-order approximate model of the third-order system $G_1(s)$. Use MATLAB or Simulink for the numerical comparison.

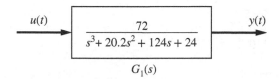

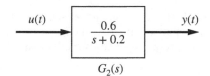

Figure P7.22

MATLAB Problems

7.23 Use Simulink to verify the solution to Problem 7.7. Plot the angular velocity response $\omega(t)$.

SS 7.24 Given the SSR

$$\dot{\mathbf{x}} = \begin{bmatrix} -0.2 & -0.6 \\ 2 & -4 \end{bmatrix}\mathbf{x} + \begin{bmatrix} 0 \\ 1.5 \end{bmatrix}u \qquad y = \begin{bmatrix} 1 & 0 \end{bmatrix}\mathbf{x}$$

    a. Compute the eigenvalues "by hand."
    b. Use MATLAB to verify your answer in part (a).
    c. Describe the *free* response of the output $y(t)$ given an arbitrary initial state $\mathbf{x}(0)$.
    d. Use MATLAB or Simulink to verify your answer in part (c). The initial state vector is $\mathbf{x}(0) = \begin{bmatrix} x_1(0) & x_2(0) \end{bmatrix}^T$ $= \begin{bmatrix} -2 & -1 \end{bmatrix}^T$.

7.25 Given the SSR

$$\dot{\mathbf{x}} = \begin{bmatrix} 0 & 1 & 0 \\ 0 & 0 & 1 \\ -12 & -20 & -9 \end{bmatrix}\mathbf{x} + \begin{bmatrix} 0 \\ 0 \\ 0.4 \end{bmatrix}u \qquad y = \begin{bmatrix} 1 & 0 & 0 \end{bmatrix}\mathbf{x}$$

    a. Use MATLAB to determine the eigenvalues.
    b. Describe the *free* response of the output $y(t)$ given an arbitrary initial state $\mathbf{x}(0)$.
    c. Use MATLAB or Simulink to verify your answer in part (b). The initial state vector is $\mathbf{x}(0) = \begin{bmatrix} x_1(0) & x_2(0) & x_3(0) \end{bmatrix}^T$ $= \begin{bmatrix} 2 & -0.5 & 0 \end{bmatrix}^T$.

7.26 Given the SSR for a third-order system

$$\dot{\mathbf{x}} = \begin{bmatrix} 0 & 1 & 0 \\ 0 & 0 & 1 \\ -340 & -88 & -9 \end{bmatrix} \mathbf{x} + \begin{bmatrix} 0 \\ 0 \\ 2 \end{bmatrix} u \qquad y = \begin{bmatrix} 1 & 0 & 0 \end{bmatrix} \mathbf{x}$$

a. Use MATLAB to determine the eigenvalues.

b. Describe the *free* response of the output $y(t)$ given an arbitrary initial state $\mathbf{x}(0)$.

c. Use MATLAB or Simulink to verify your answer in part (b). The initial state vector is $\mathbf{x}(0) = \begin{bmatrix} x_1(0) & x_2(0) & x_3(0) \end{bmatrix}^T$ $= \begin{bmatrix} -4 & 0 & 0 \end{bmatrix}^T$.

7.27 An SSR is shown below. The input is a unit-step function, and the initial state is zero, $\mathbf{x}(0) = \mathbf{0}$.

$$\dot{\mathbf{x}} = \begin{bmatrix} 0 & 1 & 0 & 0 \\ -12 & -2.6 & 12 & 2.6 \\ 0 & 0 & 0 & 1 \\ 12 & 2.6 & -32 & -2.6 \end{bmatrix} \mathbf{x} + \begin{bmatrix} 0 \\ 0.5 \\ 0 \\ 0 \end{bmatrix} u \qquad y = \begin{bmatrix} 1 & 0 & 0 & 0 \end{bmatrix} \mathbf{x}$$

a. Does the output exhibit oscillations during the transient response?

b. Estimate the time for the output to reach steady state.

7.28 Figure P7.28 shows a system defined by a transfer function. Use MATLAB to determine the characteristic roots of the system. *Describe* the nature of the system's transient response to a unit-step input, and consider issues such as the time to reach a steady-state response, and whether or not the transient response exhibits oscillations. Verify your answer with a numerical simulation using MATLAB or Simulink.

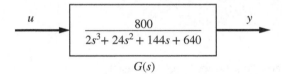

$$u \longrightarrow \boxed{\dfrac{800}{2s^3 + 24s^2 + 144s + 640}} \longrightarrow y$$

$$G(s)$$

Figure P7.28

7.29 Use MATLAB or Simulink to verify your sketch of the RC circuit response in Problem 7.8. Plot output voltage $e_O(t)$.

7.30 A simple 1-DOF mechanical system has the following transfer function:

$$G(s) = \frac{Y(s)}{U(s)} = \frac{0.25}{s^2 + 2s + 9}$$

where the position of the mass $y(t)$ is in meters. The system is initially at rest, $y(0) = \dot{y}(0) = 0$, and the applied force is a step function $u(t) = 4\,U(t)\,\text{N}$.

a. Accurately sketch the system response $y(t)$ and label all important performance criteria on your sketch.

b. Use MATLAB or Simulink to verify your sketch in part (a). Plot $y(t)$ from the numerical solution.

7.31 Repeat Problem 7.30 with the following initial conditions and input function: $y(0) = 0.04$ m, $\dot{y}(0) = 0$, and *zero* input force $u(t) = 0$ for $t \geq 0$.

7.32 Figure P7.32 shows a simple 1-DOF, frictionless, rotational mechanical system. The disk moment of inertia is $J = 0.2$ kg-m^2 and the torsional spring constant for the shaft is $k = 100$ N-m/rad. Angular displacement θ is zero when the shaft is untwisted. The disk is initially at rest (equilibrium) when the sinusoidal input torque $T_{in}(t) = 0.5\sin3t$ N-m is applied.

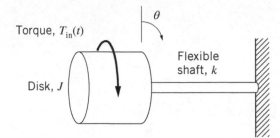

Figure P7.32

a. Determine the *general form* for the complete response $\theta(t)$ (you do not have to compute the exact solution with numerical values for the coefficients).

b. Use MATLAB or Simulink to obtain a numerical solution and plot $\theta(t)$ (use a simulation time of 10 s). Does the numerical solution match the general form of the analytical solution from part (a)? Explain your answer.

7.33 Consider again the impulse response shown by Fig. P7.17 for Problem 7.17. Develop a simple linear model of the mechanical system using the approximate values for damping ratio ζ and undamped natural frequency ω_n that were determined in Problem 7.17 (the mass is 0.2 kg). Use Simulink to simulate the impulse response, and adjust the magnitude and duration of the pulse input force (i.e., impulse) in order to match Fig. P7.17. What is the "strength" or "weight" of the impulse?

SS **7.34** A system is modeled by the following transfer function:

$$G(s) = \frac{780}{s^4 + 29.5s^3 + 119s^2 + 765.5s + 975} = \frac{Y(s)}{U(s)}$$

a. Determine a third-order transfer function that provides a relatively accurate approximation for $G(s)$.

b. Demonstrate the accuracy of your third-order transfer function in part (a) by comparing its unit-step response with the unit-step response of the original system $G(s)$. Use MATLAB or Simulink for the simulations, and plot both responses in the same figure.

c. Is it possible to determine a *second*-order transfer function that provides an accurate approximation for $G(s)$? Explain your answer.

Engineering Applications

7.35 Figure P7.35 shows the thermal system (interior office room) from Example 4.7 in Chapter 4 and Problem 6.26. The thermal system parameters are

Total thermal capacitance $C = 4.83 \left(10^5 \right)$ J/°C
North wall thermal resistance $R_1 = 0.041$ °C-s/J
South wall thermal resistance $R_2 = 0.151$ °C-s/J
East wall thermal resistance $R_3 = 0.108$ °C-s/J
West wall thermal resistance $R_4 = 0.209$ °C-s/J
Ceiling thermal resistance $R_5 = 0.240$ °C-s/J
Floor thermal resistance $R_6 = 0.159$ °C-s/J

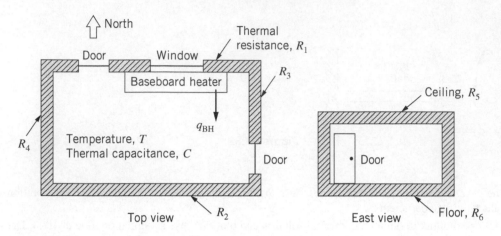

Figure P7.35

Sketch the temperature response of the room if its initial temperature is equal to the (constant) ambient temperature, that is, $T(0) = T_a = 10\,^\circ\text{C}$ (50 °F) and the baseboard heater provides a constant heat input $q_{\text{BH}} = 1000\,\text{W}$ (1 kW). Use units of hours for the time axis. Compare your sketch to the numerical solution (using Simulink) that was determined in Problem 6.26.

7.36   Figure P7.36 shows a simplified circuit for the discharge phase that produces a camera flash (originally presented in Problem 3.32). During the charging phase the capacitor C is charged by a 1.5-V battery connected to an oscillator circuit that includes a transformer for boosting voltage (not shown in Fig. P7.36). The voltage across a fully charged capacitor is about 200 V. Pressing the shutter button activates the circuit shown in Fig. P7.36 and the capacitor discharges energy to the RC circuit. A second transformer (shown in Fig. P7.36) boosts the voltage by a factor of 10 in order to ionize xenon gas and produce a flash. Ignoring the transformer component in the RC circuit, accurately sketch the capacitor voltage $e_C(t)$ if the initial voltage is $e_C(0) = 200\,\text{V}$ and the shutter is pressed at time $t = 0$. In addition, sketch the current response $I(t)$ for the RC circuit.

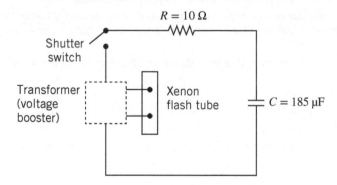

Figure P7.36

SS   **7.37**   Figure P7.37 shows the simplified, linear electrohydraulic actuator (EHA) model from Problem 5.34 in Chapter 5. The voltage input is a step function $e_{\text{in}}(t) = 0.2U(t)\,\text{V}$. The amplifier output is initially zero and the valve is initially at static equilibrium ($z = 0$).

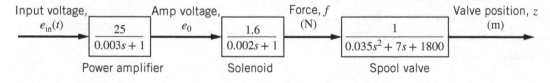

Figure P7.37

a. Accurately sketch the response of the power amplifier output.
b. Compute the steady-state position of the spool valve.
c. Accurately sketch the response of the spool-valve position $z(t)$. Label all important response characteristics on your sketch. Verify your sketch of $z(t)$ with a simulation of the EHA valve response using MATLAB or Simulink. Plot $z(t)$ from the simulation, and discuss the similarities and differences between the approximate sketch of valve position and the simulation result.

7.38 An engineer wants to develop a simple model for a DC motor, which is available in the lab with sensors to measure input voltage and shaft speed of the motor. With the motor initially at rest, she applies a constant 2-V input voltage and measures the following motor response characteristics: (a) the steady-state angular velocity of the motor is 85 rad/s, (b) the time to reach steady-state speed is 0.6 s, (c) the angular velocity response shows an exponential rise from zero to the steady-state speed without overshoot. Derive an appropriate transfer function for the DC motor based on the experimental data. Verify the engineer's experimental results with a simulation of your model using MATLAB or Simulink.

7.39 An engineer needs to select a shock-mount system for a delicate instrument's housing. The instrument and shock-mounts are modeled as a mass–spring–damper mechanical system. The total system mass is 3 kg. Two shock-mount options are available, and the composite stiffness and friction coefficients are listed below:

$$\text{Option A: } k = 11,300 \text{ N/m} \quad b = 90 \text{ N-s/m}$$

$$\text{Option B: } k = 8700 \text{ N/m} \quad b = 88 \text{ N-s/m}$$

Which shock-mount option would result in the smallest maximum overshoot for a step input? Explain your answer.

7.40 An engineer wants to develop a model of an existing mechanical system that has complicated geometry. She suspects that a low-order linear model may provide an accurate representation of the complex system. She applies a step torque input $T_{in}(t) = 0.6U(t)$ N-m and measures the angular position of the rotational system. Figure P7.40 shows the step response she obtains by experimentation.

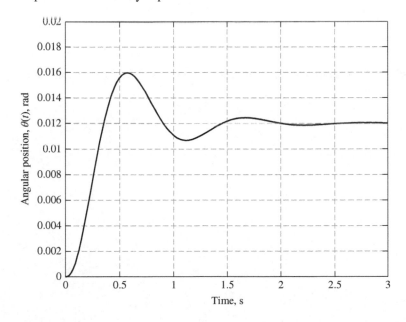

Figure P7.40

a. Develop a transfer function for the mechanical system.
b. Use MATLAB or Simulink to simulate the step response of the transfer function developed in part (a) using the torque input $T_{in}(t) = 0.6U(t)$ N-m. Compare the simulation result to Fig. P7.40. Is the low-order linear model accurate?

CHAPTER 8: PROBLEMS

Conceptual Problems

8.1 Given the following Laplace transforms $Y(s)$, compute the inverse Laplace transform "by hand" in order to determine the time function $y(t)$:

a. $Y(s) = \dfrac{-8}{s(s+2)}$

b. $Y(s) = \dfrac{s+6}{(s+2)(s+1)}$

c. $Y(s) = \dfrac{10s+4}{s(s+8)}$

d. $Y(s) = \dfrac{2s+18}{s^2+8s+20}$

e. $Y(s) = \dfrac{6s^2+4s+22}{s(s^2+2s+5)}$

f. $Y(s) = \dfrac{s+10}{s^2(s+4)}$

g. $Y(s) = \dfrac{3s+12}{s(s^2+4)}$

h. $Y(s) = \dfrac{6s+40}{(s^2+25)(s+8)}$

i. $Y(s) = \dfrac{2s+8}{(s+1)(s^2+4s+4)}$

j. $Y(s) = \dfrac{2s+8}{2s^2}$

8.2 Given the following Laplace transforms $Y(s)$ determine if a final value $y(\infty)$ exists. If $y(\infty)$ exists, find it using the final-value theorem.

a. $Y(s) = \dfrac{s+4}{s(s+2)(s+1)}$

b. $Y(s) = \dfrac{s+7}{s(s-2)(s+3)}$

c. $Y(s) = \dfrac{3s+2}{(s+3)(s-5)}$

d. $Y(s) = \dfrac{2s+14}{s^2+4s+20}$

e. $Y(s) = \dfrac{2}{s(s^2+3s+20)}$

f. $Y(s) = \dfrac{6s^2+4s}{(s^2+3s+10)(s+6)}$

g. $Y(s) = \dfrac{2s+9}{(s^2+2)(s^2+4s+16)}$

h. $Y(s) = \dfrac{3s^2+18}{(s+5)(s^2+4s+12)}$

i. $Y(s) = \dfrac{3s}{s^2+2s+18}$

j. $Y(s) = \dfrac{0.5s^2+6}{s^2(s+8)}$

8.3   Given the Laplace transforms $Y(s)$ in Problem 8.2 use the initial-value theorem to determine $y(0+)$.

SS **8.4**   The function $y(t)$ has the following Laplace transform:

$$Y(s) = \frac{6(s+2)^2}{s^3 + 5s^2 + \beta s}$$

a. Is it possible to select β so that the *initial* value of $y(t)$ is 3? If so, determine the correct value for β.
b. Is it possible to select β so that the *final* value of $y(t)$ is 3? If so, determine the correct value for β.

SS **8.5**   Given the following I/O equations, obtain the response $y(t)$ using Laplace transform methods.
a. $0.4\dot{y} + y = 0$ with $y(0) = -2$
b. $0.4\dot{y} + y = 2u(t)$ with $u(t) = 0.7U(t), y(0) = -2$
c. $2\dot{y} + 3y = u(t)$ with $u(t) = 4U(t), y(0) = 3$
d. $2\dot{y} = 3u(t)$ with $u(t) = 4\cos 6t, y(0) = 0$
e. $\ddot{y} + 3\dot{y} + 2y = 0.5u(t)$ with $u(t) = 0.2U(t), y(0) = 0, \dot{y}(0) = -1$
f. $\ddot{y} + 5\dot{y} + 6y = 0.5u(t)$ with $u(t) = 6\delta(t), y(0) = 0, \dot{y}(0) = 0$
g. $\ddot{y} + 4\dot{y} + 20y = 0$ with $y(0) = 2, \dot{y}(0) = -0.5$
h. $2\ddot{y} + 12\dot{y} + 68y = 0.5u(t)$ with $u(t) = 4U(t), y(0) = 1, \dot{y}(0) = 0$
i. $20\ddot{y} + 80\dot{y} + 260y = 0.8u(t)$ with $u(t) = 10\delta(t), y(0) = 0.2, \dot{y}(0) = 0.6$

8.6   Derive the transfer function $G(s) = Y(s)/U(s)$ for each of the following I/O equations.
a. $0.4\dot{y} + y = u(t)$
b. $2\dot{y} = 3u(t)$
c. $\ddot{y} + 3\dot{y} + 2y = 4u(t)$
d. $20\ddot{y} + 80\dot{y} + 260y = 0.8\dot{u}(t) + 3u(t)$
e. $0.1\dddot{y} + 2\ddot{y} + 18y = 7u(t)$
f. $\ddot{y} + 2y = 4\dot{u}(t)$

SS **8.7**   Figure P8.7 shows a 1-DOF mechanical system (see Problem 2.4) where the displacement of the left end, $x_{in}(t)$, is the input. When displacement $x = 0$, the spring k is neither compressed nor stretched. The mechanical-element parameters are $m = 0.2$ kg, $b = 3$ N-s/m, and $k = 10$ N/m, and the system is initially at rest; that is, $x(0) = \dot{x}(0) = 0$. The input is a weighted impulse function, $x_{in}(t) = 0.1\delta(t)$ m. Use transfer-function analysis to obtain the position response $x(t)$.

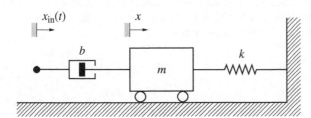

Figure P8.7

SS **8.8**   Figure P8.8 shows a series RLC circuit driven by the voltage source $e_{in}(t)$. The circuit elements are resistance $R = 0.4\ \Omega$, capacitance $C = 0.01$ F, and inductance $L = 0.002$ H. At time $t = 0$, the voltage source is stepped from zero to 1.5 V (the circuit elements initially have zero charge and zero current).

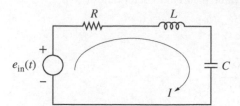

Figure P8.8

a. Determine the solution for charge $q(t)$ stored in the capacitor using Laplace transform methods.

b. Determine the solution for current $I(t)$ in the circuit.

MATLAB Problems

8.9 Rework parts (a), (b), (c), (f), (i), and (j) of Problem 8.1 using MATLAB's `residue` command to compute the residues of the partial-fraction expansion and then determine the inverse Laplace transform in order to find $y(t)$.

8.10 Rework all parts (a)–(j) of Problem 8.1 using MATLAB's Symbolic Math Toolbox command `ilaplace` to directly compute $y(t)$.

SS **8.11** Use MATLAB's `residue` or `ilaplace` command to obtain the position response $x(t)$ for the mechanical system described in Problem 8.7.

8.12 An RC circuit is modeled by

$$0.05\dot{e}_C + e_C = e_{in}(t)$$

At time $t = 0$, the capacitor voltage is $e_C(0) = 0.75$ V. A voltage source provides a 2-V step input, $e_{in}(t) = 2U(t)$ V for $t > 0$. Use MATLAB's `residue` or `ilaplace` command to obtain the capacitor voltage response $e_C(t)$.

8.13 Figure P8.13 shows a simple 1-DOF rotational mechanical system with input torque $T_{in}(t)$. The Laplace transform of the angular velocity $\omega(t)$ of the rotational mechanical system is

$$\Omega(s) = \frac{0.15s + 0.4}{s(0.01s + 0.002)}$$

where $\mathcal{L}\{\omega(t)\} = \Omega(s)$. The initial angular velocity $\omega(0)$, constant input torque, and numerical values for inertia and friction have been accounted for in the above Laplace transform. The moment of inertia of the disk is $J = 0.01$ kg-m^2 and the viscous friction coefficient is $b = 0.002$ N-m-s/rad. Use MATLAB's `residue` or `ilaplace` command to obtain the angular velocity response $\omega(t)$ for the single-disk mechanical system.

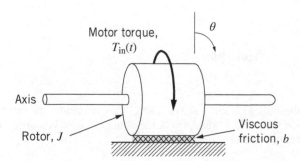

Figure P8.13

8.14 Use MATLAB's `residue` or `ilaplace` command to obtain the charge response $q(t)$ of the RLC circuit described in Problem 8.8.

8.15 A simple 1-DOF mechanical system has the following transfer function:

$$G(s) = \frac{Y(s)}{U(s)} = \frac{0.25}{s^2 + 2s + 10}$$

where position $y(t)$ is in m. The system is initially at rest, $y(0) = \dot{y}(0) = 0$, and the applied force is a step function $u(t) = 4U(t)$ N.

a. Determine the system response using Laplace methods.

b. Use MATLAB or Simulink to verify your analytical answer in part (a) and plot the analytical and numerical solutions on the same plot.

8.16 Repeat Problem 8.15 with the following initial conditions: $y(0) = -0.04$ m and $\dot{y}(0) = 0.01$ m/s. The input force is a 4-N step function.

8.17 Figure P8.17 shows a simple 1-DOF, frictionless, rotational mechanical system. The disk moment of inertia is $J = 0.2$ kg-m^2 and the torsional spring constant for the shaft is $k = 100$ N-m/rad. Angular displacement θ is zero when the shaft is untwisted. The disk is initially at rest (equilibrium) when the sinusoidal input torque $T_{in}(t) = 0.5 \sin 3t$ N-m is applied.

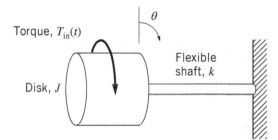

Figure P8.17

a. Use Laplace methods to determine the system response $\theta(t)$.

b. Use MATLAB or Simulink to obtain a numerical solution for the angular position response (use a simulation time of 10 s). Plot the analytical solution from part (a) and the numerical solution on the same figure.

8.18 Figure P8.18 shows the *band-stop* or *notch filter* circuit presented in Problems 3.35 and 5.32. These circuits "filter out" or attenuate input signals within a particular frequency band. The transfer function of the notch filter is

$$G(s) = \frac{E_O(s)}{E_{in}(s)} = \frac{LCs^2 + 1}{LCs^2 + RCs + 1}$$

The electrical system parameters are $R = 2\ \Omega$, $C = 0.1$ F, and $L = 0.4$ H. The voltage input is a sinusoidal function, $e_{in}(t) = 2\sin 5t$ V, and the initial output of the notch-filter circuit is zero.

a. Use MATLAB's Symbolic Math Toolbox command `ilaplace` to directly compute the output voltage of the filter, $e_O(t)$.

b. Use Simulink to obtain the output voltage response to the sinusoidal input. Plot the analytical solution from part (a) and the numerical solution from Simulink on the same figure.

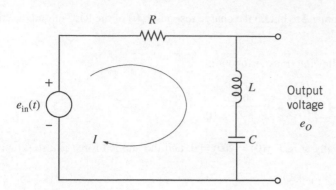

Figure P8.18

Engineering Applications

8.19 Figure P8.19 shows a *washout filter* circuit, also known as a *passive high-pass filter* (see Problems 3.34 and 5.31). The washout filter transfer function is

$$G(s) = \frac{E_O(s)}{E_{in}(s)} = \frac{RCs}{RCs + 1}$$

The electrical system parameters are $R = 2\ \Omega$ and $C = 0.1$ F. The circuit has zero current at time $t = 0$, and the constant voltage input is $e_{in}(t) = 3$ V for $t > 0$.

a. Determine the voltage output $e_O(0+)$ using the initial-value theorem.

b. Determine the steady-state output voltage using the final-value theorem.

c. Use Laplace methods to obtain the output voltage response, $e_O(t)$, for the step voltage input. Verify the initial-value and steady-state solutions from parts (a) and (b).

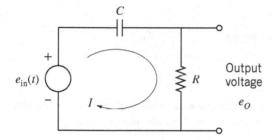

Figure P8.19

SS **8.20** Figure P8.20 shows the op-amp circuit discussed in Problem 3.21. The transfer function for the op-amp circuit is

$$G(s) = \frac{-R_2 Cs}{R_1 Cs + 1} = \frac{E_O(s)}{E_{in}(s)}$$

where e_O is the output voltage and $e_{in}(t)$ is the input voltage. The circuit parameters are $R_1 = 2\ \Omega$, $R_2 = 20\ \Omega$, and $C = 0.1$ F. This type of op-amp circuit is called an *active high-pass filter*. The output voltage is zero at time $t = 0$, and the constant voltage input is $e_{in}(t) = 3$ V for $t > 0$.

a. Determine the voltage output $e_O(0+)$ using the initial-value theorem.

b. Determine the steady-state output voltage using the final-value theorem.

c. Use Laplace methods to obtain the output voltage response, $e_O(t)$, for the step voltage input. Verify the initial-value and steady-state solutions from parts (a) and (b).

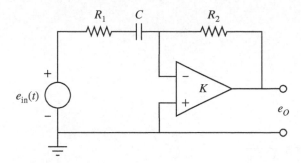

Figure P8.20

8.21 Figure P8.21 shows the simplified, linear electrohydraulic actuator (EHA) model from Problems 5.34 and 7.37. The voltage input is a step function $e_{in}(t) = 0.2U(t)$ V. The amplifier output is initially zero and the valve is initially at static equilibrium ($z = 0$).

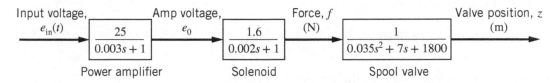

Figure P8.21

a. Use the final-value theorem to compute the steady-state position of the spool valve.

b. Use Laplace methods to determine the amplifier voltage response, $e_0(t)$, for the step input.

c. Use Laplace methods to determine the force response of the solenoid, $f(t)$, for the step input $e_{in}(t) = 0.2U(t)$ V.

d. Use Laplace methods to determine the spool-valve position, $z(t)$, for the step input $e_{in}(t) = 0.2U(t)$ V. Verify the steady-state position computed in part (a).

CHAPTER 9: PROBLEMS

Conceptual Problems

9.1 Given the transfer function

$$G(s) = \frac{4}{0.2s + 1}$$

determine the sinusoidal transfer function.

9.2 Given the transfer function

$$G(s) = \frac{3s + 1}{2s^2 + 6s + 40}$$

determine the sinusoidal transfer function.

9.3 Given the input–output (I/O) equation

$$2\dot{y} + 10y = 3u$$

compute the frequency response $y_{ss}(t)$ for the input $u(t) = 18 \sin 4t$.

9.4 Given the I/O equation

$$4\ddot{y} + \dot{y} + 30y = 0.8\dot{u} + 2u$$

compute the frequency response $y_{ss}(t)$ for the input $u(t) = 0.3 \cos 5t$.

SS **9.5** Given the I/O equation

$$2\dddot{y} + \ddot{y} + 6\dot{y} + 24y = 5\dot{u} + u$$

compute the frequency response $y_{ss}(t)$ for the input $u(t) = 2.1 \sin 4t$.

9.6 Given the transfer function

$$G(s) = \frac{2s + 3}{s(s + 8)}$$

compute the magnitude and phase angle of the sinusoidal transfer function for frequency $\omega = 2$ rad/s.

9.7 Given the transfer function

$$G(s) = \frac{0.5s + 0.4}{(s + 2)(s + 4)} = \frac{Y(s)}{U(s)}$$

compute the frequency response $y_{ss}(t)$ for the input $u(t) = 2.7 \sin 5t$.

9.8 Given the transfer function

$$G(s) = \frac{2.4}{0.6s^2 + 8s + 36} = \frac{Y(s)}{U(s)}$$

compute the frequency response $y_{ss}(t)$ for the input $u(t) = 30.2 \cos 20t$.

9.9 Given the first-order transfer function

$$G(s) = \frac{0.6}{0.25s + 1}$$

show the following facts:

a. For low input frequencies ($\omega \approx 0$), the magnitude of the sinusoidal transfer function (in decibels) is a constant that is equal to $20 \log_{10}(0.6)$ dB.

b. For very high input frequencies, the magnitude (in decibels) is a straight-line asymptote with slope -20 dB/decade when plotted on a log scale for input frequency ω.

c. The low- and high-frequency asymptotes intersect at the corner frequency $\omega_c = 4$ rad/s.

[Hint: Begin with the sinusoidal transfer function, compute the log magnitude in decibels, and evaluate at limiting values of $\omega \to 0$ and $\omega \to \infty$.]

9.10 Consider again the simple RL circuit shown in Fig. 9.5 (Example 9.1). The transfer function of the RL circuit is

$$G(s) = \frac{I(s)}{E_{in}(s)} = \frac{1}{Ls + R}$$

where the output is current $I(t)$ and the input is source voltage $e_{in}(t)$. If the system parameters are $L = 0.02$ H and $R = 1.5\ \Omega$, determine the bandwidth (in hertz, Hz) of the RL circuit.

SS **9.11** Given the I/O equation

$$2\ddot{y} + 6\dot{y} + 64y = 18u$$

compute the bandwidth and resonant frequency of this system.

9.12 Figure P9.12 shows a 1-DOF mechanical system driven by the displacement of the left end, $x_{in}(t)$, which could be supplied by a rotating cam and follower (see Problem 2.3). When displacements $x_{in}(t) = 0$ and $x = 0$, the spring k is neither compressed nor stretched. The system parameters are $m = 2$ kg, $k = 500$ N/m, and $b = 20$ N-s/m. Determine the frequency response if the position input is $x_{in}(t) = 0.04 \sin 50t$ m.

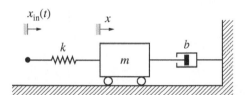

Figure P9.12

9.13 Determine the bandwidth, resonant frequency, and maximum transmissibility of the 1-DOF mechanical system in Problem 9.12.

MATLAB Problems

9.14 Use MATLAB's `abs` and `angle` commands to compute the magnitude and phase angle of the sinusoidal transfer function in Problem 9.2 with input frequency $\omega = 2$ rad/s. Verify your answer using MATLAB's `bode` command to compute the magnitude and phase angle for this input frequency.

9.15 Use MATLAB to plot the Bode diagram for the 1-DOF mechanical system in Problem 9.12 (Fig. P9.12). Estimate the frequency response for the position input $x_{in}(t) = 0.04 \sin 50t$ m by reading the Bode diagram (indicate the frequency response parameters on the plot of the Bode diagram). Obtain a more accurate answer by using MATLAB's `bode` command with left-hand-side arguments for computing magnitude and phase angle.

9.16 Use MATLAB to plot the Bode diagram of the 1-DOF mechanical system in Problem 9.12 and estimate the bandwidth, resonant frequency, and peak transmissibility.

9.17 Use MATLAB or Simulink to simulate the mechanical system in Problem 9.12 for the position input $x_{in}(t) = 0.04 \sin 50t$ m. Assume that the system is initially at rest at time $t = 0$. Plot mass position $x(t)$ and input $x_{in}(t)$ on the same plot and determine the frequency-response equation for $x_{ss}(t)$ from the simulation results.

SS **9.18** Use MATLAB to plot the Bode diagram of the second-order system in Problem 9.11 and estimate the bandwidth and resonant frequency.

9.19 A mechanical system is modeled by

$$0.6\ddot{x} + 3.4\dot{x} + 80x = f_a(t)$$

The mechanical system is initially at static equilibrium, $x(0) = 0$ and $\dot{x}(0) = 0$. The input force is a sinusoidal function, $f_a(t) = 2 \sin \omega t$ N.
  a. Determine the input frequency that results in the largest amplitude of the frequency response $x_{ss}(t)$.
  b. Use MATLAB to plot the Bode diagram and verify part (a).

SS **9.20** Figure P9.20 shows a 1-DOF mechanical system driven by the displacement of the left end, $x_{in}(t)$, which could be supplied by a rotating cam and follower (see Problem 2.4). When displacement $x = 0$, the spring k is neither compressed nor stretched. The system parameters are $m = 2$ kg, $k = 128$ N/m, and $b = 7$ N-s/m.
  a. Determine the frequency response if the position input is $x_{in}(t) = 0.04 \sin 8t$ (in meters).
  b. Use MATLAB to plot the Bode diagram of the 1-DOF mechanical system and verify the solution to part (a). Show that the amplitude of the frequency response is always less than the amplitude of the input displacement $x_{in}(t)$ for all frequencies other than $\omega = 8$ rad/s.
  c. Use the Bode diagram to determine the range of input frequencies where the amplitude of the frequency response is greater than 0.036 m if the displacement input is $x_{in}(t) = 0.04 \sin \omega t$ m.
  d. Verify the solutions to parts (a) and (c) by using Simulink to obtain the response $x(t)$ for input $x_{in}(t) = 0.04 \sin \omega t$ m. Plot $x(t)$ for the appropriate input frequency ω to verify the solutions obtained in parts (a) and (c).

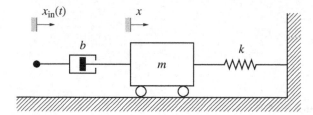

Figure P9.20

9.21 Figure P9.21 shows a 1-DOF mechanical system. Displacement z is measured from the static equilibrium position. The system parameters are $m = 0.3$ kg, $k_1 = 80$ N/m, $k_2 = 60$ N/m, and $b = 1.5$ N-s/m.

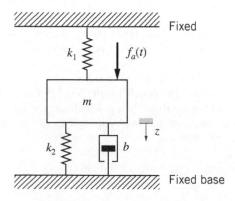

Figure P9.21

  a. Use MATLAB to create the Bode diagram using the transfer function $G(s) = Z(s)/F_a(s)$ and an appropriate SSR.
  b. Use MATLAB to estimate the bandwidth and resonant frequency.

9.22 Figure P9.22 shows the simple 1-DOF mechanical system from Problem 2.7 (in addition, see Problem 6.8). Displacement z (in meters) is measured from the undeflected spring position. External force $f_a(t)$ is applied directly to mass m. The spring force obeys a nonlinear relationship with displacement:

$$F_k = k_1 z + k_3 z^3$$

The input force is $f_a(t) = 5\sin\omega t$ N. The system parameters are mass $m = 1.5$ kg, viscous friction coefficient $b = 3$ N-s/m, linear stiffness coefficient $k_1 = 20$ N/m, and nonlinear stiffness coefficient $k_3 = 50$ N/m^3.

a. Compute the amplitude of the steady-state displacement response for a *linear* spring force (i.e., set $k_3 = 0$) for input frequency $\omega = 3$ rad/s.

b. Use Simulink to obtain the response to the sinusoidal force input for a *linear* spring force for input frequency $\omega = 3$ rad/s. Verify your answer from part (a).

c. Using Simulink to obtain the responses of the *linear* ($k_3 = 0$) and *nonlinear* ($k_3 = 50$ N/m^3) systems to the sinusoidal force input ($\omega = 3$ rad/s). Plot the linear and nonlinear responses on the same graph. Describe any differences or similarities in the steady-state amplitude and phase angle between the two responses.

d. Repeat part (c) for input frequency $\omega = 10$ rad/s.

e. Use your Simulink model and multiple trial simulations to determine the input frequency ω that maximizes the steady-state amplitude ratio (z_{ss}/f_a) for the linear and nonlinear responses.

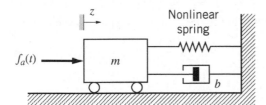

Figure P9.22

9.23 Figure P9.23 shows the two-mass mechanical damper from Problem 2.5 (in addition, see Problem 5.17). Displacements z_1 and z_2 (in meters) are measured from their equilibrium position when external force $f_a(t)$ is zero. The external force is applied directly to the rod/piston mass m_1. The system parameters are $m_1 = 0.2$ kg, $m_2 = 0.1$ kg, $k = 800$ N/m, and $b = 25$ N-s/rad.

a. Use MATLAB to obtain the frequency response of the piston (position z_1) if the applied force is $f_a(t) = 10 \sin t$ N.

b. Now, add a spring with stiffness $k_2 = 100$ N/m between the piston and the cylinder. Use MATLAB to obtain the frequency response of the piston (with added spring k_2) if the applied force is $f_a(t) = 10 \sin t$ N.

c. Using MATLAB, plot the Bode diagrams of the mechanical damper with and without the spring k_2 on the same plot (use the `hold on` command). Compare the frequency responses of the two systems at low, resonant, and high frequencies.

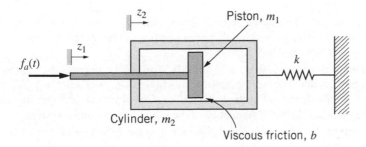

Figure P9.23

9.24 Figure P9.24 shows a block diagram for a simple second-order system that is driven by the sinusoidal input function $u(t) = 3 \sin 10t$.

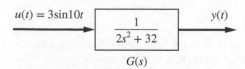

$$u(t) = 3\sin10t \quad \boxed{\dfrac{1}{2s^2 + 32}} \quad y(t)$$

$$G(s)$$

Figure P9.24

a. Compute the magnitude and phase of the sinusoidal transfer function $G(j\omega)$ for the input frequency $\omega = 10$ rad/s.

b. Use MATLAB or Simulink to obtain the response to the sinusoidal input $u(t)$. Plot $y(t)$.

c. Explain why the system response $y(t)$ obtained in part (b) *does not* match the frequency response solution presented by Eq. (9.17).

9.25 Figure P9.25 shows the magnitude Bode plot for two LTI systems. The transfer function for one of the LTI systems is

$$G(s) = \frac{576}{16s^2 + 38.4s + 576}$$

Match transfer function $G(s)$ with the appropriate magnitude plot in Fig. P9.25. Explain your answer.

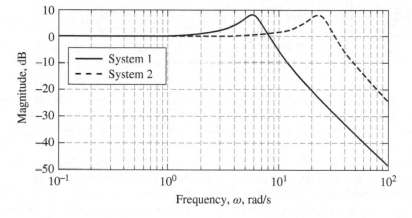

Figure P9.25

9.26 Figure P9.26 shows a two-mass mechanical system. The applied force is a sinusoidal function: $f_a(t) = 10 \sin \omega t$ N. Vertical displacements z_1 and z_2 are measured relative to their equilibrium positions. The system parameters are $m_1 = 2$ kg, $m_2 = 0.1$ kg, $k_1 = 150$ N/m, $k_2 = 360$ N/m, and $b = 20$ N-s/m.

a. The system output is the force transmitted to the base, f_t. Write a MATLAB M-file that computes the transmissibility $|G(j\omega)|$ for frequencies up to 120 rad/s, where the system transfer function is $G(s) = F_t(s)/F_a(s)$. You may use the bode command to compute the amplitude ratio. Plot transmissibility $|G(j\omega)|$ versus the input frequency ratio ω/ω_{n2}, where $\omega_{n2} = \sqrt{k_2/m_2}$ is the undamped natural frequency of the second spring–mass system.

b. Use MATLAB to obtain the Bode diagram of the system $G(s)$ and compare it to the transmissibility plot in part (a).

c. Using either plot from parts (a) and (b), explain the phenomena that occur when the input frequency is equal to ω_{n2}. Describe the motion of mass m_1 and mass m_2 when $\omega = \omega_{n2}$. A Bode diagram of a system where the output is the force transmitted to mass m_2 might help.

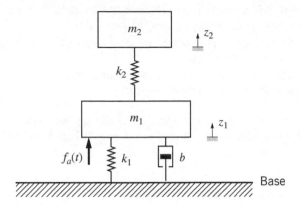

Figure P9.26

9.27 A simple model of an electromechanical actuator is

$$0.06\ddot{z} + 5.5\dot{z} + 3100z = 4.6I$$

$$0.01\dot{I} + 4I = e_{in}(t) - 2.3\dot{z}$$

where z is the position of the mechanical mass–spring–damper subsystem (in meters) and I is the current in the solenoid armature circuit. The single input to the system is source voltage $e_{in}(t)$.

a. Use MATLAB to obtain the frequency response of the mechanical actuator (position z) if the source voltage is $e_{in}(t) = 8 \sin 150t$ V.

b. Use MATLAB to compute the bandwidth of the electromechanical actuator.

Engineering Applications

9.28 Figure P9.28 shows the band-stop or notch filter discussed in Problems 3.35 and 5.32. As discussed in these prior problems, band-stop filters are used to suppress signals that contain an undesirable frequency. For example, the frames of aerospace vehicles (e.g., aircraft, missiles, rockets) vibrate at known frequencies and these frequencies can be inadvertently measured by onboard sensors such as gyroscopes and accelerometers. Hence band-stop filters are used to remove an unwanted frequency component from the measurement signal.

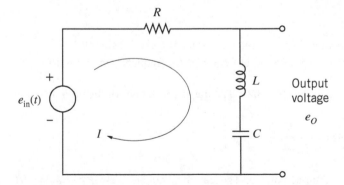

Figure P9.28

The transfer function for the band-stop filter is

$$G(s) = \frac{LCs^2 + 1}{LCs^2 + RCs + 1} = \frac{E_O(s)}{E_{in}(s)}$$

The electrical parameters for the band-stop filter are inductance $L = 0.005$ H, capacitance $C = 0.02$ F, and $R = 1\ \Omega$.

a. Using MATLAB, plot the Bode diagram of the band-stop filter. Show that the band-stop filter essentially removes an input voltage signal with a frequency equal to the "notch" frequency $\omega_N = \sqrt{1/(LC)}$ rad/s.

b. Using the Bode plot, estimate the "stop band" or input frequency range where the amplitude of the filter's output signal is reduced to less than one-half of the amplitude of the input signal.

c. If the resistance is changed to $R = 0.2\ \Omega$ describe how the performance of the band-stop filter changes and estimate the band-stop frequency range for one-half amplitude reduction [i.e., repeat part (b)].

9.29 A vibration isolation system has the transfer function $G(s) = F_T(s)/F_{in}(s)$ where $f_{in}(t)$ is the applied force from an unbalanced rotating machine (the input) and $f_T(t)$ is the transmitted force to the base (the output). Figure P9.29 shows the Bode diagram of the vibration isolation system.

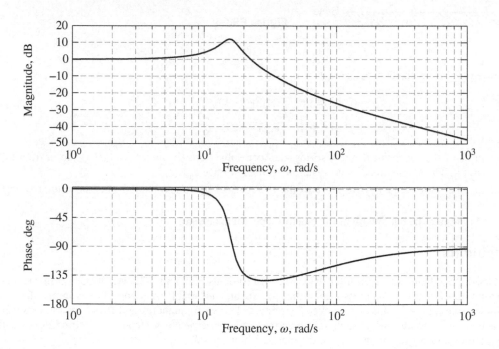

Figure P9.29

a. Compute the frequency response of the transmitted force if the input force is $f_{in}(t) = 2 \sin 50t$ N.

b. If the maximum acceptable transmissibility is 1.75, what input frequency (or frequency range) must be avoided?

9.30 Figure P9.30 shows the op-amp circuit discussed in Problem 3.21. The transfer function for the op-amp circuit is

$$G(s) = \frac{-R_2 Cs}{R_1 Cs + 1} = \frac{E_O(s)}{E_{in}(s)}$$

where e_O is the output voltage and $e_{in}(t)$ is the input voltage. The circuit parameters are $R_1 = R_2 = 2\ \Omega$, and $C = 0.002$ F. This type of op-amp circuit is called an *active high-pass filter*.

a. Using MATLAB, plot the Bode diagram of the op-amp transfer function.

b. Using the Bode plot, estimate the steady-state amplitude of the output voltage $e_O(t)$ for the input voltage $e_{in}(t) = 2 \sin t$ V. Verify your answer with a simulation using Simulink.

c. Repeat part (b) for the input voltage $e_{in}(t) = 2 \sin 100t$ V (be sure to adjust the simulation time and integration step size).

d. Repeat part (b) for the input voltage $e_{in}(t) = 2 \sin 1{,}000t$ V (be sure to adjust the simulation time and integration step size).

e. Based on the Bode diagram and the simulation results, explain why this op-amp circuit is called a high-pass filter.

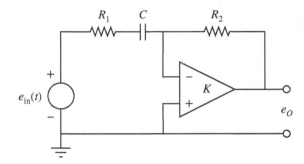

Figure P9.30

SS **9.31** The temperature of a small water tank is modeled by the simple I/O equation:

$$9250\delta\dot{T} + \delta T = \delta T_a$$

where δT is the tank's temperature deviation from its nominal value, and δT_a is the ambient temperature variation from its nominal value. Suppose the ambient temperature varies in a sinusoidal manner:

$$\delta T_a(t) = 5 \sin \omega t \; °C$$

The ambient temperature variation has matched this sinusoidal model for several days. The reference time t (measured in seconds) is defined such that the peak ambient temperature always occurs at 3 pm. Determine the maximum and minimum temperature deviations of the water tank and the respective times of day when these extremes occur.

9.32 Figure P9.32 shows the pneumatic servomechanism first studied in Problem 6.25. The total length of the cylinder is 10 cm and the piston position x is measured from the mid-point of the cylinder (hence when $x = 0$ the piston is at the middle of the cylinder as shown in Fig. P9.32). This highly nonlinear system has been linearized about a nominal pressure, volume, and mid-point piston position ($x = 0$) to produce the transfer function

$$G(s) = \frac{7.003(10^6)}{s(s^2 + 28.3s + 15{,}890)} = \frac{X(s)}{U(s)}$$

where $x(t)$ is the position of the piston-load mass and $u(t)$ is the position of the spool valve (both in meters). For additional details, see Reference [7].

a. Using Simulink, determine the piston-load mass response $x(t)$ to a sinusoidal valve position input $u(t) = 0.001 \sin 50t$ m (the input frequency is ~8 Hz). Plot piston and valve position versus time on the same figure. Describe how the simulated frequency response of the piston-load mass differs from the "standard" analytical frequency-response equation (9.17). Explain why the simulated frequency response does not match the form of Eq. (9.17).
[Hint: Reread the section on the derivation of the frequency response and the associated assumptions.]

b. Use Simulink to determine the piston-load mass response $x(t)$ to a sinusoidal valve input $u(t) = 0.001 \sin 126t$ m (the input frequency is ~20 Hz). Plot piston and valve positions versus time on the same figure to show that they are 180° out of phase. Verify this phase difference using the Bode diagram of $G(s)$.

c. Develop the transfer function with velocity of the piston-load mass (i.e., $\dot{x}$) as the output and valve position u as the input. Use Simulink to determine the velocity response to the valve input $u(t) = 0.001 \sin 50t$ m and plot $\dot{x}(t)$. Plot the Bode diagram for the velocity-output transfer function and show that it *can* be used to estimate the frequency response for piston velocity by verifying the simulated velocity response to the ~8-Hz input signal.

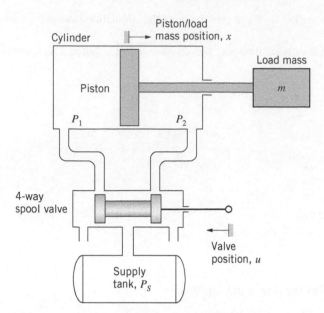

Figure P9.32

SS **9.33** A flexible spacecraft is placed on a shaker table that vibrates at a wide range of frequencies. Figure P9.33 shows the Bode diagram where the output is the spacecraft's displacement and the input is the table displacement.

a. Engineers want to develop a transfer function by modeling the flexible spacecraft as a lumped-mass mechanical system. Based on the frequency response, how many lumped masses will accurately model the flexible spacecraft? Justify your answer.

b. Using the Bode diagram, develop a transfer function for the flexible spacecraft. Explain your steps and verify your transfer function by plotting its Bode diagram and comparing it to Fig. P9.33.

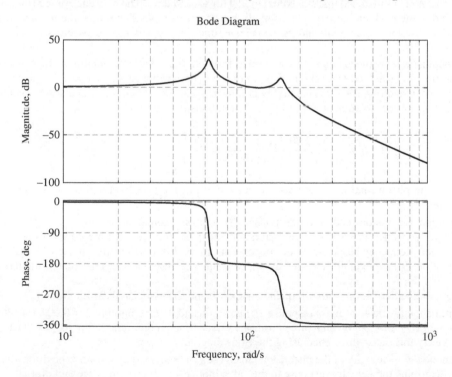

Figure P9.33

SS **9.34** An engineer has two "black boxes": one contains an RC filter circuit, and the other contains an LC circuit. The transfer functions of the two "black boxes" are

$$\text{RC filter} \quad G_{RC}(s) = \frac{1}{0.25s + 1} \qquad \text{LC circuit} \quad G_{LC}(s) = \frac{4}{s^2 + 4}$$

Capacitor voltage is the output of both transfer functions. The engineer applies a sine wave input voltage with a frequency of 4 Hz ($\omega = 25.1327$ rad/s) and an amplitude of 3 V. Figure P9.34a presents the measured output voltage of black box #1 and Fig. P9.34b presents the voltage output of black box #2. Which black box is the RC filter? Which black box is the LC circuit? Clearly, explain why each particular voltage response matches the black-box transfer function.

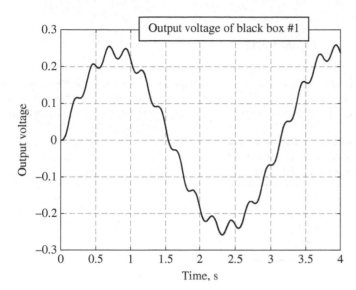

Figure P9.34a

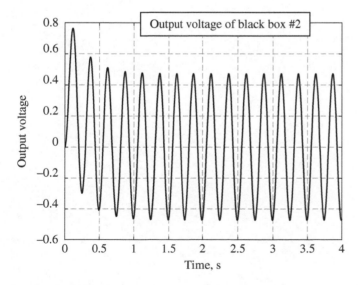

Figure P9.34b

9.35   Figure P9.35 shows a 1-DOF vibration isolation system for a sensitive instrument with mass $m = 1.4$ kg.

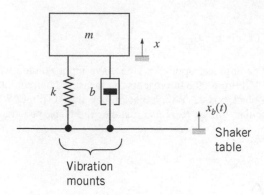

Figure P9.35

An engineer wants to determine the values of the composite stiffness k and damping coefficient b for the mounting system, so she performs a series of experiments on a shaker table. The shaker table vibration input is $x_b(t) = 0.5 \sin \omega t$ cm. The engineer measures the frequency response for a range of input frequencies and the results are tabulated below:

Input Frequency, ω, rad/s	Amplitude of Frequency Response $x_{ss}(t)$, cm
5	0.5070
30	0.7008
60	0.4975
100	0.2532

Determine the numerical values for lumped spring constant k and friction coefficient b of the vibration isolation system.

CHAPTER 10: PROBLEMS

Conceptual Problems

SS **10.1** Figure P10.1 shows a general feedback control system with forward-path transfer functions $G_C(s)$ (controller) and $G_P(s)$ (plant) and feedback transfer functions $H(s)$. Given the following transfer functions, determine the closed-loop transfer function $T(s) = Y(s)/R(s)$.

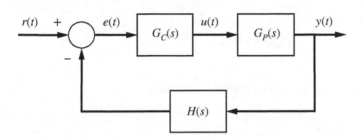

Figure P10.1

a. $G_C(s) = K_P$ $\qquad$ $G_P(s) = \dfrac{6}{s+2}$ $\qquad$ $H(s) = 1$

b. $G_C(s) = K_P$ $\qquad$ $G_P(s) = \dfrac{4}{s(s+2)}$ $\qquad$ $H(s) = 1$

c. $G_C(s) = K_P$ $\qquad$ $G_P(s) = \dfrac{8}{s^2(s+6)}$ $\qquad$ $H(s) = 1$

d. $G_C(s) = K_P$ $\qquad$ $G_P(s) = \dfrac{1}{s^2+6s+10}$ $\qquad$ $H(s) = 2$

e. $G_C(s) = K_P$ $\qquad$ $G_P(s) = \dfrac{2}{s+1}$ $\qquad$ $H(s) = \dfrac{20}{s+20}$

f. $G_C(s) = K_P$ $\qquad$ $G_P(s) = \dfrac{3}{s^2+6s}$ $\qquad$ $H(s) = \dfrac{10}{s+10}$

g. $G_C(s) = \dfrac{K_Ps + K_I}{s}$ $\qquad$ $G_P(s) = \dfrac{2}{(s+1)(s+4)}$ $\qquad$ $H(s) = 1$

h. $G_C(s) = K_Ds + K_P$ $\qquad$ $G_P(s) = \dfrac{1}{s(s+3)}$ $\qquad$ $H(s) = 1$

i. $G_C(s) = \dfrac{K_Ds^2 + K_Ps + K_I}{s}$ $\qquad$ $G_P(s) = \dfrac{1}{s^2+2s+6}$ $\qquad$ $H(s) = 1$

10.2 Again, consider Fig. P10.1 and the general closed-loop system. The system transfer functions are

$$G_C(s) = K_P \qquad G_P(s) = \dfrac{3}{s^2+8s+20} \qquad H(s) = 1$$

a. Compute the steady-state error for a unit-step input $r(t) = U(t)$ if the controller gain is $K_P = 35$.
b. Compute the steady-state error for a unit-ramp input $r(t) = t$ if the controller gain is $K_P = 35$.

10.3 Consider the general closed-loop control system illustrated in Fig. P10.1. The system transfer functions are

$$G_C(s) = \dfrac{K_Ps + K_I}{s} \qquad G_P(s) = \dfrac{3}{s^2+8s+20} \qquad H(s) = 1$$

a. Compute the steady-state error for a step input $r(t) = 0.6U(t)$ if the controller gains are $K_P = 35$ and $K_I = 60$.

b. Compute the steady-state error for a ramp input $r(t) = 0.6t$ if the controller gains are $K_P = 35$ and $K_I = 60$.

10.4 Figure P10.4 shows a simple closed-loop system. The reference input is a step function, $r(t) = 2U(t)$.

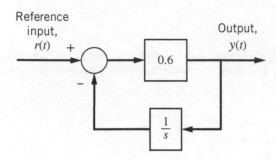

Figure P10.4

a. Compute the steady-state output, $y(\infty)$.

b. Compute the settling time for the closed-loop system to reach the steady state.

SS 10.5 Figure P10.5 shows a closed-loop control system.

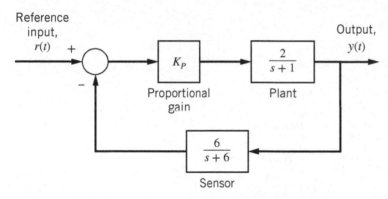

Figure P10.5

a. Compute the controller gain K_P so the undamped natural frequency of the closed-loop system is $\omega_n = 4$ rad/s.

b. Compute the controller gain K_P so that the damping ratio of the closed-loop system is $\zeta = 0.7$.

c. Compute the steady-state output for a step reference input $r(t) = 4U(t)$ and controller gain $K_P = 2$.

SS 10.6 A closed-loop control system is shown in Fig. P10.6.

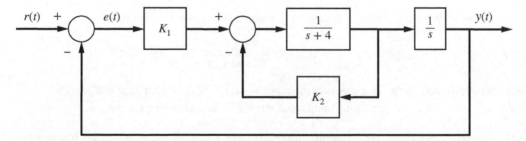

Figure P10.6

a. Compute the closed-loop transfer function for the overall system, $T(s) = Y(s)/R(s)$.

b. Two gain pairs are considered: Option 1 ($K_1 = 15, K_2 = 2$) and Option 2 ($K_1 = 30, K_2 = 3$). Which gain pair provides the greatest closed-loop damping ratio? Justify your answer.

SS 10.7 Figure P10.7 shows a unity-feedback closed-loop system. The reference input is a ramp, $r(t) = 0.2t$.

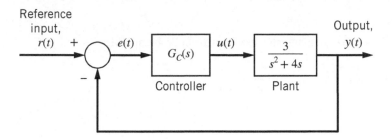

Figure P10.7

a. Compute the steady-state tracking error if the controller $G_C(s)$ is a simple proportional gain $K_P = 2$.

b. Compute the steady-state tracking error if we use a PI controller with gains $K_P = 3$ and $K_I = 1.5$.

10.8 A simple closed-loop PI control system is shown in Fig. P10.8.

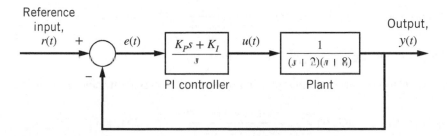

Figure P10.8

a. Show that the closed-loop system is stable for PI gains $K_P = 5$ and $K_I = 25$.

b. Determine "by hand" the closed-loop output $y(t)$ at time $t = 8$ s if the reference input is a ramp function $r(t) = 1.4t$. Use the PI gains from part (a).

10.9 Consider again the closed-loop system shown in Fig. P10.7. The controller is a lead filter

$$G_C(s) = \frac{K(s+5)}{s+15}$$

Show that the root locus has two vertical asymptotes, and determine their intersection with the real axis.

MATLAB Problems

10.10 Refer back to the general closed-loop control system shown in Fig. P10.1 and the system transfer functions presented in Problem 10.1a repeated below

$$G_C(s) = K_P \qquad G_P(s) = \frac{6}{s+2} \qquad H(s) = 1$$

    a. Use MATLAB to compute the closed-loop transfer function $T(s)$ if the P-gain is $K_P = 3$ and verify your answer in Problem 10.1a.

    b. Compute the roots of the closed-loop transfer function and estimate the unit-step response characteristics such as settling time and steady-state response. Does the closed-loop unit-step response exhibit oscillations? Explain.

    c. Create two Simulink models: one using the block diagram shown in Fig. P10.1 and another using the closed-loop transfer function $T(s)$. Simulate the unit-step response, $r(t) = U(t)$, using both Simulink models and plot output $y(t)$. Verify that both Simulink models produce the same results and verify your step-response calculations from part (b).

10.11   Repeat all parts of Problem 10.10 using the system transfer functions from Problem 10.1c

$$G_C(s) = K_P \qquad G_P(s) = \frac{8}{s^2(s+6)} \qquad H(s) = 1$$

The P-gain is $K_P = 5$.

10.12   Repeat all parts of Problem 10.10 using the system transfer functions from Problem 10.1d

$$G_C(s) = K_P \qquad G_P(s) = \frac{1}{s^2 + 6s + 10} \qquad H(s) = 2$$

The P-gain is $K_P = 4$.

10.13   Repeat all parts of Problem 10.10 using the system transfer functions from Problem 10.1g

$$G_C(s) = \frac{K_P s + K_I}{s} \qquad G_P(s) = \frac{2}{(s+1)(s+4)} \qquad H(s) = 1$$

The P-gain is $K_P = 0.3$ and the I-gain is $K_I = 2$.

10.14   Refer back to the general closed-loop control system shown in Fig. P10.1. The system transfer functions are

$$G_C(s) = K_P \qquad G_P(s) = \frac{6(s+1)}{s(s+2)(s^2 + 3s + 24)} \qquad H(s) = 1$$

    a. Use MATLAB to create the root-locus plot. Use the `rlocus` and/or `rlocfind` commands to determine the range of controller gain K_P for a stable closed-loop system.

    b. Use MATLAB and the Bode diagram to determine the gain and phase margins for the gain setting $K_P = 5$. Use the gain margin to verify the solution to part (a).

10.15   Refer back to the general closed-loop control system shown in Fig. P10.1. The system transfer functions are

$$G_C(s) = K(s+5) \qquad G_P(s) = \frac{0.4(s+1)}{s(s^2 + 4s + 36)} \qquad H(s) = \frac{30}{2s + 30}$$

    a. Use MATLAB to create the root-locus plot. Use the `rlocus` and/or `rlocfind` commands to determine the range of controller gain K for a stable closed-loop system.

    b. Use MATLAB and the Bode diagram to determine the gain and phase margins for the gain setting $K = 2$. Use the gain margin to verify the solution to part (a).

10.16   Refer back to the general closed-loop control system shown in Fig. P10.1. The system transfer functions are

$$G_C(s) = \frac{K(s+2)}{2s + 16} \qquad G_P(s) = \frac{12}{s(s^2 + 9s + 26)} \qquad H(s) = 1$$

a. Use MATLAB to create the root-locus plot. Use the `rlocus` and/or `rlocfind` commands to determine the range of controller gain K for a stable closed-loop system.

b. Use MATLAB and the Bode diagram to determine the gain and phase margins for the gain setting $K = 20$. Use the gain margin to verify the solution to part (a).

10.17 Figure P10.17 shows a closed-loop system with a "pure inertia plant," where the plant I/O equation is $\ddot{y} = 3u$.

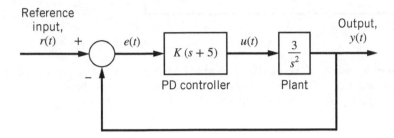

Figure P10.17

a. Compute ("by hand") the controller gain K so that the closed-loop roots have a damping ratio $\zeta = 0.7071$.

b. Verify your answer in part (a) by using MATLAB's `rlocus` and `rlocfind` commands.

10.18 A simple closed-loop system is shown in Fig. P10.18.

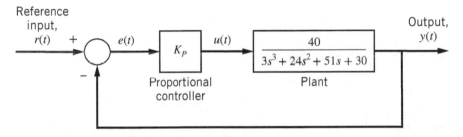

Figure P10.18

a. Use MATLAB to create the root-locus plot if the controller is a simple proportional gain.

b. Use MATLAB's `rlocus` or `rlocfind` command(s) to determine the proportional control gain K_P for marginal stability. Compute the closed-loop poles for the marginally stable case.

c. Use Simulink to simulate the marginally stable P-controller found in part (b). Let the reference input be a unit-step function. Plot the closed-loop response $y(t)$ and show that the frequency of oscillation for the marginally stable case matches the appropriate marginally stable closed-loop poles.

10.19 Consider again the PI controller in Problem 10.8 and Fig. P10.8. Use Simulink to obtain the closed-loop response $y(t)$ for a ramp input $r(t) = 1.4t$. Use the PI gains $K_P = 5$ and $K_I = 25$. Plot $y(t)$ and $r(t)$ on the same figure for a simulation time of 8 s.

10.20 Figure P10.20 shows a unity-feedback control system. Use the Ziegler–Nichols ultimate-gain method to determine a good starting point for PID gain selection. Use Simulink to obtain the closed-loop response for a unit-step reference input, and adjust the gains until a fast, well-damped response is obtained (the maximum overshoot is 10%). Plot the closed-loop response $y(t)$ and present the best PID gains.

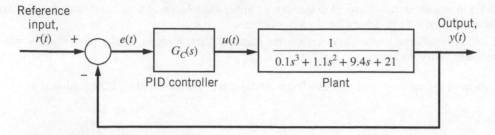

Figure P10.20

10.21 Figure P10.21 shows a closed-loop system with a PD controller. Use MATLAB and the root-locus method to design the PD controller (i.e., select gain K and zero z_C) so that the underdamped closed-loop roots have a damping ratio $\zeta > 0.65$ and undamped natural frequency $\omega_n > 1$ rad/s. Explain your steps and justify your final design with root-locus analysis.

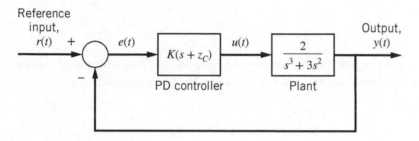

Figure P10.21

SS 10.22 Figure P10.22 shows a closed-loop system with a PID controller.

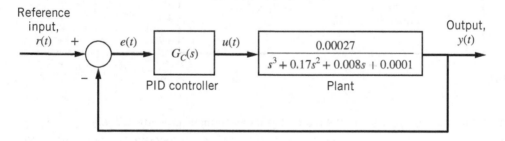

Figure P10.22

a. Use the Ziegler–Nichols reaction-curve method to select the PID control gains. Use MATLAB and Simulink as needed.

b. Use Simulink to simulate the closed-loop response to a unit-step input, $r(t) = U(t)$, with the PID controller obtained in part (a). Plot $y(t)$.

c. Use the Simulink model from part (b) and vary the PID gains in an attempt to decrease the overshoot while maintaining a fast closed-loop response (use the Ziegler–Nichols gains from part (a) as the starting point). Plot the closed-loop response $y(t)$ for an improved PID control scheme along with the output $y(t)$ obtained in part (b) using the Ziegler–Nichols gains.

10.23 Repeat all parts of Problem 10.22 using the Ziegler–Nichols ultimate-gain method to design the PID controller.

10.24 Figure P10.24 shows the mechanical position-control system from Examples 10.6, 10.8, 10.12, and 10.13. The open-loop pole of the lead controller is p_L. The actuator gain is $K_A = 2$ N/V.

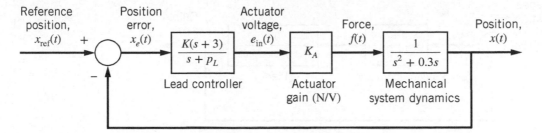

Figure P10.24

a. Use MATLAB to create the root-locus plots for four different lead controllers: $p_L = 4, 15, 25$, and 30.

b. Interpret the four root-locus plots (and their corresponding closed-loop responses) for these four lead controllers. Compare these four root-locus plots to the results from Example 10.12 (PD controller) and Example 10.13 (lead controller). What conclusions can you draw regarding the open-loop pole location of the lead controller?

SS 10.25 Consider again the unity-feedback control system shown in Fig. P10.7. The controller is a lead filter

$$G_C(s) = \frac{K(s+5)}{s+15}$$

a. Use MATLAB to create the root-locus plot. Comment on the closed-loop stability for all positive gains, $K > 0$.

b. Use MATLAB to determine the gain and phase margins for the controller gain setting $K = 20$. Does the gain margin match the solution to part (a)? Explain.

10.26 A simple closed-loop system is shown in Fig. P10.26.

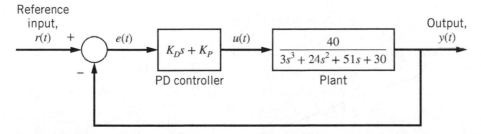

Figure P10.26

a. Use MATLAB and the Bode diagram to determine the gain and phase margins for a P-controller with $K_P = 2.5$ and $K_D = 0$.

b. Use MATLAB and the Bode diagram to determine the gain and phase margins for a PD controller with $K_P = 2.5$ and $K_D = 1.5$.

c. Use MATLAB to plot the root locus for the P-controller in part (a) and the PD controller in part (b). Use the PD controller form $G_C(s) = K(s + 1.667)$ so that the controller zero location matches the PD gains in part (b). On the basis of the results from parts (a)–(c), comment on how the PD controller changes the closed-loop response characteristics compared to a P-controller.

10.27 For the unity-feedback control system shown in Fig. P10.27, design a lead controller so that the compensated closed-loop system meets the following performance criteria: (1) phase margin is at least 50°, (2) gain margin is at least 12 dB, and (3) steady-state tracking error is less than 0.2 for a ramp input $r(t) = 0.5t$. Support your lead-controller design with the appropriate graphical analyses using MATLAB.

[Hint: first compute the static velocity error constant that is required for the steady-state tracking error constraint; next, compute the stability margins using only gain adjustment K; finally, design the lead controller to meet the stability margins.]

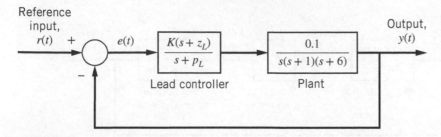

Figure P10.27

SS **10.28**   Figure P10.28 shows the Bode diagram of an open-loop transfer function $G(s)H(s)$ that is part of a closed-loop control system. The open-loop transfer function contains a lead controller with a control gain setting of $K = 2$, which corresponds to the Bode diagram in Fig. P10.28. Estimate the gain and phase margins and the control gain K_{ms} that drives the closed-loop system to the point of marginal stability.

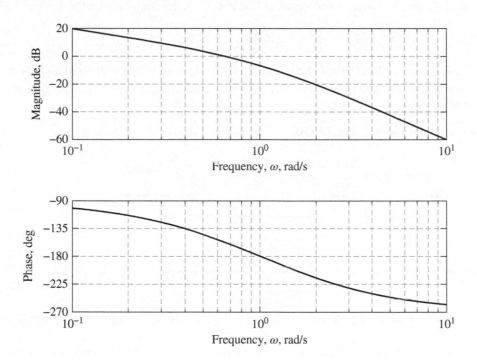

Figure P10.28

Engineering Applications

10.29   Figure P10.29 shows a closed-loop control system for the pneumatic servomechanism [8] studied in Problems 6.25 and 9.32. This highly nonlinear system has been linearized about a nominal pressure, volume, and mid-point piston position ($x = 0$) to produce the transfer function $G_P(s)$ for the pneumatic servo:

$$\text{Pneumatic servo:} \quad G_P(s) = \frac{7.003 \left(10^6\right)}{s\left(s^2 + 28.3s + 15,890\right)} = \frac{X(s)}{U(s)}$$

where $x(t)$ is the position of the piston/load mass (in meters) and $u(t)$ is the position of the spool valve (in meters). See Problems 6.25 and 9.32 for a diagram of the pneumatic servo. An electromechanical solenoid is used to actuate the valve position $u(t)$. Because the response of the solenoid–valve combination is so much faster than the pneumatic servo, we can replace it with a constant gain $K_V = 2.2(10^{-4})\,\text{m/V}$.

a. Plot the root locus using MATLAB if the controller is a simple gain, that is, $G_C(s) = K_P$. Use the root locus to show that a P-controller does not provide much flexibility for obtaining a good closed-loop response.

b. Use MATLAB's `rlocus` and/or `rlocfind` commands to compute the maximum gain K_P that results in a marginally stable closed-loop system.

c. Suppose the controller gain setting is $G_C(s) = K_P = 10$ V/m. Compute the steady-state error for the ramp input $x_{\text{ref}}(t) = 0.008t$ m.

d. Plot the root locus if we use the PD controller $G_C(s) = K(s + 5)$. Has inserting the PD controller changed either the closed-loop stability or closed-loop damping when compared to the P-controller? Explain.

e. A standard control scheme for pneumatic servomechanisms is the "position–velocity–acceleration" (PVA) controller, which has the form $G_C(s) = K_a s^2 + K_v s + K_P$ (see Reference [8]). Use MATLAB to plot the root locus using the PVA controller $G_C(s) = K(s^2 + 30s + 200)$. How has the PVA control scheme changed the stability and damping of the closed-loop system compared to the P and PD controllers?

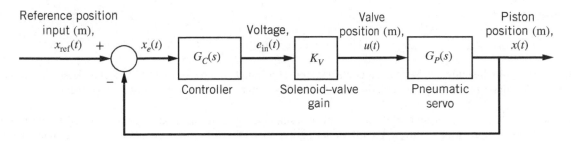

Figure P10.29

10.30 A simplified version of the Space Shuttle's Flare and Shallow Glide Slope normal acceleration command channel is shown in Fig. P10.30. The input is a flight-path angle command, γ_C (in rad), and the output is an incremental normal acceleration command, Δn_z (in units of "g"). Note that dividing the flight-path angle error γ_e by time constant τ produces the approximate angular rate $d\gamma/dt$, which is then multiplied by velocity V to produce normal acceleration. Earth gravity acceleration is $g = 32.2$ ft/s^2, and the Shuttle's velocity is $V = 420$ ft/s during the Flare and Shallow Glide Slope phase.

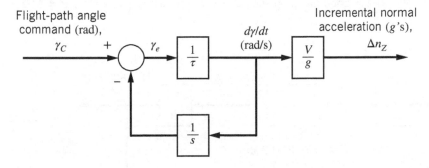

Figure P10.30

Use Simulink to simulate the closed-loop response to a step-input flight-path angle command $\gamma_C(t) = 0.09U(t)$ rad. Adjust the time constant τ so that the incremental normal acceleration $\Delta n_z(t)$ shows a peak value equal to 0.4 g. Plot $\Delta n_z(t)$ for the best design value for τ and describe the closed-loop response for the incremental normal acceleration. Relate the time constant τ to the settling time.

SS 10.31 Figure P10.31 shows the closed-loop PD control for the pitching motion of a launch vehicle during the high dynamic pressure flight phase. The combination of high dynamic pressure (i.e., large aerodynamic forces) and an offset between the aerodynamic center-of-pressure and the center of mass causes instability in the open-loop pitch dynamics. The PD controller and servo determine the rocket engine gimbal angle δ, which produces the pitch angle response $\theta(t)$. Use MATLAB and the root-locus method to determine the best gain K that maximizes the closed-loop undamped natural frequency ω_n and results in a closed-loop damping ratio $\zeta = 0.7071$.

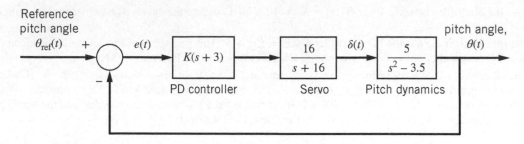

Figure P10.31

10.32 Consider again the launch vehicle pitch-control system shown in Fig. P10.31. Use MATLAB to determine the control gain K that maximizes the phase margin. Ensure that this gain results in a stable closed-loop system.

10.33 Figure P10.33a shows an autopilot for aircraft heading control. Aircraft heading ψ is adjusted by performing a banked turn, where ϕ is the aircraft's roll angle. Figure P10.33b presents the details of the roll dynamics transfer function $G(s)$.

a. Determine the roll-rate feedback gain, K_{rr}, so that the damping ratio of the inner roll-rate loop, $G_{rr}(s) = \dot{\phi}/\dot{\phi}_{ref}$, is $\zeta = 0.7$ (see Fig. P10.33b).

b. Use MATLAB and the root-locus method to determine a good roll-error gain K_r that results in good damping and fast response of the roll dynamics (see Fig. P10.33b). Use the roll-rate gain K_{rr} determined in part (a). Present the closed-loop damping ratio and undamped natural frequency of the roll dynamics $G(s)$ after determining the gain K_r.

c. If the reference roll angle is $\phi_{ref} = 0.1$ rad, determine the steady-state roll angle for the gains selected in parts (a) and (b).

d. Use MATLAB's `rlocus` and/or `rlocfind` commands to determine the heading gain K_h so that the overall closed-loop system, $\psi(s)/\psi_{ref}(s)$, has good damping (i.e., $\zeta \approx 0.7$) and fast response. Use $g = 32.2$ ft/s^2 and airspeed $V = 600$ ft/s.

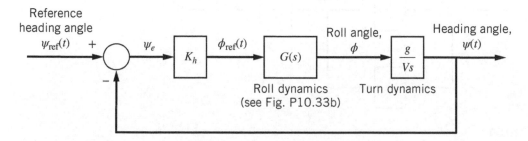

Figure P10.33a

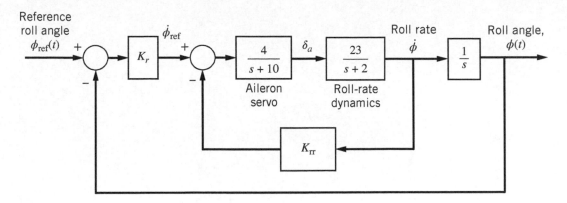

Figure P10.33b

10.34 Consider again the aircraft roll-control loop presented in Fig. P10.33b. Suppose the roll-rate feedback gain is $K_{rr} = 0.5$. If the roll-error gain is $K_r = 0.9$ s^{-1}, use MATLAB and the Bode diagram to determine the gain and phase margins.

Appendix A

Units

This textbook uses the International System of Units (SI). Table A.1 summarizes the five basic units that are used in this textbook. The reader should note that seven basic units exist; we do not use luminous intensity and amount of substance in this textbook. Angular displacement in a plane is an auxiliary unit that we use often in dynamic systems and its unit is the radian (rad).

All other units are called *derived units* as they can be expressed in terms of the basic units. One simple example is force in newtons (N), where $1\,N = 1\,kg\text{-}m/s^2$. Table A.2 summarizes the derived units used in this textbook.

Table A.1   Basic Units used in this Textbook

Quantity	Unit	Symbol
Length	meter	m
Mass	kilogram	kg
Time	second	s
Electrical current	ampere	A
Temperature	kelvin	K

Table A.2   Derived Units used in this Textbook

Quantity	Unit	Symbol	Dimension
Force	newton	N	$kg\text{-}m/s^2$
Energy	joule	J	$kg\text{-}m^2/s^2$ (or N-m)
Power	watt	W	$kg\text{-}m^2/s^3$ (or J/s)
Pressure	pascal	Pa	$kg/(m\text{-}s^2)$ (or N/m^2)
Electrical charge	coulomb	C	A-s
Voltage	volt	V	$kg\text{-}m^2/(A\text{-}s^3)$ (or W/A)
Electrical resistance	ohm	Ω	$kg\text{-}m^2/(A^2\text{-}s^3)$ (or V/A)
Electrical capacitance	farad	F	$A^2\text{-}s^4/(kg\text{-}m^2)$ (or C/V)
Magnetic flux	weber	Wb	$kg\text{-}m^2/(A\text{-}s^2)$ (or V-s)
Inductance	henry	H	$kg\text{-}m^2/(A^2\text{-}s^2)$ (or Wb/A)

Appendix B

MATLAB Primer for Analyzing Dynamic Systems

B.1 INTRODUCTION

MATLAB is a computing environment and computer language for numerical calculations, simulation, and visualization. Developed by MathWorks, MATLAB has become a universal computing platform for engineering calculations in industry and academia. In addition to being a programming language, MATLAB consists of software programs or *M-files* that can be used to perform particular functions, such as solving ordinary differential equations, curve-fitting, and statistics. *Toolboxes* are collections of M-files; one example is the Control System Toolbox. The user may write his or her own specialized MATLAB M-file to perform the desired calculations, and this user-defined M-file can access and use MATLAB's existing built-in functions and M-files as well.

This appendix provides a very basic introduction to MATLAB usage, its commands, and programming. While MATLAB is used in all areas of science and engineering, this appendix will only emphasize the commands that apply to solving problems involving dynamic systems and control.

B.2 BASIC MATLAB COMPUTATIONS

MATLAB is a command-driven computing tool. Once the user has started MATLAB, he or she will see the following prompt

```
>>
```

The user may now type in single-line commands to perform calculations. For example, the user can define variables (real or complex) by assigning numerical values. When a user defines a variable, MATLAB "echoes back" the assignment to the screen:

```
>> x = 2                          % This command defines variable x = 2
```

Upon hitting the return key MATLAB displays

```
x = 2
```

The user can suppress the print to the screen by inserting a semicolon (;) after the command. Note that in the previous simple example the character % defines the start of a "comment" line, where the user can add descriptive text that is not processed by MATLAB. After defining variables, the user can perform basic mathematical operations using + for addition, − for subtraction, * for multiplication, / for division, and ^ for power. The following simple commands illustrate the use of these math operations:

```
>> x = 2                          % define variable x = 2
>> y = 6                          % define variable y = 6
```

```
>> z = 3*x + 4.5*y;          % define variable z = 3x + 4.5y (no print to screen)
>> a = z/6 - 2.1*x^3;        % define variable a = z/6 − 2.1x³ (no print to screen)
```

After executing these commands, the four variables x, y, z, and a will be assigned numerical values in MATLAB's workspace. These four variables can be used at any time during the MATLAB session. The following commands display and clear the workspace:

```
>> whos          % list all variables in the workspace and their size
>> clear         % remove all variables from the workspace memory
```

The user can always obtain a brief on-screen tutorial on MATLAB's built-in functions by using the help 'function name' command. For example, typing help clear explains the built-in clear function.

Vectors and Matrices

All variables are stored in MATLAB as arrays or matrices (vectors are matrices with a single row or single column). In the previous example, we defined four scalar variables x, y, z, and a and upon typing the whos command we see that all four are stored in the workspace as 1x1 arrays. Vectors are used extensively in MATLAB and Simulink to store simulation data in a single row or column. We can create vectors by placing the elements between square brackets as follows

```
>> x = [ 1 -2 5 8 ]          % define 1 × 4 row vector x = [1 −2 5 8]
>> y = [ 1 ; -2 ; 5 ; 8 ]    % define 4 × 1 column vector y = [1 −2 5 8]ᵀ
```

Note that we have used the superscript T to denote transpose. We can take the transpose of a vector or matrix using the superscript symbol ' (single quote mark). For example,

```
>> z = x'          % define vector z as the transpose of vector x
```

This command displays vector z as a 4×1 column vector that is equivalent to the previously defined column vector y.

Matrices can be created by entering the elements of each row and separating the rows by a semicolon. For example, if we have the 2×3 matrix

$$\mathbf{A} = \begin{bmatrix} -1 & 0 & 2.4 \\ 3 & -0.6 & 7 \end{bmatrix}$$

it can be defined using the MATLAB command

```
>> A = [ -1 0 2.4 ; 3 -0.6 7 ]
```

which prints to the screen

```
A =
   -1.0000          0     2.4000
    3.0000    -0.6000     7.0000
```

The command whos or size(A) displays the size (dimension) of matrix A as 2x3.

A vector of sequential data can be generated by defining the starting value, an incremental step size, and the final value:

```
>> x = 0:0.25:1          % define 1 × 5 row vector x = [0 0.25 0.5 0.75 1]
```

We can also use the command linspace to generate N linearly spaced data points from a desired starting value to a desired final value:

```
>> x = linspace(0,1,20)          % row vector of 20 elements between 0 and 1
```

The command `logspace` will generate N logarithmically spaced data points between the desired decadal starting and ending values:

```
>> x = logspace(-1,3,50)          % row vector of 50 elements between 10⁻¹ and 10³
```
`% row vector of 50 elements between $10^{-1}$ and $10^3$`

MATLAB has built-in functions to construct the identity matrix, and arrays where all elements are either unity or zero:

```
>> I = eye(3)                     % define 3 × 3 identity matrix I
>> A = ones(3,2)                  % define 3 × 2 matrix where all elements are 1
>> B = zeros(4,6)                 % define 4 × 6 matrix where all elements are 0
```

Complex Variables

Complex variables can be entered by simply typing their real and imaginary parts. MATLAB assigns both i and j as the imaginary number $\sqrt{-1}$. Complex variables can defined, added, and multiplied as shown here:

```
>> x = 2 - j*5                    % define complex variable x = 2 − j5
>> y = -4 + j*8                   % define complex variable y = −4 + j8
>> z = x + y                      % add complex variables x and y
>> w = x*y                        % multiply complex variables x and y
```

The MATLAB commands `abs` and `angle` can be used to compute the absolute value (magnitude) and phase angle of a complex variable:

```
>> x = 2 - j*5                    % define complex variable x = 2 − j5
>> Mag_x = abs(x)                 % magnitude of x
>> Phase_x = angle(x)             % phase angle of x (in rad)
```

Table B.1 summarizes the basic MATLAB built-in functions described in this section as well as other common functions, such as trigonometric and logarithmic functions.

Table B.1   Basic MATLAB Built-In Functions

MATLAB Command	Description
abs(x)	Absolute value of x (magnitude if x is complex)
angle(x)	Phase angle (in rad) of complex variable x
cos(x)	Cosine of x (input angle in rad)
sin(x)	Sine of x (input angle in rad)
tan(x)	Tangent of x (input angle in rad)
acos(x)	Inverse cosine of x (answer in rad)
asin(x)	Inverse sine of x (answer in rad)
atan(x)	Inverse tangent of x (answer in rad)
exp(x)	Exponential of x (base e)
log(x)	Natural logarithm of x
log10(x)	Base 10 logarithm of x
max(x)	Maximum element of vector x
min(x)	Minimum element of vector x
size(x)	Size (dimension) of vector or matrix x
sqrt(x)	Square root of x
sum(x)	Sum of the elements of vector x

B.3   PLOTTING WITH MATLAB

A key strength of MATLAB is its extensive routines for graphing and visualizing numerical data. We only emphasize two-dimensional graphs (plots) here. The most basic command is `plot`, which plots the dependent-variable vector **y** versus independent-variable vector **x**. Axes labels, graph title, and grid lines are easily added using the following single-line commands:

Example B.1

Plot $y(t) = 2\sin 3t$ for $0 \le t \le 6$ s.

```
>> t = linspace(0,6,500);        % define time vector t from 0 to 6 with 500 points
>> y = 2*sin(3*t);               % define vector y(t) = 2 sin 3t (size is 1x500)
>> plot(t,y)                     % plot vector y versus vector t
>> title('Graph of y(t)')        % add graph title (top of graph)
>> xlabel('Time, s')             % add x-axis label
>> ylabel('y(t)')                % add y-axis label
>> grid                          % add grid lines to x and y axes
```

The `plot` command has several options for creating graphs with various line types, symbols, and colors. The user should consult the on-screen tutorial by typing `help plot` to view the details of the various plotting options. As an example, the following commands plot $y(t) = 2\sin 3t$ as a dashed red line with a circle (o) at each data point:

```
>> t = linspace(0,4,50);         % define time vector t from 0 to 4 with 50 points
>> y = 2*sin(3*t);               % define vector y(t) = 2 sin 3t (size is 1x50)
>> plot(t,y,'ro--')              % plot y(t) as red dashed line with circle symbol
```

Several curves can be plotted on the same graph by entering multiple x−y vector pairs.

Example B.2

Plot $y_1(t) = 2\sin 3t$ and $y_2(t) = -4\cos 8t$ for $0 \le t \le 4$ s.

```
>> t = linspace(0,4,200);        % define time vector t from 0 to 4 with 200 points
>> y1 = 2*sin(3*t);              % define vector y1(t) = 2 sin 3t (size is 1x200)
>> y2 = -4*cos(8*t);             % define vector y2(t) = −4 cos 8t (size is 1x200)
>> plot(t,y1,t,y2)               % plot vectors y1 versus t and y2 versus t
```

In this simple example, the two x−y pairs `(t,y1)` and `(t,y2)` are both `1x200` row vectors. The only dimensional requirement for the `plot` command is that the x−y vector pairs must have the same dimensions.

Finally, the user may use logarithmic-scaled axes using the following commands:

```
>> semilogx(x,y)                 % plot vectors y versus x with a logarithmic scale for the x-axis
>> semilogy(x,y)                 % plot vectors y versus x with a logarithmic scale for the y-axis
>> loglog(x,y)                   % plot vectors y versus x with logarithmic scales for both axes
```

These plotting commands have the same syntax rules as the `plot` command.

B.4 CONSTRUCTING BASIC M-FILES

We have seen through the previous simple examples that MATLAB is a command-driven computing tool. In many cases, we can assemble the desired single-line commands to create an algorithm or program called an M-file. When we run an M-file, these commands are executed sequentially as if we separately typed each command in the MATLAB environment.

We can create a new M-file (a blank template) by clicking on the "blank page" icon beneath the File menu in the upper left corner of the MATLAB window. When the new template is opened, the user can begin typing MATLAB commands on each line (as well as comment lines preceded by the % character). MATLAB M-file B.1 is a program that computes the function $y(x) = 2.5x^2 - \tan 1.2x$ for the independent variable $-1 \leq x \leq 1$ and plots $y(x)$. This M-file (called Ex1.m) illustrates the use of a for loop to iterate through multiple calculations. Note that we could have condensed the algorithm into a few simple command lines but instead we choose to show how to program an algorithm as an M-file. The algorithm (or script) Ex1.m begins with comment lines that document the M-file name and purpose. Next, the starting and final values of independent variable x are defined as well as the incremental step dx for 500 data points (= Npts). The for loop repeats all of the command lines between the for and end statements 500 times (the index integer begins at i=1 and ends at i=500). After the loop is repeated, the results are plotted using the plot command.

In this textbook, a user might construct a specialized M-file to perform the following tasks: (1) set the numerical values for the system parameters (such as mass, stiffness, friction, capacitance, etc.), (2) define the system objects (such as transfer functions or state-space representations), (3) obtain the system response by executing a numerical simulation tool, or (4) plot the system response(s). Simulink models could be executed to obtain the system response by using the sim command in an M-file:

```
sim model_name
```

where the Simulink model is model_name.mdl. Constructing Simulink models is explained in Appendix C. If the mathematical model is a linear, time-invariant (LTI) system, then the user might use built-in MATLAB commands such as step, impulse, or lsim to obtain the system response. These built-in commands for linear system analysis are described in the next section.

MATLAB M-file B.1

```
%
%      Example M-file Ex1.m
%
%      This M-file computes the function y(x) = 2.5*x^2 - tan(1.2x)
%      for the range -1 <= x <= 1. The function y(x) is plotted vs. x

% Define start x, final x, and step size dx
x_start = -1;
x_end = 1;
Npts = 500;
dx = (x_end - x_start)/(Npts - 1);        % compute increment for x
x_now = x_start;                          % starting value for x
% FOR loop for x
for i=1:Npts
      y(i) = 2.5*x_now^2 - tan(1.2*x_now);
      x(i) = x_now;                       % store current x
      x_now = x_now + dx;                 % increment x_now by dx
end
% Plot y(x)
plot(x,y)
title('y vs. x')
xlabel('x')
ylabel('y(x)')
grid
```

B.5   COMMANDS FOR LINEAR SYSTEM ANALYSIS

We explicitly illustrate a few of the MATLAB commands used for analyzing LTI systems. All of the following brief examples use the same LTI transfer function with output $y(t)$ and input $u(t)$

$$G(s) = \frac{3}{s^2 + 2s + 16} = \frac{Y(s)}{U(s)}$$

Defining transfer function object `sysG`

```
>> numG = 3;              % numerator of G(s) = 3/(s² + 2s + 16)
>> denG = [1 2 16];       % denominator of G(s) = 3/(s² + 2s + 16)
>> sysG = tf(numG,denG)   % define LTI transfer function G(s)
```

Note that `denG` is a row vector defining the denominator polynomial coefficients in descending powers of s. Upon hitting the return key MATLAB displays `sysG` as

```
Transfer function:

       3
---------------
s^2 + 2 s + 16
```

Therefore, the user can verify that he or she has properly defined the desired transfer function.

Roots of polynomial `denG`

```
>> roots(denG)            % roots of s² + 2s + 16 = 0
```

Poles of transfer function `sysG`

```
>> pole(sysG)             % poles of G(s); that is, roots of s² + 2s + 16 = 0
```

Undamped natural frequency and damping ratio

```
>> [Wn,zeta] = damp(sysG) % ωₙ and ζ for LTI system G(s)
```

Plot of unit-step response of `sysG`

```
>> [y,t] = step(sysG);    % unit-step response of G(s)
>> plot(t,y)              % plot the unit-step response y(t)
```

Plot of `sysG` response to an arbitrary input

```
>> t = 0:0.01:12;         % define time vector from 0 to 12 in steps Δt = 0.01s
>> u = zeros(size(t));    % define input vector u as all zeros (same size as t)
>> u(1:601) = 2;          % define u = 2 for time 0 ≤ t ≤ 6 s; u(t) is a pulse
>> [y,t] = lsim(sysG,u,t);% response of G(s) to pulse input u(t)
>> plot(t,y)              % plot the response y(t)
```

These commands, along with related commands for linear system analysis, are collected and summarized in Table B.2. Note that the built-in commands for system response in the time domain (`step`, `impulse`, `lsim`, and `initial`) and frequency-response analysis (`bode` and `bandwidth`) require an LTI system object `sys` that is created either as a transfer function (using `tf`) or state-space representation (using `ss`).

Table B.2 MATLAB Commands for Linear System Analysis

MATLAB Command	Description
`roots(denG)`	Computes roots of polynomial with coefficients defined by vector `denG`. For example, `denG` could be the denominator of transfer function $G(s)$
`sysG = tf(numG,denG)`	Creates transfer function object `sysG`. Inputs `numG` and `denG` are vectors of the numerator and denominator polynomial coefficients
`dcgain(sysG)`	Computes the DC gain of transfer function `sysG`
`pole(sysG)`	Computes the poles of transfer function `sysG`
`[Wn,zeta] = damp(sysG)`	Computes the undamped natural frequency (`Wn`, rad/s) and damping ratio `zeta` for LTI system `sysG`
`sys = ss(A,B,C,D)`	Creates the state-space representation object `sys`. Inputs A, B, C, and D are the SSR matrices
`eig(A)`	Computes the eigenvalues of square matrix A
`step(sys)`	Plots the unit-step response of the LTI system `sys` (created by either `tf` or `ss`)
`[y,t] = step(sys)`	Computes the unit-step output response `y` and simulation time `t` of system `sys` (no plot is created)
`impulse(sys)`	Plots the unit-impulse response of the LTI system `sys` (created by either `tf` or `ss`)
`[y,t] = impulse(sys)`	Computes the unit-impulse output response `y` and simulation time `t` of system `sys` (no plot is created)
`lsim(sys,u,t)`	Plots the response of the LTI system `sys` for an arbitrary input `u` corresponding to time vector `t`
`[y,t] = lsim(sys,u,t)`	Computes the output response `y` and simulation time `t` of system `sys` to arbitrary input `u` (no plot is created)
`initial(sys,x0)`	Plots the free response of the state-space representation model `sys` given the initial state vector `x0`
`[y,t,x] = initial(sys,x0,t)`	Computes the free response output `y` and states `x` of the state-space representation model `sys` given the initial state vector `x0` and time vector `t` (no plot is created)
`bode(sys)`	Creates and draws the Bode diagram for LTI system `sys` (created by either `tf` or `ss`)
`[Mag,phase] = bode(sys,w)`	Computes the magnitude and phase (in deg) of the LTI system `sys` for input frequency `w` (rad/s). No Bode diagram is drawn
`wB = bandwidth(sys)`	Computes the bandwidth frequency `wB` (rad/s) for LTI system `sys`

B.6 COMMANDS FOR LAPLACE TRANSFORM ANALYSIS

MATLAB has three very useful built-in commands for performing the Laplace transform and the inverse Laplace transform. We present examples of all three commands.

Partial-fraction expansion

The following MATLAB commands compute the residues for the partial-fraction expansion of the Laplace transform

$$F(s) = \frac{-2s + 8}{s^2 + 4s + 3}$$

```
>> numF = [ -2 8 ];              % F(s) numerator coefficients
>> denF = [ 1 4 3 ];             % F(s) denominator coefficients
>> [r,p,k] = residue(numF,denF)  % compute residues, poles, and direct term
```

Upon hitting the return key, the result is

```
r =
    -7
     5                          %  vector of residues
p =
    -3
    -1                          %  vector of poles of F(s)
k =
    []                          %  no direct term because order of numF < order of denF
```

Using the residues, the partial-fraction expansion of $F(s)$ is

$$F(s) = \frac{-7}{s+3} + \frac{5}{s+1}$$

Laplace transform using Symbolic Math Toolbox

MATLAB's Symbolic Math Toolbox can be used to compute the Laplace transform of a given time function $f(t)$. The user must define the time function $f(t)$ as the symbolic object f. The following example illustrates the built-in command `laplace`:

```
>> syms t              % define variable t (time) as a symbolic object
>> f = 6*exp(-2*t);    % define function f(t) = 6e⁻²ᵗ as a symbolic object
>> F = laplace(f);     % compute the Laplace transform F(s)
>> pretty(F)           % display  F  in a format similar to typeset mathematics
```

Upon hitting the return key, the result is

```
      6
     ----
     s + 2
```

Inverse Laplace transform using Symbolic Math Toolbox

The following example illustrates the built-in command `ilaplace` for computing the inverse Laplace transform $f(t) = \mathscr{L}^{-1}\{F(s)\}$:

```
>> syms s              % define Laplace variable s as a symbolic object
>> F = 6/(s+2);        % define Laplace transform F(s) as a symbolic object
>> f = ilaplace(F);    % compute the inverse Laplace transform f(t)
>> pretty(f)           % display  f  in a format similar to typeset mathematics
```

Upon hitting the return key, the result is

```
    6 exp(-2 t)
```

which is the inverse result of the previous Laplace transform example. The three commands for Laplace transform analysis are collected and summarized in Table B.3.

B.7   COMMANDS FOR CONTROL SYSTEM ANALYSIS

MATLAB has three very powerful built-in commands for analyzing closed-loop control systems. The first command is `feedback`, which computes the closed-loop transfer function object sysT given the forward transfer function $G(s)$ (as

Table B.3 MATLAB Commands for Laplace Transform Analysis

MATLAB Command	Description
`[r,p,k] = residue(numF,denF)`	Computes the residues r, associated poles p, and direct term k (if any) for the partial-fraction expansion of $F(s) = $ numF/denF
`F = laplace(f)`	Computes the Laplace transform of time function f (expressed as a symbolic object using `sym`)
`f = ilaplace(F)`	Computes the inverse Laplace transform of F (expressed as a symbolic object using `sym`)

`sysG`) and feedback transfer function $H(s)$ (as `sysH`). That is, `feedback` computes the closed-loop transfer function $T(s)$ using

$$T(s) = \frac{G(s)}{1 + G(s)H(s)} = \frac{Y(s)}{R(s)}$$

where $r(t)$ is the overall reference input and $y(t)$ is the plant output. Negative feedback is assumed at the summing junction. The reader should note that once the closed-loop transfer function is obtained, the LTI commands in Table B.2 (`step`, `impulse`, etc.) can be used to obtain the closed-loop system response.

The second powerful MATLAB command is `rlocus`, which plots the root locus to the screen. The user must provide the open-loop transfer function $G(s)H(s)$ as the input (as either `sysGH` or `sysG*sysH`). The closed-loop poles can be computed for a specific forward-path gain K using `rlocus`, or by using the command `rlocfind`. The interactive command `rlocfind` allows the user to place a cross-hair target on any desired root-locus branch, click, and obtain the corresponding gain K and n closed-loop poles.

The third control-system MATLAB command is `margin`, which computes the relative stability margins (gain and phase margins) and the corresponding crossover frequencies. As with `rlocus`, the user must provide the open-loop transfer function $G(s)H(s)$ as the input. The command `margin` by itself plots the Bode diagram and presents the numerical values of the gain margin and phase margin (degrees) in the graph title. Gain and phase crossover frequencies are also shown on the Bode diagram and their numerical values are noted. Gain margin Gm is computed as an absolute magnitude and therefore the gain margin in decibels (dB) must be computed using `Gm_dB=20*log10(Gm)`. Table B.4 summarizes the important MATLAB commands for analyzing closed-loop control systems.

Table B.4 MATLAB Commands for Control System Analysis

MATLAB Command	Description
`sysT = feedback(sysG,sysH)`	Creates the closed-loop transfer function `sysT` given forward and feedback transfer functions `sysG` and `sysH` (assumes negative feedback)
`rlocus(sysGH)`	Creates and draws the root-locus plot for the feedback system with open-loop transfer function `sysGH`
`CLpoles = rlocus(sysGH,K)`	Computes the closed-loop poles `CLpoles` for the feedback system with open-loop transfer function `sysGH` and forward-path gain K (no root-locus plot is drawn)
`[K,CLpoles] = rlocfind(sysGH)`	Computes gain K and closed-loop poles `CLpoles` that result from user's selection of root-locus branch with cross-hair target
`margin(sysGH)`	Plots the Bode diagram of open-loop system `sysGH` and displays the gain and phase margins
`[Gm,Pm,Wgm,Wpm] = margin(sysGH)`	Computes the gain margin Gm and phase margin Pm (in deg) and their crossover frequencies (Wgm, Wpm) given the open-loop system `sysGH`

Appendix C

Simulink Primer

C.1 INTRODUCTION

Simulink is a numerical simulation tool that is part of the MATLAB software package developed by MathWorks. It uses a graphical user interface (GUI) to develop a block diagram representation of dynamic systems. The reader should keep in mind that Simulink is used to obtain the solution to dynamic systems comprised of linear and/or nonlinear ordinary differential equations (ODEs). Simulink performs this task by numerically integrating the ODEs that the user has defined by constructing a graphical block diagram representation of the system dynamics. The user develops the desired block diagram by selecting and connecting various input–output (I/O) blocks such as transfer functions, integrators, and gains. Simulink provides the user with a GUI that contains a wide array of I/O blocks.

We can summarize the basic steps for using Simulink to obtain the system response:

1. Select the appropriate I/O blocks needed to represent the system dynamics (e.g., I/O blocks for transfer functions, state-space representations (SSRs), and integrators).

2. Select the appropriate blocks to represent the input function(s) (e.g., step, pulse, sinusoid).

3. Store the desired system output variables for plots or later analysis.

4. Select the numerical simulation parameters (e.g., run time, numerical integration method, integration step size).

5. Execute the Simulink model to obtain the system response.

Chapter 6 provides examples that demonstrate these steps for constructing a Simulink model. This appendix also explains how to use Simulink by presenting examples. We start with simple Simulink models for linear systems and progress toward more complex models. This tutorial provides some of the Simulink details that are not explicitly presented in Chapter 6. For example, we present "screen shots" of the various Simulink libraries and menus that are used to construct models. However, for the sake of completeness there is some overlap in the Simulink discussion presented in Chapter 6 and this appendix. Finally, we demonstrate Simulink in this appendix by working with "generic" dynamic systems (ODEs) that are not attached to a particular physical engineering system. We do this in order to emphasize the numerical solution of the ODEs. The development of mathematical models of physical systems and the subsequent use of Simulink to obtain the dynamic response is emphasized throughout the chapters of this textbook.

C.2 BUILDING SIMULINK MODELS OF LINEAR SYSTEMS

To start the Simulink program, open the MATLAB environment and enter the following command

```
>> simulink
```

This command will open the Simulink library browser; Fig. C.1 shows a "screen shot" of the browser. In order to create a new Simulink model, the user must left-mouse click on the "new model" icon as highlighted in the upper left corner of Fig. C.1. Doing so will create a blank screen or working template for constructing the desired block diagram.

Creates a new model

Figure C.1 Simulink library browser.

Example C.1

Let us demonstrate the construction of an extremely simple Simulink model by beginning with a system that consists of a single linear I/O equation:

$$2\ddot{y} + 6\dot{y} + 30y = 4\dot{u}(t) + u(t) \qquad (C.1)$$

with initial conditions $y(0) = 0$ and $\dot{y}(0) = 0$. The input is a step function, $u(t) = 0.2U(t)$. Because this system has zero initial conditions, we may use a transfer function to represent the system dynamics:

$$\frac{Y(s)}{U(s)} = \frac{4s + 1}{2s^2 + 6s + 30} \qquad (C.2)$$

The reader should recall that we can use transfer functions *only* in the cases where the dynamic system has zero initial conditions. Our first step is to choose an I/O block that represents the system dynamics. Highlighting the `Continuous` library in the `Simulink Library Browser` will present the standard model options for dynamic systems: transfer function, SSR, and the integrator block. Left-click and hold the `Transfer Fcn` block and "drag and drop" this icon to the blank Simulink template. Note that the `Transfer Fcn` block has a single input port and a single output port. Double clicking the `Transfer Fcn` block opens a dialog box which allows the user to set the desired numerator and denominator coefficients of the transfer function. Figure C.2 shows the dialog box with the numerator and denominator coefficients set as vectors in descending powers of *s* in order to match the transfer function in Eq. (C.2). Clicking OK will close the dialog box and display the numerical coefficients in the transfer function block.

Next, the user must provide the desired input function. Clicking on the `Sources` library (under the `Simulink Library Browser`) will display various options for input functions. Drop and drag the `Step` and `Clock` icons to the working template (the `Clock` icon will provide the simulation time *t*). The `Step` and `Clock` blocks only have output ports (they are sources). Connect the `Step` input to the `Transfer Fcn` block by left-clicking on the output port of the `Step` and dragging a signal path to the input port of the `Transfer Fcn`. Releasing the mouse button will create a signal path that connects the two blocks. Signal paths that connect the various blocks contain the time-history information of the appropriate variables and this information can be accessed or stored. Double clicking the `Step` block opens a dialog box, where the user can set the step time (zero in this case), the initial value (zero), and the final value (0.2 in this case).

The last step in constructing the Simulink model involves storing the output variables. Clicking on the `Sinks` library (under the `Simulink Library Browser`) shows various options for storing and displaying the output. Drop and drag the `To Workspace` block to the working template and connect the output of the transfer function (y) to this block. We can define the stored variable by double clicking the `To Workspace` block and changing the `Variable name` dialog box to y. It is important to save the format of the stored variable as an array so that it may be plotted from the MATLAB command line. To change the format, select the `Array` option in the `Save format` dialog box of the `To Workspace` block. Finally, we need to store simulation time t for plotting purposes. Connect the output of the `Clock` block to another `To Workspace` block (note that the user always has the option to select, copy, and paste an existing block, such as `To Workspace`, which will retain the features of the original block). Figure C.3 shows the completed Simulink model, which the user can save as `Example1.mdl` (or any desired file name). Labels for blocks and signal paths may be added for clarity (double click the block name or location in the signal path and edit the text).

Before executing the simulation, the user should set the run time and the parameters that define the numerical integration process. To do this, select the `Model Configuration Parameters` option under the `Simulation` menu in the model workspace window. Figure C.4 shows the `Configuration Parameters` dialog box for setting start/stop times and solver options. Typically, the start time is zero (default value); the stop time has been set to 4 s. While the variable-order,

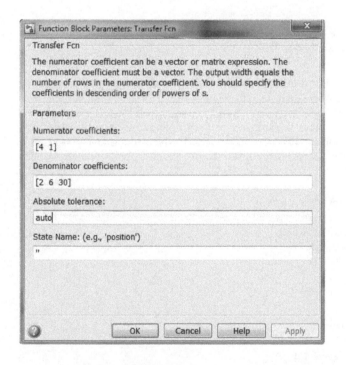

Figure C.2 Setting the transfer function parameters (Example C.1).

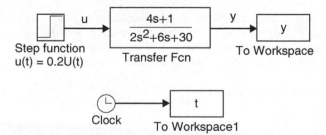

Figure C.3 Simulink model using a transfer function (Example C.1).

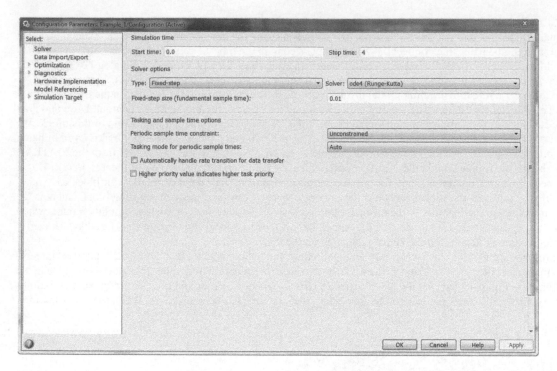

Figure C.4 Setting the numerical integration parameters (Example C.1).

variable-step solver `ode45` is a reliable and stable option for performing the numerical integration, it may take large step sizes in order to maximize computational efficiency. Large step sizes will lead to "choppy" plots of the output variable $y(t)$. One solution is to decrease the maximum step size in the `Configuration Parameters` dialog box. Another option is to select the fixed-step, fourth-order Runge–Kutta method (`ode4`) as the numerical solver. Figure C.4 shows the `Configuration Parameters` dialog box where the numerical solver `ode4` is selected and the fixed-step size is set to 0.01 s. The final step is to execute the simulation by clicking the `Run simulation` (or "play") button. A plot of output $y(t)$ can be created using MATLAB's `plot` command:

```
>> plot(t,y)                    % plot output versus time
>> grid                         % add grid lines to plot
>> xlabel('Time, s')            % add x-axis label
>> ylabel('Output, y')          % add y-axis label
```

Example C.2

Consider a second-order system governed by the linear I/O equation

$$2\ddot{z} + 8\dot{z} + 40z = 1.4u(t) \tag{C.3}$$

The system's initial conditions are $z(0) = -0.02$ and $\dot{z}(0) = 0.1$, and the input is a sinusoidal function, $u(t) = 0.8 \sin 12t$. Because this system has nonzero initial conditions, we cannot use a transfer function for the system dynamics. We have two options for representing the system dynamics when the initial conditions are nonzero: (1) SSR and (2) the integrator-block method. We solve this problem using the SSR approach and show the integrator-block method in the next example.

The reader can verify that a possible SSR of the system defined by Eq. (C.3) is

$$\dot{\mathbf{x}} = \begin{bmatrix} 0 & 1 \\ -20 & -4 \end{bmatrix} \mathbf{x} + \begin{bmatrix} 0 \\ 0.7 \end{bmatrix} u \tag{C.4}$$

$$\mathbf{y} = \begin{bmatrix} 1 & 0 \\ 0 & 1 \end{bmatrix} \mathbf{x} + \begin{bmatrix} 0 \\ 0 \end{bmatrix} u \tag{C.5}$$

where the state variables are $x_1 = z$ and $x_2 = \dot{z}$, and the *two* output variables are $y_1 = x_1 = z$ and $y_2 = x_2 = \dot{z}$. Note that we have chosen to set the output vector **y** equal to the state vector **x**, and hence the output matrix **C** is the identity matrix. The Simulink model is constructed by dragging the State-Space block from the Continuous library to a new working template. The reader should note that while the State-Space block has one input port and one output port these ports can pass scalar or vector signals. In this case, we have a single input u and a vector output **y**. Double clicking the State-Space block opens a dialog box where the user can enter the appropriate numerical values for state-space matrices **A**, **B**, **C**, and **D** (see Fig. C.5). Note that we have used the MATLAB commands eye(2) for the 2×2 identity matrix **C** and zeros(2,1) for the 2×1 null matrix **D**. The reader should carefully note that even though the direct-link term is zero in the output equation (C.5), we must enter the **D** matrix as a 2×1 column vector of zeros. Failure to heed the proper SSR matrix dimensions is a common error of beginning Simulink users. The initial conditions for the state vector, $\mathbf{x}(0) = \begin{bmatrix} z(0) & \dot{z}(0) \end{bmatrix}^T$, are also entered in this dialog box shown in Fig. C.5.

Next, the user must define the sinusoidal input function and send the outputs to storage (MATLAB workspace). The user can drag and drop the Sine Wave block from the Sources library to the Simulink template and connect it to the input port of the State-Space block. Double clicking on the Sine Wave block opens a dialog box where the user can set the amplitude equal to 0.8 and the input frequency to 12 rad/s. The output of the State-Space block is the vector $\mathbf{y}(t) = \begin{bmatrix} y_1 & y_2 \end{bmatrix}^T$. If this vector output signal is sent directly to the workspace as a *single* variable y, then plotting y versus time t will produce the two curves $y_1(t)$ and $y_2(t)$ on the same figure. An alternative method is to "split" the output vector **y** into two scalar signals y_1 and y_2 using the "de-multiplexer" or Demux block from the Signal Routing library. The user can open the dialog box for the Demux and change the number of output ports if needed (the default value of 2 outputs works for this example). Figure C.6 shows the completed Simulink model with Demux "splitting" the output vector **y** and sending y_1 and y_2 to two separate To Workspace blocks for storage. Note that we have shown the vector signal path (output **y**) as a wide, bold line by selecting this option in Signals & Ports under the Display drop-down menu.

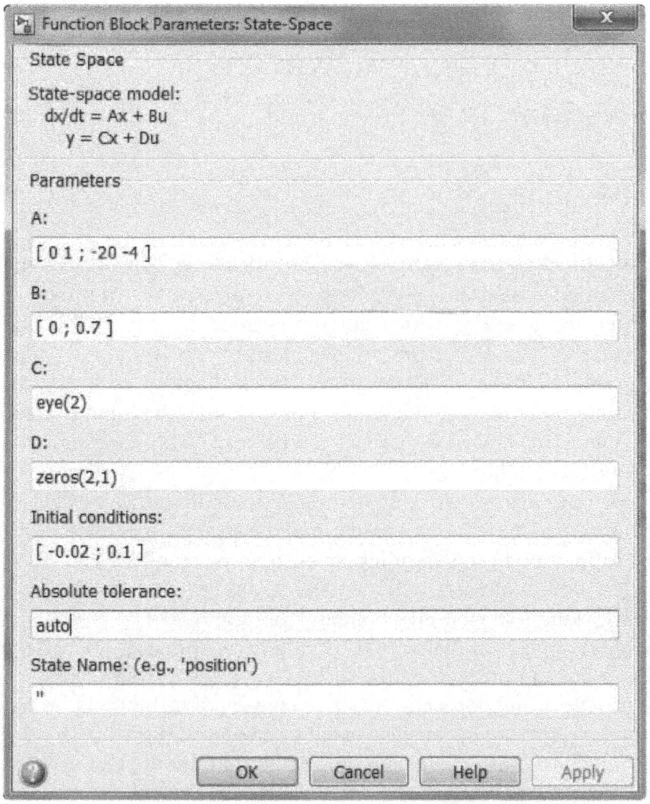

Figure C.5 Setting state-space matrices and initial states (Example C.2).

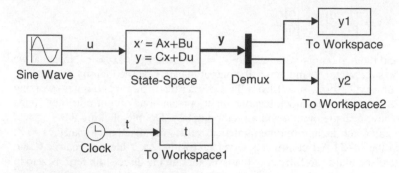

Figure C.6 Simulink model using an SSR (Example C.2).

Example C.3

Consider a dynamic system governed by two coupled, linear ODEs

$$0.5\dot{y} + 3y - 0.4z = 0 \tag{C.6}$$

$$2.4\dot{z} + 16z = u(t) \tag{C.7}$$

The system's initial conditions are $y(0) = -0.3$ and $z(0) = 0.6$, and the input is a step function $u(t) = 3.5U(t-1)$. Therefore, the input $u(t)$ is zero for time $t \leq 1$ s and has a constant magnitude of 3.5 for time $t > 1$ s. Because the system is linear, we can use an SSR; however, we demonstrate the integrator-block method here. Clearly, the system defined by Eqs. (C.6) and (C.7) is linear and second order because we have two first-order ODEs. Hence, we need two numerical integrations to solve the system, or two integrator blocks: one integrator for Eq. (C.6) and one integrator for Eq. (C.7). Let us begin by rewriting both ODEs with the respective derivative terms on the left-hand side with unity coefficients:

$$\dot{y} = \frac{1}{0.5}\left(-3y + 0.4z\right) \tag{C.8}$$

$$\dot{z} = \frac{1}{2.4}\left(-16z + u(t)\right) \tag{C.9}$$

The core of the Simulink model consists of two integrator blocks, where the right-hand sides of Eqs. (C.8) and (C.9) are the respective inputs. Figure C.7 shows a schematic diagram presenting the numerical solutions of Eqs. (C.8) and (C.9). The inputs to the respective integrators are $\dot{y}$ and $\dot{z}$, which are represented by Eqs. (C.8) and (C.9). The outputs of each integrator are $y(t)$ and $z(t)$, respectively. Note that the initial conditions $y(0)$ and $z(0)$ (i.e., the integration constants) are applied to the appropriate integrators. The key is to develop a block diagram such that the signal paths are summed in a manner to synthesize the appropriate input to each integrator. For example, the input to the $\dot{y}$ integrator must be a linear sum of $y(t)$ and $z(t)$. Consequently, these two output signals are fed back to a summing junction to create $\dot{y}$. The input to the $\dot{z}$ integrator is a linear sum of $z(t)$ and system input $u(t)$.

Let us begin the Simulink model construction by opening a new working template and dragging and dropping two `Integrator` blocks from the `Continuous` library. The reader should note that the integrator-block icon is $1/s$, which is the Laplace transform representation of integration (Simulink does not use Laplace-transform methods to obtain the response). First, we must insert a summing junction `Sum` block (from the `Math Operations` library) to the left of the $\dot{y}$ integrator. This summing junction is used to create the signal $0.4z - 3y$, which is needed for the $\dot{y}$ integrator. Double clicking the `Sum` block opens a dialog box that allows the user to set the number of input ports and their associated signs for summation. For example, setting the string '+ + −' in the `List of signs` dialog entry will create a summing junction where the top two input ports have positive signs and the bottom port has a negative sign. In this case, we enter '+ −' for the first summing junction because we wish to create the signal $0.4z - 3y$. Next, we feedback signal paths for y and z and send these new feedback paths to the summing junction. Figure C.8 shows a partially completed Simulink model where the block diagram to synthesize Eq. (C.8) has been constructed. A feedback signal for y is created by "picking off" a signal from the integrator block output signal path. This signal "pick-off" is performed by right-clicking the original path, and holding and dragging the new path backward to the left. Picking off a new signal results

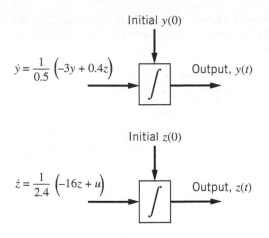

Figure C.7 Numerical integration of the two ODEs (Example C.3).

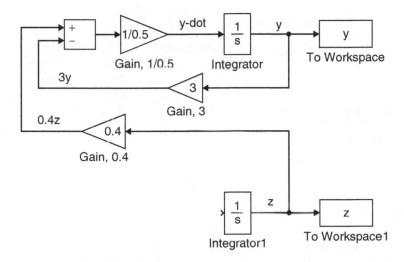

Figure C.8 Partially completed Simulink model for Eq. (C.8) (Example C.3).

in the junction point (dot) after the integrator block, as shown in Fig. C.8 (the reader should note that junction points share the same signal-path data, whereas crossing signal paths do not). A feedback signal path for z is performed in the same manner and a junction point is shown after the $\dot{z}$ integrator in Fig. C.8. The two feedback signals must be multiplied by constants (or "gained") before they are summed in order to create the signal $0.4z - 3y$. Therefore, the user must drag and drop two Gain blocks (from the Math Operations library) to the Simulink model (note that the Gain block has a triangle shape which is analogous to an op amp for boosting electrical signals). The Gain blocks show one input and output port and the default orientation is left-to-right. Because our feedback signals are from right-to-left, we need to "flip" the block. To reverse the direction of any I/O block, highlight the block and use the Flip Block command under the Diagram > Rotate & Flip menu of the Simulink model. Now the complete feedback paths for the $\dot{y}$ integrator can be created as shown in Fig. C.8 (of course it is easier to create and place the Gain and Sum blocks first and then pick off and route the feedback signals to the appropriate I/O blocks). Finally, a left-to-right Gain block is inserted to multiply the Sum output $0.4z - 3y$ by the factor $1/0.5$, as required by Eq. (C.8). The reader should be able to trace the signal paths of the Simulink model in Fig. C.8 and verify that they create the governing ODE for $\dot{y}$, that is, Eq. (C.8).

Next, we complete the Simulink model construction by inserting a second summing junction Sum block to create the signal $u(t) - 16z$ that is required for the $\dot{z}$ integrator. Figure C.9 shows the completed Simulink model. The Step block (from the Sources library) is added, and opening its dialog box allows the user to set the initial value to zero, step time to 1, and final value equal to 3.5 in order to create the desired input function $u(t)$. The user must remember to set the proper initial conditions by double clicking each Integrator block and entering the numerical values for $y(0)$ and $z(0)$. Finally, the desired numerical integration parameters can be set in the Simulation > Model Configuration Parameters dialog box.

Figure C.10 shows a plot of the simulation results for $y(t)$ and $z(t)$ after executing the Simulink model in Fig. C.9. The two dynamic variables decay to zero from their respective initial conditions. At time $t = 1$ s, the step function is applied and both dynamic variables show an exponential rise to their steady-state values.

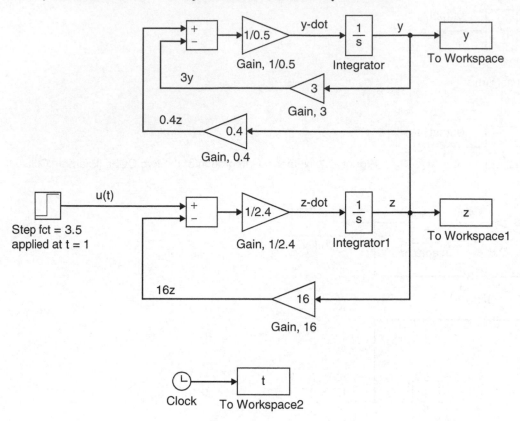

Figure C.9 Simulink model for Eqs. (C.8) and (C.9) (Example C.3).

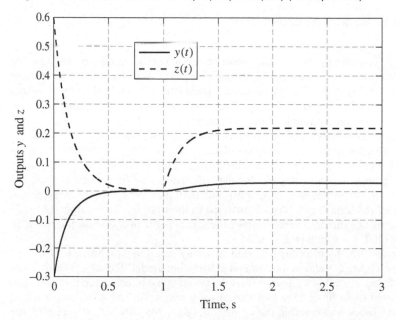

Figure C.10 System responses $y(t)$ and $z(t)$ (Example C.3).

C.3 BUILDING SIMULINK MODELS OF NONLINEAR SYSTEMS

Next we demonstrate how to construct a Simulink model of a nonlinear system. We also increase the system order so that the Simulink models become more challenging.

Example C.4

Consider a dynamic system governed by two coupled, nonlinear ODEs

$$0.5\dot{y} + 0.6\text{sgn}(y) - 1.4z = 0 \tag{C.10}$$

$$4\ddot{z} + 0.5\dot{z}^2 + 10z = u(t) \tag{C.11}$$

The system's initial conditions are $y(0) = 0$, $z(0) = 1$, and $\dot{z}(0) = 0$, and the input is a pulse function: $u(t) = 20$ for $0 < t < 1$ s and $u(t) = 0$ for $t \geq 1$ s. Equations (C.10) and (C.11) are clearly nonlinear as we have a signum (sgn or sign) term and a squared term (of course, only one nonlinear term needs to be present in order to make the entire system nonlinear). The system is third order ($n = 3$), as we have one first-order ODE and one second-order ODE. Because the system is nonlinear, we cannot use transfer functions or an SSR. The solution will require three direct numerical integrations, or three integrator blocks. To show this, we can write each ODE with the highest time-derivative term on the left-hand side with a unity coefficient:

$$\dot{y} = \frac{1}{0.5}\left(-0.6\text{sgn}(y) + 1.4z\right) \tag{C.12}$$

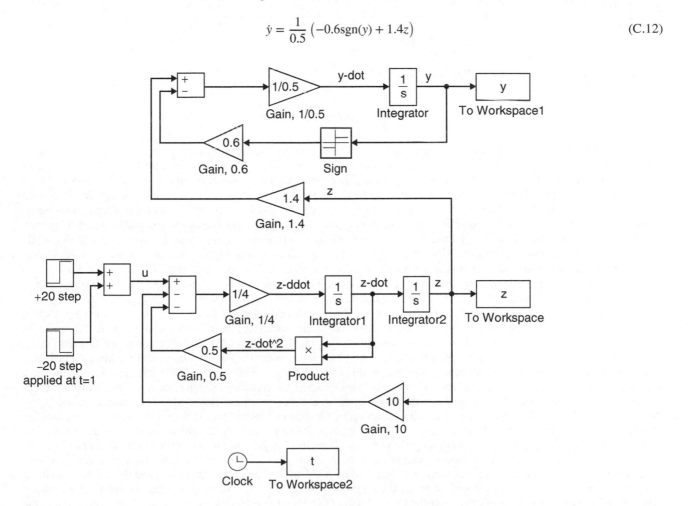

Figure C.11 Simulink model of nonlinear Eqs. (C.12) and (C.13) (Example C.4).

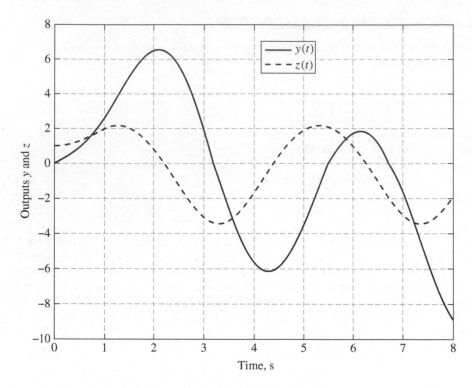

Figure C.12 System responses $y(t)$ and $z(t)$ (Example C.4).

$$\ddot{z} = \frac{1}{4}\left(-0.5\dot{z}^2 - 10z + u(t)\right) \tag{C.13}$$

It is clear that we must numerically integrate Eq. (C.12) *once* to get $y(t)$ and numerically integrate Eq. (C.13) *twice* to get $\dot{z}(t)$ and $z(t)$. Hence, the foundation of the Simulink model will be a single integrator block for Eq. (C.12) and a series of two integrators for Eq. (C.13). The input to each integrator block must be consistent with the right-hand sides of Eqs. (C.12) and (C.13). The Simulink model construction follows the same basic steps as the previous example: drag and drop Integrator blocks, Gain blocks, Sum blocks, and To Workspace blocks; pick off signals for feedback; send signal paths to "flipped" Gain blocks and ultimately to summing junctions. Figure C.11 shows the completed Simulink model, and the reader should be able to see that the top part is the block diagram for integrating $\dot{y}$ (one Integrator block), while the bottom block diagram integrates $\ddot{z}$ twice (two Integrator blocks in series). Signal paths for y, z, and $\dot{z}$ are picked off for feedback, gained, and sent to summing junctions in order to synthesize Eqs. (C.12) and (C.13). However, Eqs. (C.12) and (C.13) each have a nonlinear term that must be synthesized. The *signum function*, $\text{sgn}(y)$, in Eq. (C.12) returns the sign of the input, which in this case is y. Therefore, $\text{sgn}(y) = 1$ when $y > 0$, $\text{sgn}(y) = -1$ when $y < 0$, and $\text{sgn}(y) = 0$ when $y = 0$. The signum operation is performed by the Sign block, which is found under the Math Operations library. Hence, the user must insert the Sign block (flipped right-to-left) and Gain block (0.6) in order to create the signal $0.6\,\text{sgn}(y)$, as shown in Fig. C.11. The second nonlinear term is the squared term $0.5\dot{z}^2$ in Eq. (C.13). Simulink can create this term using various blocks. One option is to use the Math Function block found under the Math Operations library. The Math Function block allows the user to select a wide range of operations such as exponential, logarithmic, and power functions. Double clicking the Math Function block opens a dialog box where the user can select the square function from a drop-down menu. Another option for synthesizing the $0.5\dot{z}^2$ term is to use the Product block, which also resides in the Math Operations library. The Product block simply multiplies two or more signal-path variables to create an output signal. Figure C.11 shows that the signal $\dot{z}$ is the first *and* second input to the Product block, which creates $\dot{z}^2$ as the output. The reader should be able to verify that the signal paths in Fig. C.11 correctly represent the system equations (C.12) and (C.13).

The desired input function is a rectangular pulse of 20 that lasts for 1 s. One way to create this input signal is to sum together two step functions: the first step has a constant value of 20 and "steps up" at time $t = 0$, and the second step has a constant value of -20 and "steps down" at time $t = 1$ s. Of course when they are summed together we get a pulse function with value 20 for $0 < t < 1$ s and zero input for $t \geq 1$ s. Figure C.11 shows the summation of the two Step blocks from the Sources library. The desired initial values, final values, and step times can be set in the dialog boxes of each Step block.

As with any Simulink simulation, the user must set the desired integration parameters (ode4 is used here), stop time (8 s), and initial conditions for each Integrator block (recall that the default is zero initial conditions for each integrator). Recall that the initial condition are $y(0) = 0$, $z(0) = 1$, and $\dot{z}(0) = 0$, and, therefore, only Integrator2 in Fig. C.11 has a nonzero initial condition. Figure C.12 shows plots of the responses $y(t)$ and $z(t)$ obtained by executing the Simulink model. Note that $z(0) = 1$, as determined by setting the initial condition for Integrator2.

Creating Subsystem Blocks

Although the system defined by Eqs. (C.10) and (C.11) is only third order, the resulting Simulink model shown in Fig. C.11 requires four feedback paths and two summing junctions, which creates a somewhat cluttered block diagram. Therefore, higher-order nonlinear systems will result in a cluttered Simulink model with a web of signal paths. One way to create "clean" Simulink models is to condense parts of the overall block diagram into *subsystems*. Simulink has two methods for constructing subsystem blocks: (1) dragging and dropping the Subsystem block from the Ports & Subsystems library and (2) grouping an existing Simulink diagram into a subsystem. The second method involves constructing the subsystem model first and enclosing the entire model with a bounding box (click and hold outside of diagram, drag cursor across the diagram so that a dotted box encloses the diagram, release the mouse button). Figure C.13 shows the complete Simulink model from Example C.4 (i.e., Fig. C.11) with a "bounding box" around the block diagram section that represents Eq. (C.12). When the desired block diagram is selected, the user chooses Diagram > Subsystem & Model Reference > Create Subsystem from Selection and a subsystem is constructed. A subsystem for Eq. (C.13) can be created in a similar manner. Figure C.14 shows the complete Simulink model for Example C.4 using two subsystems. Note that the only external input to the $\dot{y}$ (or top) subsystem is dynamic variable z. The $\dot{y}$ subsystem, Eq. (C.12), also depends on dynamic variable y but this signal resides "inside" the subsystem. The $\ddot{z}$ (or bottom) subsystem, Eq. (C.13), has $u(t)$ as its sole input function. Some subsystems may have multiple inputs and multiple outputs. The user can double click the subsystem block to see and edit

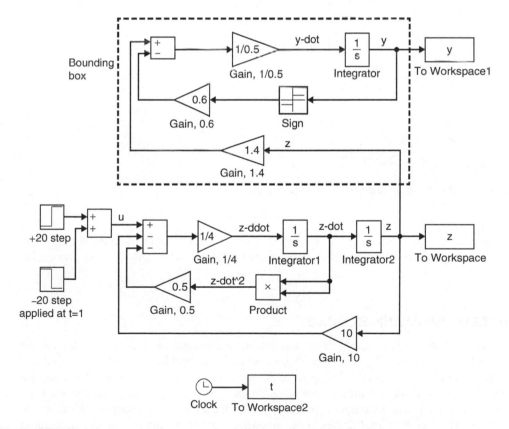

Figure C.13 Creating a subsystem for Eq. (C.12) (Example C.4).

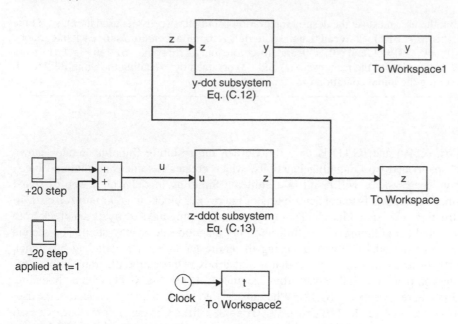

Figure C.14 Simulink model using subsystems (Example C.4).

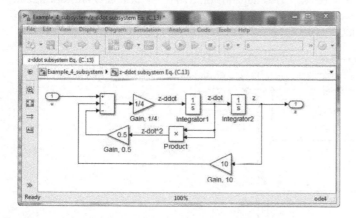

Figure C.15 Block-diagram details of the z-ddot subsystem (Example C.4).

the inner model. Figure C.15 shows a screen shot of the dialog box that appears upon double clicking the `z-ddot subsystem` in Fig. C.14. This dialog box presents the subsystem's inner block-diagram details for representing Eq. (C.13).

C.4 SUMMARY OF USEFUL SIMULINK BLOCKS

The previous examples have demonstrated how to create linear and nonlinear Simulink models. The reader can use the previous examples as guides for constructing Simulink models for essentially all problems posed in this textbook. However, there are other useful Simulink blocks that may be required for modeling linear and (in particular) nonlinear dynamic systems. Rather than attempt to illustrate their use through addition examples, we briefly describe the more common Simulink blocks that can be used to create models of physical systems that are within the scope of this textbook. For the sake of completeness, we also include the Simulink blocks described in the previous examples (such as integrators and gains). In all cases, the user sets the required parameters for each block by double clicking on the block and entering the numerical values in the dialog box. The blocks are listed in alphabetical order and categorized according to their Simulink library.

Continuous Library

`Derivative`: computes the time derivative of the input signal using a numerical method.

`Integrator`: computes the time integral of the input signal using a numerical method. The user may set the initial condition for the output.

`State-Space`: simulates the system dynamics as a linear, time-invariant SSR. The input (`u`) and output (`y`) can be either scalars or vectors as determined by the user's definitions of the **A**, **B**, **C**, and **D** matrices. The user may set the initial state vector.

`Transfer Fcn`: simulates the system dynamics as a transfer function. The user sets the numerator and denominator vectors as descending powers of s.

Discontinuities Library

`Coulomb and Viscous Friction`: computes the Coulomb (dry) friction based on the sign of the input and the viscous friction coefficient. The user defines the offset (dry friction) value and the viscous friction coefficient.

`Relay`: determines an "on" or "off" output signal that depends on the input. The user defines the desired output values corresponding to "on" and "off" and the input thresholds for the switch-on and switch-off points.

`Saturation`: limits the output to be between upper and lower values set by the user. If the input signal is between the upper/lower limits, the output is equal to the input. If the input is greater than the upper limit, the output is set to the upper value; if the input is less than the lower limit, the output is set to the lower value.

Math Operations Library

`Abs`: computes the absolute value of the input.

`Gain`: multiplies the input signal by a constant (gain). The user defines the gain.

`Rounding Function`: performs the MATLAB rounding operations such as `floor`, `ceil`, `round`, and `fix`. The output is an integer.

`Math Function`: performs various mathematical functions such as exponential, logarithmic, and power. The user selects the desired math function.

`Product`: multiplies two or more input signals together to compute their product. The user defines the number of input signals.

`Sign`: signum or sign function. The output is 1 if the input is positive, -1 if the input is negative, and zero if the input is zero.

`Sum`: summing junction. The output is the summation of two or more input signals. The user defines the number of input signals and their associated signs (+ or −) at the summing junction.

`Trigonometric Function`: the user may select from sine, cosine, and tangent trigonometric functions (and their inverses) as well as hyperbolic functions.

Ports and Subsystems Library

`Subsystem`: provides template for constructing a subsystem. The default is a subsystem with one input and one output. The user can add input or output ports by dragging and dropping `In1` (Inport) or `Out1` (Outport) from the same library. The user builds the subsystem by adding the desired blocks such as `Integrator`, `Transfer Fcn`, `Sum`, `Gain`, etc.

Signal Routing Library

Demux: de-multiplexor. "Splits" an *n*-vector input signal into *n* scalar output signal paths. The user defines the number of output signal paths.

Mux: multiplexor. Creates an *n*-vector output signal from *n* scalar input signal paths. The user defines the number of input signal paths.

Switch: passes either input signal 1 or input signal 3. Input signal 2 determines the switching by comparing it to a threshold value set by the user.

Sinks Library

Scope: plots the input signal with respect to simulation time.

To File: writes simulation time (row 1) and the input signal (row 2) to a MAT file that is defined by the user.

To Workspace: writes the input signal to MATLAB's workspace. The user defines the variable name and format (Array is the recommended format).

Sources Library

Clock: output is simulation time.

Constant: output is a constant defined by the user.

Ramp: output is a ramp function. The user defines the initial output value, start time of the ramp function, and the slope of the ramp.

Random Number: output is a random signal with a normal (Gaussian) distribution. The user defines the mean and variance of the Gaussian distribution.

Sine Wave: output is a sinusoidal function. The user defines the amplitude, frequency (in rad/s), bias (if any), and phase shift. The default sinusoid is a sine wave; the user can create a cosine wave by setting the phase shift to $pi/2$ rad.

Step: output is a step function. The user defines the initial output value, step time, and the final value of the step function.

User-Defined Functions Library

Fcn: general function of input signal u. The user defines the output function in the dialog box using standard MATLAB mathematical operations such as 4*sin(3*u), exp(−2*u), 5*u - u^3, and so on.

Interpreted MATLAB Fcn: applies the specified MATLAB function to the input(s). Allows the user to write a customized script (M-file) that computes the output(s) as a function of the input(s).

Index